Indigenous Peoples and Environmental Issues

INDIGENOUS PEOPLES AND ENVIRONMENTAL ISSUES

An Encyclopedia

BRUCE E. JOHANSEN

GREENWOOD PRESS
Westport, Connecticut • London

Library of Congress Cataloging-in-Publication Data

Johansen, Bruce E. (Bruce Elliott), 1950–
  Indigenous peoples and environmental issues : an encyclopedia / by Bruce E. Johansen.
    p.   cm.
  Includes bibliographical references and index.
  ISBN 0–313–32398–4 (alk. paper)
  1. Indigenous peoples—Ecology—Encyclopedias. 2. Human ecology—Encyclopedias.
  3. Environmental degradation—Encyclopedias. 4. Environmental policy—Encyclopedias. I. Title.
  GF50.J65   2003
  304.2—dc21   2003040582

British Library Cataloguing in Publication Data is available.

Library of Congress Catalog Card Number: 2003040582
ISBN: 0–313–32398–4

First published in 2003

Greenwood Press, 88 Post Road West, Westport, CT 06881
An imprint of Greenwood Publishing Group, Inc.
www.greenwood.com

Printed in the United States of America

The paper used in this book complies with the
Permanent Paper Standard issued by the National
Information Standards Organization (Z39.48–1984).

10 9 8 7 6 5 4 3 2 1

Copyright Acknowledgments

The author and publisher gratefully acknowledge permission to reprint the following:

Excerpts from *Teak is Torture*, by Tim Keating. Published on www.rainforestrelief.org.

Excerpts from World Rainforest Movement Bulletin. http:/www.wrm.org.uy

Excerpts from *Ecocide of Native America*, Donald A. Grinde, Jr. and Bruce Johansen. Published by Clear Light Publishers, Santa Fe, NM. www.clearlightbooks.com.

Excerpts from *The Canadaigua Treaty: A View from the Six Nations*, by Oren Lyons. Published by Clear Light Publishers, Santa Fe, NM. www.clearlightbooks.com.

Excerpts from *Treaty of Canadaigua 1794*, by Peter Jemison and Anna M. Shein, eds. Published by Clear Light Publishers, Santa Fe, NM. www.clearlightbooks.com.

Exerpts from *Treaty Making: The Legal Record*, by Paul Williams. Published by Clear Light Publishers, Santa Fe, NM. www.clearlightbooks.com.

Excerpts from *The Death of Friday Newiido*, by Nick Ashton-Jones. Published on www.shell-terror.net. Permission can be confirmed by email at NJAJA@compuserve.com.

CONTENTS

THE HAUDENOSAUNEE (IROQUOIS) ENVIRONMENTAL WORLDVIEW: DIFFERENT TREES IN A DIFFERENT FOREST *BY ROBERT W. VENABLES* 179

HONDURAS   191

INDIA   194

GUIDE TO RELATED TOPICS

ANIMAL MANAGEMENT AND POACHING

Canada: "The Ojibway: Black Bear Poaching"

Indonesia: "The Attack of the Sun Bears"

United States: "Native Peoples Line Up Against Yellowstone National Park's 'Buffalo Cull'" (Montana)

Zimbabwe and Botswana: "An Alliance for Wildlife"

BAUXITE MINING

Costa Rica: "Resisting Central America's Largest Dam, Bauxite Mining, and an Aluminum Smelter"

India: "Deaths of Bauxite Mining Protesters"

BIODIVERSITY

Biodiversity and Indigenous Environmentalism

CLIMATE CHANGE

Climate Change and Indigenous Environmentalism

COAL STRIP MINING

Colombia: "The Tabaco: Coal Strip Mining"

United States: "The Hopi and Navajo: Turning Black Mesa to Coal Slurry" (Arizona and New Mexico)

United States: "The Northern Cheyenne: Methane Gas Extraction" (Montana)

United States: "The Zuni: Sacred Waters and Coal Strip Mining" (New Mexico)

COBALT MINING

Philippines: "Mindoro Island's Mangyan, Alangan, and Tadyawan Peoples: Nickel and Cobalt Mining"

South Pacific: "New Caledonia's Kanaky Nickel Mine"

COPPER AND ZINC MINING

Canada: "The Taku River Tlingit First Nation of British Columbia: Resisting Zinc, Copper, Gold, and Silver Mining"

Fiji: "The Namosi, Serua, Nadroga, and Rewa: Fighting a Proposed Copper Mine"

Irian Jaya/Papua New Guinea: "Freeport's Grasberg Mine: Tidal Waves of Waste"

Panama: "The Ngobe-Bugle: Winning Land and Resisting Mining"

Philippines: "The Marinduque Islanders: Copper-Mining Tailing Spills"

United States: "Copper Mining at Picuris Pueblo" (New Mexico)

DAMS (HYDROELECTRIC POWER)

Argentina: "The Wichí: Opposition to Hydroelectric Development"

Brazil: "The Toll of Dam Building on Indigenous Peoples"

Bolivia: "Indigenous Peoples, Logging Concessions, Oil Exploration, and Toxic Spills"

Canada: "The Cree: Hydro-Quebec's Electric Dreams"

Canada: "The Pimicikamak Cree of Manitoba: Imperiled by Hydropower"

Chile: "The Pehuenche and Mapuche: Logging, Dam Building, and Land Rights"

Colombia: "The Embera: Conflicts with Dam Construction"

Costa Rica: "Resisting Central America's Largest Dam, Bauxite Mining, and an Aluminum Smelter"

Dam Sites and Indigenous Peoples

Guatemala: "The Achi Maya: Protesting Hydroelectric Flooding"

Honduras: "A Murder Follows Protest of Dam Construction"

India: "Saying 'No' to the Narmada Dam Complex"

Irian Jaya/Papua New Guinea: "Dam Development in Papua New Guinea"

Mexico: "The Maya: Oil Exploitation and Deforestation in Chiapas"

Philippines: "The Ibaloi: Damming a Sacred River"

Sri Lanka: "The Wanniyala-Aetto: Hydropower and Logging"

Thailand: "Indigenous Peoples in the Powerhouse"

Turkey: "The Kurds: Dam Building"

DEFORESTATION (LOGGING)

Belize: "The Mopan and Kekchi: Opposition to Industrial-Scale Logging"

Bolivia: "Indigenous Peoples, Logging Concessions, Oil Exploration, and Toxic Spills"

Brazil: "Accelerating Deforestation in the Amazon Valley"

Brazil: "The Kaiapo: Greenpeace and Mahogany Logging"

Burma (Myanmar): "Forced Labor in the World's Last Teak Forest"

Cambodia: "Introduction: Deforestation Spurs Indigenous Pressure"

Cambodia: "Protecting Resin Trees"

Cameroon: "Pygmies: Losing Their Homes to Industrial-scale Logging"

Chile: "The Pehuenche and Mapuche: Logging, Dam Building, and Land Rights"

Congo Basin: "Deforestation at 'Alarming Rates'"

Congo Republic: "Pygmies: Logging Threatens Livelihood"

Guyana: "The Akawaio Nation: Seeking a Land Base Before It Is Logged Away"

Eritrea: "The Tigre, Beni Amer, Hidareb, and Kunama Tribes: Deforestation of Their Homelands"

Forest Stewardship Council

Indonesia: "The Batak: Shutting Down Pulp-and-Paper Manufacturing in North Sumatra"

Indonesia: "Forest Management and Indigenous Peoples on Java"

Indonesia: "Logging, Pearl Harvesting, and Tourism on Togean"

Indonesia: "The Penan: Obstructing Logging"

Irian Jaya/Papua New Guinea: "The Moi: Logging and Mining in West Papua"

Malaysia (Sarawak): "Roadblocks Greet Logging Companies"

Mexico: "The Maya: Oil Exploitation and Deforestation in Chiapas"

Nicaragua: "The Mayagna (Sumo): Battling Illegal Logging and Cataloging Endangered Species"

Panama: "The Ngobe-Bugle: Winning Land and Resisting Mining"

Peru: "The Camisea Natural Gas Project"

Philippines: "Mindanao's Lumads: Logging, Mining Wastes, and Evictions"

South Pacific: "The Solomon Islands: Indigenous Peoples Relocated for Gold Mining and Logging"

Sri Lanka: "The Wanniyala-Aetto: Hydropower and Logging"

Suriname: "The Saramaka Maroons: Gold Mining and Logging"

Zambia: "Blaming 'the Poor' for Deforestation"

DIAMOND MINING

Botswana: "The Khwe (Kalahari Bushmen): End of the Line"

Canada: "The Dogrib First Nation: Diamond Mining"

Brazil: "The Yanomami: Gold Rush"

Canada: "The Taku River Tlingit First Nation of British Columbia: Resisting Zinc, Copper, Gold, and Silver Mining"

Ghana: "Gold-Mining Tailings Spills"

Guyana: "The Isseneru: Mercury Poisoning from Gold Mining"

Honduras: "The Spread of Gold-mining Concessions"

Indonesia: "The Dayak: Traditional Gold-Mining Practices"

Irian Jaya/Papua New Guinea: "Freeport's Grasberg Mine: Tidal Waves of Waste"

Irian Jaya/Papua New Guinea: "The Moi: Logging and Mining in West Papua"

Mexico: "Silver Mining and Lead Poisoning of Children"

Panama: "The Ngobe-Bugle: Winning Land and Resisting Mining"

Peru: "Indigenous Peoples, Gold Mining, and Mercury Poisoning"

South Pacific: "The Solomon Islands: Indigenous Peoples Relocated for Gold Mining and Logging"

Suriname: "The Saramaka Maroons: Gold Mining and Logging"

Tibet: "A Monk Pays the Price for Protesting Gold Mining"

United States: "The Quechan: Gold Mining" (California)

IMPORTED DISEASES

Brazil: "The Panara: Road Building, Imported Diseases, and Genocide"

India: "Imported Diseases on the Andaman and Nicobar Islands"

Peru: "The Urarina (Kachá): Oil Development, Imported Diseases, and 'Hippie' Tourists"

KITTY-LITTER MINING

United States: "The Political Economy of Kitty-Litter Strip Mining" (Nevada)

LAND TENURE

Argentina: "The Kolla: A Struggle for Land Tenure"

Botswana: "The Khwe (Kalahari Bushmen): End of the Line"

Brazil: "The Guarani and Kaiowa: Asserting Rights to Their Land amidst a Wave of Suicides"

Brazil: "The Kaiapo: Greenpeace and Mahogany Logging"

Brazil: "The Pataxó: Taking Back Their Land"

Canada: "The Lubicon Cree: Land Rights and Resource Exploitation"

Chile: "The Pehuenche and Mapuche: Logging, Dam Building, and Land Rights"

Guyana: "The Akawaio Nation: Seeking a Land Base Before It Is Logged Away"

Kenya: "The Maasai: Fighting Land Expropriation for Military Testing"

METHANE GAS EXTRACTION

NATIVE AMERICAN ENVIRONMENTALISM

NICKEL AND COBALT MINING

NIOBIUM MINING

NUCLEAR WASTE DUMPS

NUCLEAR WEAPONS TESTING

OIL AND NATURAL GAS EXPLOITATION

PHOSPHATE MINING

South Pacific: "The Mataiva, Nauru, and Banaba Islands: Sacrificed for Phosphate Mining"

POISON GAS

Iraq: "The Kurds: Poison Gas"

POLLUTION

Bolivia: "Indigenous Peoples, Logging Concessions, Oil Exploration, and Toxic Spills"

Brazil: "The Pemon: Mercury Poisoning and Amazon Gold Mining"

Canada: "The Innu of Labrador: Industrialism's Intrusions"

Canada: "The Inuit: Dioxin and Other Persistent Organic Pollutants"

Canada: "The Ouje-Bougoumou Cree: Water Pollution"

Colombia: "The Underside of U.S. Anti-drug Spraying"

Guam: "The Chamorro: Military PCB Pollution"

Guyana: "The Isseneru: Mercury Poisoning from Gold Mining"

Irian Jaya/Papua New Guinea: "Freeport's Grasberg Mine: Tidal Waves of Waste"

Irian Jaya/Papua New Guinea: "The Moi: Logging and Mining in West Papua"

Marianas Islands: "PCB Contamination"

Mexico: "The Huichole: Living with Pesticides around the Clock"

Mexico: "Silver Mining and Lead Poisoning of Children"

Peru: "Indigenous Peoples, Gold Mining, and Mercury Poisoning"

Peru: "A Lead Smelter Fouls the Air at La Oroya"

Philippines: "Mindanao's Lumads: Logging, Mining Wastes, and Evictions"

Russia (Siberia): "The Environmental Legacy of Soviet-Era Policies"

Russia (Siberia): "Siberian Indigenous Peoples and Uranium Poisoning"

Thailand: "The Karen: The Toll of Lead Poisoning"

United States: "Akwesasne: The Land of the Toxic Turtles" (New York)

United States: "The Blackfeet: Foot-High Mushrooms and Toxic Mold" (Montana)

United States: "Bombs Away in the Aleutian Islands" (Alaska)

United States: "The Cheyenne River Sioux: The Homestake Mine's Toxic Legacy" (South Dakota)

United States: "The Coeur d'Alene: Mining Waste in Idaho" (Idaho)

United States: "The Gros Ventre and Assiniboine: Gold Mining and Cyanide Poisoning in Montana" (Montana)

United States: "The Isleta Pueblo Tastes Albuquerque's Effluent" (New Mexico)

RAILWAY CONSTRUCTION

SHRIMP FISHING

SULFIDE MINING

TITANIUM MINING

TOURISM

URANIUM MINING

VIOLENCE AND FATALITIES

WATER SCARCITY

PREFACE

This encyclopedia describes the intrusion of foreign (usually destructive) elements into the environments (cultures, lifeways, and economies) of indigenous peoples around the Earth. *Indigenous people*, in this context, means groups of human beings who have occupied an area before other groups intruded. In the usual rhetorical terms, many of these intrusions have been performed by Western industrialism and its attendant tourism industries, even though, in today's world, the intruders are as likely to have home addresses in Tokyo as in London or New York City. Whatever their origins, they usually extract resources from indigenous peoples to the benefit of stockholders and consumers elsewhere.

These intrusions may take the form of oil, coal, or natural gas development; the building of dams that inundate native lands; the introduction of gold or uranium mining; or even the arrival of the first ecotourists, who often represent an indigenous people's first encounters with such pathogens as chicken-pox, to which most of us are accustomed and largely immune. Other intruders leave their condom wrappers behind in once-isolated valleys. Many times the intruders are builders of hydroelectric dams seeking to provide power to growing cities; the irony, of course, is that they inundate the lands of peoples who usually do not use electricity.

Indigenous Peoples and Environmental Issues: An Encyclopedia includes contemporary reporting worldwide from places where indigenous peoples are attempting to maintain some measure of their lands, resources, and cultural integrity in the face of rising pressure to exploit natural resources. Information gathering for this book has been aided immensely by the growing use of Internet technology, through which peoples in the most remote locations post pages on the Internet—a vast international forum, subject only to the availability of electricity and telephone connections.

The scope of this encyclopedia is global, involving peoples on all continents (except parts of Europe and Antarctica), and in all types of climates and environments. It is, to my knowledge, the first attempt to assemble under one cover a survey of indigenous peoples' contemporary struggles with environmental issues.

Entries take the reader from the Arctic, where global warming and chemical pollutants threaten the Inuit, to Borneo, where logging has devastated indigenous habitats. The real, Earth-girdling news, which most of our media largely miss, is that ways of life finely tuned to and enmeshed in natural settings are breaking down in our lifetimes, under the intrusion (some native peoples would call it an assault) of industrial-scale resource exploitation. Whether from large-scale deforestation; mining of gold, coal, uranium, bauxite, or other minerals; the expansion of monocrop, plantation-style agriculture; or even mass tourism, indigenous peoples are finding that the systems supporting their ways of life are giving way.

At the beginning of the third millennium on the Christian calendar, roughly 300 million people lived in 5,000 indigenous cultures worldwide ("Earthpulse," 2001, n.p.). Roughly 40 percent of the world's countries (72 of 184) included indigenous peoples; worldwide, "there are over 350 million indigenous people representing some 5,250 nations" (Gedicks, 2001, 15). Jason Clay, an anthropologist and former research director at Cultural Survival, has noted that the twentieth century, "considered by many to be an age of enlightenment, progress, and development—has witnessed more genocides, ethnocides, and extinctions of indigenous peoples than any other in history" (Gedicks, 2001, 16).

The advance of industrial-scale resource extraction has been paralleled, around the world, by a retreat of indigenous peoples, destruction of traditional economies, and reduction of native populations. For example, when the Amazon Basin was first sighted by European explorers and colonists during the sixteenth century, it was probably home to about six million indigenous people. Today, less than 500 years later, the indigenous population is fewer than 250,000 ("To the Ends of the Earth," n.d.). At the same time, resource extraction from indigenous lands is increasing rapidly. By the year 2010, 90 percent of all the world's new gold, about 80 percent of its nickel, more than 60 percent of its copper, and about half of its coal may be removed from indigenous peoples' lands ("Indigenous People," 2000).

These environmental changes have come with amazing swiftness, within a human generation or two in many areas. As Sir Edmund Hillary remarked in his foreword to *The Law of the Mother* (1993):

> In forty years, I have seen the transformation of the remote Khumbu area on the southern slopes of Mount Everest. In 1951, the Khumbu was a place of great beauty, with 3,000 tough and hardy Sherpas living a remarkably full and cooperative life despite their rigorous environment. Now it has become largely a tourist area, with 12,000 foreigners streaming in each year, leaving their litter and tempting the Sherpas to break their traditional forestry customs and sell hundreds of loads of firewood for luxury fires. (Kemp, 1993, xi)

Hillary reflected that he had accelerated this process, first with his world-famous mountain-climbing in the area and, secondly, by advocating construction of an airport which allowed easy access for tourists. Actions that open an area to outside contact can cut both ways. Widespread access to the Internet, for example, which has made compilation of this volume possible, also requires telephones, electricity, and other modern modes of living.

This book traces lines of conflict between traditional indigenous societies and the expansion of what is loosely called Western industrialism. Indigenous peoples have been dealing with the changes brought to them in a variety of ways, from armed resistance to accommodation. Some former headhunters in Kalimantan (Borneo), for example, have turned their longhouses into bed-and-breakfasts, mainly for Japanese tourists.

This book examines the continuing expansion of the world industrial system that one author describes this way:

> In the past few generations . . . the world's collection of highly diverse adaptations to local environmental conditions has been replaced by a world culture characterized by very high levels of material consumption, economic growth based on conversion of fossil fuels to energy, greatly expanded international trade, and improved public health. . . . [These changes] have spurred such a rapid expansion of human numbers that new approaches to resource management have been required. These powerful incentives to produce more goods have overwhelmed the conservation measures of local communities, bringing overexploitation and poverty to many rural communities and great wealth to cities and certain individuals. (McNeely, 1993, 250)

Arranged alphabetically, the book includes entries on countries from Argentina to Zimbabwe and on such general topics as dam sites and Native American concepts of ecology. When I first began to compile the work, I wanted to use tribal names as the major organizing device but I soon found that few readers would recognize any names except the most familiar. How many North American library patrons, for example, would recognize the Ogoni, referring to the people of Africa's Niger Delta whose traditional way of life has been spoiled by exploding oil wells and pipelines? Who would recognize the Khwe (aka Kalahari Bushmen) of Botswana, who now must apply for licenses to hunt a dwindling supply of bushmeat, while affluent big-game hunters from Japan, Europe, and North America are courted for their dollars, pounds, marks, or yen. Who, would recognize the Karen of Burma, who have been enlisted by the country's military government to make logs of the world's last sizable teak forest? Who would know the name of the Kaiapo of the Brazilian Amazon, who are watching the same type of logging in Earth's last mahogany forests, or the Namosi of Fiji who are battling plans to initiate a large copper mine on their island?

The book, therefore, is arranged geographically by country. Each country entry covers the peoples and environmental issues in that country. Those readers interested in specific topics, such as deforestation or oil and natural gas exploitation, will find the guide to related topics, as well as the subject index, useful.

REFERENCES

"Earthpulse: Cultural Extinctions Loom." *National Geographic*, September 2001, n.p.

Gedicks, Al. *Resource Rebels: Native Challenges to Mining and Oil Corporations*. Cambridge, Mass.: South End Press, 2001.

"Indigenous People Suffering Gross Rights Violation!" *Free Press Journal*, August 9, 2000. [http://www.indiaworld.co.in/news/features/feature495.html]

Kemp, Elizabeth, ed. *The Law of the Mother: Protecting Indigenous Peoples in Protected Areas*. San Francisco: Sierra Club Books, 1993.

McNeely, Jeffrey A. "Afterword: People and Protected Areas: Partners in Prosperity." In Elizabeth Kemf, ed., *The Law of The Mother: Protecting Indigenous Peoples in Protected Areas*. San Francisco: Sierra Club Books, 1993.

"To the Ends of the Earth: Revenge of the Lost Tribe: the Amazon's Indigenous Peoples." N.p., n.d. [http://www.channel4.com/plus/ends/tribe4.html]

ACKNOWLEDGMENTS

In the course of research and writing this book, I have received help from many people, among them my editors Cynthia Harris and John Wagner at Greenwood Press, as well as the production staff there. In addition, I received help with textual matter from Dan Plumley, Christine Kasel, John Kahionhes Fadden, and Robert W. Venables, as well as from Virginia Luling and Roger Mortimer of Survival International in London. I thank Tim Keating and Edith Mirante of Rainforest Relief for factual assistance on the Burma entry. Thanks also are due to Nick J. Ashton-Jones and the staffs of Greenpeace, the World Rainforest Movement, and *Drillbits and Tailings*, and my department chair, Deb Smith-Howell. For help with graphics, thanks are due to Survival International, Greenpeace, and the Environment News Service.

As always, the most considerable measure of thanks are due my wife Pat Keiffer, without whom I would be lonely, not to mention freezing in the dark, as well as grand-daughter Samantha Keiffer-Rose, both of whom often keep my mind on what's really important.

INTRODUCTION

I had thought of myself as reasonably well informed on matters indigenous and environmental before Cynthia Harris at Greenwood Press asked me to undertake this survey of indigenous peoples and environmental issues. I had, after all, written widely about ecological issues facing Native Americans and recently traveled in the Arctic to see the impact of global warming and persistent organic pollutants (POPs) on the Inuit.

I was not prepared, therefore, for the number and the acerbic nature of environmental conflicts that engaged indigenous peoples in other parts of the world, especially Africa and the tropical reaches of Asia. I found myself immersed suddenly in issues involving more than 170 peoples who are themselves immersed in environmental issues of life and death. Many of these issues involve tribal peoples whose names were unfamiliar to me. These included, for example, the Ogoni of Africa's Niger Delta, whose native lands have been rendered largely uninhabitable by oil spills, pipeline leaks, and explosions. Residing atop one of the richest caches of oil in the world outside the Middle East, the Ogonis, unlike the Saudis, have recouped little except misery for their wealth of oil.

I read accounts of tiny Pacific atolls, once small stretches of paradise, which have been strip-mined in near totality to supply modern industrial societies with phosphates, zinc, bauxite, or other salable minerals and metals. I also encountered stories of the Karen of Burma, forced into near-slavery to harvest the world's last sizable stands of teak. These and many other images haunted me as I wrote this work, tracing the travails endured by indigenous peoples who have found their traditional territories invaded and their lives transformed by the consumptive appetites of industrialism.

Very quickly, in our time, much of the indigenous world has become collateral damage to the appetites of Western industrialism. Although frequent consumers of print and electronic media in the United States might have a hard time believing such a thing, the world is roiling with conflict initiated by many indigenous peoples regarding the despoliation of their lands and ruination of their traditional economies by Western industrialism's extraction of natural resources.

THE ARCTIC: PERSISTENT ORGANIC POLLUTANTS AND GLOBAL WARMING

Many educated people, who still profess surprise at the substitution of the native name Inuit for Eskimo, also are unaware that the Arctic, thanks to oceanic and atmospheric circulation, has become the world's dumping ground for synthetic pollutants such as dioxin and PCBs. Many Inuit mothers are now being told not to breast-feed their children, because their milk is laced with these substances. Sheila Watt-Cloutier was raised in an Inuit community in remote northern Quebec. Unknown to her at the time, toxic chemicals were being absorbed by her body, and by those of other Inuit in the Arctic. "As we put our babies to our breasts we are feeding them a noxious, toxic cocktail," said Watt-Cloutier, a grandmother who also is Canadian president of the Inuit Circumpolar Conference (ICC). "When women have to think twice about breast-feeding their babies, surely that must be a wake-up call to the world" (Johansen, 2000, 27).

Watt-Cloutier now ranges between her home in Iqaluit (pronounced "Eehalooeet," the Baffin Island capital of the new semisovereign Nunavut Territory) to and from Ottawa, Montreal, New York City, and other points south, doing her best to alert the world to toxic poisoning and other perils faced by her people. The ICC represents the interests of roughly 140,000 Inuit spread around the North Pole, from Nunavut (which means "our home" in the Inuktitut language) to Alaska and Russia. Nunavut itself, a territory four times the size of France, has a population of roughly 27,000, 85 percent of whom are Inuit.

Many residents of the temperate zones hold fond stereotypes of a pristine Arctic largely devoid of human pollution. To a tourist with no interest in environmental toxicology, the Inuits' Arctic homeland may *seem* as pristine as ever during its long, snow-swept winters. Many Inuit still guide dogsleds onto the pack ice surrounding their Arctic-island homelands to hunt polar bears and seals. Such a scene may seem pristine, until one realizes that the polar bears' and seals' body fats are laced with dioxins and PCBs. The toxicological due bills for modern industry at the lower latitudes are being left on the Inuit table in the Canadian Arctic. Native people whose diets consist largely of sea animals (whales, polar bears, fish, and seals) have been consuming a concentrated toxic chemical cocktail. Abnormally high levels of dioxins and other industrial chemicals are being detected in Inuit mothers' breast milk.

Persistent organic pollutants (POPs) have been linked to cancer, birth defects, and other neurological, reproductive, and immune-system damage in people and animals. At high levels, these chemicals also damage the central nervous system. Many of them also act as endocrine disrupters, causing deformities in sex organs as well as long-term dysfunction of reproductive systems.

Persistent organic pollutants also interfere with the function of the brain and endocrine system by penetrating the placental barrier and scrambling the instructions of naturally produced chemical messengers. The latter tell a fetus how to develop from the womb through puberty; should interference occur, immune, nervous, and reproductive systems may not develop as programmed by the genes inherited by the embryo.

Persistent organic pollutants are only the most urgent of several environmental problems facing the Inuit above the Arctic Circle. Global warming is provoking changes in their homelands more quickly than any other place on Earth. Hunters told me of losing relatives to thin ice. While I visited Iqaluit, on Baffin Island, hungry, hot, and miserable polar bears assaulted unwary tourists in a local park, something they rarely did before, when the ice brought them food. Hunters described finding nearly hairless seals with strange welts on their skins. No one yet knows why, although depleting stratospheric ozone is suspected.

UNITED NATIONS WORLD POPULATION REPORT FOR 2001

Even as environmentalists stress the necessity to move beyond an energy-generation system based on fossil fuels to preserve the health of the planet, this same system is still expanding into formerly pristine areas of the Earth. Surveying conflicts between indigenous peoples and corporations on a worldwide scale reveals that the fingers of fossil-fueled industrialism are still spreading over the planet. According to the United Nations, the Earth is being mined, drilled, and logged at an unprecedented and unsustainable rate that needs to be curbed quickly to avoid disaster. "More people are using more resources with more intensity than at any point in human history," the United Nations warned in its annual world population report for 2001 (Lovell, 2001, 6).

According to the same study, "The costs of delaying action will increase rapidly over time. By 2050, 4.2 billion people—over 45 per cent of the global total—will be living in countries that cannot meet the daily requirement of 50 liters of water per person to meet basic needs" (Lovell, 2001, 6). The Earth's human population, which has doubled to 6.1 billion during the past 40 years, is projected to surge 50 percent to 9.3 billion within the next 50 years. Much of this growth will take place in poor countries where resources already are stressed. The same United Nations report said that by the year 2001, 1.1 billion people already did not have access to clean water. In Third World nations, as much as 95 percent of sewage and 70 percent of industrial waste was being dumped untreated into many of the same bodies of water that many people use for drinking, bathing, and cooking.

Rainforests are being destroyed at the highest rate in history, obliterating crucial sources of biodiversity and contributing to global warming, thereby boosting already-rising sea levels. The seas continue to be massively overexploited and erosion is taking a rising toll on plant species, a quarter of which could be lost forever by 2025. The previously cited United Nations report said that food production will have to double and distribution will have to improve to feed exploding populations, with most of the increase coming from higher-yielding varieties which require more environmentally dangerous chemicals for their production. The same report said that globalization of commerce has increased wealth but, at the same time, has added to global inequalities, with the world's poorest people forced to plunder their scarce natural resources simply to survive from day to day.

The world of numbers, as outlined by this report, can become numbingly abstract. As I researched, I was jolted time and again by personal stories of indigenous peoples whose lives had been ruined by intrusive resource exploitation. Some of the most graphic personal stories involved native people in the United States and Canada who were hired to mine uranium beginning in the 1940s. They were not warned that uranium could kill. As late as the 1960s, Navajos mined uranium as if it was coal, with no protection at all. They ate their lunches in the mines and slaked their thirst with water running over radioactive rock. The mines lacked toilet facilities, so Navajos wiped with radioactive yellowcake uranium ore after relieving themselves. Two to three decades later, many of the same miners died painfully of lung cancer, a disease nearly unknown to the Navajo before the advent of uranium mining.

Similar deaths afflicted the Dene of Northwest Canada. When mining began about 1950, they called uranium *the money rock*. Only later did they come to understand its lethal qualities. Regarding uranium mining, Andrew Nikiforuk of the *Calgary Herald* wrote as follows:

> Al King, an 82-year-old retired member of the steelworkers' union in Vancouver, [British Columbia] has held the hands of the dying. He recalls one retired Port Radium miner whose chest lesions were so bad that they had spread to his femur and exploded it. "They couldn't pump enough morphine into him to keep him from screaming before he died," King said. (Nikiforuk, 1998, A-1)

Such individual stories add up to a world in wrenching pain, a planet rapidly losing a rich diversity of biology and human experience.

MAJOR SOURCES OF CONFLICT BETWEEN WESTERN INDUSTRIALISM AND INDIGENOUS PEOPLES

Empowerment of Women: The Predator/Prey Relationship

The conflicts described in this book share several common attributes. One is the replacement of matriarchal or matrilineal cultures by resource-exploiting governments and industries run largely by men. We ignore the voice of women at our own peril for the future of a viable Earth. According to a report from the United Nations, "It is clear that providing full access to reproductive health services would be far less costly in the long run than the environmental consequences of the population growth that will result if reproductive health needs are not met" (Lovell, 2001, 6). According to Winona LaDuke, the women's point of view is crucial to making the changes in guiding ideologies that are required to reverse destruction of indigenous peoples and ways of life. LaDuke is White Earth Anishinabe and was a Green Party candidate for the vice-presidency of the United States in 2000 (on Ralph Nader's ticket).

LaDuke has written that:

> The origins of this problem [rampant industrialism] lies with the predator/prey relationship industrial society has developed with the Earth, and subsequently, the people of the Earth. This same relationship exists *vis à vis* women. We, collectively, find that we are often in the role of the prey, to a predator society, whether for sexual discrimination, exploitation, sterilization, absence of control over our bodies, or being the subjects of repressive laws and legislation in which we have no voice. (LaDuke, 1995)

LaDuke believes that most matrilineal societies, in which governance and decision making are largely controlled by women, have been vanquished from the face of the Earth by colonialism, and subsequently by industrialism.

The nature of sexual politics supporting the destruction of indigenous societies is clear to LaDuke. On a worldwide scale and in North America, she believes that indigenous societies historically, and today, remain in a prey/predator relationship with industrial society.

Indigenous peoples are the peoples with the land—land and natural resources required for someone else's development program and the amassing of wealth. The wealth of the United States, that nation which today determines much of world policy, easily expropriated these lands. Similarly the wealth of indigenous peoples of South African, Central American, South American, and Asian countries was taken for the industrial development of Europe, and later for settler states which came to occupy those lands (LaDuke, 1995).

Another element common to most of the conflicts described in this book is the driving power of mass consumption. LaDuke believes that mass consumption is the most visible aspect of the predator/prey relationship that is devouring indigenous cultures around the world:

> So long as the predator continues, so long as the middle, the temperate countries of the world continue to drive an increasing level of consumption, and, frankly continue to export both the technologies and drive for this level of consumption to other countries of the world, there will be no safety for the human rights of women, rights of indigenous peoples, and to basic protection for the Earth, from which we get our life. Consumption causes the commodification of the sacred, the natural world, cultures, and the commodification of children, and women. (LaDuke, 1995)

Loss of Biodiversity

Another element common to most of the conflicts described in this volume is the replacement of human and natural diversity (such as native languages and cultures and the plants and animals on which they depend) by a uniform, worldwide monoculture. At the dawn of the twenty-first century, an estimated 5,000 to 7,000 human languages were being spoken around the world, with 4,000 to 5,000 of these usually classed as indigenous. More than 2,500 of these are in danger of immediate extinction, and many more are losing their links with the natural world ("Globalization Threat," 2001). Some areas are uncommonly rich in different languages. In Papua New Guinea, 847 different tongues were being spoken, even in the year 2000. Many of these are teetering on extinction as the forests which house people speaking these languages fall to logging, mining, and plantation-style agriculture. Several hundred languages spoken by small numbers of indigenous peoples in Indonesia and the Philippines face a similar fate as the fingers of industrial capitalism crawl into the rainforests.

The United Nations estimates that up to 90 percent of the world's indigenous languages could cease to be spoken during the present century, and with their demise will disappear much valuable knowledge about nature ("U.N. Warns," 2001). This traditional knowledge describes how to manage habitats and the land in environmentally sustainable ways that has been passed down by word of mouth over generations. The United Nations, according to a British Broadcasting Corporation report, "Warns of a reciprocal loss of natural medicines and an increasing risk of crop failures. 'Nature's secrets, locked away in the songs, stories, art and handicrafts of indigenous people, may be lost forever as a result of growing globalization,'" the report said ("U.N. Warns," 2001).

COMMON ELEMENTS OF THE BOOK'S DIVERSE CASE STUDIES

Several common activities span the diverse case studies of this book: industrial-scale logging, mining and refining (most often of gold and uranium), and construction of hydroelectric dams.

Industrial-Scale Logging

Industrial-scale logging, as well as the clearing of forests for plantations, resource development, and roadways, is afflicting traditionally oriented indigenous peoples around the world. Driven by increasing demand for wood products, from toothpicks to lumber for houses, deforestation has been particularly rapid in the tropics, with the forests of Indonesia and the Amazon Valley being the two most often-cited examples. Other tropical areas (e.g., Mexico and the rainforests of Central Africa) also are experiencing rapid depletion of forests that have housed and fed indigenous peoples for many thousands of years. The burning of tropical forests is having a substantial effect on the atmosphere's carbon balance. The burning of Indonesian forests during 1997 "released as much carbon as all fossil-fuel emissions in Europe that year" (Gedicks, 2001, 6).

Mining and Refining

Many of the local indigenous struggles described in this book relate to mining. According to one observer,

> Mining activity often poses a tough and divisive dilemma for indigenous peoples and their communities. On the one hand, mining activity and its associated social and environmental impacts pose a threat to indigenous ways of life and livelihood and to cultural and spiritual sites of importance. On the other hand, in some instances indigenous communities have welcomed mining as an avenue of regaining economic self-sufficiency in the face of the enormous impacts of colonization. (Ali and Behrendt, 2001)

Advances in technology in recent years have benefited the mining industry. These advances have reduced the cost of mining at all stages, from exploration to production. Sophisticated equipment has accurately pinpointed the location of mineral deposits, efficiently analyzed the quality of these minerals, and determined the quantity of extractable ore reserves within any given site. Tele-mining, for example, allows the entire cycle of mining to be performed by remote operation.

Particular attention is paid in the entries that follow to methods of mining gold that involve the use of toxic cyanide, which bonds to the metal. During the 1960s, the Newmont Corporation of Colorado joined with the United States Bureau of Mines to invent a cyanide-based technique that was found to remove 97 percent of available gold from ore extracted from the deserts of Nevada. This low-grade ore was being extracted from the ancestral lands of the Western Shoshone. The use of this cyanide-heap leach technology has been combined with powerful explosives and massive earth-moving equipment, which allows today's gold miners to pulverize entire mountains of low-grade ore to extract gold that may comprise a fraction of one percent of overburden. The rest becomes tailings, or waste rock. Cyanide-heap leaching makes possible recovery of gold from very low-grade ores. Using massive equipment, miners dig large open pits and crush the rock. Ore is then heaped on a liner and sprayed with a cyanide solution. Miners recover the gold and recycle the cyanide solution to the next ore heap ("Global Response," 2001).

Cyanide is very toxic to all forms of life. Ingestion of a solid cyanide pellet the size of a grain of rice is lethal to a human being; small amounts also kill fish, birds, and mammals other than humans. In the leaching process, cyanide also dissolves toxic metals such as arsenic, lead, zinc, uranium, mercury, and cadmium, allowing them to contaminate water and soils. Toxic metals accumulate in living tissue and are passed through the food chain, causing several illnesses in humans and other animals. Acid formed during the extraction process often continues to drain from mines long after all available gold has been extracted, contaminating water decades after the mines are closed.

Cyanide poisoning can occur through inhalation, ingestion, or contact with skin or eyes. Cyanide blocks the absorption of oxygen by cells, causing victims to suffocate. Human beings may experience decreased respiratory and thyroid functions, cardiac pain, vomiting, headaches, and central-nervous-system toxicity from oral exposure to low levels of cyanide. Short-term exposures to high levels of cyanide compounds also may contribute to central-nervous-system toxicity and gastrointestinal corrosion. Adverse impacts of cyanide on fish have been reported at levels of 0.01 parts per million, while concentrations as low as 5 parts per billion have been found to inhibit reproduction of fish. Levels of 0.03 p.p.m. may kill fish (Chatterjee, 1998).

During 1998, the state of Montana passed a ban on all new open-pit cyanide processes for gold mines. In 2001, the State Senate of Wisconsin passed a similar measure. Also during 1998, the Turkish Supreme Court found in favor of citizens who claimed that a gold mine violated their right to a clean environment. In Canada, citizens successfully blocked the Windy Craggy Gold Mine. In the United States, activists stopped the New World Gold Mine near Yellowstone National Park and the Crown Jewel Mine in Washington State.

URANIUM AND INDIGENOUS LANDS

Approximately two-thirds of the uranium deposits on federal lands in the United States are located on Indian reservations. During the mid-1970s, almost 100 percent of all federally produced uranium was mined on Native lands. During this time, Indian nations collectively were the fifth-largest producer of uranium in the world. Historically, Native peoples were employed at low wages to work in the thousands of mines in the Southwest. Several hundred Navajo uranium miners have since died of lung cancer, a disease that was virtually unknown among them before uranium mining began.

For each ton of uranium oxide yellowcake that reaches the market, several thousand tons of tailings remain behind. Such tailings hold up to 85 percent of the ore's original radioactivity, which has been leaching into the soil, air, and drinking water of Native American homelands. On the Navajo Nation alone, more than a thousand abandoned uranium mines remain contaminated. In 1978, the U.S. Nuclear Regulatory Commission reported "uranium mining and milling are the most significant sources of radiation exposure to the public of the entire nuclear fuel cycle, far surpassing nuclear reactors and nuclear-waste disposal" ("Uranium/Nuclear," 2001).

Some of the uses to which uranium has been put on Native lands illustrate a severe disregard for human and animal life. To cite one local example of many: the Sequoyah Fuels facility, which is used to refine uranium oxide for the production of plutonium for the U.S. military, is located on Cherokee land, named by its owner, the Kerr-McGee Corporation, in honor of the Cherokee alphabet's inventor. Sequoyah Fuels also convinced the state of Oklahoma to approve a fertilizer called Raffinate, manufactured from processed uranium, which was used on local fields. Use of the material was stopped after it was implicated in the deaths of large numbers of cattle. "As a direct result of the relentless and sustained community organizing efforts of the Native Americans for a Clean Environment (N.A.C.E.)," said a commentary on the Indigenous Environmental Network, "Sequoyah Fuels was finally forced to close down in 1993" ("Uranium/Nuclear," 2001).

The Toll of Hydroelectric Dams

The building of hydroelectric dams is another common source of indigenous confrontations with industrialism. Today, on a worldwide scale, according to Winona LaDuke, 50 million indigenous peoples live in the world's rainforests, a million of whom are slated for relocation by dam projects during the next decade (LaDuke, 1995). As many as 80 new hydroelectric plants were being planned for the Amazon Valley alone during the late 1990s ("Cotingo Dam," 1998, 15). Inundation of traditional lands by dams has been a common complaint of indigenous peoples the world over, as well as a source of militant physical actions to interfere with the construction and operation of dams.

One of the most dramatic such actions has occurred in Thailand, where many hundreds of people marched from rural areas to the grounds of the national legislature in Bangkok. These people were dispersed by riot police as they poured into legislative chambers to protest the ruination of their lands by dam building. At home, many of these peoples have occupied dam powerhouses to interfere with dams that destroyed life-sustaining fish runs. Some Thai natives hung their laundry on fish ladders to illustrate just how ineffectual they are at guiding fish upstream.

According to the World Bank, the construction of 300 large dams worldwide will provoke displacement of more than 4 million indigenous people. The World Commission on Dams is leading international activities to highlight the issue of dams and their impact on indigenous peoples ("Indigenous People," 2000). In India, where more than 600 out of a planned 1,600 dams are currently under construction, 40 percent of the people displaced are indigenous Adivasis. Almost all the larger dam schemes proposed in the Philippines are on the land of the country's 6.5 million indigenous people. The majority of the 58,000 people evicted to make way for Vietnam's Hoa Binh Dam come from ethnic minority groups, as are most of the 112,000 who will be displaced by the proposed Ta Bu dam ("Indigenous People," 2000).

This book comprises a worldwide collection of such local invasions, usually in search of resources, in disregard of local lives and traditions, usually inspiring some degree of resistance from peoples who want only to continue to live in ways they know, rather than as adjuncts of an industrial system that is alien to them and dangerous to their traditions. Indigenous peoples should survive in their natural habitats not only for their own good, but because all the peoples of the world have an interest in a diversity of historical experience. Indigenous experience is an encyclopedia of vital knowledge about the animals and plants with which they live. "Enshrined in their cultures and customs are also secrets of how to manage habitats and the land in environmentally friendly, sustainable, ways," writes an observer ("Globalization Threat," 2001, n.p.). Losing a language and its cultural context is, according to the United Nations Environmental Program, "like burning a unique reference book of the natural world" ("Globalization Threat," 2001, n.p.). This book seeks to keep this natural reference book in print, and in circulation.

REFERENCES

Ali, Saleem, and Larissa Behrendt, eds. "Mining Indigenous Lands: Can Impacts and Benefits be Reconciled?" *Cultural Survival Quarterly* (spring 2001): n.p. [http://www.ienearth.org/mining_campaign_1b.html#spring, 2001] [http://www.cs. org/publications/CSQ/index.htm]

Black Elk. *Black Elk Speaks, as told to John G. Neihardt*. New York: William Morrow, 1932.

Chatterjee, Pratap. "Gold, Greed and Genocide in the Americas: California to the Amazon." *Abya Yala News: The Journal of the South and Meso-American Rights Center* (1998): n.p.

"Cotingo Dam in Brazil Is Halted, Sparing the Macuxi and the Ingarico." *Native Americas* 12, nos. 1 and 2 (summer 1998): 15.

Gedicks, Al. *Resource Rebels: Native Challenges to Mining and Oil Corporations*. Cambridge, Mass.: South End Press, 2001.

"Globalization Threat to Cultural, Linguistic and Biological Diversity." *U.N.E.P. Businessworld* (Philippines), February 15, 2001, n.p. (in LEXIS)

"Global Response: Environmental Education and Action Network." October, 2001. N.p., n.d. [http://www.globalresponse.org/gra/current.html]

Grossman, Zoltan. Personal communication with author. Wisconsin Campaign to Ban Cyanide in Mining and the Midwest Treaty Network, November 6, 2001.

"Indigenous People Suffering Gross Rights Violation!" *Free Press Journal*, August 9, 2000. [http://www.indiaworld.co.in/news/features/feature495.html]

Johansen, Bruce E. "Pristine No More: The Arctic, Where Mother's Milk Is Toxic." *The Progressive*, December 2000, 27–29.

LaDuke, Winona. "The Indigenous Women's Network: Our Future, Our Responsibility." Statement of Winona LaDuke, Co-Chair Indigenous Women's Network, Program Director of the Environmental Program at the Seventh Generation Fund, at the United Nations Fourth World Conference on Women, Beijing, China, August 31, 1995. [http://www.igc.org/beijing/plenary/laduke.html]

Lovell, Jeremy. "U.N. Rings Disaster Warning Bell over Plundered Earth." *The Advertiser*, November 8, 2001, 6.

Nikiforuk, Andrew. "Echoes of the Atomic Age: Cancer Kills Fourteen Aboriginal Uranium Workers." *Calgary Herald*, March 14, 1998, A-1, A-4. [http://www. ccnr.org/deline_deaths.html]

O'Connor, Michael. "Afghan Caves Hold More Than Bin Laden; A U.N.O. Professor Says Copper and Iron Deposits Can Be Mined and Help the War-torn Country Get Back on its Feet." *Omaha World-Herald*, November 22, 2001, 1-B.

"U. N. Warns over Indigenous Tongues." British Broadcasting Corporation, February 8, 2001. [http://news.bbc.co.uk/hi/english/sci/tech/newsid_1161000/1161406.stm]

"Uranium/Nuclear Issues and Native Communities." Indigenous Environmental Network, 2001. [http://www.ienearth.org/nuciss.html]

ARGENTINA

INTRODUCTION

Following a struggle dating from the first Spanish *conquista*, the existence of indigenous peoples in Argentina was first recognized by the Argentine legal system during 1985. By 2001, 24 indigenous nations lived in Argentina, totaling more than 1.5 million people. Despite their legal recognition, many native peoples in Argentina find their traditional economies threatened by increasing deforestation, especially by industrial-scale agricultural production. Some peoples, such as the Kolla, continue to struggle for practical ownership of their land even after provision of legal guarantees; others, such as the Mapuche, find their lands threatened by oil contamination. The Wichí continue to resist inundation of their homelands by hydropower development and industrial-scale logging.

THE KOLLA: A STRUGGLE FOR LAND TENURE

The indigenous Kolla live in the Argentine departments of Iruya, Santa Victoria, Los Andes, La Poma, Cachi, and Oran in Salta Province, as well as in parts of Jujuy Province. Some Kolla also live in neighboring southern Bolivia and northern Chile. The number of Kollas, who live in communities called *ayllus*, has been estimated at 120,000 people. The Kolla indigenous people who live in the northern Argentine provinces of Jujuy and Salta are defending their traditional residency in the *yungas*, one of the last remaining mountain forests in Argentina.

Conflicts over land-tenure rights between the Kolla communities and landowners have centered on the Santiago Estate in the department of Iruya. For several decades, Kolla living in the communities of Colanzuli, Volcan Higueras, Isla de Canas, and Rio Cortaderas, a total of about 3,000 people, have been seeking the return of their former land base (which first was lost during the Spanish invasion).

Many of the Kolla have been forced off their land and into the cash economy on the estates of the colonists, for a miserable salary ("Argentina: The Struggle," 1997). Most often, Kollas have worked for the sugar-cane company San Martin del Tabacal, which has strongly resisted the return of local indigenous land ownership. As early as 1950, when the Spanish company Manero-Quiroc bought portions of the Santiago Estate for logging, indigenous people were employed there to cut trees.

A 1993 Argentine law recognizing the Kollas' legal right to exist also returns these lands to them, in legal principle. However, as of 2002, landed interests in the area had refused to obey the law, and central Argentine authority had not enforced it. In August 1996, during the Kollas' celebrations of *Pachamama* (Mother Earth month), the Kolla blocked a road and obstructed traffic in pursuit of their rights to the land. The blockade was "violently repressed by the police," according to reports carried in the *World Rainforest Movement Bulletin*. On March 19, 1997, the Kolla took legal possession of the Santiago Estate. "Our Mother Earth was on our side," stated Festo Chausque, one of the Kolla leaders ("Argentina: The Struggle," 1997).

In a similar struggle, the Kolla communities of San Andres, Santa Cruz, and Angosto de Parani in the department of Oran have been fighting for ownership of the San Andres Estate. In 1986, San Martin del Tabacal (a subsidiary of the U.S. Seaboard Corporation) donated these lands to the Province of Salta for transfer to the Kolla. "Nevertheless," reported the *World Rainforest Movement Bulletin*, "this was never realized and still in January 1997 the company continued to exploit forests under a special authorization from the Government of Salta" ("Argentina: The Struggle," 1997).

Even as some of their land is being returned, the Kolla have faced a new intrusion on their lands by a proposed natural gas pipeline. The proposed pipeline would transport natural gas from eastern Salta to copper mines in northern Chile. During April 1998, ENARGAS, an Argentine regulatory body, approved the gas-pipeline project as proposed by Consorcio Norandino, under which the pipeline would cross Finca San Andres, which is inhabited by roughly 350 Kolla families, most of whom oppose it. The Kollas sued in Argentine courts and won a stay of construction because the company lacked an adequate environmental assessment of the project's effects on the land and its resident indigenous population.

Following that ruling, Salta's provincial government obtained the support of people in the nearby town of Oran, who had been promised jobs at a new plant of a company named Techint, which was to be built in association with the pipeline. The Argentine Federal Court of Appeal quickly revoked the federal judge's decision and reauthorized the project. Heavy machinery employed by Techint quickly opened a road 12 meters wide through areas of the forest, which included Kolla cemeteries and archeological sites. Local people worried that earth exposed by such road building could provoke landslides, further damaging their forested homelands. Landslides also could damage the pipeline and lead to gas leaks during episodes of flood-bringing rains that are common in the mountains ("Pipeline Project," 1998).

Festo Chausque, one of the Kollas' leaders, observed that "the Government has everything on its side: judges, politicians, mass media. But they have forgotten that we have fought for 500 years to recover our lands and natural resources. For the non-pollution of the water, the air, the soil. For our cultural values and cosmovision. In sum, for our existence as human beings in harmony with the surrounding nature" ("Pipeline Project," 1998).

THE MAPUCHE: OIL CONTAMINATION

The Mapuche community is settled in the Loma de la Lata zone, which contains large gas and petroleum deposits that are being exploited by Repsol. A study carried out by the undersecretary for health in Neuquen indicated that Mapuche living in the Loma de la Lata region, and particularly children and elderly people, have been victimized by high concentrations of heavy metals, mainly lead, in their blood and urine. The contamination has probably spread from contaminated drinking water, plant matter, and animals. Children in the same community also have reported problems with mental concentration, progressive loss of eyesight, painful joints, and kidney problems ("Demand Justice," 2001). On October 12, 2001, a crowd of some 30 youths protested the gradual contamination of underground water in their region, which had been poisoning people in their community. Roughly 30 police attacked these Argentine Mapuche young people, aged 6 to 17, to prevent them from painting slogans on the walls of the Spanish oil company Repsol-YPF in Neuquen, Argentina.

THE WICHÍS' OPPOSITION TO HYDROELECTRIC DEVELOPMENT

Once a fertile grassland dotted with bushes and trees, large parts of the indigenous Wichí homeland have become a dry, sandy desert. Shimmering, chest-high grasses that once spread over the Wichí homeland have disappeared, along with many of the animals the Wichí once hunted ("Wichí: Fighting for Survival," n.d.). The Wichí now find their homelands encircled by colonists as their food supplies have been reduced by desertification, which has produced "a sandy desert where a grassland ecosystem once thrived" ("Wichí: Fighting for Survival," n.d.). The lack of water is reflected in declining game and harvests.

Roughly 20,000 to 50,000 Wichís live in southeastern Bolivia and northern Argentina in villages of matrilocal clans, whose members share small houses constructed of mud, branches, and leafy boughs built to withstand summer temperatures that sometimes reach 50 degrees Celsius. Members of neighboring peoples, including the Iyojwaja, Nivaklé, Qomlec, and Tapy'y, often live among the Wichí, sometimes marrying into their society ("Wichí: Fighting for Survival," n.d.).

For almost a century, the Wichí have resisted takeover of their land by outsiders, so that they may continue to fish as they have for centuries.

Standing waist-deep in the muddy water, holding nets strung between two poles, the Wichí fisherman detects the fish by noting movements in the river's surface. Plunging the net over the fish and swinging downward, the catch is enveloped in the trap. Swiftly and with minimal impact on the aquatic environment, a natural resource yields a nutritious meal. ("Wichí: Fighting for Survival," n.d.)

Given enough moisture, the Wichí also cultivate corn, watermelons, beans, and pumpkins, which are encircled with thorny branches to keep out the colonists' roaming cattle. During the dry winter months, the Wichí subsist on fish from the Pilcomayo River. The Wichí also hunt now-dwindling numbers of deer, armadillos, and iguanas. They also harvest wild honey throughout the year.

In addition to the travails of colonization and desertification, the Wichís' Pilcomayo River may be altered by the Paraguay-Paraná Hidrovía industrial waterway project, which is being promoted by the governments of the La Plata river basin. The project would require widening and deepening the channels of the Paraguay and Paraná rivers, which are part of South America's second largest river system (after the Amazon), to allow oceangoing ships access to the port of Cáceres, Brazil, 2,100 miles upstream from the river's mouth. Plans call for draining of wetlands adjacent to the river. According to one observer, "for the indigenous peoples dependent on the rivers targeted by Hidrovía, which includes the Wichí, the environmental impacts could be devastating, worsening their already precarious living conditions" ("Wichí: Fighting for Survival," n.d.).

The Salta provincial government also has attempted to allocate land to individual owners, giving Wichís and colonists legal rights to equal-sized parcels. The Wichís assert that such allocation denies them access to large areas now used communally. Instead, the Wichís want legal guarantees to their traditional lands, roughly 162,000 hectares, which recognize traditional communal ownership patterns. In the meantime, Wichís living on these lands complain that they have been barred from hunting, sometimes at gunpoint, by nonnative colonists, as well as prevented from gathering wild fruits and berries. Some say they have been denied access to waterholes that are increasingly crucial to survival as the local climate becomes progressively drier. Settlers' cattle, lacking grass on which to feed, often invade and destroy indigenous people's gardens.

A Wichí elder was quoted in a 1994 report by Survival International as saying: "They [the colonists] threaten us saying, 'Indian, don't come around here. I own this land and I don't like Indians on it. If you want to hunt here, you must ask for my permission—or I'll kill you.' . . . They don't own those resources. The things that we Wichí live on do not belong to anyone. They belong to God" ("Wichí: Fighting for Survival," n.d.). In response to such threats, the Wichí issued a collective statement, which said, in part: "We are not animals running loose. We are not dogs to be driven away at the whims of their owner. We are the flowers of the Earth, planted by God Himself to live and thrive in these lands" ("Wichí: Fighting for Survival," n.d.).

During the late 1990s, a bridge was built across the Pilcomayo River (on the border between Argentina and Paraguay) adjacent to an Indian village called Nop'ok 'Wet ("La Paz"). The Wichí in the village were told that a frontier town would be established at the base of the bridge. To resist these changes, the Wichí formed an organization called Thaka Honat ("Our Land"), which included representatives from each village. Next, the Wichí began to consider a blockade of the new bridge.

Faced with a lack of official response before the inauguration of the new bridge, the 35 communities belonging to Thaka Honat decided to peacefully occupy the lands around the bridge to prevent its completion. A statement issued by the protesters said, "We will occupy the land until the government of Salta gives a concrete response in regards to our requests. This is an act of hope" ("Wichí: Fighting for Survival," n.d.).

As some Wichí protested development of this bridge, others fought eviction from their homes by loggers. The people in the Wichí community of Hoktek T'oi found themselves facing eviction from a tropical rainforest that had been their home for at least 12,000 years before the arrival of the first European-descended Argentines. The first visitors were quickly followed by loggers who began to cut the forests that sustain the Wichí. Shortly after that, a struggle over ownership of the forest ensued, during which Argentine authorities recognized ancestral ownership rights to only 27 of 75,000 hectares comprising the community's traditional range (Carrere, 2001).

Since 1910, legal title to the Hoktek T'oi community's land has changed hands several times. Most recently, in 1966, ownership was transferred to the agricultural company Los Cordobeses. Shortly after assuming ownership, the company tried to move indigenous people out of their traditional territory. The people resisted, asserting that the company wanted to move them to land that would flood. The company then built a fence around the village's 27-hectare reservation. The company set about clear-cutting large parts of the surrounding forest, making use of a deforestation permit granted by the government of Salta (Carrere, 2001). Wichí complaints asserting cultural ethnocide and irreparable social and environmental damage caused by the deforestation were ignored.

The *World Rainforest Movement Bulletin* described the scene as follows:

The millenary forest is being eliminated with heavy machinery and chains; the tree trunks, branches and roots are burnt. The plantations are sprayed from the air and so are the people from the community; there have even been attempts at destroying their homes and their graveyard. . . . Presently, the Hoktek T'oi community is a green island in the middle of brown fields, where the forest has definitively been destroyed and [replaced] by agricultural plantations. The company, not satisfied with what it had already devoured of the forest, attempted several times (with a bulldozer, with the police, with hired staff and with a notary public), to make the green island housing the community even smaller. They wanted to cut it down to one-third and evict the community. (Carrere, 2001)

Following intervention by the World Rainforest Movement and Argentine courts, the Wichí finally achieved legal recognition for possession of 44 hectares, "maintained," wrote Carrere, "as an oasis of life surrounded by the depredation caused by the company 'locusts'" (Carrere, 2001).

REFERENCES

"Argentina: The Struggle of the Kolla People." *World Rainforest Movement Bulletin* 5 (October 1997): n.p. [http://www.wrm.org.uy/bulletin/5/Argentina.html]

Carrere, Ricardo. "Argentina: Forest Conserved by the Wichí Destroyed by Agricultural Companies." *World Rainforest Movement Bulletin* 49 (August 2001): n.p. [http://www.wrm.org.uy]

"Demand Justice for Mapuche Youth Brutally Repressed for Protesting Against REPSOL." *Drillbits and Tailings* 6:9 (November 30, 2001): n.p. [www.moles.org]

"Pipeline Project Opposed by Argentinian Kollas." *World Rainforest Movement Bulletin* 18 (August 1998): n.p. [http://www.wrm.org.uy/bulletin/5/Argentina.html]

"Wichí: Fighting for Survival in Argentina." *Abya Yala News: The Journal of the South and Meso-American Rights Center.* n.d. [http://saiic.nativeweb.org/ayn/wichi.html]

AUSTRALIAN ABORIGINES

INTRODUCTION

When uranium mining was initiated on their lands, Australia's Aborigines were promised that it would be their ticket to the modern world. Decades later, promised jobs were nearly nonexistent, housing was substandard, and stretches of customary aboriginal homelands, piled high with waste tailings, were unusable because of residual radioactivity. In addition to problems associated with uranium mining, Australian Aborigines also have reported health problems stemming from nuclear testing in the neighboring South Pacific during the mid-twentieth century; after the tests, fallout (which the native peoples called a "black mist") was carried over their homelands by prevailing winds. Elsewhere in Australia, native peoples are resisting industrial-scale gold mining that may replace sacred sites with open pits.

URANIUM MINING

The Mirrar Aborigines of Australia's Northern Territory have been resisting development of new uranium mining within their territory. They contend that similar projects showed the proposed Jabiluka mine could destroy their way of life. The nearby Ranger uranium mine, environmentalists argue, provided a disastrous environmental precedent. Mine workers came to greatly outnumber Aborigines, having a severe impact on local Aboriginal people and depriving them of their traditional economic systems ("Australia: Aborigines Fight Mine," 2001). The Mirrar Aborigines fear that they will face a similar fate if the Australian government approves a proposal by Energy Resources, an Australian company, to develop Jabiluka. Development of the mine would leave the Mirrar with millions of tons of radioactive waste, creating the potential for several health problems.

Initially the Mirrar agreed to allow uranium mining being led to believe that approval was their only way to secure legal rights to their land. An Australian Senate inquiry and the United Nations have criticized the tactics used to obtain this agreement. Between 1979 and 1988, the Nabarlek uranium mine in West Arnhem Land, owned by Queensland Mines, extracted, stockpiled, and processed 11,000 tons of ore. This open-pit mine was constructed despite opposition by many local Aboriginal people, who staged a sit-in on the mine's access road and later took Queensland Mines to court. The mine was less than one mile from an area of special significance to Aboriginal people, the Gabo-djang (Dreaming Place of the Green Ants).

During March 1981, after heavy rain from a tropical cyclone, radioactive material from the mine's tailings dumps was released into a nearby creek. After the mine was closed, required cleanup work was not completed, leaving local Aboriginals with a pile of radioactive rubbish. Given such experiences, many Australian Aborigines have opposed new proposals to mine uranium on or near their traditional lands. According to Vincent Forester, writing under the aegis of Australia's Sustainable Energy and Anti-Uranium Service, the 'Aboriginals' reasons for opposing new mining include:

- seepage from existing tailings dams;
- concentration of radioactive contaminants in water systems;
- soil erosion;
- radon gases escaping from the tailings;
- the fact that cyclones could disperse contaminated dust from strip-mining operations;
- the fact that return of the tailings to the pit at the end of mining operations poses long-term effects in the Alligator Rivers area;
- major geological faults in the wall-rock of the pit area;
- the presence of a geological fault under the north wall of the Ranger tailings dam; [and]
- contaminated water release into Magela Creek. (Forrester, 1997)

RADIOACTIVITY REDUX? THE JABILUKA URANIUM MINE

Aboriginals and environmentalists have called upon uranium miner Energy Resources of Australia to rescind plans for the Jabiluka mine, which would adjoin the Kakadu National Park, an area made famous by the Crocodile Dundee films. The Kakadu Park houses an extraordinary ecosystem that the Aboriginals endowed with spiritual significance from ancient times. Along with some of the richest uranium deposits in the world, the area also is home to communities of Australian Aboriginals who comprise one of the world's oldest surviving indigenous populations.

Energy Resources of Australia (ERA) is majority-owned by global mining corporation Rio Tinto, whose chairman said during the late 1990s that development of the mine was only a remote possibility. The company refused to back away from the project altogether, however. Rio Tinto's chairman of the board, Sir Robert Wilson, said that the company was not pursuing the mine at present, because current and foreseeable market conditions indicated that investment in Jabiluka was "economically unattractive" ("Hotspots: Australia," 2001). Wilson left open the possibility that market conditions could change.

The Mirrar people regard the area as their ancestral home and point to the damage done over the years by the Ranger mine, which has left 20 million tons of radioactive tailings in spoil heaps around its operations. According to Friends of the Earth, there have been 120 breaches of the mine's operational guidelines, most recently in May 2000, when 2 million gallons of radioactive liquid contaminated with manganese, uranium, and radium was released. Some of this contamination escaped into the Kakadu wetlands (Brown, 2001).

Ed Matthew, from Friends of the Earth in London, said: "This [Jabiluka] mine is on land unjustly wrested from the aboriginal people and inside a World Heritage site. If Rio Tinto proceeds with this mine, it will be telling us that there is no place on Earth the company is not prepared to plunder" (Brown, 2001).

Extralegal means have been used by the Mirrar to protect their country and sacred sites, including a blockade during 1998. Despite the blockade, construction work at the mine was delayed for little more than a few hours. Energy Resources of Australia used helicopters to fly its workers into the mine compound for several weeks during April and May, when blockaders cemented cars into place, blocking a mine gate. Police later cleared the obstruction. Subsequently, the blockade was cleared with bulldozers.

Protesters then resorted to mass trespass on the mineral lease area, occasionally locking themselves to trucks, gates, and mining equipment. Similar protests also took place at the Ranger mine. Mass trespass actions often resulted in large numbers of arrests. During these 1998 protests, as many as 118 people were taken into custody on one June morning alone. The last of several protests took place during the week preceding the federal election on October 3, 1998, producing more than 90 arrests, as protesters walked onto the lease area wearing masks depicting Australian Prime Minister John Howard. A few days later, after Howard won reelection as prime minister, the blockade camp was dismantled as the monsoon season set in.

Yvonne Margarula, a leader of the Mirrar people, has been active in the antimining movement, along with Jacqui Katona, another Mirrar leader. Both have said that the Mirrar people oppose the mine for two reasons: firstly, it will devastate a broad area that includes many sacred places; secondly, the Mirrar have a very real fear of the mine's potential for radioactive poisoning of their land, which is likely to result from the release of radon gas into the atmosphere. Radon is a heavy gas that stays close to the Earth's surface, contaminating all life in its path, according to the mine's opponents (Daters, 1997).

By the year 2000, most Australian Aborigines were united against further uranium mining on their traditional lands. This opposition was reflected in a statement by senior traditional owners of the Jabiluka mineral lease, the Gundjehmi people, and the Alliance Against Uranium, a global coalition of numerous Aboriginal and environmental groups from all over Australia. Katona, who works for the Mirrar people in Kakadu National Park, said:

> [The] Mirrar Gundjehmi, Mirrar Erre, Bunitj, and Manilakarr clan leaders have many concerns about mining in their homelands. A new mine will make our future worthless and destroy more of our country. We oppose any further mining development in our country. We have no desire to see any more country ripped up and further negative intrusions on our lives. ("Traditional Owners Statements," 1997)

Katona, supported by the Gundjehmi Aboriginal Corporation, said that the Aborigines' Northern Land Council "was told [by federal negotiators] that if approval was not given by traditional owners the Land-rights Act would be dismantled. . . . In the words of one of the Land Council members, the Ranger [uranium mine] agreement was signed through lies and trickery. . . . [which] condemned our people as passive recipients of the consequence of resource development" ("Traditional Owners Statements," 1997). Katona continued: "We have watched our people die in the shadow of industrial gain. Our community is regarded as fringe dwellers in our country. As soon as the ink was dry on the agreement, our citizenship rights were withdrawn. . . . No other citizens are asked to make this sacrifice" ("Traditional Owners Statements," 1997).

Katona's beliefs are similar to Aboriginal points of view expressed to the Australian federal government after it created an official inquiry to anticipate the social and economic impacts of uranium mining on Aboriginal communities. The inquiry concluded that Aboriginal people have a clear and unquestionable right to successfully claim their lands; it also found that a large majority of Australian Aboriginal people have been strongly opposed to the mining of uranium.

"There were many promises made to our people about the benefits of mining," Katona said. "School, housing, employment, health services and investments. Well, we have no aboriginal graduates of secondary education. Housing is substandard. The vast majority of the community is unemployed. Health services are minimal. And those strategic investments are losing value each year" ("Traditional Owners Statements," 1997). Instead, development of uranium mining has come with increased Aboriginal consumption of alcohol and dependence on welfare.

Traditional landowners among the Aboriginals argue that ERA, the leaseholder for the Ranger and Jabiluka uranium deposits, "releases contaminated water from the Ranger mine into our wetlands every year" ("Traditional Owners Statements," 1997). Katona declared that "nothing can replace our country when it is mined. Nothing can reverse the damage to our water system and our food sources. Our culture can't be replaced by money. Inherent in our laws and culture is an obligation to protect and preserve our homeland for future generations. It isn't negotiable. It isn't a matter of convenience" ("Traditional Owners Statements," 1997).

Katona believes that "stopping the Jabiluka mine is the first step in changing the future for our community. We have a responsibility to our children, and grandchildren and their children, to strengthen their heritage by acting now. This is our future. Without industrial domination. Without aggression. With meaningful positive change. For us, by us" ("Traditional Owners Statements," 1997).

According to Vincent Forrester, chairperson of the Northern Territory National Aboriginal Conference, "There is simply no proper information given to Aboriginal people living in the area about the effects of uranium mining on the land. The monitoring scientists have made no attempt to interpret their findings to the affected Aboriginal people" (Forrester, 1997).

Forrester continued:

> The local aboriginal community has no involvement in this and must depend on the government or on statutory bodies dependent on royalties from uranium mining. . . . This dependency, I believe, is a form of ransom. We must break this dependency on mining activity for money for essential services. It is morally bankrupt. No aboriginal community should be put in the position of deciding on development that is tied to the uranium industry. Until all aboriginal service needs are met by direct grants from the federal treasury, our people have little choice in this matter. (Forrester, 1997)

Forrester also said that no substantial study had been done of radiation levels in Aboriginal people's diets in the uranium-mining regions. Because they lack crucial information, he said, "We can only guess what amount of radiation they have in their bodies or in the food chain" (Forrester, 1997). Closed uranium mines also pose problems for Aboriginal people. One such mine, Rum Jungle, was abandoned in 1971. Its tailings dam has been breached by monsoon rains, which have polluted the Finniss River with radioactive materials. Aboriginal people who live in the area can no longer safely use the affected land.

Some Australian Aboriginal peoples also may have sustained health damage following dumping of radioactive materials from other mines. Fifteen thousand gallons (60,000 liters) of radioactive liquid was sprayed onto the ground at the Beverly uranium mine in South Australia about 300 miles (520 kilometers) north of the city of Adelaide early in 2002. The spill was one of 24 spills of radioactive liquid at the mine during the previous two years. Australian environmental groups are calling for the closure of the mine, which is run by U.S.-based Heathgate Resources.

The closest community to the mine is an Aboriginal settlement about 37 miles (60 kilometers) away. The company and the government deny any environmental damage or harmful exposure to mine workers. The Australian Conservation Foundation (ACF), an environmental group opposed to *in situ* leach uranium mining, a technique used at the Beverly mine, said that the mine should not be restarted until there has been an independent assessment. The ACF has also said that the *in situ* leaching process has caused serious pollution in Eastern Europe ("Hotspots: Australia," 2002).

Anti–uranium mining demonstration at Jabiluka mining lease, led by Yvonne Margarula and Peter Garret. Protesters believe the mine will endanger the neighboring Kakuda National Park in northern Australia. (Painet)

Radioactive leaks have become a constant problem on Mirrar lands. Early in 2002, a uranium leak from the Jabiluka and Ranger uranium mines contaminated Swift Creek in Kakadu National Park. Resident Mirrar people said that ERA, the company that owns and operates the two mines, waited six weeks to notify them of the leak ("Hotspots: Australia," 2002). According to a report in *Drillbits and Tailings,* elevated uranium levels also were detected at three checkpoints within the two different mines during January. *Drillbits and Tailings* reported that, according to Andy Ralph, executive officer of the Gundjehmi Aboriginal Corporation (which represents the Mirrar traditional owners), a retention pond at Jabiluka was too small. Because of its elevated position on the Arnhem land escarpment, water was sure to run off into the wetlands that surround the mines and connect to the Magela river system used by the Mirrar ("Hotspots: Australia," 2002). Tests revealed that uranium levels in the creek had reached almost 2,000 parts per billion, 4,000 times the drinking-water standard.

During spring 2002, yet another large leak was revealed at the Ranger mine, raising renewed protests from the Mirrar people. Australia's Office of the Supervising Scientist released a report that said the internal management at the scene in charge of the mine had failed when a uranium leak occurred earlier this year.

Andy Ralph from the Gundjehmi Aboriginal Corporation, which represents Mirrar traditional owners, said that the latest leak was seven times larger than the one discovered earlier in the same year. "There is a concern that a lot have bypassed the Magela Creek and bypassed their wetland filter and did not actually get filtered," he said ("Traditional Owners Concerned," 2002).

Early in September 2002, Rio Tinto effectively abandoned its efforts to open the Jabiluka mine without aboriginal consent. Speaking on the British Broadcasting Corporation's "World Hardtalk," Rio Tinto Chairman Sir Robert Wilson said, "There would be no development of that project without the consent of the traditional landowners, the Mirrar people. . . . We won't develop it without their consent, full stop" ("Uranium Mine," 2002).

A senior traditional leader of the Mirrar, Yvonne Margarula, reaffirmed her long-standing opposition to the mine. Margarula said, "It doesn't matter how many times they ask, I'm not going to agree to this mine, whatever money they ask for it. Mining ruins the land. Just like the way the other Rio Tinto uranium mine, Ranger, has destroyed my land. My mind is firmly set" ("Uranium Mine," 2002).

The Mirrar challenged Rio Tinto to rehabilitate the Jabiluka mine site, where a 1.2 kilometer underground tunnel was drilled before the site was acquired by Rio Tinto. Wilson committed his company to rehabilitation of the mine site, stating on the BBC that Rio Tinto will "rehabilitate that area" and "block off the adit [mine tunnel], but this is not a very large area, nor in any way is it a threat to the environment" ("Uranium Mine," 2002).

NUCLEAR TESTING

Australian Aborigines also have suffered health problems following tests of atomic bombs in the neighboring South Pacific. The Pitjantjatjara and Yaknunytyara peoples believe that many deaths among their peoples during the 1950s and early 1960s can be associated with fallout from the tests.

Following some of the tests, clouds of fallout carried by prevailing winds passed over and adjacent to Australian Aboriginal communities. According to Vincent Forrester, the Pitjantjatjara Council called for a Royal Commission to inquire into the circumstances surrounding the nuclear tests in South Australia in the 1950s and 1960s. Council representatives went to London to lobby over the issue. Leading the delegation was Yami Lester, who lost his sight after a fallout cloud from the first Emu test descended on him and his people (Forrester, 1997).

An Australian Royal Commission, established during July 1984 under the leadership of Justice Jim McLelland, reported in November 1985 that an Aboriginal community at Wallatinna had been exposed to a *black mist* (radioactive fallout). The mist could have caused harm to the people's health, according to the Royal Commission's report. The same inquiry also found that Aboriginal people had been denied access to their traditional lands and that plutonium-contaminated areas at Maralinga should be cleaned up. During 1994, the British government agreed to a limited cleanup, in which plutonium-contaminated soil would be gathered into existing pits of radioactive rubbish and fused into a solid mass.

"No one told my people about the tests at the time and only now, after a barrage of leaks and statements, is the Australian government considering holding a full inquiry into the matter. But the full extent of cancers and other illnesses being suffered by my people may never be known. Aboriginal land in the immediate test area may not be useable for 50,000 years" (Forrester, 1997).

GOLD MINING

Aboriginal landowners and environmentalists in New South Wales, Australia have been battling Canadian mining giant Barrick Homestake to prevent exploratory drilling for gold near Lake Cowal, 47 kilometers northeast of West Wyalong, in Wiradjuri County. The Mooka Traditional Owners Council filed a lawsuit after an Australian court dismissed an injunction that would restrain the subsidiary company, Homestake Australia Limited, from drilling ("Aboriginal Landowners," 2002). The Council hopes that the lawsuit will prevent further exploratory drilling until current National Parks and Wildlife Service inspections have been completed and a report issued in court.

Barrick Gold acquired Homestake, the former parent company of Homestake Australia Limited, during June 2001. The company plans to mine 2.7 million ounces of gold from 76 million tons of ore, creating a 1-kilometer-long by 825-meter-wide and 325-meter-deep open pit on Lake Cowal. According to *Drillbits and Tailings*, "Some traditional owners of the Wiradjuri Nation say that exploratory drilling for the mine is already threatening federally registered sites and that continued development will jeopardize the community's sacred sites and cultural artifacts" ("Aboriginal Landowners," 2002).

"They are tearing up the very fabric of the sacred land. That is our sacred heartland. It is the heartland of the Wiradjuri people. We are not going to give up," said Neville Williams, a representative of the Mooka Traditional Owners Council of the Wiradjuri Nation ("Aboriginal Landowners," 2002). Oshlack, a traditional leader, said the drilling rigs on the site were breaking up the ground and destroying a number of rare stone artifacts, including stone hammers and axes.

Lake Cowal has been proposed as a site for gold mining since 1981, when Australia-based North Limited began exploration there. The company's application to develop the mine was denied in 1996 on environmental grounds, following opposition by environmental groups. Cowal is the largest lake in the state of New South Wales and one of the state's most important wetlands. According to *Drillbits and Tailings*, the lake supports significant numbers of threatened and migratory birds, as well as other animals, fish, and plant species ("Aboriginal Landowners," 2002)

REFERENCES

"Aboriginal Landowners Fight off Barrick Homestake and Gold Mining." *Drillbits and Tailings* 7:1 (January 31, 2002): n.p. [englishdrillbits@topica.com]

"Australia: Aborigines Fight Mine." Survival International Update, July 2001. [http://www.survival.org.uk/about.htm]

Brown, Paul. "Gift of Life." *The Guardian* (London), February 14, 2001. [http://www.urg.org.au/jabiluka/index.htm]

Daters, Michaela Reuss. "Taking a Job at Jabiluka." *Sydney City Hub,* December 31, 1997. [http://www.big.com.au/film/jab31-12-97.html]

Forrester, Vincent. "Uranium Mining and Aboriginal People." The Sustainable Energy and Anti-Uranium Service, Inc., Australia, 1997. [http://www.sea-us.org.au/black uranium.html]

"Hotspots: Australia." *Drillbits and Tailings* 6:7 (October 31, 2001): 3. [englishdrill bits@topica.com]

"Hotspots: Australia." *Drillbits and Tailings* 7:1 (January 31, 2002): n.p. [englishdrill bits@topica.com]

"Hotspots: Australia." *Drillbits and Tailings* 7:3 (March 29, 2002): n.p. [www.moles.org]

"Nabarlek Uranium Mine." N.d. [http://www.urg.org.au/other_mines/nabarlek_intro. htm]

The Parliament of the Commonwealth of Australia. *Jabiluka: The Undermining of Process: Inquiry into the Jabiluka Uranium Mine Project.* Report of the Senate Environment, Communications, Information Technology and the Arts References Committee, Canberra, June 1999.

The Parliament of the Commonwealth of Australia. *Unlocking the Future: The Report of the Inquiry into the Reeves Review of the Aboriginal Land Rights (NT) Act 1976.* House of Representatives Standing Committee on Aboriginal and Torres Strait Islander Affairs, Canberra, August 1999.

"Traditional Owners Concerned at More Leaks from Ranger Mine." Australian Broadcasting Corporation Indigenous News, April 24, 2002. [http://abc.net.au/news/newsitems/s538750.htm]

"Traditional Owners Statements: Statement from the Gundjehmi Aboriginal Corporation." Sustainable Energy and Anti-Uranium Service, Australia, April 1997. [http://www.sea-us.org.au/trad-owners.html]

"Uranium Mine in Australian National Park Dead." Environment News Service, September 6, 2002. [http://ens-news.com/ens/sep2002/2002-09-06-01.asp]

BANGLADESH

GAS WELL EXPLOSION

On June 15, 1997, 12 days into initial drilling, Occidental Petroleum's Moulavi Bazar #1 gas well in northern Bangladesh experienced a major blowout. The gas explosion grew into a 300-feet-high inferno that spread to surrounding rainforest, cropland, and villages. The explosion initiated a political chain of events that brought the Clinton administration to the defense of U.S. corporations that had ravaged an indigenous village.

According to a report from the scene in *Drillbits and Tailings*, the indigenous village of Magurcharra Khasi Punji, which was only 600 feet from the drill site, lost seven homes, and many trees on which the village had relied for basic survival. The gas leak that initiated the explosion then continued to seep gas for six months. "Our lives have come to a total halt," said Danis Khongla, a local resident ("Clinton/Gore Administration," 2000).

Occidental Petroleum failed to secure environmental permits before beginning drilling on June 3, 1997. The company said that, based on seismic studies, the pocket of gas that caused the blowout was hit by accident, about a mile above the gas deposit that Occidental had meant to tap. "It was gross negligence on the part of the operator" that caused the fire, said Professor Badrul Imam, chairman of the geology department of Bangladesh's Dhaka University ("Clinton/Gore Administration," 2000).

After the accident, and after the Bangladeshi government expressed reluctance to renew Occidental's permit to explore for natural gas in the area, the U.S. envoy to the United Nations, Bill Richardson, accepted an invitation from Occidental lobbyist Robert McGee to meet with the Bangladeshi Commerce Ministry. Richardson also traveled to Bangladesh during April 1998.

According to *Drillbits and Tailings*, a confidential memo from Richardson to the Bangladeshi government said, "I'm troubled by reports that the future of the joint venture between Occidental Petroleum and Unocal . . . is in jeopardy." He asked the government to "move swiftly to grant the extension . . . so they can get on with their important work. Nothing would please me more than to inform President Clinton that the U.S. companies were awarded the blocks they are seeking" ("Clinton/Gore Administration," 2000). The extension was granted, and Occidental continued to explore for gas in Bangladesh until a business deal in 1999 gave Unocal sole control of the project, including U.S. $125 million in compensation claims from the explosion and resulting fires. This incident was part of a broader campaign by the Clinton administration to increase foreign aid to Bangladesh while advocating for development of gas reserves there.

REFERENCES

"Clinton/Gore Administration Backs Occidental Petroleum (Part 2)." *Drillbits and Tailings* 5:8 (May 31, 2000): n.p. [http://groups.yahoo.com/group/graffis-l/message/11105]

Silverstein, Ken. "Bangladesh: The Ambassador from Big Oil." *Earth Island Journal* (fall 1998): n.p.

Silverstein, Ken. "Gore's Oil Money." *The Nation,* May 22, 2000, n.p.

BELIZE

THE MOPAN AND KEKCHI: OPPOSITION TO INDUSTRIAL-SCALE LOGGING

Like many indigenous peoples in Central America, the Mopan and Kekchi Maya of southern Belize have no legal title to the lands that sustained them for many centuries. As in so many other cases, these two Mayan peoples have found themselves competing with intensifying industrialization—in this case, logging—as they seek clear title to their homelands.

During 1995, the Mopan and Kekchi learned that Belize's government had, without their consent, granted a portion of their traditional hunting range to Atlantic Industries, a Malaysian logging company. Soon thereafter, the leaders of 36 Mayan villages in the area discovered that 17 other logging concessions had been granted by the same government across their traditional hunting range, comprising more than 550,000 acres. In addition, they learned that Atlantic Industries was assembling plans for a large sawmill capable of processing 150,000 board feet of lumber a day, to be connected to the outside by a new road. The government's planners also revealed plans for an 18,000-acre industrial park and a toxic-waste incinerator in the traditional lands of the Mopan and Kekchi.

Julian Cho, chairman of the villages' common council, told the Belize Audubon Society that "we are not against development, but neither do we support developments that would have lasting negative impacts on the social structure, culture, and economic lives of the Maya people" (Schaff, 1996, 11). Cho said that efforts to secure legal title in the area are now more necessary than ever given the fact that rapid invasion of their homelands by outsiders will soon confront the Mopan and Kekchi with demands to sell their lands for money. With aid from geographers and cartographers at the University of California at Berkeley, the Maya are developing an atlas to document their land claims.

REFERENCES

Schaff, Deborah. "Belize: Rainforest Destruction Threatened Mayan People." *Native Americas* 13:3 (fall 1996): 10–11.

BIODIVERSITY AND INDIGENOUS ENVIRONMENTALISM

Indigenous peoples comprise less than 4 percent of the world's population, but they also constitute 95 percent of its cultural diversity and more than 50 percent of the population in areas of high biodiversity (Sabaratnam, 2001, 1).

Biodiversity benefits everyone. For example, usage of medicinal plants depends on the traditional knowledge systems of the indigenous people and rural communities, who are quickly being absorbed into the industrial web of capitalism. The same rainforests are rich in biogenetic resources. Recent advances in biotechnology have increased scientists' ability to investigate organisms at a genetic level and to develop commercial products as a result. Hence, many scientists have expressed a need to conserve rainforests for the sake of worldwide humanity, as well as for their indigenous occupants. "Without their botanical knowledge, scientists will be testing blindly the medicinal properties of the world's estimated 250,000 plant species," wrote Professor Gurdial Singh Nijar, a consultant to the Third World Network, who presented a paper at a roundtable discussion on biodiversity and indigenous peoples' knowledge (Sabaratnam, 2001, 1).

Most indigenous peoples are not benefiting from the profits that are realized from the utilization of their knowledge. "This knowledge has been exploited by multinationals from developed countries without sharing of profits and improving the lives of the local communities," said Dr. Nadzri Yahaya, principal assistant director of the Conservation and Environmental Management Division, Ministry of Science, Technology and the Environment (Sabaratnam, 2001, 1).

"States often treat indigenous peoples as backward, unproductive, or even destructive. They frequently deny them land rights, seek forcible relocation and promote the takeover of indigenous territories by national colonists and foreign companies," reported the International Alliance of the Indigenous Peoples of the Tropical Rainforests (IAIPTR.) (Sabaratnam, 2001, 1). According to the IAIPTR, indigenous peoples have nurtured species variations for thousands of years, making possible their current range of biodiversity. Indigenous peoples have actively sought ways to create a wider diversity of plant species by practicing cross-breeding over long periods of time, especially with medicinal and food plants.

"Indeed, so innovative have indigenous peoples been in developing and encouraging species diversity, whether under agricultural cultivation or not, that the distinction [between] 'domesticated' and 'wild' is somewhat meaningless," according to the IAIPTR statement. "Further, there must be national legislation in place governing access to our biodiversity: to protect from bio-theft, bio-piracy and bio-prospecting" (Sabaratnam, 2001, 1).

REFERENCES

Sabaratnam, Sarah. "Plunder of Indigenous Knowledge." *New Straits Times* (Malaysia), May 17, 2001, 1.

BOLIVIA

INDIGENOUS PEOPLES, LOGGING CONCESSIONS, OIL EXPLORATION, AND TOXIC SPILLS

A large percentage of Bolivia's people share indigenous heritage, although a declining proportion live on the land. Those indigenous people who do still live on the land are facing increasing pressure from commercial deforestation and oil exploration, as well as toxic-waste spills associated with mining of antimony, gold, silver, and zinc. Roughly 1,000 tons of mining waste was dumped into Bolivian rivers every day by the middle 1990s.

Logging Concessions

By the year 2000, the Bolivian government had allocated more than a million hectares of primary rainforest as part of new logging concessions, none of which were negotiated with local indigenous peoples' consent. These new logging concessions included 27 on lands recognized under the Bolivian constitution as indigenous territories. At the same time, growing reaches of indigenous lands in Bolivia have been facing increasing pressure from oil exploitation and also from the mining of several metals and minerals, provoking environmental damage from cyanide and arsenic associated with breaches of mining-waste reservoirs.

On July 31, 1997, Bolivia's forest superintendent granted 85 new forest concessions, "effectively eliminating large stretches of primary forest, which constitute zones of traditional and cultural usage that are indispensable to the survival of the Indigenous People" ("Bolivian Rainforests," 1999). Appeals of these decisions were denied in a number of administrative channels, then appealed to the Bolivian Supreme Court of Justice, which refrained from blocking them.

According to local observers, "The concessions effectively eliminate 500,000 hectares of Guarayo territory, more than 140,000 hectares of Chiquitano de Monte Verde territory, more than 15,000 hectares of Yaminahua

Machineri territory, more than 17,000 hectares of indigenous multiethnic territory, and more than 28,000 hectares of indigenous territory [in the] Isiboro Secure National Park" ("Bolivian Rainforests," 1999). In all, a total of more than 700,000 hectares of legally recognized indigenous territory was assigned for exploitation by transnational lumber businesses. Following the allocation of these concessions, conflicts increased between displaced indigenous peoples and logging companies, whose interests often were supported by local police forces, Bolivian army troops, and agents of the national superintendent of forests.

The logging concessions came at a time when the area of Bolivia's remaining forests was steadily declining. During the late 1990s, Bolivia's national territory included roughly 440,000 square kilometers of rainforests, comprising 57 percent of the country's lowlands. At the same time, the rate of deforestation had reached 168,000 hectares per year as log exports increased steadily. Deforestation in Bolivia also was aggravated by oil exploration. Inhabitants of San Ignacio de Moxos reported that Repsol, an oil company owned mainly by Spanish nationals, had advanced 90 kilometers into the Multiethnic Indigenous Territory in the Amazon forest, using a road previously opened by loggers. The Multiethnic Indigenous Territory is inhabited by several indigenous peoples, including the Trinitary, Mojeño, and Chimán. Repsol was reported to have drilled two exploratory wells without legally required environmental permits ("Bolivia: Indigenous Peoples' Forests," 2000). The wells are said to have affected an area inhabited by the Quichua and Aymara indigenous peoples.

Mining and Pollution

Areas of Bolivia's remaining forests have been increasingly threatened by industrial-scale pollution. For example, a rupture in a dike at a Compania Minera del Sur (Comsur) Bolivian mine—which is owned by Gonzalo Sanchez de Lozada, the president of Bolivia—caused 235,000 tons of pollutants (including arsenic and cyanide) to be spilled into the Yana Machi River and other tributaries of the Pilcomayo River. More than 8,000 indigenous Mataco and Chiriguano peoples live along the Bolivian portion of the Pilcomayo (which also traverses parts of Paraguay and Argentina), sustaining themselves principally on the sabalo, a fish that is rapidly becoming extinct due to arsenic and other toxic residues ("Major Toxic Spill," 1996).

Comsur is one of several mines in Bolivia that have dams holding waste from mining operations (another is the Inti Raymi mine). Approximately three dozen other mines that extract antimony, gold, silver, and zinc dump their wastes directly into rivers feeding the Pilcomayo. Studies estimate that by the middle 1990s, an average of 1,000 tons of mining waste were being dumped into Bolivian rivers every day ("Major Toxic Spill," 1996).

In the southern Bolivian town of Tarija, near the Argentine border, indigenous peoples assembled a protest march "in defense of life and the environment," demanding that Comsur compensate them for damage caused to crops irrigated by water from the Pilcomayo. "We want to avoid more deaths, pollution of crops and the displacement of people living along the Pilcomayo due to the contamination of its waters," said the president of the Tarija civic committee, Julio Rodriguez ("Major Toxic Spill," 1996).

A study by the state university of Tarija reported that arsenic poisoning of the Pilcomayo by an accident at the Comsur mine probably had caused the deaths of three miners who drank water and ate fish from the river. A government commission that visited the area of the spill downplayed the significance of the accident, however, claiming that thanks to quick action by Comsur, no plants or animals were harmed ("Major Toxic Spill," 1996).

REFERENCES

"Bolivia: Indigenous Peoples' Forests Menaced by Oil Exploration." *World Rainforest Movement Bulletin* 35 (June 2000): n.p. [http://www.wrm.org.uy/bulletin/35/Bolivia.html]

"Bolivian Rainforests Allocated without Indigenous Consent." Worldwide Forest/Biodiversity Campaign News, August 1, 1999. [http://www.wildideas.net/forest/alerts/1999-08-09-bolivia.html]

Centro de Estudios Juridicos y Investigacion Social (Center for Legal Studies and Social Investigation) in Bolivia. [email: cejis@scbbs-bo.com].

"Major Toxic Spill at Mine Owned by Bolivian President." *Drillbits and Tailings* (November 7, 1996): n.p. [http://www.moles.org/ProjectUnderground/drillbits/1101/96110101.html]

BOTSWANA

THE KHWE (KALAHARI BUSHMEN): END OF THE LINE

Introduction

The last Kalahari Bushmen are being forced to leave the Central Kalahari Game Reserve in Botswana. The government of Botswana, having forced the Bushmen to become dependent on its services, has eliminated access to food, water, and basic health care. As of January 2002, Survival International asserted that "this latest move is part of a long-standing drive by the government of Botswana to evict the Khwe Bushmen from their land to make way for tourism and diamond mining" ("Last Kalahari Bushmen," 2001).

Africa's Original Inhabitants

The Khwe Bushmen are Southern Africa's original inhabitants. During the 1960s the Central Kalahari Game Reserve was created as a haven for them. By 1997, however, the government of Botswana had evicted many of the Bushmen. They were forcibly removed from the Reserve, and placed in bleak resettlement camps. Following a campaign by Survival International and organizations advocating the Khwe's rights to a land and lifestyle of their own, the removals ceased for a time. By 2001, however, pressure was building again to remove the Khwe from their homeland. By 2002, most had been forced off the land as the government withheld their access to food, water, and medical services.

The hunt is central to the religious and ritual lives of the Bushmen. Their view of themselves in relation to their environment and spiritual world is intimately bound up with animals and their dependence on them ("Bulletin: Botswana," 2001). Botswana authorities have forbidden the Bushmen, who formerly sustained themselves by hunting and gathering, from capturing more than a few animals a year. The Bushmen are forbidden to hunt without a license that allows each hunter only three large antelope per year. The game quotas also forced the Bushmen to depend on outside authorities for basic necessities, including food, water, and shelter, which then were manipulated to

A woman of the Khwe and her children, the Bushmen of the Kalahari. (Painet)

force their relocation from the mineral-rich Kalahari Game Reserve. While Bushman hunting is severely restricted, the government encourages hunting for sport by tourists elsewhere in the country.

Beginning in 1997, government officials ordered police and troops to invade Bushman villages. The police demolished their homes and trucked inhabitants to resettlement camps. Several hundred Bushmen managed to resist relocation and remained on their ancestral lands. Survival International, a organization based in London devoted to supporting tribal peoples, then initiated a vigorous international protest campaign.

Survival International reported during May 2001 that "several Bushmen have recently been tortured by wildlife officials and local police for supposedly exceeding their hunting allowance. The government is forcing the Bushmen to choose between starvation and leaving the land they have lived on for 20,000 years" ("Last Kalahari," 2001). The Central Kalahari Game Reserve, originally created to shelter the Bushmen and the game that comprises their traditional economy, has been under increasing pressure from mining and other development interests since the 1980s.

According to a Survival International report,

Wildlife Department officials invaded the Khwe Bushman community of Molapo in August 2000 and, forcing their way into several Bushman homes, physically bullied at least two dozen men and women. One of the men, Mathambo Sesana, died a few days later. At least 12 of the men were then driven away and subjected to repeated interrogations, beatings, and torture. Some were tied to trees and threatened with fire; most had their feet tied to vehicles, forcing them into a "press-up" position, whilst being kicked and beaten. The terrified Bushmen were finally released after about six days. . . . Many now say they are too scared to hunt, so that they and their families are going hungry. ("Bulletin: Botswana," 2001)

In the opinion of Survival International, "Recent statements from senior government officials that the Khwe Bushmen inside the reserve should move out 'for their own good,' 'join the mainstream,' and 'become civilized' display a backward and colonialist mentality. It is time for the government to realize that the Bushmen way of life is not 'primitive' or 'uncivilized' but a different, modern and sustainable way of living. Integration or assimilation of the Bushmen, the stated aim of the Botswana government, will lead to the destruction of these people" ("Botswana Persecutes Bushmen," 2001).

Pressure on the Last of the Bushmen

By the end of 2001, only a few Bushman communities remained in the Central Kalahari Game Reserve, refusing to move from land containing the graves of their ancestors. The government maintained considerable pressure on them, using threats and intimidation to persuade the roughly 700 who remained in the Reserve to leave. By late 2001, the government announced that all water and food supplies to Bushmen living inside the Reserve would be cut off by the end of January 2002. Restricted by the government from hunting at a level required to feed themselves and their families, the Khwe found themselves increasingly dependent on outside supplies.

De Beers Diamond Co. (which maintains a near-monopoly over the world's diamond mining and marketing) and AngloAmerican have been engaged in mining operations in the area from which the last of the Bushmen are being removed. The companies have been cooperating with the government of Botswana to ensure that the Bushmen will have no claims to mineral wealth underlying the Central Kalahari Game Reserve. De Beers thus far has identified two potential diamond mines within the reserve, at Gope and Xaxe ("De Beers and Anglo Linked," 1997).

As the last of the Kalahari Bushmen were being punished for hunting without permits, a touring company offering "Untamed Wildlife Safaris" advertised for the "Bushman Experience," described as "a safari sharing moments with the Real People" in Xai Xai, a village of . . . Bushmen and Babanderu people . . . one of the last *very remote* parts of Botswana," which, it is said, "offers a unique and exclusive possibility to experience the life of hunters and gatherers in the beautiful setting of the northern Kalahari" ("Untamed Wildlife Safaris," n.d.). Advertising from a Web site on the Internet continued:

This experience might offer you a thoroughly peaceful satisfaction while you sleep in traditional grass huts, walk in the blazing sun, stalk a *kudu*, perhaps eat porcupine meat at the open fire, dig up potato-like roots for breakfast, eat loads of sweet berries to fill your stomach and just be there to share and be part of a sacred unity, clapping and singing into the early hours while the men tirelessly dance a well-worn groove into the earth circle formed by their tracks." ("Untamed Wildlife Safaris," n.d.)

This version of an authentic "Survivor" is said in its advertising to take place in a location "well-known for its huge giraffe and springbok population . . . located north of the Maun-Nata road in Northern Botswana [in] one of Botswana's less-frequented national parks, Nxai Pan National Park, a major transit point for migratory zebra and elephant. . . . " ("Untamed Wildlife Safaris," n.d.). Wildlife highlights include the Nxai lions, "who are often seen hunting at the park's water holes during the dry season. The park also hosts large herds of springbok and impala, which attract visits from roaming desert cheetahs and packs of wild dogs. Gemsbok abound in the adjacent Kalahari sand-veld and one may catch sight of the elusive desert giraffe, roaming bull elephants, brown hyena and bat-eared foxes. Birdlife is exceptional during the rains" ("Untamed Wildlife Safaris," n.d.).

In the meantime, on February 1, 2002, Botswana's government cut off water supplies to remaining Kalahari Bushman communities in an attempt to drive the surviving remnants of the Bushman tribes off the land on which they had lived for about 20,000 years. The Bushmen struggled against governmental plans to place them in resettlement camps where they were not allowed to hunt or gather. The Bushmen in the camps became dependent on government handouts and subject to alcoholism and despair. One Bushman described the camps as a place of death ("Botswana: Government Plans," 2002). The government asserted that it could not afford to provide the remaining bush camp with water at a cost of about U.S. $5 per person per week, despite the fact that Botswana is the world's largest producer of diamonds. The government also failed to respond to offers by the European Union to pay for the water ("Botswana: Government Plans," 2002).

Some high-ranking ministers in the government have described the Bushmen as primitive and stone-age creatures. Survival International's director, Stephen Corry, said," "The Botswana government has spent 16 years harassing the . . . Bushmen. This latest move—cutting the water—risks destroying them once and for all. The international community must speak out now to halt this racist crime against humanity" ("Botswana: Government Plans," 2002).

In mid-February 2002, Survival International relayed word that Botswanan authorities were prosecuting 13 additional Bushmen for hunting without a license after they had been tortured by wildlife officers. The 13 hunters, from the village of Molapo, went on trial February 18 in the Botswanan town of Lethlakane. Testimony collected by Survival International researchers "described how wildlife officers came into the village in August 2000 and subjected them to a six-day ordeal of violence and torture. Some were tied to a tree which was set alight, [as] others were tied to the front of the officers' vehicle for three days and beaten" ("Botswana Tortures Bushmen," 2002).

By the third week of February, as threatened, the Botswana government left the last of the resisting Bushman families in the Central Kalahari desert to die without water. Officials removed parts from the Bushmen's only pump, making it impossible for them to get any water of their own. The same officials also deliberately emptied tanks containing remaining water supplies. Bushmen who protested were beaten. As it cut off the Bushmen's water supplies, the Botswana government also stripped them of solar-powered radio transceivers provided by Survival International, their only link with the outside. Two Bushmen from outside who brought food and water were told that after these deliveries they would need permits purchased from the government to visit the reserve.

"PEOPLE ARE LEFT HOPELESS"

Lacking access to water, most of the surviving independent Bushmen surrendered to relocation during February, boarding government trucks. Others were reported by Survival International as determined to stay. A Survival International press release said, "A Bushman spokesperson said yesterday [February 21, 2002], 'People are left hopeless, they don't know what to do or where to go'" ("Botswana Leaves Bushmen in Desert," 2002).

Early in 2002, the Botswana government was criticized by the United Nations Human Rights Commission, which said that the Kalahari Bushmen had been victims of discriminatory practices and were being dispossessed of their traditional lands. The criticism was contained in a report by the United Nations' special rapporteur on indigenous peoples, Rodolfo Stavenhagen, who visited Botswana during February 2002. His report also says that the Bushmen's "survival as a distinct people is endangered by official assimilationist policies" ("U.N. Condemns," 2002).

As the Bushmen were taken from their homes during the spring of 2002, protest vigils began in London, Madrid, Paris, Milan, and Zurich. A vigil also started outside Botswana's honorary consul in Rome. Indigenous people in Canada also expressed support for the Bushmen. An article by Innu Nation president Peter Penashue appeared in a Botswana daily newspaper urging Botswana to "learn from Canada's mistakes, and end the misguided policy of trying to forcibly integrate the Bushmen into your cultural mainstream. Canada has shown the world that this doesn't work" ("Bushmen Campaign Spreads," 2002).

During mid-February 2003, Kalahari Diamonds Limited—formed at the initiative of, and partly owned by, BHP Billiton—secured U.S. $2 million funding from the International Finance Corporation (IFC), part of the World Bank, to explore for diamonds in the Central Kalahari Game Reserve (CKGR), Botswana. By the same time, as the last of the Bushmen were being forced from their homelands, most of the reserve was parceled out in diamond exploration Concessions, according to Survival International ("World Bank Funds," 2003).

REFERENCES

"Botswana: Government Plans to Destroy Bushman Tribes." Survival International, January 30, 2002. [http://www.survival.org.uk/about.htm]

"Botswana Leaves Bushmen in Desert Without Water." Survival International, February 22, 2002. [http://www.survival.org.uk/about.htm]

"Botswana Persecutes Bushmen." Survival International, August 15, 2001. [http://www.survival.org.uk/about.htm]

"Botswana Tortures Bushmen, then Prosecutes Them." Survival International, February 14, 2002. [http://www.survival-international.org]

"Bulletin: Botswana—Bushmen Tortured for Hunting." Survival International, May 2001. [http://www.survival.org.uk/about.htm]

"Bushmen Campaign Spreads." Survival International news release, May 10, 2002. [http://www.survival-international.org/ad.htm]

"De Beers and Anglo Linked to Forced Removal of Last Kalahari Bushmen." *Drillbits and Tailings*, October 7, 1997. [http://www.moles.org/ProjectUnderground/drill bits/971007/97100704.html]

"Last Kalahari Bushmen Tortured and Facing Starvation." Survival International, May 2001. [http://www.survival.org.uk/about.htm]

"U.N. Condemns Botswana's 'Dispossession' of Bushmen." Survival International, April 2, 2002. [http://www.survival-international.org]

"Untamed Wildlife Safaris: South Africa—Botswana—Namibia—Zimbabwe—Zambia—Malawi—Tanzania—Uganda—Cameroon—Mozambique." N.d. [http://www.untamedwildlife.com/program3.html]

"World Bank Funds Controversial Diamond Project on Bushmen's Land." Survival International Press Release, February 17. 2003. [st2@survival-international.org]

BRAZIL

INTRODUCTION

Brazil, most notably the Amazon Valley, has become a leading hot spot for indigenous assertions of environmental rights in the face of intrusion by outside interests. The fingers of industrialism are reaching into the interior of Brazil, including dam building, gold mining, road building, power transmission, and logging, with a speed and intensity not unlike the westward movement in North America's interior during the mid- to late-nineteenth century. Nonnative populations in the Amazon grew from about 2.5 million during the 1960s to more than 20 million by 2002 ("Population, Highways," 2002).

Dams, logging, and road building in the Amazon threaten some of the most diverse biological areas on Earth, including many animals, plants, and insects yet to be identified by Western science. More than half of all modern medicines are developed from plants and animals, many in the tropics. "Year by year," commented the newspaper *Indian Country Today*, "new cures are found and developed from the combined knowledge of indigenous medicine and scientific research. . . . New studies by Brazil's National Institute of Amazon Research and the University of Michigan estimate that if the current pace of development is sustained, only 5 per cent [of] the original forest will survive by 2020" ("Oil, Mega-development Plans," 2001, A-4).

THE TOLL OF DAM BUILDING ON INDIGENOUS PEOPLES

As many as 80 new hydroelectric plants were being planned for the Amazon Valley by the late 1990s ("Cotingo Dam in Brazil," 1998, 15). Plans to construct this web of dams could sound the death knell of native peoples' aspirations to demarcate (establish) legal rights to their homelands. Struggles against construction of dams often are tied in with indigenous land tenure as well as the economic base. The Apinaje, Kraho, Xerente, Tapuia, and Karaja, for example, have objected that the changes in the Mortes, Araguaia, and Tocantins rivers by proposed dam construction will increase mortality among fish and other animals on which they depend on for survival. The proposed Tocantins-Araguaia waterway and adjacent dams (with hydroelectric power plants) would connect the central-western region of Brazil with Atlantic ports in Brazil's northeast, promoting agricultural development in Brazil's heartland ("Brazil's Indigenous People Resist," 2001).

An electricity shortage in the urban areas of Brazil has intensified pressure to build hydroelectric dams in and around the Amazon Valley, with attendant flooding and clearing of land for power lines and service roads. The landless poor inevitably follow the roads from Brazil's urban areas into indigenous lands, where they attempt to establish farms and ranches. Logging operations also follow the expanding web of roadways. Lobbyists for farmers have been pressing the government to change Brazil's Forest Code (which determines how much of protected areas may be used for human endeavor). The proposed changes would allow up to 50 percent of private property to be deforested.

Resisting dam development in Brazil can he hazardous to one's health. On August 25, 2001, at about 2:30 A.M., one of the coordinators of the Movement for the Development of the Trans-Amazon and the Xingu was assassinated. Ademir Alfeu Federicci, known as "Dema," was shot in the head following a struggle with an armed man who entered his house in Altamira. Dema, a leader of the Federation of Agricultural Workers (FETAGRI) had worked to organize community leaders, labor unions, cooperatives, and associations along the Trans-Amazon Highway to provide sustainable development in the region. Dema also had made several influential enemies by speaking out against corruption in local governments. With his associates, Dema also had opposed the Belo Monte dam. According to plans of its developers, Belo Monte, if constructed, would be the first dam on the Xingu River in Vitória do Xingu, in the Pará state in the Brazilian Amazon, and the world's second-largest hydroelectric installation ("Co-ordinator of Movement," 2001).

The Belo Monte dam was first proposed during the 1980s (it was then called the Kararaô dam), but construction was repeatedly postponed following pressure from local, Brazilian national, and international environmental movements. During 2000, Eletronorte renewed work on the dam project with an intensive publicity campaign financed with public funds to convince the public of the importance of the dam.

Dema and his colleagues sought development "based upon the rational use of our natural resources and the preservation of our rivers and forests, sharing the wealth with all men and women" ("Co-ordinator of Movement," 2001). In addition to their opposition to the Belo Monte dam, Dema and his allies also opposed industrial-scale agriculture (usually involving soybeans and grains), which requires intensive use of chemicals. The same alliance also opposed industrial-scale cattle ranching and mining.

Large, vocal protests by indigenous peoples have become part of daily life at many of Brazil's public agencies that supervise the building of dams. During March 2001, roughly 1,500 indigenous people and nonnative environmentalists occupied the Brazilian Ministry of Mines and Energy's head office in

Brasilia. "These men, women, and children are here to demand a halt to new dam construction until already existing problems caused by dams in Brazil are solved," said Sadi Baron, one of the coordinators of the Movement of Dam-Affected People (MAB) ("Dam Protesters," 2001). Protesters also occupied nearby offices of the Inter-American Development Bank, which finances many of the new dams. The group has threatened to occupy dam sites and force construction to stop.

In the meantime, Mining and Energy Minister Josa Jorge said Brazil needs all the hydroelectric power that it can generate. Jorge asserted that the government might have to adopt electrical energy rationing due to the low water levels behind Brazil's existing hydroelectric dams, which produce more than 90 percent of the electricity produced in Brazil ("Dam Protesters," 2001).

At the turn of the millennium, several Brazilian indigenous groups were actively protesting plans to build dams on their lands. For example, a large delegation of Apinaj protesters traveled to Brasilia from Tocantins state in central Brazil where the Lajeado dam is located and others are planned. "Lajeado Dam is bringing disease, prostitution, hunger, and alcohol to the Xerente people, and causing disrespect for our culture and an increase in violence on our lands," leaders of this group declared ("Dam Protesters," 2001). The Apinaj indigenous people are fighting the Serra Quebrada dam on the Tocantins River, which will flood more than five percent of their land. "We Apinaj people have the river as our source, because our culture is the mother earth, the river, nature, and animals. We do not accept this dam. We will fight to the death so that our children may live in peace," their leaders said ("Dam Protesters," 2001). The protesters are calling for the democratization of Brazilian energy policy and implementation of renewable energy generation and conservation measures to avoid the social and environmental impacts caused by large dams.

On October 30, 2001, about 350 men, women, and children from across Brazil flooded into the headquarters of the Belgian transnational company Tractebel in Rio de Janeiro to protest plans to build several dams in the Amazon Valley. Tractebel is a partial owner of the electric utility Gerasul, which is constructing some of the dams. The takeover is part of a national mobilization by the Brazilian Movement of Dam-Affected People (MAB). The demonstrators asserted that the company had been unfair in its dealings with people who will be displaced by its dams, including the Ita dam in Rio Grande do Sul and Santa Catarina states, and the Cana Brava dam on the Tocantins River in the central-western region of Brazil. Indigenous activists charged that Tractebel had failed to address outstanding resettlement and compensation issues for 200 families whose problems are as yet unresolved at Ita despite the fact that construction of the dam had been completed ("Dam Affected People," 2001).

At Cana Brava, a dam financed by the Inter-American Development Bank, Glenn Switkes, speaking for the Latin America Campaigns section of the International Rivers Network, said that hundreds of families had not received compensation for losses sustained during construction of the dam years earlier. Additionally, most families who had been compensated found the payments too small to buy new homes and land. "Most of the sharecroppers, renters, fishermen, and . . . gold miners who worked along the river are being ignored by the company, as well as many families who will be isolated by the formation of the reservoir," said Switkes ("Dam Affected People," 2001).

National MAB coordinator Helio Mecca said that "Tractebel has refused to consider the needs of populations who will lose everything when the floodgates on Cana Brava are closed" ("Dam Affected People," 2001). The dam project affects the homelands of several indigenous peoples, including the Kayapó, Parakanã-Apiterewa, Araweté do Igarapé, Ipixuna, Asurini do Xingu, Arara do Pará, Juruna, Xipaia, and Curuaia.

ACCELERATING DEFORESTATION IN THE AMAZON VALLEY

During early 2002, a team of scientists from the United States and Brazil disclosed in the journal *Environmental Conservation* that forest destruction in the Brazilian Amazon had accelerated during the 1990s. The team, led by William Laurance of the Smithsonian Institution's Tropical Research Institute, analyzed deforestation estimates produced by Brazil's National Space Agency that were based on detailed satellite images of the Amazon since 1978 ("Amazonian Deforestation," 2002). According to a report by the Environment News Service, this report is contrary to assertions by the Brazilian government that threats to Amazonian forests had declined in recent years because of improved environmental laws and public attitudes.

Deforestation in Brazil, as in many other tropical regions of the world, threatens indigenous peoples who use the forests for food and shelter. To cite one of several examples, the hunter-gatherer Awa, one of the last nomadic tribes in Brazil, are being killed by loggers, ranchers, and settlers. By the year 2000, only a few hundred Awa survived. The World Bank gave the Brazilian government money 19 years ago to recognize and protect the Awas' land.

The Awa, a tribe of nomadic hunter-gatherers, are one of the most acute examples of what deforestation can do to a traditional people. Since the 1950s, according to a report by the British Broadcasting Corporation, "they have been viciously persecuted, hunted down and murdered by gunmen employed by cattle ranchers and loggers who are invading and plundering their land" ("B.B.C. Exposes," 2002). According to Survival International, "The massive Carajas development project, funded by the World Bank and European Union, has opened their lands to colonists who are pouring in, in ever-greater numbers. For 20 years the Brazilian government has failed to demarcate and protect Awa land, despite the fact World Bank funding is available to do so, and demarcation was a condition of its loan to Brazil in 1982" ("B.B.C. Exposes," 2002).

The contacted Awa number about 230, and roughly 60 to 100 others live in isolation, hiding in fragments of remaining forest in small family groups, fleeing from the loggers and settlers. Many Awa have tragic stories. Karapiru was the survivor of a massacre by ranchers who wiped out his entire family. Kamara, a hunter featured in the BBC piece, narrowly escaped death when a settler shot at him recently ("B.B.C. Exposes," 2002).

By March 2003 the demarcation—legal recognition, mapping out and marking the ground—of the Awas' land was completed. Only 300 Awa remain; about 60 still live uncontacted in small nomadic groups.

"Forest destruction from 1995 to 2000 averaged almost two million hectares a year," said Laurance. "That's equivalent to seven football fields a minute, and it's comparable to the bad old days in the 1970s and 1980s, when forest loss in the Amazon was catastrophic" ("Amazonian Deforestation," 2002).

The research team released its findings as the Brazilian government announced plans to invest more than $40 billion in new highways, railroads, hydroelectric reservoirs, power lines, and gas lines in the Amazon Valley. According to these plans, about 5,000 miles of highways will be paved ("Amazonian Deforestation," 2002). "There's no way you can criss-cross the basin with all these giant transportation and energy projects and not have a tremendous impact on the Amazon," Laurance said. "When you build a new road in the frontier, you almost always initiate large-scale forest invasions by loggers, hunters, and slash and burn farmers" ("Amazonian Deforestation," 2002).

New environmental laws in Brazil have been implemented to slow forest loss, but the research team asserted that most of these laws were rarely enforced. That, combined with growing population and expanding logging and mining industries, has increased deforestation in Amazonian forests despite legal prohibitions. "The scariest thing is that many of the highways and infrastructure projects will penetrate right into the pristine heart of the Amazon,"

The charred remains of logging slash in the Brazilian rainforest, 1991. (Painet)

said Laurance. "That could increase forest loss and fragmentation on an unprecedented scale" ("Amazonian Deforestation," 2002).

Deforestation of the Brazilian Amazon was greater during 2000 than at any time since 1995, according to satellite data. Brazil's National Institute for Space Research, which monitors deforestation by satellite, issued a provisional estimate for the period August 1999 to August 2000, based on a sampling. According to the satellite survey, the mean annual rate of gross deforestation during that period was 19,836 square kilometers (7,658 square miles). Between August 1998 and August 1999, the mean annual rate of forest cut down was 17,259 square kilometers (6,663 square miles), according to the same survey ("Brazil's Amazon Rainforest," 2001). The institute said that increased road building was proceeding apace with acceleration of deforestation.

The satellite survey may be understating actual deforestation because it lacks resolution in areas less than 6.4 hectares. According to the Environment News Service, "This means that the impacts of hundreds of thousands of small-scale farmers and selective logging of lucrative species are not included ("Brazil's Amazon Rainforest," 2001). "The new figures clearly show that efforts by the Brazilian government have failed to stop, or even to slow, deforestation of the Amazon," Greenpeace Amazon campaigner Paulo Adário said from Manaus, where he monitors illegal logging in cooperation with the Brazilian environment agency (2001). Adário continued:

> This loss of forest cover in the Amazon is unacceptable and unsustainable. Scientific studies have repeatedly shown that Amazon soil is not suitable for agriculture and cattle ranching. The biological richness of the region lives only in the standing forest. To continue unchecked deforestation means to condemn the Amazon to inevitable environmental and social crises. ("Brazil's Amazon Rainforest," 2001)

Greenpeace has called upon the government of Brazil to reduce deforestation to zero by the year 2010. In 1970, according to Greenpeace, only one percent of the Brazilian Amazon had been deforested. In 1978, 13 million hectares of the Amazon Valley were deforested. That figure later tripled, to 37 million hectares by 1988 and 41 million in 1990 ("Ten Years," 1998). By 2000, almost 15 percent of the valley's land area had been cleared; a forested area the size of France had been deforested in 30 years.

At present rates of deforestation, as little as five percent of the Amazon's rainforest may remain in a pristine state by 2020, according to a team of Brazilian and U.S. scientists who analyzed how the Amazon ecosystem will respond to a new road-development project that will cost U.S. $40 billion, promoted by Avanca Brasil (Advance Brazil). Avanca Brasil's proposals include construction of hydroelectric dams as well as roads. The team used satellite pictures to develop computer models that forecast the course of forest destruction based on road building and other forms of development during the previous 20 years.

According to an account by Steve Connor for Amazon Watch, "The scientists accuse the Brazilian government of fast-tracking the project by ignoring environmental agencies (including its own Environment Ministry) thus accelerating logging and deforestation" (Connor, 2001). "Once a road or highway is built, a Pandora's box is opened which is almost impossible for a government to control," said Laurance. "Once you build a road into a pristine forest you start an inevitable process of illegal colonization, logging, land-clearing and forest destruction" (Connor, 2001). "We used the past as a guide to the future," wrote Connor. "We looked at the entire network of roads and highways in the Amazon to see how deforestation occurs in the region of a new road," said Laurance. "There's really nothing that has been done that approaches the scale of what we've done. Our computer model is very comprehensive" (Connor, 2001).

The scientists' study provides two possible scenarios for the future of Amazon forests: optimistic and nonoptimisitic. According to Connor,

> both suggest that the Amazon will be drastically altered by current development schemes. Under the less optimistic scenario, more than 95 per cent of the Amazon will lose its untouched status and 42 per cent of the forest will be totally denuded or heavily degraded by 2020. Even under the more optimistic view, well over half of Amazonia will no longer be in a pristine state and about 30 per cent will be lost forever. (Connor, 2001)

By the year 2000, according to this study, Brazil's portions of the Amazon basin were losing almost 5 million acres of forest a year. The Avanca Brasil plan would increase this rate of loss by between 14 percent and 25 percent each year, according to the study. "At stake is the fate of the greatest tropical rainforest on Earth," the scientists wrote (Connor, 2001).

CHICO MENDES AND THE INDIGENOUS RUBBER WORKERS

As in many other parts of the Amazon Valley, destruction of indigenous ways of life in Acre, part of the western Brazilian Amazon, began late in the nineteenth century, as peasants from Brazil's Northeast colonized the area, escaping drought and poverty. They immigrated to the Amazon rainforest to harvest rubber for the *seringalistas,* powerful Brazilian and foreign owners of rubber plantations. Local people who were displaced from their homelands and traditional lifeways accepted employment on the rubber plantations, where they became known as *seringueiros.* During the ensuing century, only 10 of the 60 indigenous peoples who had lived in the Jurua Valley of Acre survived. Even the surviving peoples' populations dropped sharply as large parts of their homelands became industrial-scale rubber plantations.

During the 1960s, the Brazilian government began to promote development of Amazonia more intensely, attracting investors from industrialized southern Brazil with incentives for colonization. Land in Acre was advertised in Brazil's urban areas as plentiful and cheap. By the early 1970s, the Amazon Valley was linked to the outside by the Trans-Amazon Highway. Between 1970 and 1975 the *fazendeiros* (big landowners) purchased 6 million hectares of land in Acre, with support from the Brazilian government, expelling the families of the earlier immigrants. Local people's homes were set on fire, their cattle were killed, and their women abused. At the same time, the forests were rapidly destroyed, turned into agricultural land ("Ten Years," 1998). By 1975, 180,000 rubber trees (*seringueiras*) and 80,000 chestnut trees (*castanheiras*) had disappeared because of logging and fires, as the land was cleared for other types of commercial agriculture and cattle ranching. Many newcomers from Brazil's burgeoning urban areas settled illegally on territories that had been utilized for many years by the *seringueiros* or other indigenous peoples ("Ten Years," 1998).

The *seringueiros* then began to organize resistance against the newest wave of invaders, creating the *empate* movements. Francisco (Chico) Mendes, who was born in 1944 at Porte Seco, became one of the best-known leaders of this movement. Groups of *seringueiros* and their families moved into locations slated for logging or burning and occupied them in a peaceful manner. Some of these actions were widely publicized around the world, prompting popular pressure to preserve the forests and their indigenous residents ("Ten Years," 1998). Mendes became the best known of the extractivist *seringueiros* who advocated sustainable use of the forests in a collective manner.

Violence increased in Acre, centered on disputes over ownership and use of the land. Chico Mendes was one of a number of prominent *seringueiros* who advocated establishment of collective reserves in Cachoeira and Sao Luis do Remanso. On December 22, 1988, he was assassinated at his home in Xapiru. During a speech a few days before his assassination, Mendes anticipated his death when he said:

> I only hope that my death contributes to a halt in the impunity of the killers, who count on the protection of the police of Acre, and which have already killed 50 persons like me, *seringueiro* leaders, [who are] committed to save the Amazon forest and to show that progress without destruction is possible. ("Ten Years," 1998)

Ten years after Mendes was assassinated, a longtime ally of his, Jorge Viana, was elected governor of Acre. With the world price of rubber continuing to decline (due in large part to mass-produced synthetics), some of the *seringueiros* cleared their land and introduced farming of rice, corn, and beans.

THE APURINA, PAUMARI, DENI, AND JUMA: PROTESTING OIL AND GAS PIPELINES

Oil and gas pipelines are another frequent intrusion into the lives of many Amazonian indigenous peoples. Heretofore isolated peoples living in the Brazilian Amazon, including the Apurina, Paumari, Deni, and Juma, expect to experience acute social and environmental pressures following the construction of two proposed oil and gas pipelines that would link their homelands with some of the largest population centers in the area. The indigenous residents of the area believe that illegal logging and colonization by land-hungry farmers will ruin their forests and their traditional ways of life.

According to a report by Amazon Watch, the two new pipelines would expand oil and gas production from the Urucu and Jurua gas fields in the heart of the Brazilian Amazon. The report details prospective detrimental social and environmental consequences of the projects. The pipelines, roads, and related infrastructure would link the two largest cities in the Brazilian Amazon (Manaus and Porto Velho) to the homelands of the Apurina, Paumari, Deni, and Juma, in areas described by Brazilian and international scientists as "the highest priority for bio-diversity conservation" ("Amazon Watch," 2001).

The two proposed pipelines would extend 325 miles (550 kilometers) from Urucu to Porto Velho (Rondonia) and 245 miles (420 kilometers) from Coari to Manaus (Amazonas). Key participants in the project include GasPetro, a 100 percent subsidiary of Petrobras; El Paso Energy Internacional; Halliburton; Japan Exim Bank; Schlumberger; Techint Engineering; and the state-owned Brazilian National Development Bank (BNDES) ("Amazon Watch," 2001).

Local indigenous and environmental groups have demanded extensive public hearings in all major communities along the proposed pipelines' routes. International environmental groups also have responded to local calls for international support to pressure international financial organizations to withdraw funding from the projects. Environmentalists and local activists assert that oil and gas extracted from these areas should be transported by barges on local rivers rather than through new pipelines. According to Amazon Watch, "The Urucu gas reserves will be exhausted in about 15 years, yet the impacts of the [pipeline] projects [would] be irreversible" ("Amazon Watch," 2001).

THE GUARANI AND KAIOWA: ASSERTING RIGHTS TO THEIR LAND AMIDST A WAVE OF SUICIDES

When the first European colonizers of Brazil arrived from Portugal, roughly 1 million Guarani lived in what is now western Paraguay, extreme southern Brazil, and Uruguay, as well as parts of Bolivia and northern Argentina. Until the 1940s, the Kaiowa and the related Guarani tribe roamed freely over about a quarter of Mato Grosso, which means "thick forest" in Portuguese. Today, the state has been divided in two and most of the forest has been cut down for pasture and grain fields. The 60,000 Indians in the two tribes are confined to small, hard-to-farm lots (Astor, 2002).

By the 1990s, only about 30,000 Guaranis survived. Many of the survivors were being forced into resettlement camps, removed from their traditional social and economic moorings. Some of the last surviving Guaranis moved deeper into the forests to escape forced removal as their traditional range was restricted by expansion of monocrop agriculture, logging, and non-Guarani intrusion. Some individual Guaranis have been removed from their lands as many as six times. As Jennifer Hanna wrote in *Native Americas*, "Each time these people were evicted from the land, they turned around and came back" (Hanna, 1997, 38).

The Tupinikim and Guarani of Espirito Santo, in southern Brazil, have been struggling for years against corporate giant Aracruz Celulose in defense of traditional lands into which the company moved, beginning in 1967. Indigenous peoples' marches and land occupations have provoked intimidation and violence at the hands of company security agents, backed by the Brazilian military.

Indigenous rallies in support of the Tupinikim and Guarani for demarcation (legal recognition) of their traditional lands began on March 6, 1998. Within two weeks, the Brazilian government had launched a military operation to snuff out the movement. Roads that provide access to indigenous villages in the area were occupied by Brazil's federal police. Leaders from CUT (Central Unica dos Trabalhadores, the Central Workers Union), who participated in the demarcation rallies, were arrested and, according to a report in the *World Rainforest Movement Bulletin*, "treated like criminals" ("A Dictatorship-type Action," 1998). Similarly, members of the Movimento dos Sem Terra (Landless Peasants Movement), who also had supported the indigenous action, were removed by force from their homes.

During April 1998, the Tupinikim and Guarani were forced to sign an agreement, valid for 20 years, by which the indigenous peoples accepted corporate jurisdiction over large eucalyptus plantations in exchange for promised financial assistance. The agreement was signed under the gun-muzzles of armed police. One local witness said that the conditions under which the agreement was signed "offend seriously the fundamental rights and liberties guaranteed in the Federal Constitution [which says that native lands may not be sold] valid for any individual—including the Indians" ("A Dictatorship-type Action," 1998).

The Tupinikim and Guarani then collectively decided to resist the coerced agreement by refusing to plant eucalyptus trees on their lands. The *World Rainforest Movement Bulletin* reported that "the main reason for this decision was that the communities realized that their main struggle was and is against eucalyptus monocultures, which are the symbol of the invasion of their land by Aracruz and which have resulted in major social and environmental impacts affecting their lives and livelihoods. As a Tupinikim leader said during a workshop in Espirito Santo: 'Eucalyptus forests are dead forests that kill everything.' The indigenous people decided to assist in the regeneration of local native forests instead of the company's plantations" ("Aracruz: Indigenous Peoples," 2000).

As some Guarani engaged in militant assertions of their rights to the land and traditional ways of life, others were killing themselves out of hopelessness. The Guaranis' wave of suicides was a response to eviction from their homelands (and destruction of their culture) during the mid-1990s. More than 330 in a population of 30,000 have killed themselves during the last 17 years. Twenty percent of the suicides were children under the age of 14. Dona Marta Silva Vito Guarani, president of the Kaguateca Association for Displaced Indians, said that "the suicides are a form of protest. This problem is a direct result of a forced and very violent integration." She continued:

Politics, land issues, poverty, abandonment—what's left for us [but] death? . . . People get worn out, they don't want to go hungry, they don't want violence. . . . Indians are the same as plants. If you pull off their leaves [and] their roots they become sad. . . . This is why young Guarani are killing themselves, why they are searching for the end, hanging themselves. The women from the community . . . told me that they will kill their children and kill themselves afterwards if they [squatters] try to take their lands away again. I cannot stand with my arms folded while my people are being massacred. (Hanna, 1997, 34, 41)

During early June 2002, several newspapers in the United States carried accounts of 24-year-old Kaiowa Ramao da Silva's suicide on March 5, 2002 in a "grass-roofed hut [with a doorway] so low that he had to kneel to hang himself." He strangled himself with a dirty shirt. When da Silva hung himself, federal police were preparing to evict him and 300 other Kaiowa Indians from an area they had reclaimed three years earlier to pressure the government into declaring it an Indian reservation. A last-minute court injunction later postponed the eviction for 90 days, long enough for the Indians to harvest their meager crops (Astor, 2002).

According to the report, "Neighbors recalled how da Silva had agonized over the prospect of being forced off his tribe's traditional lands. No matter that it was 74 rocky acres unsuitable for farming, that it had no game or fresh water, that it legally belonged to a cattle rancher. He would die before he left it" (Astor, 2002). Since 1986, according to this report, 348 Kaiowa and Guarani Indians had committed suicide, 42 of them in 2001 alone, in the Mato Grosso do Sul state of Brazil.

During the last few years a number of Guarani and Kaiowa have run away from settlement camps and returned to their traditional territories, often contesting local farmers and ranchers for use of some areas. Cacique (chief) Otavio Vera, from the village of Taquaraty, noted that suicides had virtually disappeared among indigenous people who returned to the land. "In the few years since returning to our land," he said, "our people have a renewed joy in life and are no longer committing suicide. It's a very moving and beautiful sight to behold." Emilia Romero Avavera, age 97, said, "We have gone back to dancing, healing, and praying" (Funari and Hanna, 2001, 40).

While many of the families' returns to the land have been peaceful, some have met with violence. On March 24, 2001, the Guarani-Kaiowa *tehoka* (extended family) of *Ka'a Jari*—men, women, and children—tried for a third time to reclaim their ancestral land, "only," according to a report in the news magazine *Native Americas*, "to be met by a spray of bullets from the local *fazendeiros* [hired gunmen]. One death and numerous wounded were reported" (Funari and Hanna, 2001, 40).

During March 2000, Federal Indian Bureau anthropologists identified two sacred Indian cemeteries and marked the boundaries of 24,200 acres around the invaded area as Kaiowa land, the first step in the long bureaucratic process of creating a reservation. Much of the proposed Cerro Marangatu Indian reservation crosses Pio Silva's 3,344-acre Fronteira ranch. Under Brazilian law, Silva is entitled to government compensation for improvements to the land, such as houses, fences, and bridges, but not for the land itself. In December 2001, Silva obtained a court order to evict the native people who were occupying land they insisted was theirs by ancestral right.

During November 2002, a group of Guarani-Kaiowa in the Cerro Marangatu region of Brazil won title to their land after 50 years, as Brazil's Minister of Justice signed a bill approving the demarcation of 9,300 hectares of land, covering the Cerro Marangatu area that was taken from them during the 1950s. Roughly 400 Guarani-Kaiowa were thus allowed to move off of a nine-hectare reservation on which lack of land had led to food shortages and social breakdown, as young children died of malnutrition and some committed suicide.

For other Guarani-Kaiowa communities, the struggle for land rights continued. According to a press release from Survival International, "Guarani-Kaiowa from Taku·ra were camped by the side of a road, having been evicted from their land by armed police and soldiers. When living in Taku·ra, leader Marcos Veron said, 'This here is my life. My soul. If you take me away from this land, you take my life.' His son now reports that Marcos Veron has talked of committing suicide" ("Guarani-Kaiowa," 2002).

Veron himself was murdered during January 2003, after his community tried to move back onto its ancestral land, stolen by ranchers in the 1950s. He was shot and beaten. He was the third Brazilian Indian to be killed during 2002.

THE KAIAPO: GREENPEACE AND MAHOGANY LOGGING

During October 2001, in a highly unusual move, the Brazilian national government suspended all trade in mahogany, following a stringent domestic crackdown on illegal logging of this valuable wood. Scarcity of mahogany has enticed many loggers into illegal operations deeper than ever into the rainforest, where they often bribe indigenous peoples (notably the Kaiapo) to let them harvest trees for as little as $30 each. Illegal mahogany loggers frequently bribe government officials, as well.

The mahogany trade is nearly as old as European colonization of the Amazon; the wood is prized for its appearance, durability, and malleability. "Only cocaine has a multiplier effect on such a scale," said Paulo Adario, chief of Greenpeace in the Amazon, who was supplied with governmental bodyguards after his life was threatened for authoring reports critical of illegal mahogany logging. Brazil supplies about 95 percent of mahogany imports to the United States (Jordan, 2001, B-1).

The same trees may eventually sell as parts of finished furniture and other products for as much as $130,000 each. The Indians often need the little money they receive for food and medicines. Recognizing the crimp that the ban placed on their cash requirements, some of the Kaiapo painted themselves red in traditional warrior fashion to protest the government crackdown on mahogany logging on their territories. In the Amazon Valley, mahogany is known widely as *green gold*.

Brazil's mahogany belt covers about 80 million hectares of the Brazilian Amazon, stretching from the south of Para to Acre, and crossing the northern regions of Mato Grosso, Rondonia, and southern Amazonas. Mahogany prospectors often fly small aircraft hundreds of kilometers over dense forest in search of isolated mahogany trees, usually fewer than one tree per hectare. "To gain access to a single mahogany tree," according to the *World Rainforest Movement Bulletin,* "loggers often bulldoze illegal access roads sometimes stretching over hundreds of kilometers, criss-crossing previously untouched forest" ("Brazil: Mahogany Loggers," 2001). Logs are sometimes are taken as far as 500 kilometers from the nearest sawmill, resulting in widespread forest destruction, affecting indigenous lands in Para State.

Fifteen parcels of Indian lands covering 16,243,000 hectares of forest have been materially affected by illegal mahogany logging, despite legal protections from logging ("Brazil: Mahogany Loggers," 2001) An unknown number of Brazilian Indians have been murdered because of their opposition to illegal mahogany logging, according to a report in the *World Rainforest Movement Bulletin* ("Brazil: Mahogany Loggers," 2001).

Although it is mainly illegal, the mahogany trade has flourished, according to the *World Rainforest Movement Bulletin,* using forged documents to make it appear that the logs were harvested legally. The legal mahogany is then exported to overseas buyers. Furniture-manufacturing companies then buy the logs for products marketed by reputable companies such as LifeStyle Furnishings International, Furniture Brands International, Stickley, and Ethan Allen (in the United States); Gibbard Furniture Shops in Canada; International Timber, Timbmet, James Lathams, and Vincent Murphy in the United Kingdom; and others ("Brazil: Mahogany Loggers," 2001).

Brazil has cracked down on illegal logging by utilizing hundreds of federal police, who cooperate with Greenpeace activists and use satellite surveillance, airplanes, helicopters, and boats to raid illegal logging sites as well as clandestine lumber mills. Sometimes the activists and police cross paths with hired gunmen associated with the mahogany loggers. "Those operating with illegal mahogany are working against the forest and its indigenous peoples," Hamilton Casara, president of Ibama, a Brazilian environmental group, told Miriam Jordan of the *Wall Street Journal* (Jordan, 2001, B-1). An unknown number of indigenous people have been murdered in Para State for resisting mahogany logging that is legally prohibited on Indian lands.

Between September and October 2001, Greenpeace documented a large number of illegalities in the mahogany industry on Indian lands in the Amazon and issued a report, "Partners in Mahogany Crime," which was delivered to Brazil's federal prosecutor and environmental authorities. The report described a mahogany mafia tied to the international timber trade. Based on this report, Brazilian authorities seized the largest volume of illegal mahogany logs in Brazil's history. During five days, Greenpeace seized 7,165 cubic meters of illegal mahogany worth almost U.S. $7 million on the international market ("Brazil Cracks Down," 2001).

On May 8, 2000, according to an account by the Environment News Service (ENS), during a routine flight of a Greenpeace Cessna aircraft, illegal logging of a tree that is even rarer than mahogany was discovered—Samauma, which is called the "queen of the forest" by many indigenous people and rubber tappers. According to the ENS, "Because of its vast size, the felling of one Samauma tree may damage as many as 30 surrounding trees. In Amazonas State, the Samauma tree is one of the most used species in the production of plywood" ("Brazil Cracks Down," 2001). Not only were the trees cut illegally, but their removal left vast trails of destruction in the forest. Some of these logs measure more than 1.8 meters (5.8 feet) in diameter, and were very old trees.

By the end of March 2002, the European Commission had advised all European Union countries to reject importation of Brazilian mahogany. On March 26, a "Note to the Management Authorities" of the 15 E.U.-member states from Christoph Bail, head of global and international affairs in the environment directorate, advised "member states not to accept export permits for specimens of *Swietenia macrophylla* [mahogany] from Brazil until further notice, without first obtaining from the Brazilian authorities a statement that those specimens were legally acquired" ("Europe Rejects," 2002).

THE PANARA: ROAD BUILDING, IMPORTED DISEASES, AND GENOCIDE

The 200 surviving Panara Indians of Mato Grosso and Para states in Brazil won, during the year 2000, a U.S. $500,000 judgment from a Brazilian court, which implicated Brazil's government in genocide that killed 80 percent of the Panara people. The cause of the genocide: a road, built during the early 1970s, which brought rapacious settlers and outlaw gold-miners who spread deadly diseases among the Panara. This was the first time a Brazilian court had held the government liable for failure to enforce its Indian-protection laws.

During 1991, some of the Panara returned to their traditional territories to find that the land, according to one report, had been "reduced to wasteland by the gold rush and cattle ranching." However, according to the same report, "The headwaters of the Iriri River remained intact, signaling a potential for restoration" ("Brazil: Panara," 2000, 9). In 1996, the government set aside a reservation of 1.2 million acres on which logging and mining were legally prohibited. Laurie Parise, executive director of the Rainforest Foundation, was quoted as saying: "This decision represents a major victory and illustrates how legal initiatives can be used to create precedent-setting changes in the protection of the rights and lands of indigenous peoples. . . . After many years of struggle, a terrible injustice has finally been righted" ("Brazil: Panara," 2000, 9).

The Panaras' example has inspired other indigenous peoples in Central Mato Grosso State to take direct action in protection of their lands against road-borne invaders. For example, during mid-2001, more than 100 Terena, wearing war paint and feathers, held nine journalists hostage, demanding land that had been promised by the government, but not delivered. The journalists, mainly television crews from the area, had been sent to cover the Terenas' blockade of a highway. A spokesman for FUNAI (the State Agency on Indigenous Peoples' Issues, a Brazilian government agency charged with protecting native rights) told Reuters that negotiations were underway. According to FUNAI, government officials were offering land to the Indians for immediate occupation. Officials from FUNAI maintained that the Incra land agency had simply been slow in handing over the land promised under an accord signed the previous year, after the Panara took several journalists hostage over the same issue.

"They are doing exactly the same thing now as they did last year, when they got the deal for land. They are using the press to pressure the government for the quickest solution," the spokesman said, adding that such situations usually ended favorably for the Indians. Brazil's Globo Television showed the hostages calmly sitting and walking around a hut, guarded by Indians with wooden spears. "Indians like the Terena, the Xingu, the Xavante—they learned the tactics of keeping people hostage without harming them just to get what they want," the spokesman said ("Brazil Indians Take Reporters," 2001).

On October 5, 2001, having held the journalists for three days, the Indians freed them. More than 100 Terena Indians gathered with Brazilian officials to negotiate the land claim that had provoked the hostage-taking. "The land issue has not been resolved and the chances are the Indians will resume their protests if the land is not handed over soon," said a spokeswoman for the Agrarian Development Ministry's Incra land agency ("Brazil Indians Free Reporters," 2001). The Terena Indians demanded that the government hand over land it promised as part of an accord signed last year when the tribe took several journalists hostage over the same issue. The tribe has about 700 members and has a reservation in the sparsely populated area.

THE PATAXÓ: TAKING BACK THEIR LAND

During 1999, the Pataxó-Hã-Hã-Hãe indigenous peoples reoccupied a section of their traditional territory in the Brazilian state of Bahia. "Since then," reported the *World Rainforest Movement Bulletin*, "they have been struggling to have their rights recognized by the government, with little support from environmental organizations, many of [which] seem to deny them their capacity to manage the forest that rightly belongs to them" ("Brazil: The Struggle," 2000).

The Pataxós' traditional territories, demarcated (legally recognized) in 1936, consist of 53,000 hectares that are now occupied, for the most part, by almost 400 ranchers who received their titles illegally from successive governments of Bahia after 1960. These lands, which also include remnants of the Atlantic forest (*mata atlántica*), have been largely converted into pastures and cacao plantations ("Will There Be Justice," 2000).

The Brazilian Anthropologic Society criticized Brazil's Ministry of Environment for promoting projects in the area before the demarcation of the Pataxós' lands was finalized. Anthropologist Silvio Coelho dos Santos, coordinator of the Commission for Indigenous Affairs of the Anthropologic Society, wrote in a letter addressed to the government that "there appears to be a systematic movement against the indigenous presence within conservation areas" ("Brazil: The Struggle," 2000).

The Pataxós were violently evicted from their traditional territory during 1961, after which the land was declared part of the Monte Pascoal National Park. After the Pataxós reoccupied their land in August 1999, the government established a Technical Working Group to carry out demarcation of the Pataxó areas. Political pressures kept the committee from completing its mandate. In the meantime, the Brazilian government failed in its repeated efforts to extract the remnant of Pataxós from the park. In Coelho's opinion, "There is no evidence to believe that this natural heritage is now especially threatened or vulnerable as a result of the occupation of the park by the Pataxó 14 months ago; on the contrary, the opposite appears to be true" ("Brazil: The Struggle," 2000).

During December 1999, military police expelled about 150 Pataxó families who had camped within the municipality of Prado in the State of Bahia. The police action followed a legal complaint from two cattle ranchers who earlier had evicted them. The Pataxó had camped there as they awaited finalization of their land's demarcation at Barra do Caí. The Pataxó later decided to leave the area peacefully in order to avoid a confrontation and immediately organized a demonstration in town against the police action and the "lack of will by FUNAI (the State Agency on Indigenous Peoples' Issues) to finalize the demarcation of the indigenous lands in Barra do Caí" ("Brazil: The Struggle," 2000).

THE PEMON: MERCURY POISONING AND AMAZON GOLD MINING

The Pemon, an Amazonian indigenous people, are being poisoned by gold mining in which mercury is used to remove gold from ore. In areas where metallic mercury has been used as an amalgam to separate particles of gold from river sediment, mercury poisoning occurs, also known as Mad Hatter's or Minamata disease. The mercury has been bioaccumulating in the local food chain in an area where most Pemon men have been gold miners at one time or another. Local indigenous women and children also are being poisoned from eating contaminated fish and other foods.

Metallic mercury escapes from gold mines, leaching into local rivers, where it is transformed into the highly toxic and persistent methyl mercury, a persistent organic pollutant that accumulates in the food chain, ultimately reaching unhealthy concentrations in fish, the staple diet of local people. In its most advanced form, mercury poisoning can cause birth defects and brain damage, and eventual death. Less severe poisoning may result in tunnel vision and neurological disorders.

Greenpeace estimates that between 1,000 and 2,000 tons of mercury were released into the Amazon's ecosystem during the 1990s. Mercury levels have been increasing as larger volumes of gold are mined. For every kilogram of gold produced, four times that amount of mercury is released into the environment ("Quicksilver for Gold," n.d.).

International health guidelines state that tolerable levels of mercury should not exceed five micrograms per gram of creatine (as measured in fingernails). A seven-year-old Pemon boy, Jose Blanco, was tested at 17.29 micrograms. Emilio Palacio, a miner, was tested at 24.0 micrograms. For hair, international guidelines state that tolerable levels should not exceed 2.0 micrograms of mercury per gram of hair. The hair of Antolin Bolivar, a Pemon aged nine, tested at 17.88 micrograms. Miner Eliseo Yepez, aged 32, tested at 18.68 ("Mercury, Mining, and Mayhem," 1999). A 1989 study in the mining district of El Callao in Bolivar State, Venezuela, indicated that in a random sampling of 51 individuals, 72.5 percent showed symptoms of mercury contamination. Of this sample, 6 people suffered from clouding of the cornea with a complete loss of vision, 15 had hypertension, and 27 showed severe problems with memory loss and disorders of the nervous system ("Mercury, Mining, and Mayhem," 1999).

During 1996, 15 inhabitants of El Vapor, an indigenous community in the Amazon, had their hair and fingernails tested for mercury contamination. Three years later, results were delivered. Of the 15 tested, only 3 had normal levels in their hair (as determined by international health standards) and only 2 had normal levels in their fingernails ("Mercury, Mining, and Mayhem," 1999). Analysis of the sediment from the river Caroni, between the Guri dam and the city of Puerto Ordaz in Bolivar State, revealed mercury levels of 3,679 micrograms of mercury per kilogram, approximately 183 times the natural level of the river ("Mercury, Mining, and Mayhem," 1999).

A similar situation is suspected on the Tapajos River in the neighboring Brazilian Amazon. Early in 1998, two Japanese scientists, Maszumi Harada of Kumamoto University and Junko Nakanishi from Yokahama National University, published a study of mercury contamination around San Louis de Tapajos. They tested 50 villagers from various parts of the region: all had high levels of mercury and 3 had symptoms of severe mercury poisoning ("Mercury, Mining, and Mayhem," 1999).

Studies in the Brazilian Amazon have shown that levels of mercury in Tapajos river fish in 1995 were 3.8 parts per million (p.p.m.), almost eight times the permitted U.S. federal maximum of 0.5 p.p.m. In 1989, fish in the Madeira River tested as high as 2.7 p.p.m. (Chatterjee, "Gold, Greed," 1998). Small-scale miners who used mercury were evicted from the Yanomami territory in January 1998 by the Brazilian army. The Macuxi peoples of Roraima, Brazil blockaded roads in 1997 to successfully demand the removal of gold miners from their territory.

In 2003, the Evandro Chagas Research Institute, working with the Brazilian Health Ministry, found high levels of mercury contamination among 60 percent of the newborns at three hospitals in the city of Itaituba, in the Brazilian Amazon. The institute tested the blood of all the 1,666 babies born during 2002 in the three hospitals of the city and found 1,000 of them to be contaminated. Some of the children had 80 parts per million of mercury in the blood. The highest acceptable level, according to the World Health Organization, is 30 p.p.m. The contamination is due to gold mining activities in the region's rivers during the 1980s.

THE YANOMAMI: GOLD RUSH

More than 100,000 Yanomami roamed the watershed of the Rio Branco and Orinoco rivers in the Northern Amazon basin when the first Spanish colonizers reached the New World. They had lived in the tropical rainforests of the Amazon Valley for about 40,000 years by the time the first Spanish eyes greeted them. By the later years of the twentieth century, only 22,000 Yanomami survived, roughly 9,400 in Brazil and the rest in Venezuela (Donnelly, 1998).

The Yanomami language is linguistically isolated, with many dialects, leading anthropologists to believe that they once occupied a much larger area than at present. Their word for disease and epidemics is *xawara*, signifying an evil spirit that lives in the bottom of the world. They use the same word for gold, because its appearance, usually in the hands of outsiders, has been associated with the arrival of disease and death. The Yanomami perceive the *nabebe* (white men) as having an insane desire to bring disease and gold to them from the bottom of the world (Donnelly, 1998).

For several hundred years of colonization, the Yanomami resisted integration with modern capitalism. According to an analysis by R. Donnelly in Great Britain's *Socialist Standard*, "Portuguese exploiters, who attracted [other] indigenous peoples into their settlements and into slavery, failed to lure the Yanomami from their traditional communal culture. Likewise, early missionaries failed to convert them" (Donnelly, 1998). The Yanomami preferred their own culture and religious rituals, remaining one of the last remaining societies on Earth that entered the twenty-first century still living in kinship groups and inhabiting *malocas* (communal huts). The Yanomami retained their staple diet of cassava gathered from manioc plants and wild game, notably monkeys and turtles, from the jungles of their traditional homelands.

By the later years of the twentieth century, however, the tentacles of the industrial state came calling, in the form of various gold-mining enterprises, whose workers brought diseases and claims to the land, which were enforced by Brazil's military. Many Yanomami were killed during the 1970s, as the Brazilian military government, in an attempt to open the Amazon to gold speculators and cattle ranchers, built the first highways through the Yanomamis' terrain. Thousands of Yanomami died from imported illnesses, such as yellow fever, introduced by the road builders.

A 1998 photo of a Yanomami woman and her child, residents of the Yanomami reservation in the Brazilian State of Roraima. Home to some 9,000 Yanomami, the reservation is being threatened by mining and destruction of the forest. (Painet)

A large reservation was set aside in the Amazon Valley for the Yanomami during the early 1990s. Before its first birthday, however, much of the reserve was overrun by several thousand gold miners, leading to the massacre of an entire Yanomami village. The Kayapo of Brazil's Para State and the Yanomami both have expelled thousands of gold miners (locally called *garimpeiros*) who invaded their lands during the 1990s, not only because they disrupt the land, but because they bring increased incidence of diseases, especially malaria. More than 40,000 miners invaded Yanomami lands looking for gold in 1990; most of them had been expelled by 1992, when a fresh surge of 8,000 miners inundated the area. Following the miners' invasion in 1993 of Yanomami lands, 2,000 were expelled by Brazilian federal authorities during 1994. Local observers expected the miners to return, utilizing clandestine aircraft-landing strips in the jungle ("Kayapo and Yanomami," 1998, 14).

Shortly after the miners' invasion of 1993, Davi Yanomami was delegated by his people to seek the assistance of the United Nations to protect his people's human rights. He described ferryboats carrying loads of illegal miners over the Calaburi River on the western edge of the Yanomami reservation, bringing with them malaria and other diseases. "We are getting ill," Yanomami told the United Nations. "Malaria is out of control. My relatives are dying. We are fearful of another epidemic, like the one between 1987 and 1991, that killed so many of us" (Ewen, *Voices of Indigenous Peoples*, 1994, 109–10). In that epidemic 1,500 of fewer than 10,000 surviving Yanomami died.

According to Alexander Ewen, writing in *Native Americas*, "With the election of President Fernando Enrique Cardoso in 1994, the juggernaut of ranchers and colonists, mining companies and the military once again began to slowly grind up Indian people" (Ewen, 1996, 22). By 1996, 160 of 210 Indian reserves in the Amazon, including the Yanomamis, were under some sort of pressure to diminish their land tenure. During the 1990s, the Yanomamis' 9,000-square-kilometer reservation was reduced to 2,000 square kilometers. Of that, in 1990, only 256 square kilometers had been set aside for gold mining. Very suddenly, 45,000 gold miners surged into the Yanomamis' homeland, polluting their rivers with mercury, blowing up villages, and shooting children (whom some of them called monkeys) out of the trees for sport (Donnelly, 1998).

In addition, a changing climate has devastated large areas of the Yanomamis' lands. During the El Niño years of the late 1990s, forest fires destroyed or severely damaged large swaths of the Yanomamis' remaining forests. Many of these fires were first started deliberately to clear land for cattle before they raged out of control due to unusually dry weather conditions. The Yanomami, whose economy depends on the forest, found their land base shrinking once again.

During 2000, Brazil's Higher Court of Justice upheld a judgment by a lower court finding a group of outlaw gold miners guilty of genocide against 16 Yanomami for the 1993 Haximu massacre in the state of Roraima. The 22 defendants were sentenced to between 19 and 22 years each in prison. According to an Environmental Defense Fund report from Mato Grosso, Brazil, released during November 1996:

> A group of loggers and miners near the town of Pontes e Lacerda ambushed and violently assaulted at least 14 Katitaulhu Indians in the Sarare reserve. The loggers subsequently looted the Indians' village, damaging a health post and school and stealing money, tools and vehicles belonging to the Indians. Supporters of the Indians, who have attempted to mobilize federal officials to comply with court orders to remove the illegal loggers and miners from the reserve subsequently received death threats and intimidation. The Katitaulhu were also threatened with further violence by the invaders. Medical reports state that 14 Indians were wounded, many by having been tied up and beaten. (Chatterjee, 1998)

The Katitaulhu are 1 of 12 Nambikwara subgroups whose lands were first invaded during the 1970s when the World Bank–funded road from Cuiaba in Mato Grosso to Porto Velho in Rondonia was opened by Brazil's military government. Reduced by epidemics and forcibly relocated to make way for the road, the Nambikwara died in great numbers as they attempted to reestablish lives on their traditional lands. Roughly 6,000 gold miners invaded the Sarare reservation in the 1990s, seriously polluting major watercourses in the area, disrupting local fishing and hunting, and spreading malaria and viral diseases.

The incident described above is only one of many attacks on the Nambikwara in the last two decades (Chatterjee, 1998).

According to Davi Yanomami, "The biggest problem for the Yanomami now are the *garimpeiro* (gold miners) . . . and the illnesses they bring with them. Among them some have illnesses like flu, tuberculosis and venereal diseases, and contaminate my people. Now we are afraid they will bring measles and also AIDS, this illness which is so dangerous that we do not want it among us. But the worst illness for us is malaria, which comes in with the gold miners. The government's National Health Foundation say that 1,300 Yanomami had got malaria up until May this year" (Chatterjee, 1998).

Facing budget constraints during 2002, the Brazilian government threatened to curtail health care for the Yanomami, forcing them to face the prospect of higher death rates from malaria and tuberculosis imported by gold miners. According to Survival International, "Without a full health care program, the Yanomami could again suffer appalling epidemics like those which recently [reduced] their population by more than 20 per cent" ("Brazil: Yanomami," 2002). Following international protests, health care was restored in April 2002.

REFERENCES

"Amazonian Deforestation is Accelerating, U.S. Study Finds." Environment News Service, January 15, 2002. [http://ens-news.com/ens/jan2002/2002L-01-15-09.html]

"Amazon Watch Launches Mega-Project Report: New Pipelines Threaten Intact Amazon Rainforest in Brazil." Amazon Watch, August 10, 2001. [http://www.amazon watch.org/newsroom/newsreleases01/aug1001_br.html]

"Aracruz: Indigenous Peoples Refuse to Plant Eucalyptus." *World Rainforest Network Bulletin* 32 (March 2000): n.p. [http://www.wrm.org.uy]

Astor, Michael. "Indian Land Conflict Hits Brazil." Associated Press, June 1, 2002.

"B.B.C. Exposes Scandal of Brazilian Government Neglect: Last Amazonian Hunter-Gatherers Face Extinction." Survival International, August 29, 2002. [http://www.survival-international.org]

"Brazil Cracks Down on the Illegal Mahogany Trade." Environment News Service, December 12, 2001. [http://ens-news.com/ens/dec2001/2001L-12-12-03.html]

"Brazil Indians Free Reporters Taken Hostage." Reuters News Service, October 5, 2001. (In LEXIS)

"Brazil Indians Take Reporters Hostage to Get Land." Reuters News Service, October 5, 2001. (In LEXIS)

"Brazil: Mahogany Loggers Destroying the Amazon Forest." *World Rainforest Movement Bulletin* 53 (December 2001): n.p. [http://www.wrm.org.uy/deforestation/logging.html]

"Brazil: Panara Win Damages for 1970s Genocide." *Native Americas* 17:4 (winter 2000): 9.

"Brazil's Amazon Rainforest Shrinking Fast." Environment News Service, May 15, 2001. [http://ens.lycos.com/ens/may2001/2001L-05-15-03.html]

"Brazil's Indigenous People Resist Large River Modifications." Environment News Service, May 30, 2001. [http://ens.lycos.com/ens/may2001/2001L-05-30-01.html]

"Brazil: The Struggle of the Pataxó Indigenous Peoples in Bahia." *World Rainforest Movement Bulletin* 31 (December 2000): n.p. [http://www.wrm.org.uy/bulletin/41/Brazil.html]

"Brazil: Yanomami to Lose Another 20 Per Cent?" Survival International, January 28, 2002. [http://www.survival.org.uk/about.htm]

Chatterjee, Pratap. "Gold, Greed and Genocide in the Americas: California to the Amazon." *Abya Yala News: The Journal of the South and Meso-American Rights Center* (1998): n.p. [http://saiic.nativeweb.org/ayn/goldgreed.html]

Connor, Steve. "Death Sentence for the Amazon: Scientists Say $40 Billion Project is Set to Destroy 95 per cent of Rainforest by 2020." Amazon Watch, 2001. [http://www.amazonwatch.org/newsroom/mediaclips01/010119iuk.html]

"Co-ordinator of Movement Against Xingu Dams is Murdered." Amazon Alliance [amazon@amazonalliance.org] Movimento Pelo Desenvolvimento de Transamazonia. [fvpp@amazoncoop.com.br] August 25, 2001. [http://www.amazonwatch.org/newsroom/mediaclips01/braz/010825feab.html]

"Cotingo Dam in Brazil is Halted, Sparing the Macuxi and the Ingarico." *Native Americas* 12:1+2 (summer 1998): 15.

"Dam Affected People Occupy Tractebel Headquarters in Rio." Environment News Service, October 31, 2001. [http://ens-news.com/ens/oct2001/2001L-10-30-03.html]

"Dam Protesters Occupy Brazil's Ministry of Energy." Environment News Service, March 14, 2001. [http://ens.lycos.com/ens/mar2001/2001L-03-14-02.html]

"A Dictatorship-type Action Gives Aracruz a Spurious Victory." *World Rainforest Movement Bulletin* 11 (April 1998): n.p. [http://www.wrm.org.uy/bulletin/11/Brazil.html]

Donnelly, R. "The Curse of Xawara." *Socialist Standard* (Great Britain), June 1998. [http://www.worldsocialism.org/xawara.htm]

"Europe Rejects Brazilian Mahogany Imports." Environment News Service, March 29, 2002. [http://ens-news.com/ens/mar2002/2002L-03-29-01.html]

Ewen, Alexander. "Indians in Brazil: Is Genocide Inevitable?" *Native Americas* 13:4 (winter 1996): 12–23.

———. *Voices of Indigenous Peoples: Native People Address the United Nations.* Santa Fe, N.M.: Clear Light, 1994.

Funari, Ricardo, and Jennifer Hanna. "Recovering Lands; Recovering Futures." *Native Americas* 18:1 (spring 2001): 38–41.

"Guarani-Kaiowa of Brazil Win Land Rights." Survival International Press Release, November 13, 2002. [http://www.survival-international.org/enews.htm]

Hamilton, Dominic. "Pemon Indians Paralyze Powerlines." *Native Americas* 14:3 (fall 1997): 9.

Hanna, Jennifer. "The Guarani Are Committing Suicide: Mato Grosso de Sul, Brazil." *Native Americas* 14:1 (spring 1997): 32–41.

Jordan, Miriam. "Brazilian Mahogany: Too Much in Demand; Illegal Logging, Exports Are Lucrative for Criminals, Disastrous for Rainforest." *Wall Street Journal*, November 14, 2001, B-1, B-4.

"Kayapo and Yanomami Battle Miners." *Native Americas* 12:1+2 (summer 1998): 14.

"Mercury, Mining, and Mayhem: Slow Death in the Amazon." *Drillbits and Tailings* 4:10 (June 24, 1999): n.p. [http://www.moles.org/ProjectUnderground/drillbits/4_10/4.html]

Muggiati, André. "One Thousand Brazilian Babies Poisoned by Mercury." Environment News Service, May 20, 2003.

"Oil, Mega-development Plans Would Destroy Amazon." *Indian Country Today*, October 10, 2001, A-4.

"Partners in Mahogany Crime: Amazon at the Mercy of 'Gentlemen's Agreements.'" Greenpeace, October 2001. [http://www.greenpeace.org/%7Eforests/forests_new/html/content/reports/Mahoganyweb.pdf]

"Population, Highways Lead to Amazon Deforestation." Environment News Service, July 8, 2002. [http://ens-news.com/ens/jul2002/2002-07-08-09.asp#anchor3]

Quicksilver for Gold: The Poisoning of the Amazon. Greenpeace, n.d. [http://www.greenpeace.org/~thoml/mercury.html]

"Ten Years Without Chico Mendes." *World Rainforest Movement Bulletin* 18 (August 1998): n.p . [http://www.wrm.org.uy/bulletin/18/Brazil2.html]

"Will There Be Justice for the Pataxó-Hã-Hã-Hãe?" *World Rainforest Movement Bulletin* 32 (March 2000): n.p. [http://www.wrm.org.uy/bulletin/32/Brazil.html]

BURMA (MYANMAR)

FORCED LABOR IN THE WORLD'S LAST TEAK FOREST

Forest-dwelling peoples in Burma, notably the Karen, have been impressed into slave labor to harvest the world's last sizable forests of teak by the country's military rulers. The Burmese military government, which has thrown out democratic election results for the country as a whole, also has been using roads originally built for gas exploration to allow logging access in indigenous homelands comprising the panhandle in the country's far south. Boycotts of Burmese teak have been organized by environmental activists in the United States and Europe, where the valuable wood is most often used in Scandinavian-style furniture. Large amounts of teak also are sold (much of it harvested in Burma, but processed in Thailand and China) for luxury yachts and other pleasure boats. In some cases, Burmese teak also is being used as flooring and outdoor furniture.

In addition to the world's last teak forests, Burma's territory once included some of the largest virgin rainforests that remained in mainland Asia. Many of these forests were home to rare species, such as the Asian rhino and the Asian elephant, among others, as well as the aforementioned forest-dwelling peoples.

HISTORY OF BURMA'S JUNTA AND THE BURMESE TEAK HARVEST

Burmese teak logging began in earnest during the British colonial period. British demand for ships made of the durable wood consumed most of the commercially viable teak in India and eventually in Thailand. Destruction of teak forests also has provoked flooding and drought in parts of these countries. By the late 1990s, commercial-scale teak harvesting was restricted almost entirely to Burma. In 1994, it was estimated that Burma held 80 percent of the world's remaining natural teak (Stevens, 1994).

At the beginning of the twentieth century, forests covered 80 percent of Burma. In 1948, 72 percent of the country was forested. By 1988, Burma's forest cover had decreased to about 47 percent of its land area. By the early 1990s, a decade later, Burma's forest cover was estimated to be roughly 36 percent. By the turn of the millennium, Burma was experiencing the third-highest rate of deforestation in the world, after Brazil and Indonesia, at roughly 8,000 square kilometers a year (Rainforest Relief, 1997).

A military coup in Burma during 1962 initiated a reign of terror and oppression that continues to this day. In 1988, after millions of Burmese rallied for democracy, the military junta formed the SLORC (State Law and Order Restoration Council) to strengthen its domination of Burmese government. The SLORC, composed of several high-level generals, ordered the killing of at least 3,000 (other estimates are much higher) dissident Burmese demonstrators in 1988, during widespread unrest.

The junta later called general elections, during which its opponents won more than 82 percent of the seats in Parliament. The military then ignored these results, and refused to yield power. The SLORC generals consolidated their rule with forced labor, rape, torture, forced relocation, and intimidation (Rainforest Relief, 1997). The same methods were used against the Karen, as well as the Shan, Karenni, Mon, and Chin indigenous peoples as the junta sought to raise foreign exchange by harvesting the world's last sizable stands of teak, as well as other valuable hardwoods across Burma.

Until the late 1990s, large areas of southern and eastern Burma remained relatively free of military rule due to resistance of numerous indigenous ethnic groups such as the Mon, Karen, and Karenni. However, with a massive infusion of new capital, largely from selling natural gas concessions offshore, a so-called ethnic cleansing operation was initiated by the junta in an attempt to consolidate its rule in rural areas of Burma. Much of this capital came from large American energy corporations (notably Unocal and Texaco); the French energy giant Total; and a Thai company, PTT. Teak was harvested in east-central Burma. The hardwood forests of southern Burma were concessioned, with indigenous peoples often supplying unwilling forced labor. According to local observers, this cleansing involved "burning of villages, raping and torturing of villagers, forced labor, and forced relocations" (Rainforest Relief, 1997).

The SLORC's minister for forestry, Lieutenant General Chit Swe, stated during 1997 that Burma's teak forests would be logged to increase economic development with support from the private sector (ASIA, Inc., 1996, cited in Keating, 1997).

In practical terms, this meant that forest-products exports were exempted from commercial tax after May 1996. Daw Aung San Suu Kyi, a leader of the Burmese National League for Democracy, called this kind of economic development crony capitalism. According to the Burmese National League, "The generals and their friends get rich, while the Burmese populace starves" (Rainforest Relief, 1997).

Some of the largest buyers of Burmese teak in Europe are Scandinavian furniture manufacturers that supply several retail chains in the United States and Europe, including Scandinavian Design, Happy Viking, Scan Design, and

Dania. Some of the manufacturers and retailers assert that their sales of Burmese teak are helping that nation achieve economic development (Rainforest Relief, 1997). During the late 1990s, the Burmese democratic government-in-exile called for an international boycott of teak from Burma. The boycott included teak exported from Thailand, Singapore, and Taiwan that is Burmese in origin.

According to an analysis by Rainforest Relief, teak logging, like most tropical deforestation, causes extreme degradation of tropical forests. Teak trees grow in dense stands throughout the forest. Logging roads plowed to gain access to these stands play a fundamental role in allowing further deforestation (as well as lower water tables) for primary forests in Burma, Laos, Cambodia, and Thailand.

Roads also serve as conduits for other invasions of indigenous lands, including prospectors for various metals and minerals, squatters looking for land, and (especially in Burma's case) troops charged with forcing the indigenous population into a slave-labor force for the harvesting of saleable teak.

Intense logging in indigenous areas of Burma has aggravated flooding during seasonal monsoons. According to a report by Tim Keating of Rainforest Relief,

> As forests are cleared, rain runs off instead of being absorbed by the forest and recirculated into the environment and atmosphere slowly. This also leads to periods of drought, both of which adversely affect local peoples' ability to grow food. Erosion is exacerbating flooding and [has] caused silting of rivers, affecting fish populations . . . having a negative impact on local people. (Keating, 1997)

Similar effects are evident for other hardwoods, as well as for evergreen trees in Burma's far north.

By the early 1990s, the sale of teak and other hardwoods had become the second-largest legal moneymaker for the Burmese military (not counting its illegal trade in heroin). In 1992 and 1993, Burma extracted nearly one million cubic tons of teak logs with state-owned or contracted operations, up from 700,000 in 1983 (Keating, 1997). China is the largest importer of teak logs from Burma, followed by Thailand. A large proportion of this teak is processed for export as lumber, furniture, and other consumer items (including luxury yachts and other pleasure boats) for sale in the United States and Europe.

TESTIMONIES OF TORTURE

Indigenous activists and environmental organizations have collected many testimonies of torture related to forced labor by indigenous peoples in the Burmese teak harvest. A few of these testimonies follow.

A teak-forested hill behind Moung Pyin town was selected for the newly arrived Light Infantry Battalion 360. People from Moung Pyin and surrounding villages were ordered to clear it, and the felled logs were sold in Keng Tung town, with the money shared by the brigade commanders of LIB 43 and LIB

360. The labor was unpaid; workers were subject to a fine and jail for refusing to work. Those who complained sometimes were subjected to torture. Nearby forests were cleared for firewood and charcoal for the army. Construction materials for barracks were supplied by local people without compensation (Keating, 1997).

In Keng Tung, Number 244 Battalion forced people to cut all the trees in a nearby forest. They also used porters from other places to clear the trees. The cut logs were then taken to an army base. The army ordered all trees harvested, including the natural forest, as well as trees planted by the villagers for their own use. When the villagers protested, the soldiers refused to listen. As a result, all of the local mountains became barren, contributing to flooding during monsoon rains (Karen Human Rights Group, 1994; also cited in Keating, 1997).

Everyone in an unnamed town and its surrounding area were forced to work in rotating shifts. Each village and section of the town was directed to send people to work two of each three months. "Each day," reported one conscript, "my section of town has to send [as many as] 20 people, depending on how many the soldiers demand. There are 60 houses in my section. I've had to go twice to cut the trees, for one day each time. We had to take all our own tools, machetes and saws. They make us cut everything down, even the bamboo trees. Then we have to dig out the stumps too, and give them to the Army. It's all taken away by Army trucks" (Karen Human Rights Group, 1994; also cited in Keating, 1997).

The Burmese Na Sa Ka forces of Buthidaung township ordered the Muslim villages on the west bank of the Mayu River to supply 5,000 people to the Na Sa Ka for road construction, building embankments and logging in the Mayu hills. These laborers were further ordered to work continuously for 10 days and to cut 6,000 trees (Rohingya Solidarity Organization, 1994; also cited in Keating, 1997).

Another conscript told an observer for the Burma Students' Union: "There are three concentration labor camps in the Babow valley of the Tamu township in the Saggaing Division. . . . Wet Shu camp was the first to be built and the main camp of the three. The camp was constructed by the porters from nearby villages. The prisoners were divided into six teams for sawing. The sawing stands were one to two hours' walk from the camp in the forests. During 1994, three prisoners [broke] their legs in accidents with logs. No compensation was provided for the prisoners (All Burma Students' Democratic Front (ABSDF), 1995; also cited in Keating, 1997).

In the Toungoo district, most of the forced labor was used for construction of roads, and most of these roads directly benefited military efforts or powerful business interests taking part in teak logging and oil exploration. Laborers also were forced to build a road between Papun to Parheik, to a large military base. From Parheik, this road was extended northwards to Kyauk Nyat on the Salween River, on the border with Thailand. The river itself has always served as the major transportation artery in the area, both for military purposes and logging interests. Soldiers and timber traders ply the river from dawn to dusk in long-tail boats. When the access road is completed, it will greatly facilitate movement of troops through the area, as well as the transport of logs and timber to Rangoon (Images Asia, 1996; also cited in Keating, 1997).

The price of resistance in the teak forests can be painful. When soldiers dislike a villager, reports indicate that they beat him with sticks. "Sometimes," according to one firsthand account, "they beat them so hard that they had to be carried to hospital" (Karen Human Rights, 1994; also cited in Keating, 1997).

During the first week of March 1997, the villagers from Min Tha village were subjected to forced labor. Men, women, and even aged people were directed to remove teak from the jungle for the Burmese Army. Forced laborers were beaten by soldiers. One villager was severely beaten and his leg was broken, and the army denied him medical treatment. Relatives brought the beating victim to the Tamu Government Hospital, where treatment was denied. After a trip of many miles on foot, medical treatment was administered at Kalay Myo (Free Trade Unions of Burma, 1997; also cited in Keating, 1997).

The Karen Human Rights Group linked the cutting of trees to changes in local climate.

> They have been cutting so many trees that the climate is now changing here and it has become drier, so every year the rice harvest is worse. Usually the traders hire villagers to cut the trees. They pay them 250 Kyat for a ton but then they sell the logs for 12,000 Kyats a ton. The traders get permits to cut the trees by bribing the military. (Karen Human Rights Group, 1994; also cited in Keating, 1997)

THE JUNTA DISOWNS ITS OWN POLICIES

A U.S. human rights group has accused the Myanmar (Burma) military government of maintaining forced labor despite its own declaration stating that coerced labor is illegal. At one point, the military junta even issued a decree making the requisitioning of forced labor punishable by a maximum of one year in jail and a fine under section 374 of its Penal Code.

Human rights groups consider this decree to be public-relations window dressing that has no practical effect on real-world conditions in Burma's teak forests. Human Rights Watch, based in New York City, has said that it had clear evidence of continuing forced labor indicating that the aforementioned decree was passed only to avoid international criticism, not to change the junta's behavior ("Human Rights Group," 2001). The allegation by Human Rights Watch was issued a week before the governing body of the International Labor Organization was scheduled to meet in Geneva to review progress by Burma toward ending forced labor. "Either the Burmese government thought it could avoid international pressure by a sham decree or it just has made no effort to enforce the ban," Sidney Jones, Asia director for Human Rights Watch, was quoted as saying ("Human Rights Group," 2001).

Myanmar has long been assailed by the United Nations and Western countries for suppression of democracy and its human rights record, including its use of unpaid civilian labor on infrastructure projects. The junta's officers have maintained that civilians contribute their labor voluntarily to promote development of their nation. The Human Rights Watch statement called on the junta to enforce the decree and grant access to independent observers who will monitor compliance. Human Rights Watch said it conducted interviews in Thailand's Chiang Mai Province in late February 2001 with many Myanmar people "who had been recently subjected to forced labor" ("Human Rights Group," 2001).

Human Rights Watch cited the case of one ethnic Shan farmer who said that in January, a local unit of the Myanmar military forced him to dig trenches and fence-post holes for a military base in Ton Hu in Shan State's Nam Zarng Township. The farmer and about 20 other villagers said that they were forced to travel to the site five times during the month for two to three days at a time. Villagers were compelled to provide their own food and to sleep at the work site. None of them were compensated for their labor ("Human Rights Group," 2001).

OIL CORPORATIONS AND FORCED LABOR

According to *Drillbits and Tailings*, eyewitnesses in Burma—including the international human rights group EarthRights and the U.S. embassy in Rangoon—have reported that coerced labor has been used to build a Burmese natural gas pipeline. The consortium building the Yadana pipeline includes the French oil company Total, the U.S. firm Unocal, the Petroleum Authority of Thailand (which is run by the Thai government), and MOGE, Burma's state-owned oil and gas company.

The Yadana and Yetagun pipelines cross the southeastern Tenesserim region of Burma, pumping gas from offshore wells in the Andaman Sea to power plants in neighboring Thailand ("New Evidence," 2000). The pipelines have been built from the Yadana gas field off the coast of southern Burma, through the Tenesserim rainforest, then into Thailand.

Deserters from the Burmese army have reported that they were ordered on a primary mission to secure the area on behalf of the oil companies. "Securing" the area, according to the eyewitnesses, "included the construction of a string of military bases, intimidation and terror against villagers, and forcible relocation of entire villages" ("New Evidence," 2000). According to the same reports, military housing was built with forced, unpaid labor, including young children. These reports were denied by the government.

Unocal employee Joel Robinson told the U.S. embassy in Rangoon that Unocal had hired members of the Burmese military as security guards for the Yadana pipeline. Soldiers also were said by Robinson to have been hired as supervisors for construction of helicopter landing sites. The military, in turn, forced unpaid laborers to build the helipads. The Total company later provided some of the laborers with small amounts of money, much of which subsequently was taken from them by soldiers ("New Evidence," 2000).

On May 22, 2000, the same day that the U.S. embassy's report was released, more than 100 demonstrators rallied at Unocal's annual meeting in suburban Brea, California to protest the treatment of the Burmese villagers. According to an account in *Drillbits and Tailings,* as drummers in skull masks performed with mock bones on upturned Unocal oil barrels outside, a shareholders' resolution sought to scale Chief Executive Officer Roger Beach's salary according to the company's ethical performance, including its behavior on its Burma project. The proposal won approval of 16.4 percent of the company's shareholders, which is high for a human rights–based measure on a corporate ballot ("New Evidence," 2000). The demonstration was organized by the Burma Forum of Los Angeles and endorsed by the Los Angeles County Federation of Labor.

A state appeals court in California early during 2003 ruled that Unocal should face trial on allegations that it was complicit in human rights abuses in Burma during the construction of a 39-mile (62 kilometers) natural gas pipeline. The suit was filed by refugees from Burma and alleges that Unocal was aware of, and took no steps to prevent, the Burmese military from raping, murdering, and enslaving villagers within the vicinity of the pipeline ("Hotspots: Burma," 2003).

Activism in the United States in opposition to coerced labor in Burma has been spearheaded by Ka Hsaw Wa, 29, director of EarthRights International. His name means "White Elephant," a nom de plume that he uses to avoid reprisals from Burma's military regime. Ka Hsaw Wa was born in Burma and was tortured by the junta for his environmental and human rights convictions before his subsequent escape to the United States. Ka Hsaw Wa, who is Karen, in 1988, at the age of 18, joined massive demonstrations in Burma demanding human rights, democracy, and an end to military rule. He was arrested and tortured for three days ("Earth Day," 2000). He was then forced to flee his home and go into hiding in the forests near the Thai border.

Slipping anonymously back and forth across the border between Burma and Thailand, Ka Hsaw Wa has documented environmental and human rights abuses associated with construction of the Yadana natural gas pipeline. He has called the pipeline "the most notorious project in Burma" because it is damaging the Tenesserim rainforest, which is home to many ethnic groups as well as many rare plants and animals ("Earth Day," 2000). Ka Hsaw Wa was a 1999 winner of the $125,000 Goldman Prize for Asia. He also has been honored with the $20,000 tenth-annual Conde Nast Traveler Environmental Award, and in 1999 he received the Reebok Human Rights Award.

"The Tenesserim rainforest is one of the largest intact rainforests in mainland Southeast Asia," said Ka Hsaw Wa. "The people of Burma do not want the pipeline, but Unocal wants it and builds it anyway. They have sent the brutal Burmese military dictatorship to protect their investment" ("Earth Day," 2000). The Tenesserim rainforest is on the Isthmus of Kra, between the

Andaman Sea and the Gulf of Thailand—land that has long been occupied and used by the Karen and other indigenous peoples.

Ka Hsaw Wa described how the local people are forced to work for the pipeline, to carry ammunition and grow food for the soldiers. "Many villagers were forced to move from their land. Women have been raped and many have been killed," he said ("Earth Day," 2000). "The same soldiers hired by the Unocal company who are oppressing people in the pipeline regions are also destroying the environment. They are catching wild animals, they are cutting down trees to sell, to grow food and other projects," Ka Hsaw Wa continued ("Earth Day," 2000).

Corporate representatives of Unocal maintain that the company is improving the lives of the local people. "Unocal does not defend the actions and policies of the government of Myanmar [Burma]. We do defend our reputation and the integrity of the Yadana project. Our hope is that Myanmar will develop a vital, democratic society built on a strong economy. The Yadana project, which has brought significant benefits in health care, education, and economic opportunity to more than 40,000 people living in the pipeline area, is a step in the right direction," Unocal said in a statement on its Web site ("Earth Day," 2000).

REFERENCES

All Burma Students' Democratic Front (ABSDF). "Wet Shu Concentration Labour Camp." New Delhi, India, April 20, 1995. [http://www.rainforestrelief.org/reports/teak_tort.html]

ASIA, Inc.: David DeVoss Talks to Lt. Gen. Chit Swe, the Burmese Forestry Minister. BurmaNet, April 1996.

"Earth Day with Ka Hsaw Wa: Everybody Belongs." Environment News Service, April 7, 2000. [http://ens.lycos.com/ens/apr2000/2000L-04-07-01.html]

Free Trade Unions of Burma. "Forced Labour at Indo-Burma Border." BurmaNet, March 1997. [http://www.rainforestrelief.org/reports/teak_tort.html]

Gentile, Gary. "Total Denial Continues." EarthRights International, May 2000.

"Hotspots: Burma." Drillbits and Tailings 8:3 (April 11, 2003): n.p. [project_underground@moles.org]

"Human Rights Group Says Myanmar Still Using Forced Labor." Associated Press, March 8, 2001.

Images Asia. "Forced Labor and Human Rights Abuses." March 1996. [http://www.rainforestrelief.org/reports/teak_tort.html]

Karen Human Rights Group. "Interviews About Shan State." July 27, 1994. [http://www.rainforestrelief.org/reports/teak_tort.html]

Keating, Tim. "Forced-labor Logging in Burma. Draft: Second in the Rainforest Relief Reports Series of Occasional Papers; In Cooperation with the Burma UN Service Office of the National Coalition Government of the Union of Burma." June 1997. [http://www.rainforestrelief.org/reports/teak_tort.html]

"New Evidence Reveals Unocal's Complicity in Abuses in Burma." Drillbits and Tailings 5:8 (May 31, 2000): n.p. [http://groups.yahoo.com/group/graffis-l/message/11105]

Rainforest Relief. "Teak is Torture and Burma's Reign of Terror; Mon, Karen and Karenni Indigenous peoples Threatened in Burma." Teak Week of Action Press Release, New York City, March 29, 1997. [http://nativenet.uthscsa.edu/archive/nl/9703/0119.html]

Rohingya Solidarity Organization. "The Newsletter." Bangladesh, February 15, 1994.

Shan Human Rights Foundation Report. Shan Human Rights Foundation Report, April 16, 1992.

Stevens, Jane. "Teak Forests of Burma Fall Victim to Warfare." *The Oregonian (Portland, Oregon)*, March 16, 1994.

CAMBODIA

INTRODUCTION: DEFORESTATION SPURS INDIGENOUS PRESSURE

By the 1990s, deforestation caused by large-scale industrial logging had become the main environmental problem experienced by Cambodia's rural, indigenous peoples. Nearly all remaining forested lands in the country, except for a few protected areas, had been allocated as logging concessions to foreign companies, according to the *World Rainforest Movement Bulletin* ("Cambodia: Timber Concessions," 2001). The World Rainforest Movement (WRM) estimated that 90 percent of all logging activities in Cambodia were illegal during the late 1990s.

A review conducted during 1999 by the Asian Development Bank described Cambodia's forestry situation as a "total system failure" ("Cambodia: Timber Concessions," 2001). The report said that "the scenario is clear: the industry wants to cover its investment costs rapidly and continue earning as long as the resource lasts. In permitting this level of forest exploitation, Cambodia displays a classic example of unwise forest resource utilization. The country may soon turn from being a net exporter of timber to a net importer" ("Cambodia: Timber Concessions," 2001). Deforestation has become such a severe environmental threat that a logging moratorium is being suggested, with an emphasis on decision making by local indigenous peoples under the rubric of community forestry. According to the *World Rainforest Movement Bulletin*, "Facing destruction and loss of their livelihood, communities are starting to organize themselves with petitions, demonstrations and direct confrontations with loggers and the military, with sometimes surprisingly successful outcomes" ("Cambodia: Timber Concessions," 2001).

PROTECTING RESIN TREES

Cambodia's existing forest laws prohibit the cutting of trees that villagers traditionally tap to collect resin. These laws are evaded, because many logging concessionaires and their subcontractors (as well as some military units) violate them. Resin trees are cut, or villagers are coerced into selling their resin trees. Such cutting undermines the local indigenous economy, which for a long time has been based on the production of resin. Because of these conflicts, local people have become more involved in policing the resin harvest and protecting trees that supply it.

The *World Rainforest Movement Bulletin* observed that "in almost all areas of Cambodia that still have forests, people obtain their family income from collecting resin. Those forest areas are subject to highly developed systems of community-based management. Villagers own resin trees privately; when one person has tapped a tree, usually no one else collects resin from that tree. Resin can be collected from a given tree for many years, with resin trees passed on to children at the time of marriage. The forest is effectively divided into plots that are managed by individual families" ("Cambodia: Villagers," 2002).

In February 2001, according to the *World Rainforest Movement Bulletin*, "seeing that the Pheapimex Fuchan Concession Company was cutting their resin trees, 17 villagers from O Lang village went together to guard their resin trees." When the company workers arrived, the villagers showed them copies of the forest law that forbids cutting of resin trees. They stopped cutting for a time. About a month later, according to the *World Rainforest Movement Bulletin*, Pheapimex workers claimed villagers' trees by placing tags on them. Villagers then informed the commune chief, who went with 53 people (from 3 villages) to the area, and "met the workers putting up tags, and explained that the law forbids cutting resin trees. The company took down the tags, and stopped cutting trees. Since then, there has been no cutting of resin trees in the area" ("Cambodia: Villagers," 2002).

Several similar incidents have been reported across Cambodia. In the Kampong Damrei commune, for example, the Casotim Concession Company was found to be cutting trees. Around May 2001, villagers placed stickers reading "Save Resin Tapping" on their trees. They also attached the Department of Forestry legal notices regarding resin-tree protection to their trees. Those trees were not cut ("Cambodia: Villagers," 2002).

REFERENCES

"Cambodia: Timber Concessions vs. Community Forests." *World Rainforest Movement Bulletin* 53 (December 2001): n.p. [http://www.wrm.org.uy/deforestation/logging.html]

"Cambodia: Villagers Defend Their Resin Trees." *World Rainforest Movement Bulletin* 54 (January 2002): n.p. [http://www.wrm.org.uy]

Hardtke, Marcus. "Cambodia: The Forestry Sector Reform and the Myth of a Sustainable Logging Industry." *Global Witness*, Phnom Penh, 2000. [http://www.oneworld.org/globalwitness/reports/credibility/credibility.htm]

CAMEROON

PYGMIES: LOSING THEIR HOMES TO INDUSTRIAL-SCALE LOGGING

In Cameroon, timber is the second most valuable export, after oil. With the country facing high levels of international debt (to France, Germany, and Austria, among others), pressure from the World Bank and International Monetary Fund has intensified to increase logging in areas that have provided longtime habitats to several indigenous peoples, including the Pygmies. Governmental corruption also has provided an atmosphere for unsustainable levels of logging.

Three quarters of remaining forest lands in Cameroon, or about 75,000 acres, have been identified by the government as productive forest, meaning that they are ripe for assignment as concessions for commercial logging ("Cameroon: Unsustainable," 2001). New logging roads are being built in previously inaccessible forests, pushing wildlife to extinction and driving out native peoples who depend on the flora and fauna of the forests to survive ("Cameroon: Unsustainable," 2001).

According to the 'World Rainforest Movement Bulletin, Cameroon's government enacted a new forestry law in 1994, setting a basic framework for industrial-scale logging. The country has weak environmental monitoring and lax enforcement, and the lack of political will to interfere with illegal logging, allowing it to flourish. According to WRM estimates, the country's commercial-grade timber may be gone by the year 2020 ("Cameroon: Unsustainable," 2001).

European companies dominate the forest industry in Cameroon as concessionaires and also as subcontractors to concessions allocated to Cameroon nationals. Most of Cameroon's timber production goes to European processing plants and consumers: in 1998, Italy and France accounted for more than 61 percent of Cameroon's log exports, followed, in order, by China, Spain, and Portugal. Other major harvesters include companies based in Germany, the Netherlands, and the United Kingdom.

Logging companies are legally required to pay local taxes, but indigenous peoples who live in the forests only rarely see any benefits from the taxes, nor do they reap benefits from industrial-scale logging. Instead, according to the *World Rainforest Movement Bulletin*, "the intrusion of the cash economy into the forests has disrupted, for example, [the] Pygmies' traditionally close-to-nature lives. Now, they often capture bushmeat for commercial traders who follow the logging roads or find commercially exploitable trees for loggers, thereby accelerating the end of their livelihood" ("Cameroon: Unsustainable," 2001).

According to the *World Rainforest Movement Bulletin*, industrial-scale logging in Cameroon threatens the survival of many large mammals there, since excessive, unsustainable logging goes hand-in-hand with overhunting. The *World Rainforest Movement Bulletin* said that "experts foresee that some of the large mammals in Cameroon will shortly disappear if this hunting is not put to a stop" ("Cameroon: Social," 2001). Gorilla and elephant meat is sold at high prices in the best restaurants. "Theoretically," according to the *World Rainforest Movement Bulletin*, "it is forbidden to hunt these species and sell the meat or other parts of gorillas and chimpanzees, but this prohibition is widely transgressed" ("Cameroon: Social," 2001).

In Cameroon, according to the *World Rainforest Movement Bulletin*, Baka Pygmies often are employed briefly by the company as prospectors to indicate the species of trees of commercial interest. "In this way," commented the *Bulletin*, "they are unconsciously participating in the destruction of their own environment" ("Cameroon: Social," 2001). The local workers also may be shortening their own lives, because they work with little or no protective clothing (such as gloves, helmets, or masks against the dust), in areas that are treated with toxic products against parasites and fungi.

Due to their dependency on primary forests, according to the WRM, "the Pygmies are the main victims of forestry exploitation in Cameroon. According to estimates, at the end of the 1990s, approximately 3,400 Bakolas lived in the southwest [of Cameroon] and 40,000 Bakas [lived] in the Equatorial forests in the south and southeast. In the Yokadouma-Moloundou region, the Bakas are even more numerous than the Bantu. As their territorial rights are not recognized by the authorities, they cannot defend themselves against the present expansion of industrial logging in Eastern Cameroon" ("Cameroon: Social," 2001).

REFERENCES

"Cameroon: Social and Environmental Impacts of Industrial Forestry Exploitation." *World Rainforest Movement Bulletin* 53 (December 2001): n.p. [http://www.wrm.org.uy/deforestation/logging.html]

"Cameroon: Unsustainable Forestry for European Benefit." *World Rainforest Movement Bulletin* (November 2001): n.p. [http://www.wrm.org.uy]

CANADA

INTRODUCTION

Canada often prides itself on an allegedly humane civil-rights record regarding indigenous peoples (First Nations), but it has become a major source of indigenous environmental contamination and conflict. These conflicts span the country from east to west and north to south. The Innu of Labrador have been afflicted with sulfide mining, aluminum smelting, and noise pollution from squadrons of military aircraft. Along the Arctic Circle, the native Inuit suffer major problems from contamination by persistent organic pollutants such as dioxins, PCBs, and others.

Some of the most intense resource exploitation in Canada takes place in remote locations, such as among the Lubicon Cree of northern Alberta, whose lands were so inaccessible in 1900 that treaty makers completely missed them. Today, roads have opened their lands to massive oil drilling and logging. The lands of the Cree in Quebec have been scarred by widespread dam building near James Bay that has contaminated large areas with toxic methyl mercury, as well as other pollutants. Uranium mining has decimated the Dene in Canada's Northwest Territories, much as it has ravaged the Navajo in the Southwestern United States. The catalog of indigenous environmental issues in Canada spans a wide range of resources—hydropower, diamonds, uranium, gold, silver, sulfide, aluminum, oil, and natural gas.

THE CREES: HYDRO-QUEBEC'S ELECTRIC DREAMS

During the 1970s, planners and engineers at Hydro-Quebec indulged themselves in dreams of an electric empire that would make their utility the biggest supplier of electricity in the world. The utility planned to harness dozens of rivers flowing into James Bay, reshaping the ecology of an area the size of Iowa in the service of electrical generation. At first, the planners at Hydro-Quebec paid very little attention to what might become of the flora and fauna of the James Bay region during what they envisaged as the largest earthmoving project in the history of humankind. Although it is a public entity (a Crown corporation in Canada), Hydro-Quebec found itself unrestrained even by a public board of directors. Hydro-Quebec's electric dreams met problems other than the Crees' resistance. By 1990, when the first phase had been completed, the utility was $26 billion in debt, mainly due to expenses associated with planning and construction of its James Bay projects.

An Unwelcome Guest Comes to Stay

The James Bay area is home to 10,000 Cree and 6,000 Inuit people who were not consulted before construction began on James Bay I. James Bay hydroelectric development was first interjected into the Crees' lives on April 30, 1971, when Quebec Premier Robert Bourassa (after whom part of the complex later was named) announced its inception to the people of the province.

The Cree occupy parts of northern Quebec as far north as Whapmagoostui. Inuit communities dot the eastern shores of Hudson Bay in Quebec, north from Kuujjuarapik to Ivujivik and Salluit. In the Northwest Territories, Inuit communities extend from Arviat on the western shore of Hudson Bay to Coral Harbour on Southampton Island. The Inuit community of Sanikiluaq is located on the Belcher Islands in southeastern Hudson Bay, about 100 kilometers from the mouth of the Great Whale River.

Hunting and fishing comprise the heart of the Cree and Inuit economies. The Cree have traditionally hunted migratory birds, particularly during the spring, as well as terrestrial mammals such as moose. The Cree fish the rivers in the region and trap fur-bearing mammals such as muskrats and beavers. Traditionally, Inuit harvested fish and marine mammals such as seals, walruses, and whales. Some communities also depend heavily on caribou.

The Cree and Inuit, who at first were scarcely even recognized as longtime occupants of the land by Hydro-Quebec, soon found themselves challenging the construction project in court. It was a matter of life and death not only for traditional cultures, but also for the peoples themselves. The construction ruined many traditional hunting areas, and replaced them with poisoned earth laced with life-threatening mercury compounds.

The Cree and Inuit of northern Quebec signed a comprehensive claims agreement (formally named the James Bay and Northern Quebec Agreement) with the Canadian federal and Quebec provincial governments in 1975. The Cree entered agreements (in essence, treaties) with the governments of Quebec and Canada, but they found the terms of these treaties being violated by the ecological consequences of hydroelectric construction.

The 1975 agreement provided the Cree with $22 million (Canadian) and the right to continue to hunt, fish, and trap in their usual territories. Not even Hydro-Quebec anticipated that its plans would make a significant portion of the territory unsustainable for any of those activities. Hydro-Quebec's lawyers were more interested in other stipulations of what came to be called a treaty between the Cree and Quebec: the clauses that gave their clients the right to undertake construction of the James Bay project's first phase. In 1979, the first of four projected power stations at La Grande (in Hydro-Quebec shorthand, LG-1) came on line. With its advent, roughly 4,200 square miles of Cree hunting and trapping grounds were flooded. Three hundred black bears drowned as the reservoirs filled (Bigert, 1995). Ignoring natural seasonal cycles, Hydro-Quebec's engineers filled their reservoirs just after the bears had put themselves to sleep for the winter, drowning them. The fact that the black bear is the most sacred animal in Cree mythology did not bother them.

Transformation of the Crees' Homeland

During the James Bay project's first phase, large areas of the Crees' homelands were transformed. Rivers that once spawned large numbers of fish were reduced to trickles or stopped entirely by the creation of reservoirs behind energy-generating dams. Forests were clear-cut and burned, adding greenhouse gases to the atmosphere. More than 10,000 caribou drowned after Hydro-Quebec's rearrangement of the landscape spilled deep water across their migration routes. Hydro-Quebec called the deaths an "act of God" (Bigert, 1995). This incident was yet another reminder that while hydroelectricity is often touted as clean energy that does not emit pollution or directly increase the atmosphere's overload of greenhouse gases, it is not environmentally benign.

Hydro-Quebec's James Bay II (the Great Whale Project) proposed to dam eight major rivers in northern Quebec at a cost of more than $170 billion to provide electricity to urban Canada in the St. Lawrence valley, and also to the northeastern United States. Little is said these days of the original James Bay project's third proposed stage, the most ambitious electric dream of all. As originally envisaged by the engineers of Hydro-Quebec, the project's second phase (which was successfully impeded by the Cree) was to be followed by an even more ambitious third phase, a $100 billion proposal to build a 100-mile dike across the mouth of James Bay, "separating it from Hudson Bay so that the now fresh water from James Bay can be pumped (possibly using nuclear-powered pumps) to the Great Lakes and thence to the Midwestern and Southwestern United States" (Native Forest Network, 1994). Given free range, the empire builders of Hydro-Quebec were planning to create a utility that would become a power merchant not only to Quebec, Ontario, and New England, but also to the eastern half of the United States, with the raw material, electricity, provided by clean hydroelectric energy.

Formally called the Grand Canal, or Great Recycling and Northern Development proposal, the third phase of the James Bay Project would have transformed James Bay into a gigantic freshwater lake. According to Hydro-Quebec's plans, "Fresh water from this reservoir would be diverted south through existing watercourses and new canals to the Great Lakes and thence to the American Midwest and the thirsty, booming Southwest, as well as to the Canadian prairies" (Sierra Club, n.d.).

According to now-abandoned plans, once phase two was completed, Hydro-Quebec estimated that its generating capacity would rise to 27,000 megawatts, as much electricity as would be made available by fully harnessing the power of Niagara Falls 13 times (on both sides of the U.S.-Canadian border) ("Sustainable Development," n.d.). Following construction of the James Bay project's first phase, Hydro-Quebec was generating one-quarter of the electricity consumed in the United States and Canada. Hydro-Quebec engineers estimated that, by the time James Bay itself was dammed at the end of phase three, their utility would be able to meet the demands of two-thirds of the same market.

The Toll of Methyl Mercury

Just as Hydro-Quebec's electric dreams reached a point of ecocidal fantasy, the Cree and Inuit peoples got in the way. As a matter of life and death, the Crees were called to protest the James Bay project during the 1980s, as they faced methyl mercury contamination caused by the earthmoving of the James Bay project's first phase.

Mercury contamination occurs when vast areas of land are disrupted. Plant decay associated with the James Bay project's earthmoving caused large amounts of ordinary mercury to become methyl mercury, a bioaccumulative poison. Rotting vegetation accelerates microbial activity that converts elemental mercury in previously submerged glacial rocks to toxic methyl mercury. By 1990, many Crees were carrying in their bodies 20 times the level of methyl mercury considered safe by the World Health Organization. Hydro-Quebec did not anticipate the accelerated release of methyl mercury into the waters of the region, contaminating the entire food chain for the Cree and Inuit peoples and the birds, fish, and animals. This type of mercury poisoning can cause loss of vision, numbness of limbs, uncontrollable shaking, and chronic brain damage. By the late 1980s, the Quebec government's health ministry was telling the Cree not to eat fish from their homeland.

Fishing represents more than sustenance for the Cree. Fishing activities are important in knitting family and community. Mercury contamination thus disrupted an entire way of life. Before 1978, the concentration of mercury in 700-millimeter pike was approximately 0.6 microgram per kilogram. After the completion of phase one, the concentrations increased gradually. In 1988 concentrations were 3 milligrams per kilogram, five times the original concentration and six times the maximum permissible concentration in Canada for commercial fish (e.g., for human consumption) (Dumont, 1995).

By 1984, the concentration of mercury in the hair of Crees at all ages was much higher than in surveys conducted during the 1970s. During 1993 and 1994, the Cree Board of Health completed an assay of mercury levels in the hair of the Cree. This survey revealed a wide variation in exposure levels between different communities. "If the 6 mg/kg maximum hair concentration recommended by the World Health Organization is used, at least half of the population of several communities is over that limit. In 1984, when Whapma-goostui (Great Whale) was surveyed, 98 per cent of the population surveyed had mercury concentrations above 6 mg/kg" (Dumont, 1995). Mercury levels increased generally with a person's age. Persons recognized as trappers in their communities consistently tested for higher mercury concentrations. "During the past ten years, the number of individuals with high mercury concentrations has decreased considerably," due in large part to programs persuading the Cree to avoid eating tainted fish (Dumont, 1995).

The Crees, wrote Charles Dumont, have expressed great concern for the effects of mercury on the coming generations: "Mercury is presumably toxic at lower doses on the fetus and can cause delays in walking and speaking, and abnormal reflexes. Boys are probably more sensitive than girls. In higher doses it can cause cerebral palsy" (Dumont, 1995). Dumont and colleagues examined changes in mercury levels within the Cree population between 1988 and 1993–1994. Hair-sample assays for mercury content were taken by the Cree Board of Health and Social Services during 1988 and again during 1993 and 1994 in all nine Cree communities of northern Quebec. The studies found that

> the proportion of the Cree population with mercury levels in excess of 15.0 mg/kg declined from 14.2 per cent in 1988 to 2.7 per cent in 1993/94. Wide variations in mercury levels were observed between communities: 0.6 per cent and 8.3 per cent of the Eastmain and Whapmagoostui communities respectively had mercury levels of 15.0 mg/kg or greater in 1993/94. Logistic regression analyses showed that significantly higher levels of mercury were independently associated with male sex, increasing age and trapper status. There was a correlation between the mercury level of the head of the household and that of the spouse. . . . Mercury levels in the Cree of James Bay have decreased in the recent past. Nevertheless, this decrease in mercury levels may not be permanent and does not necessarily imply that the issue is definitively resolved. (Dumont et al., 1998)

In 1984, two of every three people in Chisasibi, a community of 2,500 at the mouth of the La Grande river, had unhealthy levels of mercury in their bodies; some elders who exhibited symptoms of mercury poisoning had twenty times the usual levels ("Sustainable Development," n.d.).

The Crees Organize against James Bay II

When James Bay II was proposed, the Cree and other Native American peoples living in the area not only took the case to court, but also sought to organize the customers of the electric utilities that would receive the power, mostly in Eastern Canada and New England. This activism ultimately convinced an activist nucleus of customers to pressure their utilities to refuse Hydro-Quebec's power on moral grounds.

The Crees became very effective at addressing public forums in New York and Vermont, where a large part of the electricity would be sold, telling people that they shared complicity with Hydro-Quebec in the devastation of northern Quebec. The Crees also urged electricity consumers and utilities to conserve, and to consider other sources of supply. Several non-Indian environmental groups joined with the Grand Council of the Cree against the James Bay projects. In New York and New England, electricity users were urged to conserve energy to eliminate a need for more generating capacity.

The James Bay projects did not envisage the construction of traditional dams in fast-flowing waterways bordered by mountains, the usual situation in the North American West. The plan was to rework the dozens of rivers that traversed the rolling countryside on their way into James Bay into a network of short dams and shallow reservoirs. In 1971, at its inception, Prime Minister Bourassa called Hydro-Quebec's designs for the James Bay region "the project of the century," as he bemoaned the fact that "every day, millions of potential kilowatt hours flow downhill and out to sea. What a waste!" (Bigert, 1995). Bourassa made his support of Hydro-Quebec a matter of provincial pride, and sometimes questioned the patriotism of the project's opponents. He dreamt out loud that northern Quebec would become a perfect location for magnesium and aluminum smelters that would consume the newly generated power.

In replies to Bourassa and Hydro-Quebec, the Crees (often through their spokesman, Cree Grand Chief Matthew Coon-Come) asserted that "beavers are the only ones who should be allowed to build dams in our territory" (Bigert, 1995). The Crees came to this conclusion as scientists from Quebec's health ministry warned everyone (especially women of childbearing age) not to eat local fish, which were becoming contaminated with methyl mercury. Predatory fish, such as the pike, had been accumulating a great deal of methyl mercury in their bodies since Hydro-Quebec began its ambitious plans to reduce the natural landscape of northern Quebec into a shape and a function suitable for generating electricity. Until then, fish had been a major constituent of the Cree diet for thousands of years.

Said Coon-Come: "Nobody asked us or told us, but three or four times from 1670 on, our people and our lands were handed between kings and companies and countries. We found out about all of this when they came to build the dams and when the courts told us that we did not have any rights, that we were squatters on our own land" (Coon-Come, 2001).

On land, the depredations of the earthmovers were matched by another incident in the south: hunters in orange suits from Canada's urban areas, arriving in late-model sports utility vehicles along freshly paved roads, left behind the lifeless, rotting bodies of moose and caribou, their trophy heads removed. Such behavior profoundly offended the Cree.

Coon-Come recalled:

In the early 1970s, I was a young student in Montreal when I read in the newspaper that Hydro-Quebec . . . was going to build a hydroelectric mega-project of the century in our territory, by diverting and damming the rivers and flooding our traditional lands. I looked at a map and saw that my family's trap-line and our community was going to be underwater! I immediately returned home, and as a result of a speech I made from the back of a community hall I was launched on my political career. (Coon-Come, 2001)

"Twenty-five years later," Coon-Come recalled in the year 2000, "our people know that the treaty we signed was more broken than honored. Many of the most important promises, such as for economic involvement and development, for protection of the environment and our traditional economy and way of life, and for housing, community development, and infrastructure have been twisted, ignored, or broken" (Coon-Come, 2001).

James Bay I involved 9 dams, 206 dikes, 5 major reservoirs, and the diversion of 5 major rivers over an area roughly the size of Connecticut. Rotting vegetation in the area had released about 184 million tons of carbon dioxide and methane into the atmosphere by 1990, adding to global warming. In the meantime, James Bay I had saddled each of Hydro-Quebec's ratepayers with an average of $3,500 in debt (LaDuke, 1993, A-3). James Bay I uprooted 2,000 Cree from Chisasibi Island (also known as Fort George Island). "It's painful to look at young people caught between two worlds. I've seen their faces, knowing that their heritage is under water. It was a whole disintegration of the spirit—all that was drowned," said Coon-Come ("Canadian Indians," 1990, A-2).

In addition to its toll on the Cree, hydroelectric development ruined wetlands and coastal marshes that were important staging grounds for migratory waterfowl, including several species of duck, teal, and goose. The area also was home to rare and endangered species of freshwater seals, beluga whales, polar bears, and walruses. Anadromous fish such as the brook trout and lake whitefish entered the waters of James Bay to spawn from rivers that would have been disrupted or destroyed by additional extensive hydroelectric development under James Bay II. According to the Sierra Club, "Estuaries, heath-covered islands, salt marshes, freshwater fens, sub-tidal eelgrass beds, and ribbon bogs in the James Bay region provide nourishment for huge flocks of geese, ducks, and loons. Tundra on either side of the bays provides habitat for caribou, moose, otter, muskrat, beaver, lynx, and polar bear. This wildlife, in turn, supports the traditional Cree trappers and fishers of the James Bay region and, along Hudson Bay, the Inuit and Naskapi. The Cree regard their part of the eco-region as a 'garden' providing for all their needs" (Sierra Club, n.d.).

The Cree and Inuit have adapted to this world of forest, flowing water, and marsh since the end of the last ice age about 10,000 years ago. At the end of the twentieth century, Hydro-Quebec proposed to alter the entire ecosystem, destroying large parts of it to provide electricity to urban areas in more temperate reaches. Commented one observer:

Hydro projects could have far-reaching and negative impacts if, for example, breeding waterfowl habitat in Quebec river estuaries was damaged, thereby affecting migratory bird populations wintering in the eastern and southern United States. . . . Approximately 75 per cent of the global population of

Atlantic brant geese are concentrated on the eel grass beds of the Quebec coast and parts of the Ontario coast of James Bay, and almost the entire North American population (up to 320,000) of black scoters use southern James Bay as a staging area. Other waterfowl species that utilize inshore, inter-tidal and brackish coastal habitats in the Hudson Bay/James Bay bio-region include black duck, pintail, mallard, wigeon, green-winged teal, and scaup. Mergansers and loons make extensive use of offshore water for feeding, and a significant number of common eider pass the winter in James Bay and the Belcher Islands." ("Sustainable Development," n.d.)

For several years during the early and middle 1990s, Cree leaders and elders utilized the lecture circuit with the story of how Hydro-Quebec's plans would damage their world. During July 1993, Bill Namagoose, a James Bay Cree, was quoted as saying:

> We're up against the perception that Hydro-Quebec is the engine [of progress] and they've used it to whip up the nationalism of Quebecers against the Cree and the Inuit. [However], if you cut off the market there's no point in doing the financing. If there is no financing there is no construction. And so far we have managed to cancel two American contracts. There is no economic justification for the project, no environmental, no energy reason why these projects should be built.
>
> We've been living up there in harmony with the environment. But when the invasion began in 1970, we were totally overwhelmed. . . . We've survived as a nation for many thousands of years and we [would] like to arrive in the future on our own terms and traditions . . . living on an environmentally sound land, with all of our rivers intact for thousands and thousands of years. Two-fifths of our land has already been impacted by the first phase. We know what it's like to live in the middle of a mega-project. That's why we're trying so hard to protect the parts of our lands we still have left. (Vance and Shafer, 1995)

Albert Meascum, a Cree trapper living in the village of Ouje-Bougoumou, said that he had asked the company not to cut certain parts of his trapping territory, but they paid no attention. "Before the forest companies came," he said, "there was lots of game. By cutting down the trees they have chased away the animals." The companies also pollute, he said, as they spill oil and gas on the land and poison the young willows that feed the moose. "When the snow melts," he said, "the gas and oil go into the lakes and pollute the water and the fish" (Native Forest Network, 1994).

The future of the James Bay project was litigated for several years. During September 1991, a Canadian federal judge in Ottawa ordered a new assessment of the James Bay II project under a process that could give federal authorities the right to stop it. "I conclude that the Crees' right to an independent parallel federal review must be honored," wrote Justice Paul Rouleau ("Cree Indians Win Battle," 1991, 4). At the time, the total cost of James Bay I and II was estimated to be $62 billion.

During February 1994, the Supreme Court of Canada ruled that the National Energy Board had the right to examine the environmental effects of the Great Whale Project, essentially the same issue that had been decided by Judge Rouleau two-and-a-half years earlier. Armand Couture, president of Hydro-Quebec, denied at the time that the Court's order would have any impact on plans to throw the switch on the second phase of the Great Whale Project by the year 2003.

In the meantime, the Crees continued to stoke popular pressure against Hydro-Quebec's efforts to market power from the James Bay projects. Ten Cree and Inuit activists paddled a 25-foot combination kayak-canoe from Ottawa to New York City during the spring of 1990. They crossed the Quebec-Vermont border at the northern end of Lake Champlain and timed their arrival in Central Park to coincide with Earth Day. During the first week of October 1991, Cornell University's American Indian Program organized a forum on James Bay development, with Matthew Coon-Come, grand chief of the Crees, as keynote speaker. In mid-April 1993, native leaders and their supporters bought a full-page advertisement in the *New York Times* opposing James Bay II. Robert Kennedy Jr. and Coon-Come held a joint press conference, also in New York City, yet another high-profile action that turned more New Yorkers against the project. New York State then yielded to growing consumer protests and canceled a $17 billion export contract with Hydro-Quebec.

On November 18, 1994 (not too many months after Hydro-Quebec's president Armand Couture had asserted that electricity would begin to flow from James Bay II in 2003), Quebec Premier Jacques Parizeau said that the project was shelved indefinitely. He told the press: "We're not saying never, but that project is on ice for quite a while" ("Quebec Will Shelve," 1994, 7). The decision was made after two governmental review committees said that Hydro-Quebec "would have to go back to the drawing board to correct 'major inadequacies'" in its environmental assessment of the project." Hydro-Quebec had been stopped by a blizzard of paper; company officials complained that its environmental-impact study, which ran to 5,000 pages and took eleven years to compile, had cost $190 million ("Quebec Will Shelve," 1994, 7).

Great Whale Redux

The Great Whale (James Bay II) hydroelectric project was revived on a reduced scale during 2001 after having been shelved seven years earlier. The project was revived with a new twist: it now contained a proposal that the Crees, who had played a large role in opposing the project, would eventually become its sole owners. The idea was proposed to the Crees by Toronto-based Amec, a large engineering and construction firm that had worked on Hydro-Quebec's original James Bay hydroelectric project, as well as China's Three Gorges dam and Colombia's Urra dam. The company's name was Agra until it was bought by Amec P.L.C., a British-based conglomerate with 50,000 employees in 40 countries (Roslin, 2001).

The band councils of Whapmagoostui (the Cree name for the community of Great Whale) and neighboring Chisasibi held quiet discussions addressing the idea of reviving Great Whale in 2001. "It's just an idea, it's not official," said David Masty, a chief at Whapmagoostui. "It's at the discussion stage. We haven't taken a position on it as a community." He continued: "Some Cree officials are furious that the talks have gone on with virtually no community involvement. They say the discussions are a slap in the face to those who fought to kill the earlier variants of the project. They also point out that Crees have repeatedly voted—in regional assemblies, band-council votes and a community referendum—against new hydro projects" (Roslin, 2001).

The new proposal involved diversion of the Great Whale River at its head-water, Lac Bienville. At an estimated cost of $350 million, the river's water would be redirected through 10 kilometers of canals into Hydro-Quebec's La Grande hydroelectric complex, increasing the water flow through the turbines. In 1997, Hydro-Quebec had proposed to divert the Great Whale and, for the first time, sought the Crees' consent and offered a minority partnership in the project. This proposal was shelved after 92 percent of Whapmagoostui Crees voted against any development projects on the river (Roslin, 2001). The new proposal retained the diversion idea, but proposed that the Crees would own the facility once its debt was paid off. Amec would design the facility, arrange financing, and retain an ownership share while the debt was paid off.

Hydro-Quebec, which earned a net profit of more than $1 billion in 2000, wanted to construct the dams to increase its profits by selling surplus power outside Quebec, which already produced enough power to meet its own needs. Hydro-Quebec planned to allow small private energy producers to build and operate the dams. In turn, the owners of the smaller dams would sell their electricity to Hydro-Quebec. According to an account in the *Cleveland Plain Dealer*, "Quebec has turned its back on the environment, suggesting that even the Rouge River in the beautiful Laurentians could be a site for a dam. Thousands visit the area every year to paddle the white water rapids and pristine pools" (Egan, 2001, D-15).

On October 23, 2001, the Quebec government and natives in the James Bay region signed an agreement in principle to end a bitter legal dispute over hydroelectric development in the area and open the door to further resource development in northern Quebec. Under the agreement, the Cree would receive Canadian $3.5 billion (roughly U.S. $2.2 billion) in natural-resources royalties from the province over 50 years in return for giving up all legal action against a major power project in the region. "This agreement constitutes, I am sure, the basis of a great peace between Quebec and the Crees," Quebec Premier Bernard Landry said before a signing ceremony with Cree grand chief Ted Moses. "This is an historic turning point and a truly profound revolutionary vision for the Cree and aboriginal peoples generally," Moses said (White, 2001). The project, which had been on hold for several years because of opposition from the Cree, planned to produce 1,300 megawatts, or about 15 percent of the power from entire James Bay area, and create about 8,000 jobs during 6 years of construction.

Canadian Cree leader Dr. Ted Moses signs the Quebec Electrical Industry Association honorary book after speaking about a new deal with the Quebec government that was signed on October 23, 2000. (NewsCom.com/Zuma Press)

The agreement, which later was ratified by the Cree communities of northern Quebec, allowed the 15,000 natives an annual revenue flow and direct participation in any economic development on native land in northern Quebec. The agreement pledged to pay the Cree $70 million a year for 50 years. It also included Hydro-Quebec jobs for Crees, an important issue in communities where more than 80 percent of the young people under 25 years of age were unemployed. The agreement also promised remediation of mercury contamination, funding for start-up business programs, job training, health and social services, electricity, sanitation, and fire services for Cree communities (Matteo, 2002).

In return, the Crees promised access to resources (including diversion of the Rupert River for hydroelectric development) and cession of $3.6 billion in environmental lawsuits. The proposed diversions of the Rupert River were subject to environmental-impact reviews, but the government of Quebec was allowed to make the final decisions. "In effect," according to one analysis, "the Cree will not be able to protest, stop, inhibit, or litigate" environmental outcomes, including the anticipated drying up of parts of the Eastman River and the flooding of trap lines (Matteo, 2002).

The Cree agreed to drop their environmental lawsuits in part because legal fees were costing them $9 million a year, but also because many of the cases were faring badly in the Canadian court system. Courts in Quebec had ruled against the Cree on forestry issues, as well as on their opposition to hydroelectric development on the Eastman River (Matteo, 2002). The agreement also was important because it saved roughly 8,000 square kilometers of land from being flooded, according to Coon-Come. "We want jobs," he said after the new agreement was negotiated. "We want a say in where development takes place [and] what happens in our own backyard" (Matteo, 2002). "It isn't a blank check," said chief Cree negotiator Abel Bosum. "Quebec is taking a big risk here. We consent to Eastmain-Rupert but there is no guarantee that it will be approved." Bosum explained that the project had to pass impact studies by Quebec, Ottawa, and the Crees. "It's not a done deal," he said (Dougherty, 2001).

Cree spokesman Romeo Saganash recalled that in 1975 the original James Bay agreement also was greeted with hope. Instead, the Crees faced 26 years of frustration, fighting Quebec and Ottawa in the courts. "Today my only hope is that my children will not be in front of your children in 25 years' time saying: 'Here's another agreement. After 25 years of frustration, we couldn't implement the deal we signed in 2001'" (Dougherty, 2001).

Once the Crees' tentative agreement with Hydro-Quebec was negotiated, an intense debate began among the Crees as to whether it should be accepted, with two Cree leaders taking opposite sides. On December 18, 2001, Grand Chief Ted Moses denied that the tentative deal was unraveling. "Sure, there is criticism and suspicion," Moses told the Quebec Electricity Association. "Why should this be surprising? And why in any society should this democratic debate be interpreted as a sign of failure? We have been in a long and difficult struggle to achieve recognition and respect for the Cree people." Moses said that opponents of the agreement were "a small group of people trying to make a lot of noise" (Macafee, 2001).

"I stand firm in my opposition to the deal," Matthew Mukash, deputy grand chief, said shortly after arriving in Montreal for a Cree Grand Council meeting. "The Cree Nation is clearly divided on this. It's difficult at this time to know which way they're going to go" (Macafee, 2001). Mukash said he was concerned that aboriginal rights could eventually be extinguished under the deal. There was also an implication the agreement would reaffirm Quebec's position that it owned the resources on the land, not the Cree. "I think it's very dangerous when you look at it over the long term," said Mukash (Macafee, 2001).

During late January 2002, Quebec Crees voted in favor of the $3.4 billion deal with the provincial government. Of the 6,500 eligible voters, 69.35 percent, or 3,106, voted in favor of the deal, with 30.65 percent, or 1,373, against. The decision, which Cree Grand Chief Ted Moses called historic, meant that the Cree Nation was ready to complete a process it started in October when it signed an agreement in principle with Quebec Premier Bernard Landry to allow hydro installations along the Eastmain and Rupert rivers subject to environmental approval (Authier, 2002). One Cree leader, George Wapachee, said the Cree who opposed hydroelectric development had sat on the fence too long while the world passed them by. In his community, the population had jumped to 500 by 2001 and would hit 1,000 in 20 years. The land could only support so much traditional trapping and logging activity, he said. "There's a lot of young people who have nothing to do," said Wapachee. "We have to look forward, move ahead. That's the way life is" (Authier, 2002).

As Cree Grand Chief Ted Moses and Quebec Premier Bernard Landry signed the hydroelectric power deal in the Cree village of Waskaganish on February 7, 2002, an elderly Cree chief, Henry Diamond, broke into the assembly hall, shouting his objections. He was tackled by seven police officers and hauled away, bleeding from the head, to be charged with disturbing the peace and resisting arrest. Another dissident, Chisasibi Band Councilor Larry House, tried to voice his objections at a press conference following the signing. He was arrested for shoving an officer and impeding pedestrian traffic, for which he was fined $25 (Taylor, 2002). Opposition to the agreement was centered in the Chisasibi community, which sits at the foot of a reservoir that has been plagued with technical problems. The reservoir would rise at least six feet once the Rupert River is diverted, causing floods.

"Today," said Cree deputy chief Matthew Mukash, "Native communities across Canada are calling us sellouts." Mukash believes that the deal is a major mistake that will eventually lead to "the gradual and progressive takeover of the Cree by Canada and Quebec" (Matteo, 2002). He pointed out that while 72 percent of the Cree who voted supported the agreement, only 53 percent of those eligible to vote did so. Many who did not vote no did so from fear, Mukash believes (Matteo, 2002).

Nevertheless, by 2002 Hydro-Quebec was preparing to pursue its electric dreams once again, although on a much-reduced scale than a decade earlier, when a gaggle of corporate engineers nearly forgot that nature and the Crees existed.

THE PIMICIKAMAK CREE OF MANITOBA: IMPERILED BY HYDROPOWER

The Pimicikamak Cree assert that, for 25 years, the Manitoba Hydro Power Company has been destroying their traditional lands and society with a series of dams and canals cut through wilderness to divert the flow of rivers upon which their traditional economy depends. The Crees signed a treaty in 1977 that said they would receive full compensation for damage to their homeland. They assert, however, that this treaty has not been honored by Manitoba Hydro. Now, the Pimicikamak Crees, who have been experiencing 70 to 80 percent unemployment, have lost their traditional economic base and their cultural and spiritual foundations, leading to an epidemic of suicides (Peterson, 2000).

Dams built during the 1970s have reengineered Manitoba's two largest rivers and largest lake to serve the energy needs of Manitobans and American export customers. For 25 years, unnaturally fluctuating waters in Pimicikamak Cree territory have eroded shorelines. According to advocates for the Cree, "Burial grounds have been exposed, islands eroded right off the map, [the] ecological balance [has been] disrupted, and large areas of Cree fishing and trapping grounds have become despoiled and inaccessible" ("Hydroelectric Production," 2001). Cross Lake, immediately downstream of the Jenpeg dam, has lost as much as 300 square kilometers, about half of its surface area, because the water has been held in Lake Winnipeg to be released during times of peak electrical demand in winter. Although Manitoba Hydro constructed a rock weir in 1991 (14 years after committing to build it), "navigation remains treacherous and receding waters still expose muddy flats at certain times" ("Hydroelectric Production," 2001).

The debate over James Bay II continued in tandem with an environmental analysis of seven similar dams that were built during the early 1970s in the Nelson and Churchill river systems of Manitoba. These rivers drain into the west coast of Hudson's Bay, of which James Bay is a part. By the early 1990s, studies showed that one in six people in the area was suffering from mercury contamination (LaDuke, 1993, A-3). A large percentage of game had disappeared from the affected area, making subsistence hunting very difficult. Several hundred native people were displaced into housing projects as their homes were flooded.

The environmental-impact assessment for the Churchill-Nelson project was an after-the-fact matter; it was initiated after Manitoba Hydro had fixed the configuration, operating regime, and timing of construction for the diversion of the Churchill River into the Nelson River. The unforeseen environmental impacts of the Manitoba projects included:

> Severe shoreline erosion, caused by impoundment of Southern Indian Lake as part of the diversion, led to increased turbidity. This resulted in the collapse of the commercial whitefish industry. Whitefish populations in Cross Lake, another large lake in the Churchill-Nelson river basin, fell 65 per cent. In addition, walleye and northern pike in all flooded lakes along the diversion route accumulated mercury levels that exceeded Canadian limits for the protection of human health. Some commercial fisheries were permanently closed, and local residents were encouraged to avoid consumption. The collapse of the commercial fishery placed a severe strain on the social fabric of northern aboriginal communities, forcing them to move and/or to rely increasingly on compensation payments for income. ("Sustainable Development," n.d.)

Methyl mercury contamination was discovered in northern Manitoba during the early 1970s by scientists from the Freshwater Institute in Winnipeg. Similar contamination had not been anticipated by Hydro-Quebec as it began construction of James Bay I at the end of the decade, despite the experiences of a similar utility, doing similar construction work, geographically adjacent. Apparently, no one was paying attention.

The Pimicikamik Cree held a "Hands Across the Border" conference in 2001 to strengthen its campaign to build a coalition of native nations, farmers, consumers, and environmentalists in the United States and Canada to force the cancellation of utility contracts with Manitoba Hydro requiring a new megadam project. Five existing dams in Manitoba have submerged 3.3 million acres of land, including large tracts of boreal forest and extensive animal habitat. The methyl mercury created by construction of these dams has contaminated fish, a staple of the Cree diet, in the Nelson River.

The Pimicikamak Crees of Manitoba adopted a strategy similar to that of the Quebec Crees, who forced Hydro-Quebec to shelve its James Bay II project during the middle 1990s. They traveled to the United States to convince consumers of the utilities who might purchase the electricity generated by the dams to advocate cancellation of their contracts on moral grounds. As part of this effort, the Pimicikamak Cree lobbied Minnesota public utilities to weigh damage to the native peoples' ecosystems when purchasing power from Manitoba.

Late in the year 2000, the Minnesota Public Utilities Commission agreed to include ecological damage in its decision making, a decision that Andy Orkin, one of the Crees' attorneys, called "a big win for the environment" ("Cree Claim," 2000). The decision required the commission to assess how energy projects would affect the environment and the people who live near them. For years, electricity generated in Manitoba had been sold in Minnesota, but the state's public utilities commission never had investigated the human impact of power generation. This decision was intended to create a methodology that would compare coal, nuclear, hydroelectric, wind, and other generation methods with each other in terms of socioeconomic and environmental impacts.

Orkin said the commission's decision meant that the impact of Manitoba Hydro's hydroelectricity projects on native peoples would come under closer scrutiny. The renewal of Manitoba Hydro's contract with Minnesota utilities was upheld, however. "What the Pimicikamak Cree Nation was much more concerned about was protecting the Lake Winnipeg, Churchill, [and] Nelson Rivers' environment, particularly in and around [the] Pimicikamak Cree Nation's traditional territory, against the impacts of further large-scale hydroelectric development by Manitoba Hydro," said Orkin ("Cree Claim," 2000).

Pimicikamak Cree Nation Chief John Miswagon traveled to Minneapolis to attend an annual shareholders' meeting of Xcel Energy, which purchases approximately 40 percent of Manitoba Hydro's power output, supplying power to 12 states south of the border. Miswagon addressed 1,800 shareholders and utility managers as they voted on a resolution demanding that the company withdraw from Manitoba Hydro and increase its power supply from renewable resources that do not negatively impact human rights and the environment. The motion was organized by the As You Sow Foundation, a U.S. organization that advocates corporate responsibility. The resolution was rejected, but 8.8 percent of shareholders voted in favor of it. A 3 percent vote was required to place the resolution on Xcel's agenda for the following year ("Cross Lake," 2001). To promote its cause, the Crees bought a full-page advertisement in the *New York Times*, purchased for about $100,000, with funding from an anonymous donor.

Michael Passoff, a corporate responsibility expert with As You Sow, said the vote indicated a high level of support for change among Xcel shareholders.

"The vote . . . sends a clear message to management that this issue is likely to grow unless the company addresses shareholder concerns about the destructive environmental and human rights impact of increasing energy imports to the U.S. from Manitoba Hydro's mega-projects" ("Cross Lake," 2001). Miswagon said the vote was clearly a success for the first nation. "This is a major victory . . . in our struggle for the environment and human rights," Miswagon said. "We are blown away by this level of compassion and support from American shareholders" ("Cross Lake," 2001).

Manitoba Hydro maintains that it had spent $400 million (Canadian) over 20 years to compensate northern communities, but there appeared to be little to show for it. The Crees and other native peoples asserted that poverty, unemployment, and suicide among the Pimicikamak Cree may be attributed to hydroelectric development ("Cree Claim," 2000). George M. Ross, a Cree elder, said, "The south is benefiting at the expense of our misery" ("Hydroelectric Production," 2001).

THE INUIT: DIOXIN AND OTHER PERSISTENT ORGANIC POLLUTANTS

To environmental toxicologists, the Arctic by the 1990s was becoming known as the final destination for a number of manufactured poisons, including, most notably, dioxins and polychlorinated biphenyls (PCBs), which accumulate in the body fat of large aquatic and land mammals (including human beings), sometimes reaching levels that imperil their survival. Thus the Arctic, which seems so clean, has become one of the most contaminated places on Earth—a place where mothers think twice before breast-feeding their babies, and where a traditional diet of country food has become dangerous to the Inuits' health.

Most the chemicals that now afflict the Inuit are synthetic compounds of chlorine; some of them are incredibly toxic. For example, one-millionth of a gram of dioxin will kill a guinea pig (Cadbury, 1997, 184). To a tourist with no interest in environmental toxicology, the Inuits' Arctic homeland may seem as pristine as ever during its long, snow-swept winters. Many Inuit still guide dogsleds onto the pack ice surrounding their Arctic-island homelands to hunt polar bears and seals. Such a scene may seem pristine, until one realizes that the polar bears' and seals' body fats are laced with dioxin and PCBs.

"As we put our babies to our breasts we are feeding them a noxious, toxic cocktail," said Sheila Watt-Cloutier, a grandmother who also is president of the Inuit Circumpolar Conference (ICC). "When women have to think twice about breast-feeding their babies, surely that must be a wake-up call to the world" (Johansen, 2000, 27).

Watt-Cloutier was raised in an Inuit community in remote northern Quebec. Unknown to her at the time, toxic chemicals were being absorbed by her body and also by those of other Inuit in the Arctic. As an adult, Watt-Cloutier ranged between her home in Iqaluit (pronounced "Eehalooeet," capital of the new semisovereign Nunavut Territory) to and from Montreal, New York City, and other points south, doing her best to alert the world to toxic poisoning and other perils faced by her people. The ICC represents the interests of roughly 140,000 Inuit who live around the North Pole from Nunavut (which means *our home* in the Inuktitut language) to Alaska and Russia. Nunavut itself, a territory four times the size of France, has a population of roughly 25,000, 85 percent of whom are Inuit. Some elders and hunters in Iqaluit have reported physical abnormalities afflicting the seals they catch, including some seals without hair, "and seals and walruses with burn-like holes in their skin[s]" (Lamb, n.d.).

Persistent organic pollutants (POPs) have been linked to cancer, birth defects, and other neurological, reproductive, and immune-system damage in people and animals. At high levels, these chemicals also damage the central nervous system. Many of them also act as endocrine disrupters, causing deformities in sex organs as well as long-term dysfunction of reproductive systems. Persistent organic pollutants also can interfere with the function of the brain and endocrine system by penetrating the placental barrier and scrambling the instructions of the naturally produced chemical messengers. The latter tell a fetus how to develop in the womb and postnatally through puberty; should interference occur, immune, nervous, and reproductive systems may not develop as programmed by the genes inherited by the embryo.

Pesticide residues in the Arctic today may include some used decades ago in the Southern United States. The Arctic's cold climate slows the natural decomposition of these toxins, so they persist in the Arctic environment longer than at lower latitudes. The Arctic acts as a cold trap, collecting and maintaining a wide range of industrial pollutants, from PCBs to toxaphene, chlordane, and mercury, according to the Canadian Polar Commission (P.C.B. Working Group, n.d.). As a result, "many Inuit have levels of PCBs, several forms of D.D.T., and other persistent organic pollutants in their blood and fatty tissues that are five to ten times greater than the national average in Canada or the United States" (P.C.B. Working Group, n.d.).

During the late 1990s, ecologist Barry Commoner and his colleagues used a computer model to track dioxins released from each of 44,091 sources in North America, a list that included trash-burning facilities and medical-waste-burning plants. For one year, the scientists followed dioxins as weather patterns scattered them from their sources. Winds took some of the pollution north in a hurry. Riding strong air currents, dioxin molecules can travel 400 kilometers in one day, according to atmospheric scientist Mark Cohen of the National Oceanic and Atmospheric Administration in Silver Spring, Maryland, who adapted the model for the study (Rozell, 2000).

Dioxins can travel from a smokestack in Indiana, for example, to the breast milk of a woman in Coral Harbour, Nunavut. After riding air currents northward, dioxins drop with snowflakes into Hudson Bay. During the summer, heat may promote evaporation of pesticides in the fields of the American South, feeding a molecular trickle of toxaphene, chlordane, and other compounds that makes its way to the Arctic, then falls to Earth with precipitation. In water, algae absorb the dioxins. A fish eats the algae; a bearded seal eats the fish, and dioxins build up in the animal's fatty tissue. The woman in Coral Harbour eats the seal meat, and her body transfers the dioxins to the fatty molecules of her breast milk (Rozell, 2000; Schneider, 1996, A-15). "The Arctic is more than myth and dreams. . . . The fish and whales carry scary amounts of contaminants," said Canadian Environment Minister Sergio Marchi (Schneider, 1996, A-15). "This is an important issue for indigenous people in the Arctic," said Commoner. "There's no way of protecting [areas from dioxin fallout]. You can't put an umbrella over Nunavut" (Rozell, 2000).

Persistent organic pollutants have been taking a toll on Arctic peoples in Russia as well as in Canada. With regard to POPs, "The peoples of the Arctic of Russia are at the edge of an abyss, physical disappearance," said Yeremei Danilovich, president of an association representing the more than two dozen native groups in Russia's Arctic. "The reforms have mercilessly hit the people of the North. The oil and gas companies, the logging companies, the gold and silver companies, have given nothing to the indigenous people. . . . I hope the world understands how important this entire area is" (Schneider, 1996, A-15).

Welcome to ground zero on the road to environmental apocalypse: a place, and a people, who never asked for any of the travails that industrial societies to the south have brought to them. The bevy of environmental threats facing the Inuit are entirely outside their historical experience.

"We are the miner's canary," said Watt-Cloutier. "It is only a matter of time until everybody will be poisoned by the pollutants that we are creating in this world" (Lamb, n.d.). "At times," said Watt-Cloutier, "we feel like an endangered species. Our resilience and Inuit spirit and of course the wisdom of this great land that we work so hard to protect gives us back the energy to keep going" (Watt-Cloutier, 2001).

The toxicological "due bills" for modern industry at the lower latitudes are being left on the Inuits' table in Nunavut. Native people whose diets consist largely of sea animals (whales, polar bears, fish, and seals) have been consuming a cocktail of concentrated toxic chemicals. Abnormally high levels of dioxins and other industrial chemicals are being detected in Inuit mothers' breast milk.

One may scan the list of scientific research funding around the world and add up what ails the Arctic. In addition to a plethora of studies documenting the spread of POPs through the flora and fauna of the Arctic, many studies aim to document the saturation of the same area by levels of mercury, lead, and nuclear radiation in fish and game ("P.D. 2000 Projects," 2001). Below is a sampling of titles of research projects directed at documenting ongoing ecological troubles in the Arctic. The listed studies have been generated by scientists in the United States, Canada, Sweden, Norway, and Russia.

"Assessment of Organochlorines and Metal Levels in Canadian Arctic Fox"

"Concentrations and Patterns of Persistent Organochlorine Contaminants in Beluga Whale Blubber"

"Contaminants in Greenland Human Diet"

"Effects of Metals and POPs on Marine Fish Species"

"Effects of Prenatal Exposure to Organochlorines and Mercury in the Immune System of Inuit Infants"

"Effects and Trends of POPs on Polar Bears"

"Endocrine Disruption in Arctic Marine Mammals"

"Estimation of Site-Specific Dietary Exposure to Contaminants in Two Inuit Communities"

"Follow-up of Pre-school-aged Children Exposed to PCBs and Mercury Through Fish and Marine Mammal Consumption"

"Heavy Metals in Grouse Species"

"Lead Contamination of Greenland Birds"

"Metals in Reindeer"

"New Persistent Chemicals in the Arctic Environment"

"Persistent Toxic Substances, Food Security, and Indigenous Peoples of the Russian North"

"Retrospective Survey of Organochlorines and Mercury in Arctic Seabird Eggs"

"Temporal Trends of Persistent Organic Pollutants and Metals in Ringed Seals of the Canadian Arctic"

"Ultraviolet (UV) Monitoring in the Alaskan Arctic"

"UV-Radiation and Its Impact on Genetic Diversity, Population Structure, and Foodwebs of Arctic Freshwater" ("P.D. 2000 Projects," 2001)

The bodies of some Inuit on the northernmost islands of Nunavut, thousands of miles from sources of pollution, have the highest levels of PCBs ever found, except for victims of industrial accidents. Some native people in Greenland have several dozen times as much of the pesticide hexachlorobenzene (HCB) in their bodies as temperate-zone Canadians.

Generation of POPs has become an issue in Watt-Cloutier's present residence, Iqaluit, on Baffin Island, where the town dump burns wastes that emit dioxins. The dump's plume provides only a small fraction of Iqaluit residents' POP exposure, but it has become enough of an issue to provoke a three-month shutdown of the dump that caused garbage to pile up in the town. The dump was reopened after local public-health authorities warned that the backlogged garbage could spread disease; that "the hazard posed by the rotting piles of garbage outweighed the risks of burning it" (Hill, 2001, 5). In 2001, Iqaluit's government was asking residents to separate plastics and metals from garbage that can be burned without adding POPs to the atmosphere.

Inuit Infants: "A Living Test Tube for Immunologists"

Eric Dewailly, a Laval University scientist, accidentally discovered that the Inuit were being heavily contaminated by PCBs. During the middle 1980s,

Dewailly first visited the Inuit as he sought a pristine group to use as a baseline with which to compare women in southern Quebec who had PCBs in their breast milk. Instead, Dewailly found that Inuit mothers' PCB levels were several times higher than the Quebec mothers in his study group. Dewailly and his colleagues then investigated whether organochlorine exposure is associated with the incidence of infectious diseases in Inuit infants from Nunavut (Dewailly et al., "Breast Milk Contamination," 1993; Dewailly et al., "High Organochlorine," 1993; Dewailly et al., 1994; Dewailly et al., 2000).

Dewailly and his colleagues reported that serious ear infections were twice as common among Inuit babies whose mothers had higher-than-usual concentrations of toxic chemicals in their breast milk. More than 80 percent of the 118 babies studied in various Nunavut communities had at least one serious ear infection in the first year of their lives. The three most common contaminants that researchers found in Inuit mothers' breast milk were three pesticides (dieldrin, mirex, and DDE) and two industrial chemicals, PCBs and HCB. The researchers could not pinpoint which specific chemicals were responsible for making the Inuit babies more vulnerable to illnesses because the chemicals' effects may amplify in combination.

Inuit infants have provided what one observer has called "a living test tube for immunologists" (Cone, 1996, A-1). Due to their diet of contaminated sea animals and fish, Inuit women's breast milk by the early 1990s contained six times more PCBs than that of women in urban Quebec, according to Quebec government studies. Their babies have experienced strikingly high rates of meningitis, bronchitis, pneumonia, and other infections compared with other Canadians. One Inuit child out of every four has chronic hearing loss due to infections. Born with depleted white-blood cells, the children suffer excessive bouts of diseases, including a twentyfold increase in life-threatening meningitis compared to other Canadian children. These children's immune systems sometimes fail to produce enough antibodies to resist even the usual childhood diseases.

"In our studies, there was a marked increase in the incidence of infectious disease among breast-fed babies exposed to a high concentration of contaminants," said Dewailly (Cone, 1996, A-1). A study published September 12, 1996 in the *New England Journal of Medicine* confirmed that children exposed to low levels of PCBs in the womb grow up with low IQs, poor reading comprehension, difficulty paying attention, and memory problems.

According to the Quebec Health Center, a concentration of 1,052 parts per billion (ppb) of PCBs has been found in Arctic women's milk-fat. This compares to a reading of 7,002 ppb in polar bear fat, 1,002 ppb in whale blubber, 527 ppb in seal blubber, and 152 ppb in fish. The U.S. Environmental Protection Agency (EPA) safety standard for edible poultry, by contrast, is 3 ppb, and in fish, 2 ppb. At 50 ppb, soil is often classified as hazardous waste by the EPA. Research by the Canadian Federal Department of Indian and Northern Affairs indicates that Inuit women throughout Nunavut experience DDT levels nine times more than the average of women in Canadian urban areas. The milk of

An Inuit woman photographed in Nunavut, Canada, in 1997. (Painet)

Inuit women of the Eastern Arctic has been found to contain as much as 1,210 ppb of DDT and its derivative, DDE, while milk from women living in southern Canada contains about 170 ppb (Suzuki, 2000).

The Arctic Monitoring and Assessment Programme, a joint activity of the Arctic nations and organizations of indigenous Arctic people, found in its study *Pollution and Human Health* that "P.C.B. blood levels, while highest in Greenland and the eastern Canadian Arctic, were high enough (over 4 micrograms of P.C.B.s per liter of blood) that a proportion of the population would be in a risk range for fetal and childhood development problems" (P.C.B. Working Group, n.d.).

"The last thing we need at this time is worry about the very country food that nourishes us, spiritually and emotionally, poisoning us," Watt-Cloutier said. "This is not just about contaminants on our plate. This is a whole way of being, a whole cultural heritage that is at stake here for us" (Mofina, 2000, A-12). "The process of hunting and fishing, followed by the sharing of food—the communal partaking of animals—is a time-honored ritual that binds us together and links us with our ancestors," said Watt-Cloutier (P.C.B. Working Group, n.d.)

Toxic Contamination of Traditional Inuit Foods in Alaska and Russia

Toxic contamination of traditional foods has become an issue in Alaska, as well as in Nunavut. Some Alaska natives are avoiding their traditional foods out of fear that wild fish and game species are contaminated with pesticides, heavy metals, and other toxins, native delegates told an international conference on Arctic pollution in Anchorage on May 1, 2000.

"I have a son who has quit eating seal meat altogether," said St. Paul Island resident Mike Zacharof, president of the Aleut International Association. The association was formed because indigenous people in western Alaska and Russia are worried about pollution that crosses their countries' boundaries, Zacharof said. According to a report in the *Anchorage Daily News*, "Fear that seal livers may contain mercury has made many islanders wary of eating the staple of their diet, though most still do" (Dobbyn, 2000). "Many people in Prince William Sound no longer eat their traditional foods because of the [Exxon Valdez] oil spill. This impacts not only our physical well-being but our emotional and spiritual lives as well," said Patricia Cochran, director of the Alaska Native Science Commission (Dobbyn, 2000). According to a report in the *Anchorage Daily News,* Cochran said that "natives from every region of Alaska have been noticing more tumors, lesions, spots and sores on land and sea animals" (Dobbyn, 2000). Indigenous reports are being compiled in a report for the Alaska Native Science Commission, called the Traditional Knowledge and Contaminant Project, started in 1997. The report was funded by the U.S. Environmental Protection Agency.

Research by the multinational Circumpolar Arctic Monitoring and Assessment Program has revealed high levels of the pesticide DDT in the breast milk of Russian mothers in Arctic regions. Another study, by the University of Alaska at Anchorage, indicates that pregnant Alaska native women who eat subsistence foods may be exposing their fetuses to potentially dangerous pollutants. "There's no question that people are concerned not only about what it's going to do to them but to their unborn children," Cochran said (Dobbyn, 2000). The Tanana Chiefs' Conference, a tribal social services agency, has detected high levels of DDT in salmon.

Because of toxic contamination, Alaskan Inupiat hunters now more closely examine animals they are preparing to butcher. It is called "playing doctor," wrote David Hulen, reporting for the *Los Angeles Times,* who described one hunter as he "slips on a pair of rubber gloves, wipes clean a titanium-blade knife and begins cutting tissue samples from the seal to be sent off and tested for PCBs, DDT and nearly 50 other industrial and agricultural pollutants" (Hulen, 1994, A-5). The hunters, in this case, were accompanied by Paul Becker, a Maryland-based scientist from the National Marine Fisheries Service. Becker has been working with native hunters since 1987, collecting samples for a national marine mammal tissue bank. He has trained several groups of villagers in Nome, Barrow, and other settlements to take samples as part of their regular hunts. Hulen reported that "all of this has caused no small amount of anxiety in the villages, where oil from rendered seal fat is a staple consumed like salad dressing and where people routinely dine on dishes such as dried seal meat, walrus heart and whale steaks" (Hulen, 1994, A-5).

Charlie Johnson, executive director of the Eskimo Walrus Commission, a native group that works with the government to help manage Pacific walrus populations, said: "We've seen a huge increase in cancer rates. People hear about these things turning up and wonder if that has anything to do with it. I don't think there's enough known yet to say there's a problem here in Alaska . . . but I think we have to be aware that we could have a problem eventually" (Hulen, 1994, A-5). Because of their role in handling body wastes, the kidneys of mammals are especially vulnerable to accumulation of toxic chemicals. Natives continue to eat the kidneys of the bowhead whale, however, because the kidney is one of the best-tasting parts of the animal.

The Inuit: The Intrusion of the Money Economy in the Arctic

Inuit elders use nature metaphors to describe the coming of *qallunaat*, the non-Inuit money economy; it was a time when the old world shifted, "they say, like a storm they could not read in the clouds. . . . Against the wind, the people first had to brace themselves, then they had to adapt, then they had to try to stop the wind." The change has come with a stunning swiftness, in some cases within a generation or two. According to one report,

> "Jamaise Mike was born in 1928 in an igloo, but now he is sitting in a Kentucky Fried Chicken restaurant in Pangnirtung, Nunavut, [population 1,300, on Baffin Island about 100 miles north of Iqaluit.] "Most elders were born and raised on the land," Mike said, through an interpreter, speaking in the Inuktitut language. "Everything people needed to survive surrounded them on the land. Today it is different, living in a community with a store. Everything is different from when you had to do everything yourself to survive. You depended on yourself. Now, you need money." (Brown, 2001, A-1)

As in many other parts of the world, the story of indigenous environmentalism is shaped, first and foremost, by the collision of the money economy with traditional, environmentally based economies. Many Inuit are the first generation off the land, having moved to towns at the behest of the Canadian federal government. In some cases, the government's agents "systematically killed many of the dogs that pulled their sleds, giving people no choice but to come in from the land. Children were put in Christian boarding schools. "'White people from the south,'" said Mike, "'were more terrifying than polar bears'" (Brown, 2001, A-1).

The collision of cultures can be tragic. A few decades ago, consumption of alcohol was rare, but Baffin Island Inuit now have Canada's highest rates of drunkenness, as well as suicide. In 1999, 58 of Nunavut's 27,000 people committed suicide: 52 were by hanging, 6 were by firearm, and 57 of them were Inuit (Brown, 2001, A-1).

As the economic system of the south becomes rooted, the Inuit sense a more acute need to preserve traditions that are slipping away. "We live in wooden houses, drive Jeep Cherokees, and fly in jumbo jets all over the world. But we are still Inuit. It is our spirit, our inner being, that makes us Inuit," said John Amagoalik, former chief commissioner of the Nunavut Implementation Commission (Brown, 2001, A-1).

THE LUBICON CREE: LAND RIGHTS AND RESOURCE EXPLOITATION

Despite the fact that almost $10 million worth of oil has been pumped out of their land since 1980, the 500-member indigenous Lubicon Cree community of Little Buffalo, in far northern Alberta, has no running water, inadequate housing, no sewage, and no public infrastructure (Green, n.d.). The Lubicon have been faced with a choice: rely on an overtly hostile provincial government or move in with another First Nation that can offer them social services. In the words of Chief Bernard Ominayak, the proposed change could "tear our people apart" ("Unions Back Lubicons," 2000). In 1990, after six years of deliberation, the United Nations charged Canada with human rights violations under the International Covenant on Civil and Political Rights, stating that "recent developments threaten the way of life and culture of the Lubicon Lake Cree and constitute a violation of Article 27 so long as they continue" ("Resisting Destruction," n.d.).

In 1971, the province of Alberta announced plans for construction of an all-weather road into the Lubicons' traditional territory to provide access for oil exploitation and logging. Road-building plans were undertaken without Lubicon consent, and were resisted by the native band in Canadian courts. At one point, the federal government asserted that the Lubicon were "merely squatters on Provincial crown land with no land rights to negotiate" ("Resisting Destruction," n.d.).

Having cleared legal challenges, Alberta built the all-weather road into Lubicon territory during 1979. Construction of the road was followed by an explosion of resource-exploitation activity that drove away moose and other game animals, causing the Lubicons' traditional hunting-and-trapping economy to collapse. Within four years, by 1983, more than 400 oil wells had been drilled within a 15-mile radius of the Lubicon community. From 1979 to 1983, the number of moose killed for food dropped 90 percent, from 219 to 19, as trapping income also dropped 90 percent, from $5,000 to $400 per family. At the same time, the proportion of Lubicon Cree on welfare shot up from less than 10 percent to more than 90 percent. During 1985 and 1986, 19 of 21 Lubicon pregnancies resulted in stillbirths or miscarriages. "In essence the Canadian government has offered to build houses for the Lubicon people and to support us forever on welfare—like animals in the zoo who are cared for and fed at an appointed time," said Chief Bernard Ominayak ("Resisting Destruction," n.d.).

The Lubicon: A Long Search for a Land Base

The Lubicon Lake Cree have a claim to about 10,000 square kilometers of land in northern Alberta, east of the Peace River and north of Lesser Slave Lake. The land was regarded as so remote in 1900 that Canadian officials seeking to negotiate treaties completely ignored it. The Lubicon did not sign Treaty 8, which was negotiated in 1899 and 1900, providing the band no legal title to its land—title they were still seeking to negotiate with Canadian officials a century later.

During the century that the Lubicon have sought legal title to their land base, it has been scarred by oil and gas production as well as industrial-scale logging. While Canadian officials have refused to set aside land for the Lubicon, large oil, gas, pulp, paper, and logging companies have moved into the area to exploit its natural wealth (Green, n.d.).

The Lubicon were first promised a reserve by the Canadian federal government in 1939, before oil was discovered under their lands. Oil companies flooded into the area in the 1970s, all but destroying the Lubicon society and economy. The story of the Lubicon is a case study of how modern resource exploitation can ruin a natural setting and the indigenous people who once lived there.

The Lubicon engaged in negotiations with the federal and provincial governments three times during the 1990s, but the talks broke off because the two sides failed to agree on the size of the land mass and monetary compensation that should be allotted to the Lubicon (Green, n.d.). "It is very worrisome when you are at the table year in, year out . . . with government sponsored and supported resource development . . . subverting the rights you are at the table to negotiate. You have to wonder if there is any sincerity about achieving a settlement," said band advisor Fred Lennarson (Guerette, 2001).

During the mid-1980s, after the provincial government enacted retroactive legislation to prevent the Lubicon from filing legal actions to protect its traditional territory from booming oil and gas development, the Lubicon launched a protest campaign against petroleum companies sponsoring the 1988 Calgary Olympics (Guerette, 2001). On the eve of the international event, Premier Don Getty established a personal dialogue with Lubicon Chief Bernard Ominayak that led to an agreement committing the province to transfer to Canada the 95-square-mile reserve the Lubicon had been seeking (Guerette, 2001).

Oil and gas revenues from Lubicon ancestral lands had continued at about $500 million a year. Not a penny went to the Lubicon. During 1988, after 14 years getting nowhere in the courts, the Lubicon asserted active sovereignty over their land. The peaceful blockade of access roads into their traditional territory stopped all oil activity for six days. Later the barricades later were forcibly removed by the Royal Canadian Mounted Police. When Alberta's Premier Don Getty met with Lubicon leaders in Grimshaw, Alberta, the result was the Grimshaw Accord, an agreement on a 243-square-kilometer (95.4-square-mile) reserve area.

Twelve years later, the provincial government abandoned the agreement. In the meantime, environmental assessments disclosed that more than 1,000 oil- and gas-well sites had been established within a 20-kilometer radius of Lubicon Lake, on land that had been promised to the indigenous people. During 2001, Marathon Canada announced plans for a natural gas compressor and pipeline in the area.

Logging on Lubicon Land

By 2001, forest industries held concessions from the provincial government that covered nearly all of the land claimed by the Lubicon not already leased for oil and gas production. The first of the industrial loggers was the Japanese paper company Daishowa, which began logging on Lubicon land during the 1980s. In 1988, Daishowa announced plans for a pulp mill near the proposed Lubicon territory that would have daily processed lumber equaling the area of 70 football fields. The province of Alberta also granted Daishowa timber rights to an area including the entire Lubicon traditional territory.

An international boycott of Daishowa was launched to protest the company's clear-cutting of Lubicon land. In response, Daishowa stayed off Lubicon land during the 1991–1992 winter logging season. Logging later resumed, as the company sought to lift the boycott by suing the Lubicons' nonnative allies in Canadian courts. Daishowa's legal action was thrown out of court during the late 1990s, as the boycott intensified. Daishowa, which manufactures paper bags, newsprint, and other paper products, retracted its plans to log Lubicon land only after the boycott began to reduce its revenues. Daishowa then again pledged to stay out of Lubicon forests until the natives' land rights were delineated. At this point, the boycott ended. The company asserted that the boycott had cost it $20 million in lost sales (Guerette, 2001).

THE DENE: DECIMATED BY URANIUM MINING

At the dawn of the nuclear age, Paul Baton and more than 30 other Dene hunters and trappers who were recruited to mine uranium called it the money rock (Nikiforuk, 1998, A-1). Paid $3 a day by their employers, the Dene hauled and ferried burlap sacks of the grimy ore from one of the world's first uranium mines at Port Radium across the Northwest Territories to Fort

McMurray. Since then, according to an account by Andrew Nikiforuk in the *Calgary Herald*, at least 14 Dene who worked at the mine between 1942 and 1960 have died of lung, colon, and kidney cancers, according to documents obtained through the Northwest Territories Cancer Registry (Nikiforuk, 1998, A-1). The Port Radium mine supplied the uranium to fuel some of the first atomic bombs. Within half a century, uranium mining in northern Canada had left behind more than 120 million tons of radioactive waste, enough to cover the Trans-Canada Highway two meters deep across Canada. By the year 2000, production of uranium waste from Saskatchewan alone occurred at the rate of over one million tons annually (LaDuke, 2001).

The experiences of the Dene were similar to those of Navajo uranium miners in Arizona and New Mexico at the same time (see "The Navajo: The High Price of Uranium," United States of America). Since 1975, a 30-year latency period after uranium mining began in the area, hospitalizations for cancer, birth defects, and circulatory illnesses in northern Saskatchewan had increased between 123 and 600 percent. In other areas impacted by uranium mining, cancers and birth defects have increased, in some cases, to as much as eight times the Canadian national average (LaDuke, 1995).

"Before the mine, you never heard of cancer," said hunter Paul Baton, 83. "Now, lots of people have died of cancer" (Nikiforuk, 1998, 1). The Dene were never told of uranium's dangers. Declassified documents have revealed that the U.S. government, which bought the uranium, and the Canadian federal government, at the time the world's largest supplier of uranium outside the United States, withheld health and safety information from the native miners and their families. A 1991 federal aboriginal health survey found that Deline, a Dene community, reported twice as much illness as any other Canadian aboriginal community.

While mining, "many Dene slept on the ore, ate fish from water contaminated by radioactive tailings and breathed radioactive dust while on the barges, docks and portages. More than a dozen men carried sacks of ore weighing more than 45 kilograms for 12 hours a day, six days a week, four months a year. . . . Children played with the dusty ore at river docks and portage landings. And their women sewed tents from used uranium sacks" (Nikiforuk, 1998, A-1).

While many of the Dene blame uranium mining and its waste products for their increased cancer rates, some Canadian officials compiled statistics indicating only marginal increased mortality from uranium exposure. André Corriveau, the Northwest Territories' chief medical officer of health, noted that high cancer rates among the Dene do not differ significantly from the overall territorial profile. He said that the death rate was skewed upward by high rates of smoking (Nikiforuk, 1998, 1). The Dene, in the meantime, maintain that the fact that almost half the workers in the Port Radium mine (14 of 30) died of lung cancer cannot be explained by smoking alone.

Until his death in 1940, Louis Ayah ("Grandfather"), one of the North's great aboriginal spiritual leaders, repeatedly warned his people that the waters in Great

Bear Lake would turn a foul yellow. According to Grandfather, the yellow poison would flow toward the village, recalls Madelaine Bayha, one of a dozen scarfed and skirted so-called uranium widows in the village (Nikiforuk, 1998, A-1).

The first Dene to die of cancer, or what elders still call the incurable disease, was Old Man Ferdinand in 1960. He had worked at the mine site as a logger, guide, and stevedore for nearly a decade. "It was Christmastime and he wanted to shake hands with all the people as they came back from hunting," recalled Rene Fumoleau, then an Oblate missionary working in Deline. After saying goodbye to the last family that came in, Ferdinand declared, "'Well, I guess I shook hands with everyone now,' and he died three hours later" (Nikiforuk, 1998, A-1).

According to Nikiforuk's account, others died during the next decade. Victor Dolphus's arm came off when he tried to start an outboard motor. Joe Kenny, a boat pilot, died of colon cancer. His son, Napoleon, a deckhand, died of stomach cancer. The premature death of so many men has not only left many widows, but also interrupted the handing down of culture. "In Dene society it is the grandfather who passes on the traditions and now there are too many men with no uncles, fathers or grandfathers to advise them," said Cindy Gilday, Joe Kenny's daughter, and chair of the Deline Uranium Committee (Nikiforuk, 1998, A-1)

"It's the most vicious example of cultural genocide I have ever seen," Gilday said. "And it's in my own home" (Nikiforuk, 1998, A-1). According to Nikiforuk, "Watching a uranium miner die of a radioactive damaged lung is a job only for the brave." He described Al King, an 82-year-old retired member of the Steelworkers Union in Vancouver, British Columbia, who has held the hands of the dying. King described one retired Port Radium miner whose chest lesions were so bad that they spread to his femur and it exploded. "They couldn't pump enough morphine into him to keep him from screaming before he died," said King (Nikiforuk, 1998, A-1).

In exchange for their labors in the uranium mines, the Dene received a few sacks of flour, lard, and baking powder. "Nobody knew what was going on," recalled Isadore Yukon, who hauled uranium ore for three summers in a row during the 1940s. "Keeping the mine going full blast was the important thing" (Nikiforuk, 1998, A-1).

The Dene town of Deline was described by one of its residents as

practically a village of widows. Most of the men who worked as laborers have died of some form of cancer. The widows, who are traditional women were left to raise their families with no breadwinners, supporters. They were left to depend on welfare and other young men for their traditional food source. This village of young men, are the first generation of men in the history of Dene on this lake, to grow up without guidance from their grandfathers, fathers and uncles. This cultural, economic, spiritual, emotional deprivation impact on the community is a threat to the survival of the one and only tribe on Great Bear Lake. (Gilday, n.d.)

THE KANESATAKE MOHAWKS: DEBATING NIOBIUM MINING

The Kanesatake Mohawks have challenged a decision to allow a $102-million niobium-mining project on land zoned for agricultural use. A Quebec provincial commission on the protection of agricultural lands decided during May 2001 to allow Niocan of Montreal to develop a niobium mine in the town of Oka, on land claimed by the Mohawks of Kanesatake. Late in 2001, the Mohawks began an appeal of that decision before the Tribunal Administratif du Quebec (Lalonde, 2001). Experts commissioned by the Mohawks told the tribunal that the mine will affect properties within three kilometers of the mine, not only 1.5 kilometers as claimed by Niocan (Lalonde, 2001). In 1990, Oka was the site of a prolonged standoff over land claims between Mohawks, Quebec police, and Canadian army troops that electrified the land-claims debate across Canada.

Canada is the world's second-largest producer and exporter of niobium, which also is known as columbium, an alloy used in production of material for the aerospace, automobile, construction, and pipeline industries. The United States imports nearly all of its niobium from Canada. The Mohawks are concerned that niobium, which contains high levels of uranium, polonium, and thorium, could contaminate groundwater that is used for drinking and irrigation.

Many Mohawks at Kanesatake also assert that the mining operation may disturb radon in the ground, thereby producing radioactive byproducts that could leach into the water table, contaminating farm produce and destroying the area's potential for tourism and agriculture. The Mohawks also asserted that the radioactive waste generated by the mine will not be disposed of safely. Grand Chief Steven Bonspille said that the health and livelihood of several Mohawk families who farm in the area are at stake. "This region is an agricultural zone and tourist area and those industries will suffer the effects of the stigma of having a mine right in the middle of it all," Bonspille said (Lalonde, 2001).

The Mohawks also have challenged Niocan's pledge that the mine will create 160 jobs in the region. "Mining is not a pick-and-shovel industry any more and I don't see them hiring people from the region and training them when there are trained miners in the province looking for work," Bonspille said (Lalonde, 2001).

On February 14, 2002, the Mohawk Council of Kanesatake passed a resolution demanding that the mine not open. According to a report in *Drillbits and Tailings*, the Mohawk Council, the government of Kanesatake, joined with the Union of Agricultural Producers "to demand a full environmental assessment on the federal and provincial levels to expose the health risks as well as the agricultural and environmental detriment that the proposed mine will bring to the surrounding communities. At press time, there has been no response from the federal or provincial governments" ("Mohawk of Kanesatake," 2002).

THE HALFWAY RIVER FIRST NATION OF BRITISH COLUMBIA: RESISTING OIL AND GAS PIPELINES AND TOURIST DEVELOPMENT

Tensions between natives in northeastern British Columbia and the provincial government regarding oil and gas development escalated during 2001 as an Indian band established a blockade to halt construction of a Petro-Canada natural gas pipeline. Chief Bernie Metecheah of the Halfway River First Nation said that he ordered the blockade—a serious decision made with the consent of his 200-member band—because the proposed 23-kilometer right-of-way crossed sacred hunting areas (Cattaneo, 2001).

A blockade of North Road, located near Fort St. John, included about 50 people protesting proposed oil and gas development. According to a news account in Canada's *National Post*, "All oil-industry personnel were being turned away by late afternoon, although people were allowed to leave the area" (Cattaneo, 2001).

The blockade was viewed by some in Canada as a localized response to a pledge made during July by Matthew Coon-Come, national chief of the Assembly of First Nations, to set up coast-to-coast blockades in a dispute with the federal government over changes to Canada's Indian Act. On August 1, Coon-Come and the government had agreed to a 30-day cooling-off period. This specific blockade also was related to Petro-Canada's use of lands that the Halfway River band deems sacred. In addition, four of the band's seven major hunting camps lie in the path of the proposed pipeline. The Halfway River band said "it is planning to continue the closure into the winter, or as long as it takes to get a moratorium on resource development until completion of a cumulative environmental impact assessment" (Cattaneo, 2001).

"We are inundated with 30 resource companies and a huge Petro-Canada pipeline that is poised to wipe out four of seven main hunting camps in the traditional area," said Gray Jones, a spokesman for the band. "For years, they have tried to negotiate fairly with governments and companies, and they get baseball caps and rides in helicopters, but nothing substantive has happened. The chief is turning up the heat" (Cattaneo, 2001). Elders in the area also fear that one pipeline may later be joined by other feeder pipelines, as well as to roads that will attract non-Indian hunters.

The area has recently experienced increased exploration for natural gas, as native peoples assert that the British Columbia government, while reaping a windfall in energy royalties from soaring oil and gas activity, is ignoring a treaty with the First Nations in northeastern British Columbia, signed with Canada in 1899, that guarantees them the right to enjoy and live off the land.

As the Halfway River band resisted oil and gas development, Native Americans elsewhere in British Columbia established blockades to impede development of tourist facilities that they assert infringe on their ancestral lands. On August 24, 2001, members of the Native Youth Movement blocked access to the Sun Peaks resort. Amanda Soper, a spokesperson for the Native Youth

Movement, said that the group set a spiritual fire in the middle of the road blocking access of all vehicles coming into the village. "Whether they [were] workers, tourists, or business owners, or residents, we stopped everybody and halted all traffic," said Soper. She continued:

> This was a message to everybody's involvement and their collaboration in the destruction of our lands, natural resources, and for the total disregard of our aboriginal title and rights that we have on this mountain. This area lies within our 1862 Neskonlith Douglas Reserve, and within Tsekuplet title territory that is legally recognized by the Supreme Court of Canada. But the provincial and federal governments continue to ignore their own laws, and to continue to not uphold their own laws in spite of our constitutionally recognized rights to this land. (Cattaneo, 2001)

The blockade, which began about 7:30 A.M., delayed about 100 vehicles, the drivers of which received handbills "advising against continuing any further business, or governmental transactions, or involvement within our traditional territory, and the consequences of further actions will result in necessary sponsors permitted by our natural laws and our right to defend [that] are put here by our creators. That's the laws that we go by. And we must protect the land, and every human being on this planet must raise their consciousness and awareness that the land is what sustains our very survival" ("Protecting Knowledge," 2001). The blockade was dismantled during the late morning of the same day, after the Royal Canadian Mounted Police threatened arrest of the participants.

According to Soper, the blockade was a direct action against clear-cutting for ski runs in an area where the Tsekuplet build winter homes. "We're up there to prevent any further destruction on that mountain, and to show our president, [by our] presence that we will never extinguish our right to this land. This land will always be under the control and title of the Tsekuplet people" ("Protecting Knowledge," 2001).

On September 24, 11 people surrendered to the Royal Canadian Mounted Police on arrest warrants for their actions August 24 at Sun Peaks. Chief Stewart Phillip stated, on behalf of those arrested:

> We are agreeing to be arrested on behalf of all elders and native people who depend on their territory for their traditional livelihood and for their medicines. . . . The people arrested today are the Elders, the youth and land-users of Skwelkwek'welt who are protecting their aboriginal title interests in the Sun Peaks area. The criminalization and harassment of community people does not settle the fundamental issues of Neskonlith's unresolved aboriginal title and of their Douglas Reserve specific claim. . . . I fully agree with Chief Arthur Manuel's call for the governments of British Columbia and Canada to enter into good-faith negotiations based on the recognition of Aboriginal Title and Rights. (Phillip, 2001)

During March 2002, local natives became enraged when Sun Peaks posted a *No Indians Allowed* sign at the entrance to its resort, meanwhile bulldozing two native homes and two sweat lodges. A rally in opposition was held March 23 at the entrance to the resort.

THE TAKU RIVER TLINGIT FIRST NATION OF BRITISH COLUMBIA: RESISTING ZINC, COPPER, GOLD, AND SILVER MINING

The largest salmon-producing watershed in southeast Alaska could be threatened by a mining proposal that depends on the construction of a 160-kilometer (100-mile) road. The future of the watershed, in turn, depends on a proposal to reopen the Tulsequah Chief mine, on a tributary of the Taku River 100 kilometers south of Atlin, British Columbia.

The Tulsequah Chief mine was operated initially by Cominco. In 1957, the company stopped mining at the site on the Tulsequah River, upstream from its confluence with the Taku River. Redfern Resources, a mining company based in Vancouver, acquired the property in 1992. The British Columbia government approved Redfern's proposal to reopen the mine in March 1998 after a three-and-a-half-year environmental review. Redfern had invested more than $8 million (Canadian) since 1995 to meet the technical, environmental, and public consultation requirements of the British Columbia Environmental Assessment Act and the Canadian Environmental Assessment Act. Questions about potential habitat damage from the mine led American Rivers, a Washington, D.C.–based conservation group, to list the Taku River fourteenth among the United States' 20 most-endangered rivers in 1998 ("Mine Approved," 2000).

The Taku watershed is an 18,000-square-kilometer (4.5-million-acre) roadless tract of land in northwestern British Columbia. The area includes habitats representing five biogeoclimatic zones, ranging from high plateaus to lush coastal temperate rainforests. The area supports some of the richest wildlife habitat on the west coast of North America. Grizzly bears, moose, caribou, black bears, mountain goats, salmon, and migrating birds thrive in large numbers due to the area's isolation and the fact that (until and unless an access road is built) it is accessible to humans only by float plane, river boat, or on foot ("Mine Approved," 2000).

The area also is the traditional home of the Taku River Tlingit First Nation. In 1999, the Taku River Tlingit asked the British Columbia Supreme Court for a judicial review of the provincial government's approval of the mine. The case was heard in March on the issue of whether government has a duty to protect the environment and the fishing and hunting rights of native peoples. If and when it is reopened, the Tulsequah Chief project could produce zinc, copper, lead, gold, and silver for at least nine years, with the potential for a much longer mine-life based on as yet unproven reserves. Each year, more than two million salmon return to the Taku and its tributaries on both sides of the border. Water quality and salmon habitat—and the communities that depend on both—could be threatened by mining operations.

THE INNU OF LABRADOR: INDUSTRIALISM'S INTRUSIONS

Introduction

The Innu of Labrador are among some of the most remote peoples in the world from industrialism. Distance is often irrelevant, however, at a time when the Innu have found themselves facing hydroelectric development, aluminum smelting, a sulfide mine, and frequent overflights by military aircraft.

Military Test Flights Ruin Innu Hunting

The Innu Nation launched a court challenge against the Canadian Department of National Defence (DND) and the Royal Netherlands Air Force to prevent supersonic test flights over its hunting territory. In a motion filed in Canadian Federal Court on August 1, 2000, the Innu Nation asserted that the proposed flights are potentially harmful and provide little useful information on the potential effects of future supersonic fighter training presently being considered by Canada and other countries flying out of the Canadian Forces' Goose Bay airbase. The court challenge followed a decision by the DND to conduct a series of planned supersonic tests. The tests were originally scheduled to take place on July 28–29, 2000, but were suspended following the announcement of the Innu Nation's intention to seek an injunction against the tests in court ("Innu Nation," 2000). The Innu asserted that inadequate consultation took place prior to the announcement of the tests, that Innu hunters may be present in the area, and that shock waves resulting from the supersonic flights pose a high risk of irreparable damage to people who may be in the area.

In 1980, when the Canadian Air Force started flying at Goose Bay with Phantom II aircraft, the Innu first voiced concerns about potential environmental impacts of military flight training over their hunting grounds in letters to the Canadian federal minister of environment. The DND began an environmental screening of the flights the following year that produced a 37-page Initial Environmental Evaluation, essentially an in-house study by Major G. Landry, which gave an environmental so-called green light for low-level training activities in Innu airspace over parts of Quebec and Labrador ("Innu Nation," 2000).

Since the early 1980s, military flight training from Goose Bay over 130,000 square kilometers of Labrador and northeastern Quebec has been conducted by European countries under a Multinational Memorandum of Understanding. The majority of approximately 6,000 annual training flights are conducted by fighter jets flying at low levels, less than 100 feet above ground level. According to a statement by Innu elders, "Innu people have long opposed military activities over their homeland, primarily due to concerns about the impacts of jet noise and pollution on humans and wildlife" ("Innu Nation," 2000).

A statement issued by the Innu assessed problems encountered by indigenous people who live within range of constant low-level jet aircraft flights:

> It is well known that shock waves can permanently damage human lungs, viscera, ears and can cause brain damage. Low-level supersonic fly-overs . . . generate shock waves which can produce irreversible damage with respect to a plane flying supersonically at 15,000 feet. . . . A human being directly under the flight path of this aircraft would likely be exposed to sound in excess of 130 decibels (the threshold for pain), and with respect to a plane flying supersonically at 5,000 feet . . . a human underneath the flight path of that aircraft would likely be exposed to sound around or in excess of 153 decibels, which is the threshold for the types of irreparable damages. ("Innu Nation," 2000)

Innu communities affected by high noise levels from the training flights include about 13,000 people who live in eastern Quebec and Labrador (an area the Innu call Nitassinan). They reside in 13 communities: Utshimassit (Davis Inlet), Sheshatshiu, St. Augustin, La Romaine, Natashquan, Mingan, Sept-Illes, Maliotenam, Betsiamites, Les Escoumins, Pointe-Bleue, Schefferville, and Kawawachikamach. Of these, two Innu communities—Sheshatshiu and Utshimassit—are in Labrador. While nonnative Canadians have labeled the Innu "Montagnais" and "Naskapi," they are very closely related to the Cree linguistically and culturally ("Innu Nation," 2000).

Tanien (Daniel) Ashini, an Innu hunter, described how military jet overflights ruin hunting, as well as a hunter's peace of mind:

> We go into the bush for the kind of spiritual tranquility that many others associate with their churches. But now the noise of low-flying jet bombers has destroyed our peace of mind in the bush. The jets startle us, terrify our children, frighten our animals and pollute the waterways. They fly into river valleys, over lakes and marsh areas—places that are also best for hunting, trapping, and fishing. (Interpress Service, 1993, 16)

Many Innu have protested the overflights physically at the practice target area and in Ottawa. Some have been arrested. Despite these Innu protests, the North Atlantic Treaty Organization (NATO) during 2001 increased the number of flights from 7,000 per year to 18,000, or an average of 50 per day. Plans formulated by NATO also called for opening of a second bombing range and increased usage of live bombs. New dogfight training would require a 30 percent increase in the amount of airspace being used. As the number of flights increased, the Innu asked whether NATO needs to continue the testing at all, given the demise of the Cold War. One independent analysis, cited in the hemispheric indigenous newsmagazine *Native Americas*, asserts that NATO's exercises are "of questionable value and excessive cost" ("NATO and Innu," 1995, 12).

The Innu Battle a Sulfide Mine

The Innu Nation and Labrador Inuit Association have been contending with proposed development of an open-pit nickel mine and mill complex at Voisey's Bay, amidst a site of what could become the world's largest such mine, estimated at 150 million tons of reserves. The Voisey's Bay Nickel Company (VBNC), a subsidiary of Inco, the largest nickel producer in the world, applied to the Newfoundland provincial government in late May 1997 for permission to construct a road and airstrip at the Voisey's Bay site. Innu and Inuit residents, in response, took their case to the Newfoundland Supreme Court where they sought an injunction to prevent the construction of infrastructure pending an environmental assessment.

Innu president Katie Rich said, "It's too bad we had to go to court. This action is about forcing VBNC to live up to their public statements about being committed to the environment. The company can't claim to be environmentally responsible when it is trying to detour around the environmental review process." ("Innu Begin," 1997). The Innu request for an injunction against the mine and mill was rejected by the Supreme Court on July 18, 1997. In the meantime, the province already had issued construction permits. Machines and materials were imported, buildings constructed, and four kilometers of roadway cut at Voisey's Bay ("Innu Begin," 1997).

On August 20, 1997, their legal appeals exhausted, the Innu Nation and the Labrador Inuit staged a demonstration at Voisey's Bay. That morning, the first Inuit protesters arrived from Nain to establish a protest camp, as more than 250 Innu and Inuit arrived to participate. "Development at the site has gone far enough. Inco is now building a road and an airstrip without Innu and Inuit consent. Without our consent there will be no project!" asserted Rich ("Innu Begin," 1997). Rich expressed her hope that the protest would remain peaceful. "This is about standing our ground, not about a standoff. The company has never asked for our permission to be here, or asked for our consent for the mine. We have clearly outlined what our consent requires and it is up to them to respond to it. We have lived here for thousands of years and we plan to stay here for thousands more, but we don't want to be living in the mess they will make here by rushing ahead. By standing together with the Labrador Inuit we are saying to Inco and to Brian Tobin [the provincial premier] that we will not be bulldozed over on our own land" ("Innu Begin," 1997).

In addition to the proposed nickel mine, Innu territory has been examined for possible siting of an aluminum smelter. During late July 2001, the Innu Nation warned potential developers of Lower Churchill hydroelectric projects that a proposed aluminum smelter would require Innu consent. The Innu Nation was responding to announcement of a new feasibility study for the construction of a Lower Churchill hydroelectric project and an aluminum smelter. Peter Penashue, president of the Innu Nation, said that "we would like to . . . again remind the politicians and promoters of this project that the consent of the Innu Nation is required for any developments in our territory. For the Innu to seriously consider major developments in our territory, there must be a settled land-claims agreement. . . . Innu will only accept projects that meet our sustainable benefits to our people" ("Innu Respond," 2001).

THE OUJE-BOUGOUMOU CREE: WATER POLLUTION

Ouje-Bougoumou, a Cree community in northern Quebec about 350 miles north of Montreal, has experienced severe metal contamination in its food and water, a decades-old legacy of pollution related to logging and mining. During the early 1990s the Canadian federal government created a new community for the Ouje-Bougoumou Cree Nation and forced its members to relocate. The move hasn't solved the pollution problem in fish, their main food source.

"I was kind of shocked," said Cree Margo Miascum Cooper. She fears the water pollutants killed her father, Albert Cooper, who died of cancer during 2000. "I was really concerned, especially for my family" ("Northern Quebec," 2001). Cooper said that the fish Albert caught and consumed were deformed by pollution from a nearby gold and copper mine. The fishes' eyes were covered in scabs; many of them were missing fins.

A report commissioned by the Grand Council of the Cree examined soil and water samples, which revealed unhealthy levels of toxic heavy metals, also found in hair samples of nearly two dozen Cree in the same area. The Cree samples were three times higher than average for selenium, four times higher for lead, and five times higher for aluminum. Hair samples also revealed usually high levels of arsenic and cyanide. "The levels we have observed particularly of cyanide in fish are killing people and will continue to kill people until something is done," said Chris Covel, the report's coauthor ("Northern Quebec," 2001).

In October 2001, the government of Quebec admitted it knew that the Cree of Ouje-Bougoumou were being poisoned by toxic waste that had fouled their water and poisoned their fish. Government reports documented what the Cree had known for years: the fish they had been catching were deformed, missing eyes and fins, poisoned by three nearby mines. Covel's study found high levels of cyanide, arsenic, mercury, and other heavy metals in the water that yielded the deformed fish. Environment Minister Andre Boisclair at first tried to dismiss the findings, but on October 25 he admitted the Ouje-Bougoumou Cree were, in fact, being poisoned. Furthermore, he said that the ministry had known about the severity of the pollution since at least 1999. "There is a problem," he said. "Their results confirm ours, so there is a problem" ("Quebec Admits," 2001).

"They just dumped it [pollution] into the water," said Chief Sam Bossum. Bossum couldn't get the Environment Ministry to do anything about it, so the

Cree commissioned their own study. United States researcher Chris Covel sampled mine tailings, lake water and sediment, fish, and human hair. The analysis showed high levels of several toxic chemicals. "The levels that we observed, particularly of cyanide in fish, are killing people," he said. As of early 2002, the government of Quebec had not ordered or funded any actions to remediate the water pollution. Boisclair merely told the Cree "to cut down on their fish intake while he orders more studies" ("Quebec Admits," 2001).

THE DOGRIB FIRST NATION: DIAMOND MINING

Diavik Diamonds and the Canadian Northwest Territories Chamber of Mines have been lobbying the Canadian federal government to approve a Diavik mining project on the Dogrib First Nation's territory in the Northwest Territories near Yellowknife. The Dogrib have responded with caution because earlier mining, for uranium and gold, has scarred their homelands.

John Zoe, chief negotiator for the Dogribs' land claims and self-government agreements, accused the company, a subsidiary of Rio Tinto, of rushing the project proposal through governmental regulatory processes without adequately addressing the mine's impact on the environment and the livelihood of the Dogrib First Nation ("Dogrib First Nation," 1999). "What Diavik wants to do with this mine has never been done before . . . the potential negative consequences are terrifying to us and to all downstream users of the water," said Zoe ("Dogrib First Nation," 1999).

The company proposes to dam and drain a substantial portion of a large Arctic lake and then begin open-pit mining on the lake's bed. The proposed mine site is a summer grazing area for the Bathurst caribou herd; is also lies on the same herd's twice-yearly migratory route. The Dogrib First Nation depends heavily upon this herd for basic subsistence, describing it as "the lifeblood of our people" ("Dogrib First Nation," 1999).

The Dogrib people assert that Diavik ignored the risks its proposed mine would pose to the caribou. Furthermore, the Dogrib criticize the company for ignoring its complaints. The Dogrib have urged the Canadian federal government to slow the approval process for Diavik's mine to ensure that environmental and community concerns are properly addressed ("Dogrib First Nation," 1999). "The diamonds aren't going anywhere. We have been harvesting on this land and the Bathurst Caribou have been traversing it since before human memory. How can the demands of Diavik for an immediate approval be balanced against this history and these needs today?" asked Zoe ("Dogrib First Nation," 1999).

The Dogrib are not entirely opposed to a diamond mine, according to a report in *Drillbits and Tailings*. During 1999, agreement was reached with B.H.P., a large Australian mining company, which now operates a diamond mine within First Nation territory, and which, according to Zoe, ensured that the development "was done right, for wildlife and the environment and from our perspective as a thriving First Nation" ("Dogrib First Nation," 1999).

According to *Drillbits and Tailings*, "Traditional lands of the Dogrib First Nation have already been plundered by government-approved mining operations, leaving hazardous uranium waste sites, tailings ponds set to overflow with cyanide and heavy metals, and massive arsenic deposits at Royal Oak's Giant Mine in Yellowknife" ("Dogrib First Nation," 1999). The Dogrib First Nation has asked the Canadian government to "take the time to get this decision right." In the words of John Zoe, "When the mining executives are long gone, and the diamonds are used up, we know we will be left to bear the consequences" ("Dogrib First Nation," 1999).

THE OJIBWAY: BLACK BEAR POACHING

The future of Canada's roughly 300,000 black bears has been threatened by a voracious demand for their gall bladders, bile, and paws from Asia. Indigenous peoples are working, along with environmentalists, to protect the bears. "Demand flourishes not only in Asia, but also in growing Asian communities in Canada and elsewhere," said the Investigative Network, which advocates internationally for animal-welfare societies. The Investigative Network said that 40,000 bears are killed legally in Canada each year, and another 40,000 are poached (killed illegally) for bear body parts (Borst, 1996).

The Canadian federal government and several provinces passed legislation during the early and mid-1990s to stop the trade in bear gall bladders and paws. Gall bladders, which are shaped like figs, are used in some traditional Chinese medicines. Paws are prized as a culinary treat in some Asian communities. The new laws call for prison terms of up to five years and fines of as much as U.S. $112,000 for illegal trafficking in bear body parts. In reality, imprisonment has been rare.

Paul Hollingsworth of the Native-Animal Brotherhood said that his group is uniting Bear Clan elders and chiefs from native bands across North America to oppose hunting and illegal trade in black bear parts. "The native people are extremely interested in preserving this species," said Hollingsworth, a member of the Ojibway Nation (Borst, 1996).

Gail Anderson, a forensic entomologist at Simon Fraser University near Vancouver who helps determine the time of death when a body is found, assists the province of British Columbia in its antipoaching efforts. She called bear poaching "the second most profitable crime in North America after drug dealing." She has reported seeing "truck loads of galls" in the United States, which has lagged behind Canada in efforts to stem the trade in bear body parts (Borst, 1996).

REFERENCES

Authier, Philip. "Cree Pact Clears Final Hurdle." *Montreal Gazette*, February 5, 2002. [http://www.canada.com/montreal/montrealgazette/story.asp?id = {43F9B38E-7577-4579-AEED-90A8F9638936}]

Bigert, Claus. "A People Called Empty." In Rainer Wittenborn and Claus Bigert, eds., "Amazon of the North: James Bay Revisited." Program for presentation, Santa Fe Center for Contemporary Arts, August 4 through September 5, 1995, n.p.

Borst, Barbara. "Poaching Threatens Bear Population." Interpress Service, May 7, 1996. [http://www.nativenet.uthscsa.edu/archive/nl/9605/0084.html]

Brown, DeNeen L. "Culture Corrosion in Canada's North; Forced Into the Modern World, Indigenous Inuit Struggle to Cope." *Washington Post,* July 16, 2001, A-1.

Cadbury, Deborah. *Altering Eden: The Feminization of Nature.* New York: St. Martin's Press, 1997.

Calamai, Peter. "Chemical Fallout Hurts Inuit Babies." Toronto *Star,* March 22, 2000. [http://irptc.unep.ch/pops/newlayout/press_items.htm]

"Canadian Indians Paddle to New York City to Protest Quebec Power Plant." *Syracuse Post-Standard,* April 5, 1990, A-2.

Cattaneo, Claudia. "B.C. Natives Blockade Petro-Can Pipeline Site." *National Post* (Canada), August 14, 2001. [http://www.nationalpost.com/]

Cone, Marla. "Human Immune Systems May be Pollution Victims." *Los Angeles Times,* May 13, 1996, A-1.

Coon-Come, Matthew. *Remarks of the National Chief Matthew Coon-Come: Peoples Summit of the Americas Environment Forum.* April 18, 2001. [http://www.afn.ca/Press%20Realeses%20%20speeches/april_18.htm]

"Cree Claim Victory in Hydro Hearings." Canadian Broadcasting Corporation, December 1, 2000. [http://www.northstar.sierraclub.org/HYDRO_Cree_Claim%20_%20Victory.htm]

"Cree Indians Win Battle in Their Struggle to Stop Canadian Hydropower Plant." *Omaha World-Herald,* September 12, 1991, 4.

"Cross Lake Takes Hydro Battle to U.S." *First Perspectives: News of Indigenous Peoples of Canada* (May 29, 2001): n.p. [http://collection.nlc-bnc.ca/100/201/300/first_perspective/2001/05-29/story7.html]

Dewailly, E., J. J. Ryan, C. Laliberte, S. Bruneau, J.-P. Weber, S. Gringras, and G. Carrier. "Exposure of Remote Maritime Populations to Coplanar PCBs." *Environmental Health Perspectives* 102, supp. 1 (1994): 205–9.

Dewailly, E., P. Ayotte, S. Bruneau, S. Gingras, M. Belles-Isles, and R. Roy. "Susceptibility to Infections and Immune Status in Inuit Infants Exposed to Organochlorines." *Environment Health Perspectives* 108 (2000): 205–11.

Dewailly, E., S. Bruneau, C. Laliberte, M. Belles-Iles, J.-P. Weber, and R. Roy. "Breast Milk Contamination by PCB and PCDD/Fs in Arctic Quebec. Preliminary Results on the Immune Status of Inuit Infants." *Organohalogen Compounds* 13 (1993): 403–6.

Dewailly, E., S. Dodin, R. Verreault, P. Ayotte, L. Sauve, and J. Morin. "High Organochlorine Body Burden in Breast Cancer Women with Oestrogen Receptors." *Organohalogen Compounds* 13 (1993): 385–88.

Dewailly, Eric, P. Ayotte, C. Blanchet, J. Grodin, S. Bruneau, B. Holub, and G. Carrier. "Weighing Contaminant Risks and Nutrient Benefits of Country Food in Nunavik." *Arctic Medical Research* 55, supp. 1 (1996): 13–19.

Dobbyn, Paula. "Contaminated Game Has Natives Worried." *Anchorage Daily News,* May 2, 2000. [http://www.adn.com] [http://www.ienearth.org/food_toxic.html]

"Dogrib First Nation Fights Back Diavik's Land Grab." *Drillbits and Tailings* 4:18 (November 9, 1999): n.p. [http://www.moles.org/ProjectUnderground/drillbits/4_18/3.html]

Dougherty, Kevin. "Deal Is No Blank Cheque, Crees Say." *Montreal Gazette*, October 24, 2001. [http://www.canada.com/montreal/montrealgazette/story.asp?id = {4FCA2E44-DDE4-4675-88C5-DEFD94486151}]

Dumont, C. *Proceedings of 1995 Canadian Mercury Network Workshop. Mercury and Health: The James Bay Cree*. Montreal: Cree Board of Health and Social Services, 1995. [http://www.cciw.ca/eman-temp/reports/publications/mercury95/part4.html]

Dumont, Charles, Manon Girard, François Bellavance, and Francine Noël. "Mercury Levels in the Cree Population of James Bay, Quebec, from 1988 to 1993/94." *Canadian Medical Association Journal* 158 (June 2, 1998): 1439–45. [http://www.cma.ca/cmaj/vol-158/issue-11/1439.htm]

Egan, D'Arcy. "Dam the Ecology if Quebec Goes Full-speed Ahead." *Cleveland Plain Dealer*, July 29, 2001, D-15.

Gilday, Cindy Kenny. "A Village of Widows." *Arctic Circle*. N.d. [http://arcticcircle.uconn.edu/SEEJ/Mining/gilday.html]

"Great Whale Still Dead." January 22, 2001. [Portfolio@newswire.ca]

Green, Sara Jean. "Fighting a GIANT." *Windspeaker*, n.d. [http://www.ammsa.com/classroom/CLASS3Lubicon.html]

Grinde, Donald A., Jr., and Bruce E. Johansen. *Ecocide of Native America: Environmental Destruction of Indian Lands and Peoples*. Santa Fe, N.M.: Clear Light, 1995.

Guerette, Deb. "No Clear-cut Answer: Timber Rights Allocation on Lubicon Land a Worrisome Development." *Grande Prairie Daily Herald-Tribune*, March 5, 2001. [http://www.tao.ca/~fol/Pa/negp/ht010305.htm]

Guin, Rhal. "Quebec, Crees Sign Deal." *Toronto Globe and Mail*, October 23, 2001. (in LEXIS)

Hill, Miriam. "Iqaluit's Waste Woes Won't Go Away; City Sets Up Bins Where Residents Can Dump Plastics, Metal." *Nunatsiaq News*, July 27, 2001, 5.

Hulen, David. "Hunt Is on for Pollutant Traces in Bering Sea; Alaska Villagers, Scientists Wonder if Toxic Substances Are Endangering Animals and People Who Eat Them." *Los Angeles Times*, August 15, 1994, A-5.

"Hydroelectric Production Is Anything but Cheap." *Virtual Circle: First Nations' Chronicles*, March 21, 2001. [http://www.vcircle.com/journal/archive/010.shtml]

"Innu Begin Occupation To Halt Construction At Voisey's Bay." *Drillbits and Tailings* (August 21, 1997): n.p. [http://www.moles.org/ProjectUnderground/drillbits/970821/97082104.html]

"The Innu Nation: Background Information." August, 2000. [http://www.ienearth.org/military_impacts.html]

"Innu Nation Launches Court Challenge to Military Plans for Supersonic Test Flights Over Innu Lands." *Indigenous Environmental Network*, August 8, 2000. [http://www.ienearth.org/military_impacts.html

"Innu Respond to N.F. Hydro/Alcoa Announcement." *Cultural Survival News Notes*, July 26, 2001. [http://www.cs.org/main.htm]

Interpress Service. *Story Earth: Native Voices on the Environment*. San Francisco: Mercury House, 1993.

Jacobson, Joseph L., and Sandra W. Jacobson, "Intellectual Impairment in Children Exposed to Polychlorinated Biphenyls *in Utero*," *New England Journal of Medicine* 335:11(September 12, 1996):783–89.

Johansen, Bruce E. "Pristine No More: The Arctic, Where Mother's Milk is Toxic." *The Progressive*, December 2000, 27–29.

LaDuke, Winona. "The Indigenous Women's Network: Our Future, Our Responsibility." Statement of Winona LaDuke, Co-Chair Indigenous Womens Network, Program Director of the Environmental Program at the Seventh Generation Fund, at the United Nations Fourth World Conference on Women, Beijing, China, August 31, 1995. [http://www.igc.org/beijing/plenary/laduke.html]

———. "Insider Essays: Our Responsibility." Electnet/Newswire, October 2, 2001. [http://www.electnet.org/dsp_essay.cfm?intID=28]

———. "Tribal Coalition Dams Hydro-Quebec Project." *Indian Country Today*, July 21, 1993, A-3.

Lalonde, Michelle. "Natives to Appeal Mining Decision." *Montreal Gazette* in *National Post* (Canada), December 17, 2001. [http://www.nationalpost.com/]

Lamb, David Michael. "Toxins in a Fragile Frontier." Transcript, Canadian Broadcasting Corporation News, n.d. [http://cac.ca/news/indepth/north/]

Ljunggren, David. "Effects of Global Warming Clear in Canada Arctic." Reuters News Service. Environmental News Network, Thursday, April 20, 2000. [http://www.enn.com/enn-subsciber-news-archive/2000/04/04202000/reu_arctwarm_12170.asp]

Macafee, Michelle. "Cree Grand Chief Defends Hydro Deal with Quebec as Negotiations Wind Down." Canadian Press in *Ottawa Citizen*, December 18, 2001. [http://www.canada.com/news/story.asp?id = {565BD3CB-8DAB-4CD2-878F-EE2F 7807666B}]

Matteo, Enzo di. "Damned Deal: Cree Leaders Call Hydro Pact Signed in Secret a Monstrous Sellout." *Now Magazine* (Toronto), February 2002. [http://www.now-toronto.com/issues/2002-02-14/news_story.php]

McKeown-Eyssen, G. E., and J. Ruedy. "Methyl-mercury Exposure in Northern Quebec: Neurologic Findings in Adults." *American Journal of Epidemiology* 118 (1983): 461–69.

"Mine Approved for Rich Salmon River Pits Alaska against B.C." Environment News Service, June 14, 2000. [http://ens.lycos.com/ens/jun2000/2000L-06-14-05.html]

Mofina, Rick. "Study Pinpoints Dioxin Origins: Cancer-causing Agents in Arctic Aboriginals' Breast Milk Comes from U.S. and Quebec." *Montreal Gazette*, October 4, 2000, A-12.

"The Mohawk of Kanesatake Resist Nobium Mining." *Drillbits and Tailings* 7:3 (March 29, 2002): n.p. [www.moles.org]

Native Forest Network. "Quebec-Hydro Project May Destroy James Bay." Friends of the Earth, Victoria, British Columbia, 1994. [http://jinx.sistm.unsw.edu.au/~greenlft/1994/137/137p14.htm]

"NATO and Innu Set for Showdown in Eastern Canada." *Native Americas* 12:3 (fall 1995): 10–11.

Nikiforuk, Andrew. "Echoes of the Atomic Age: Cancer Kills Fourteen Aboriginal Uranium Workers." *Calgary Herald*, March 14, 1998, A-1, A-4. [http://www.ccnr.org/deline_deaths.html]

"Northern Quebec Cree Poisoned by Mining Pollutants: Report." Canadian Broadcasting Corporation News On-line, October 20, 2001. [http://cbc.ca/cgi-bin/view?/news/2001/10/19/cree_quebec011019]

P.C.B. Working Group, I.P.E.N. "Communities Respond to P.C.B. Contamination." N.d. [http://www.ipen.org/circumpolar2.html]

"P.D. 2000 Projects." Arctic Monitoring and Research—Project Directory. April 11, 2001. [http://amap.no/pd2000.htm]

Peterson, Diane J. "Two Twin Cities Churches Protest Hydro Power Injustices." *Earth-keeping News* 9:5 (July/August 2000): n.p. [http://www.nacce.org/2000/manitoba.html]

Phillip, Chief Stewart (President, Union of British Columbia Indian Chiefs). "The Criminalization of Skwelkwek'welt Defenders," September 4, 2001.

"Protecting Knowledge: Native Youth Movement Protest at Sun Peaks Resort; Interview with Amanda Soper, spokesperson for the Native Youth Movement, and Chris Rogers, spokesperson for Sun Peaks Resort. CBC Radio (British Columbia), B.C. Almanac, August 24, 2001.

"Quebec Admits Toxic Waste Poisons Cree." Canadian Broadcasting Corporation News On-line, October 26, 2001. [http://cbc.ca/cgi-bin/view?/news/2001/10/26/cree_tox011026]

"Quebec Will Shelve Huge Hydroelectric Project Indefinitely." *Omaha World-Herald*, November 19, 1994, 7.

"Resisting Destruction: Chronology of the Lubicon Crees' Struggle to Survive." N.d. [http://www.lubiconsolidarity.ca/resisting.html]

Roslin, Alex. "Crees Revive Hydro Project." *Montreal Gazette*, January 21, 2000. [http://www.montrealgazette.com/news/pages/010121/5036705.html]

Rozell, Ned. "Alaska Science Forum: Dioxins: Another Uninvited Visitor to the North." Paper presented at Geophysical Institute, University of Alaska–Fairbanks, November 9, 2000. [http://www.gi.alaska.edu/ScienceForum/ASF15/1515.html]

Schetagne, R., J.-F. Doyon, and R. Verdon. *Summary Report: Evolution of Fish-mercury Levels in the La Grande Complex, Quebec (1978–1994)*. Hydro-Quebec, 1997. [http://www.hydroquebec.com/environment/activites/pdf/doc_c2.pdf]

Schneider, Howard. "Facing World's Pollution in the North." *Washington Post*, September 21, 1996, A-15. [http://www.washingtonpost.com/wp-srv/inatl/longterm/canada/stories/pollution092196.htm]

Sierra Club. "Hudson Bay/James Bay Watershed Eco-region." N.d. [http://www.sierraclub.org/ecoregions/hudsonbay.asp]

"Sustainable Development in the Hudson Bay: James Bay Bio-region Canadian Arctic Resources Committee." Paper presented at Environmental Committee of Sanikiluaq, Rawson Academy of Aquatic Science, n.d. [http://www.carc.org/pubs/v19no3/2.htm]

Suzuki, David. "Science Matters: POP Agreement Needed to Eliminate Toxic Chemicals." December 6, 2000. [http://www.davidsuzuki.org/Dr_David_Suzuki/Article_Archives/weekly12060002.asp]

Taylor, Robert. "Cree Leaders Jailed in Protest over Power Deal." *Indian Country Today*, February 17, 2002. [http://www.indiancountry.com/?1013955409]

"Unions Back Lubicons." Indigenous Environmental Network, August 16, 2000. [http://www.ienearth.org/lubicon.html#canada]

Vance, Chris, and John Shafer. "Voices of Resistance." Radio transcript, CFUV 102 FM (community radio station, University of Victoria, British Columbia), 1995; transcribed in 1995 and reprinted in the magazine *All That's Left Is Struggle*. [http://www.finearts.uvic.ca/~vipirg/SISIS/sov/allcree.html]

Watt-Cloutier, Sheila. Personal communication with author, March 28, 2001.

Wheatley, M. A. "The Importance of Social and Cultural Effects of Mercury on Aboriginal Peoples." *Neurotoxicology* 17 (1996): 251–56.

White, Patrick. "Quebec, Cree Reach [Canadian] $3.5 Billion Land-Claim Settlement." Reuters News Service, October 23, 2001.

Zoe, John B. "Diamonds Aren't Our Best Friend: A Negotiator for the Dogrib Nation Says a New Mine in [the Northwest Territories] Will Threaten the Lifeblood of His People." *Toronto Globe and Mail,* November 2, 1999, n.p.

CHAD

PYGMIES: ENCOUNTERING AN OIL PIPELINE

Introduction

An international consortium consisting of Exxon, Shell, and E.L.F. announced plans for a multibillion-dollar oil exploitation project in Chad, a sub-Saharan African country. Indigenous peoples in the area, most notably the Pygmies, and their environmentalist allies have expressed concern that the ecological and social risks of large-scale oil-related development in Chad could cause their homelands to resemble the oil-scarred lands of the Ogoni, in the Niger Delta, to their south (Horta, 1997) (see also chapter on Nigeria, this volume).

Pipeline Construction Begins

During the year 2000, the oil consortium started production in the Doba oil fields of southern Chad, along with construction of a 600-mile pipeline through Cameroon that would transport oil to an Atlantic port for export.

According to the Environmental Defense Fund (EDF), the World Bank has asserted that the project will alleviate poverty, because revenue from the oil for the government of Chad and royalties from use of the pipeline for the government of Cameroon would be invested in poverty programs. "This argument has little credibility," according to one critic, "in view of the demonstrated lack of commitment by either government to alleviate poverty" (Horta, 1997).

In addition, the Doba region in southern Chad is the seat of an ethnic and regional struggle against Chad's central government that is closely associated with similar conflicts in neighboring Sudan. This conflict could explode into violence, imperiling the safety of the pipeline, its workers, and the surrounding environment. Chad has experienced a 30-year history of destabilizing civil war in which the north of the country, mainly Muslim, is arrayed against the country's main agricultural region in the south, home to a majority of Christians and

Animists. According to an EDF analysis, the oil wealth in the south is likely to increase demands among people living in that area for greater autonomy. "The fragile truce at present is at great risk from acts of sabotage and the secessionist movement fueled by resentment of northern control over the oil income," wrote Korinna Horta for EDF (Horta, 1997).

The proposed pipeline would pass through ecologically fragile rainforest areas, including land that is the home of a Pygmy minority of traditional hunters and gatherers. The EDF fears that

> an uncontrollable influx of people in search of work will gather at the construction sites. As a result, deforestation, wildlife poaching, and the loss of farmland of local villages to the construction activities will create a destructive environmental legacy. The pipeline itself, even with state-of-the-art equipment, poses the danger of groundwater contamination and pollution of important regional river systems as crude oil containing heavy metals leaks into the environment. (Horta, 1997)

The oil consortium plans to develop three oil fields in the Doba Basin. Approximately 300 wells would be drilled and production potential is estimated at about 225,000 barrels of oil per day. The pipeline through neighboring Cameroon also would require related infrastructure such as pump and storage stations, about 500 kilometers of road upgrades, and the building of a floating storage and off-loading facility on Cameroon's Atlantic coast at Kribi. Total project costs are estimated at about U.S. $3.5 billion (Horta, 1997).

According to the EDF, by 1997 recent visitors to the area found that several local people already had lost houses and fields to oil-prospecting work, construction of a base camp, and other project-related activities. The newly landless asserted that compensation for what they had lost ranged from nothing to woefully inadequate.

Politics Complicates the Situation

Oil-related politics in Chad are very unstable, raising questions about the proposed pipeline's safety. President Idriss Deby captured 67 percent of the vote in a presidential election May 27, 2001, as six opposition candidates asserted that they lost due to fraud. A meeting of those candidates the next day was broken up by government forces, which detained the candidates as they shot and killed four opposition activists. Deby's political strategy centered around the U.S. $3.7 billion Chad-Cameroon Petroleum Development Project, including a 600-mile-long pipeline and new oil extraction efforts in Chad's Doba region. The project is a joint venture of U.S.-based ExxonMobil and Chevron and Petronas, a Malaysian state-owned company. The financing is being managed by U.S.-based Citigroup ("Bank Funding," 2001). Opponents of Deby, including indigenous residents near the path of the proposed pipeline, were prohibited from airing programs of a political nature on the radio, and two private newspapers that had been critical of Deby were shut down ("Bank Funding," 2001).

On May 30, the 6 opposition candidates and about 40 of their supporters were arrested as they prepared to lead a funeral procession for a man killed during a political demonstration. The same day, the government announced a ban on political gatherings of more than 20 people. As reports of the detentions and the torture of one of leading opposition candidate, Ngarledjy Yorongar, reached solidarity activists, protest marches assembled in Paris and Washington, D.C. Chadian exiles occupied the grounds of their country's embassy to France, while American campaigners flooded the World Bank with calls ("Bank Funding," 2001). Deby then released the other opposition candidates after a call from World Bank President James Wolfensohn. On June 11, a protest group of about 100 women was prevented by soldiers from delivering a letter to the French embassy. Fourteen of the women were injured, four seriously, when the soldiers opened fire with tear-gas canisters and grenades.

REFERENCES

"Bank Funding for Oil Project and Government Repression Both on the Rise in Chad." *Drillbits and Tailings* 6:5 (June 30, 2001): n.p. [http://www.moles.org/Project Underground/drillbits/6_05/1.html]

Horta, Korinna. "Questions Concerning the World Bank and Chad/Cameroon Oil and Pipeline Project: Makings of a New Ogoniland? Corporate Welfare Disguised as Aid to the Poor?" Environmental Defense Fund, New York City, 1997. [http://www.edf.org/pubs/Reports/c_chadcam.html]

CHILE

THE PEHUENCHE AND MAPUCHE: LOGGING, DAM BUILDING, AND LAND RIGHTS

Introduction

During the 1990s, the Spanish corporation ENDESA (Empresa Nacional de Electricidad) began to develop six hydroelectric dams, with a total generating capacity of 2,300 megawatts, on the Biobío River in Chile. The dam-building proposal brought to a head issues related to several intrusions for the Pehuenche and Mapuche, who have long contested logging on their lands, as they seek legal guarantees of their land tenure.

Resisting Dam Building

The Biobío originates in the Icalma and Galletue lakes in the Andes in southern Chile and flows roughly 380 kilometers through forests, agricultural lands, and cities to the Pacific Ocean, supplying water for drinking, irrigation, and fishing to more than a million people. The first dam, the Pangue, was completed in 1996.

Construction of the Pangue dam began in 1990, with financing from the World Bank's private-sector arm, the International Finance Corporation. According to a report in the *World Rainforest Movement Bulletin*, the indigenous Pehuenche, who live in the river's watershed, resisted relocation. During April 1998, during a visit to Santiago, James Wolfensohn, president of the World Bank, admitted that the Bank's support of the Pangue hydroelectric project had been a mistake, and that the Bank had performed bad work during its evaluation of the environmental impact of the project because the Pehuenche had not been consulted.

Wolfensohn's admission did not impede construction of the next dam, the 570-megawatt Ralco, which began during the late 1990s. If it is completed, this dam, 155 meters high with a 3,400-hectare reservoir, would displace more than 600 people, including 400 indigenous Pehuenches. The dam would flood more than 70 kilometers of the river valley, inundating the richly diverse forest and

Chilean President Ricardo Lagos greets Santos Milenao, leader of the Mapuche, during the inauguration of a meeting aimed at outlining a proposal to solve the problems of the Chilean indigenous peoples in 50 days. (NewsCom.com/Notimex)

destroying its biodiversity ("Chile: The Struggle," 2001). The Pehuenche, with support from the Biobío Action Group, sought a court injunction against the dam's construction, enjoining a legal confrontation between the Indigenous Law of 1993, designed to protect the lands of Chile's indigenous peoples, and the Electricity Law passed during Pinochet's regime, which promotes energy generation projects above indigenous rights.

Direct Action to Protect Land

While the legal battle stalled in the courts, the Pehuenche took direct action at the dam site to impede its construction. They refused to abandon their ancestral lands, and rebuked the resettlement plans of ENDESA, which called for the Pehuenche to move high into the Andes, to Huachi and El Barco, where harsh winter conditions would make their survival nearly impossible. A few families who had agreed to relocate complained that promised compensation was late in arriving, as their livestock perished in the harsh highland winter. Firewood and medical assistance also were not supplied as promised ("Chile: The Struggle," 2001).

As the Pehuenche took direct action to protect their land base, the Mapuche, also in Chile, staged a national march for the recognition of their indigenous land rights. During 1999, more than 100 Mapuche walked 700 kilometers from Wallmapuche in Mapuche Territory to Chile's capital, Santiago. Mapuche of all ages, including boys and girls, as well as elderly men and women, braved harsh weather, including heavy rain and near-freezing temperatures, at the beginning of the march. The conditions caused many of the participating Mapuche to fall ill, requiring medical attention. The marchers lodged in schools, churches, and communal centers, sleeping on cement floors and bathing in cold water. Long miles of marching on worn-out shoes injured many of the marchers' feet. Despite these and other problems, the marchers arrived in Santiago during late June.

March participants drew attention to the Mapuches' efforts to recover ancestral lands occupied by national and transnational forestry corporations, non-Mapuche individuals, and the Chilean government. The Mapuche sought not only constitutional guarantees of their land rights, but also rights to political self-determination, including establishment of an Autonomous Mapuche Parliament. The Mapuche activists also sought "to denounce the massive presence of transnational forestry corporations which operate under neoliberal policies which impinge the collective rights of the Mapuche People and to demand their withdrawal from the Mapuche Territory" ("Chile: Mapuche," 1999).

On March 8, 2002, Pehuenche Indians interrupted a ceremony attended by Chilean President Ricardo Lagos to protest the Ralco dam. The dam site also was the focus of violent clashes between police and Pehuenche Indians earlier the same week. Three days before the protest in Santiago, police arrested 55 protesters near Ralco in what their lawyer said was a heavy-handed operation reminiscent of the security forces' tactics during ex-dictator Augusto Pinochet's 17-year rule. According to Reuters, "A group of Pehuenches, a sub-group of the larger Mapuche Indian population believed to number around 1 million, had blocked a road to stop a transformer from reaching Ralco, which is 310 miles (500 kilometers) south of Santiago. Protesters hurled rocks at police in riot gear who had fired pellets at them" ("Indians Take," 2002).

At the ceremony, according to a Reuters report, "a woman dressed in Indian garb grabbed the microphone from Minister for Women's Affairs Adriana Delpiano as she was making a speech on a platform outside La Moneda presidential palace to mark International Women's Day. Witnesses said the protester shouted, 'No to Ralco,' as Lagos, sitting cross-legged, looked on calmly two yards away. Presidential security officers carried the protester away as Indian women continued to shout slogans" ("Indians Take," 2002).

REFERENCES

"Chile: Mapuche Indigenous Peoples' March to the Capital City." *World Rainforest Movement Bulletin* 23 (May 1999): n.p. [http://www.wrm.org.uy/bulletin/23/Chile.html]
"Chile: The Struggle of the Pehuenche against the Ralco Dam." *World Rainforest Movement Bulletin* 42 (January 2001): n.p. [http://www.wrm.org.uy/bulletin/42/Chile.html]
"Indians Take Dam Protest to Chilean President." Reuters, March 8, 2002.

CLIMATE CHANGE AND INDIGENOUS ENVIRONMENTALISM

INTRODUCTION

Many indigenous peoples are very directly affected by global warming wrought by increases in atmospheric levels of carbon dioxide, methane, and other trace gases. Ironically, indigenous peoples often live outside the industrial structure that produces the gases. The following analysis includes a general statement by a worldwide consortium of indigenous peoples addressing global warming, followed by two case studies. The first study concentrates on the Inuit of Nunavut, in the Canadian Arctic, who live in one of the most swiftly warming areas on Earth. Climate observations early in the twenty-first century confirmed the predictions of many climate models: warming will take place most dramatically in the Arctic. Inuit hunters confirm that their lives are being changed quickly. The second case study, "Waiting to Drown," looks at how global warming alters the lives of indigenous peoples who live on small islands, particularly in the Pacific Ocean, which are imperiled by waters that are rising because of melting ice elsewhere on Earth.

THE FIRST INTERNATIONAL FORUM OF INDIGENOUS PEOPLES ON CLIMATE CHANGE

The Declaration of the First International Forum of Indigenous Peoples on Climate Change was signed by indigenous peoples' local community representatives present at the United Nations Framework Convention on Climate Change Subsidiary Bodies meetings in Lyon, France on September 8, 2000. The Declaration, partially excerpted below, provides a sense of worldwide consensus regarding global warming from indigenous peoples.

At long last, the international community has been forced to recognize that climate change threatens the very survival of humanity. Despite the recognition of our role in preventing global warming, when it comes time to sign international conventions like the United Nations Framework Convention on Climate

Change, once again, our right to participate in national and international discussions that directly affect our Peoples and territories is denied.

Our active opposition to oil exploration, logging and mining helps prevent the accelerated deterioration of the climate. Nonetheless, our territories have been handed over to national and multinational corporations that exploit our natural resources in an indiscriminate and unsustainable fashion.

Before the signing of the Kyoto Protocol, we had already made concrete political contributions to mitigating climate change. For example, indigenous peoples of the Amazon forged a mutually beneficial alliance with European Cities in [a] joint program of the Climate Alliance, the Coordinator of Indigenous Organizations of the Amazon Basin (COICA) and the International Alliance of Indigenous and Tribal Peoples of the Tropical Forests.

Our intrinsic relation with Mother Earth obliges us to oppose the inclusion of sinks in the Clean Development Mechanism [C.D.M.] because it reduces our sacred land and territories to mere carbon sequestration which is contrary to our cosmovision and philosophy of life. Sinks in the C.D.M. would constitute a worldwide strategy for expropriating our lands and territories and violating our fundamental rights that would culminate in a new form of colonialism. Sinks in the C.D.M. would not help to reduce greenhouse-gas emissions, rather it would provide industrialized countries with a ploy to avoid reducing their emissions at [their] sources. We emphatically oppose the inclusion of sinks, plantations, nuclear power, megahydroelectric and coal. Furthermore, we oppose the development of a carbon market that would broaden the scope of globalization.

We have historically and continue to play a fundamental role in the conservation and protection of the forests, biological diversity and the maintenance of ecosystems crucial for the prevention of severe climatic change. Long ago, our elders and our sciences foretold of the severe impacts of Western "development" models based on indiscriminate logging, oil exploitation, mining, carbon-emitting industries, persistent organic pollutants and the insatiable consumption patterns of the industrialized countries. Today, these unsustainable models threaten the very life of Mother Earth and the lives of all of us who are her children.

We denounce the fact that neither the [United Nations] nor the Kyoto Protocol recognizes the existence or the contributions of Indigenous Peoples. Furthermore, the debates under these instruments have not considered the suggestions and proposals of the Indigenous Peoples nor have the appropriate mechanisms to guarantee our participation in all the debates that directly concern the Indigenous Peoples been established. In this declaration, we address the Parties and other participants of this Conference to present the conclusions of our Forum.

We have the obligation to inform the international community about our grave concern regarding the social, cultural, economic and security threats posed by climate change to indigenous peoples and local communities living in small island states. Given the extreme urgency of the need for adaptation activities in small island states, we urge that an Adaptation Fund be immediately established and activated with the full participation of Indigenous Peoples and Local Communities, even if the Kyoto Protocol is not ratified. . . . [W]e are also particularly concerned about the emergence of "biocolonialism" and "environmental racism" that Indigenous Peoples and Local communities of the world continually confront. (Declaration of the First International Forum of Indigenous Peoples on Climate Change, 2000)

CLIMATE CHANGE AND THE INUIT (ARCTIC)

Climate change in the Arctic arrived swiftly and dramatically during the 1990s, including such surprises as warm summer days, freezing winter rain, thunderstorms, and invasions of Inuit villages by heretofore unseen insects and birds.

While George W. Bush said he could find no sound science to support the theory of global warming, the temperature hit 26 degrees Celsius on July 28, 2001 in Iqaluit (pronounced "Eehalooeet"), a Baffin Island community that nudges the Arctic Circle, following a string of days that were nearly as warm. It was the warmest summer anyone in the area could remember. Travelers joked about forgetting their shorts, sunscreen, and mosquito repellent—all now necessary equipment for a globally warmed Arctic summer.

In Iqaluit, a warm, desiccating westerly wind raised whitecaps on nearby Frobisher Bay and rustled carpets of purple Saxifrage flowers as people emerged from their overheated houses (which have been built to absorb every scrap of passive solar energy), swabbing their foreheads with ice cubes wrapped in hand towels. The temperature, at 78 degrees Fahrenheit, was 25 degrees above the July average of 53, comparable to a 105- to 110-degree day in New York City. The wind raised eddies of dust on Iqaluit's gravel roads as residents swatted slow, corpulent mosquitoes.

Around the Arctic, in Inuit villages now connected by the oral history of traveling hunters as well as by e-mail, weather watchers are reporting striking evidence that global warming is an unmistakable reality. Sachs Harbour, on Banks Island, above the Arctic Circle, is sinking into the permafrost as its 130 residents swat mosquitoes. Summer downpours of rain with thunder, hail, and lightning have swept over Arctic islands for the first time in anyone's memory. Swallows, sand flies, robins, and pine pollen are being seen and experienced by people who have never known them. Shishmaref, an Inuit village on the far-western lip of Alaska, 60 miles north of Nome, is being washed into the newly liquid (and often stormy) Arctic Ocean as its permafrost base dissolves.

By the turn of the millennium, global warming was felt most intensely in the Arctic, where a world based on ice and snow was melting away. In Iqaluit, capital of the semisovereign Nunavut Territory, people spoke of a natural world turned upside down in startling ways. "We have never seen anything like this. It's scary, very scary" said Ben Kovic, Nunavut's chief wildlife manager. "It's not every summer that we run around in our T-shirts for weeks at a time" (Johansen, 2001, 18–19). The weather reports in Iqaluit and other parts of the Arctic read like the projections of the Intergovernmental Panel on Climate Change—on fast-forward.

At 11:30 A.M. on a midsummer Saturday, Kovic was sitting in his backyard, repairing his fishing boat, wearing a T-shirt and blue jeans in the warm wind, with many hours of Baffin Island's 18-hour July day remaining. On a nearby beach, down the hill, Inuit children were building sandcastles with plastic shovels and buckets, occasionally dipping their toes in the still-frigid seawater.

Warming has extended to all seasons. In Iqaluit, for example, thunder and lightning used to be an extreme rarity. Thunderstorms are now much more common across the Arctic; the day the high hit 78 degrees Fahrenheit in Iqaluit, the forecast for Yellowknife, in the Northwest Territories, called for a high of 85 degrees Fahrenheit with scattered thunderstorms. The previous winter in Iqaluit had been notable for liquid precipitation (freezing rain) in December. Snows in the area also had become heavier and wetter than previously. Winter cold spells, which still occur, have generally become shorter, according to longtime residents of the area.

As a wildlife officer, Kovic has seen changes that scare him. Polar bears, for example, are "often becoming shore dwellers rather than ice dwellers," sometimes with dire consequences for unwary tourists (Johansen, 2001, 19). The harbor ice at Iqaluit did not form in the year 2000 until December, five or six weeks later than usual. The ice also has been breaking up earlier in the spring—sometimes in late May in places that once were icebound into early July.

Polar bears usually obtain their food (seals, for example) from the ice. Without it, they can become hungry, miserable creatures, especially during spells of unaccustomed warmth. During Iqaluit's weeks of record warmth in July 2001, two tourists were hospitalized after they were mauled by a bear in a park south of town. On July 20, a similar confrontation occurred in northern Labrador as a polar bear tried to claw its way into a tent occupied by a group of Dutch tourists. That time, the tourists escaped injury but the bear was shot to death. "The bears are looking for a cooler place," said Kovic (Johansen, 2001, 19).

Until recently, polar bears had their own food sources and usually went about their business without trying to steal food from humans. Beset by late freezes and early thaws, however, hungry polar bears are coming into contact with people more frequently. In Churchill, Manitoba, polar bears waking from their winter's slumber have found Hudson Bay's ice melted earlier than usual. Instead of making their way onto the ice in search of seals, the bears walk along the coast until they get to Churchill, where they block motor traffic and pillage the town dump. Churchill now has a holding tank for wayward polar bears that is larger than its human jail.

Canadian Wildlife Service scientists reported during December 1998, that polar bears around Hudson Bay were 90 to 220 pounds lighter than 30 years ago, apparently because earlier melting of ice has given them less time to feed on seal pups. When sea ice fails to reach a particular area, the entire ecological cycle is disrupted. When the ice melts, the polar bears can no longer use it to hunt for ringed seals, many of which also have died, having had no ice on which to haul out. The offshore, ice-based ecosystem is sustained by upwelling nutrients that feed the plankton, shrimp, and other small organisms, which in turn feed the fish, which feed the seals, which feed the bears. The native people of the area also occupy a position in this cycle of life. When the ice is not present, the entire cycle collapses.

The Arctic's rapid thaw has made hunting, never a safe or easy way of life, even more difficult. Hunters in and around Iqaluit said that the weather has been seriously out of whack since roughly the mid-1990s. Simon Nattaq, an Inuit hunter, fell through unusually thin ice and became mired in icy water long enough to lose both his legs to hypothermia, one of several injuries and deaths reported around the Arctic recently due to thinning ice.

Pitseolak Alainga, another Iqaluit-based hunter, said that climate change compels caution. One must never hunt alone, he said (Nattaq had been hunting by himself). Before venturing onto ice in fall or spring, hunters should test its stability with a harpoon, he said. Alainga knows the value of safety on the water. His father and five other men died during October 1994, after an unexpected late-October ice-and-snow storm swamped their hunting boat. The younger Alainga and one other companion barely escaped death in the same storm. He believes that more hunters are suffering injuries not only because of climate change, but also because basic survival skills have not been passed from generation to generation as they were when most Inuit lived on the land.

Within two or three generations, many Inuit have become urbanized as the tendrils of industrial life infiltrate the Arctic. Where visitors once arrived by dog sled or sailing ship, they now stream into Iqaluit's busy airport on Boeing 727s in which half the passenger cabin has been sequestered for freight. Because there exist no road connections to the outside, freight as large as automobiles is sometimes shipped to Iqaluit by air.

The population of Iqaluit jumped from about 3,500 to 6,500 in less than three years. Substantial suburban-style houses with mortgages worth hundreds of thousands of dollars have sprung up around town, rising on stakes sunk into the permafrost and granite hillsides. In other areas, ranks of walk-up apartments march along the high ridges above Frobisher Bay. Every ounce of building material for both has been imported from thousands of miles away. Inuit have been moving into the town from the backcountry, and the town is awash in a sea of children. People in Iqaluit subscribe to the same cable television services (as well as the Internet) available in the so-called South. Bart Simpson and Tom Brokaw are well-known personages in Iqaluit, where some homes have sprouted satellite dishes. Iqaluit also now hosts a large supermarket of a size that matches stores in larger urban areas, except that the prices are three to four times higher than in Ottawa or Omaha. If one can afford it, mango-grapefruit juice and ready-cooked buffalo wings (as well as many other items of standard Southern fare) are available.

Climate change has been rapid, and easily detectable within a single human lifetime. "When I was a child," said Sheila Watt-Cloutier, Canadian president of the Inuit Circumpolar Conference, "we never swam in the rivers of my homeland, Nunavik [northern Quebec]. Today, kids swim in those rivers every summer" (Johansen, 2001, 19–20).

Some of the rivers to which Arctic Char return for spawning have dried up, according to Kovic; the summer of 2001 brought drought, as well as record warmth to Iqaluit and its hinterland. Flying above glaciers in the area, Kovic has noticed that their coloration has changed. "The glaciers are turning brown," he said, speculating that melting ice may be exposing debris, and that air pollution may be a factor. Some ringed seals have been caught with little or no hair, said Kovic. Asked why seals have been losing their hair, Kovic answered: "That is a big question that someone has to answer" (Johansen, 2001, 20).

In 1993, 3,000 Peary caribou lived on Bathurst Island in the High Arctic. By 1997, all but 75 of them had died, after three unusual winters during which heavy, wet snows and frequent freezing rain sealed the caribous' food under "a nearly impenetrable crust" (Jaimet, 2000, A-4). During ordinary winters, the Peary caribou, which are notable for their small size and large hooves, use their hooves to scatter snow in search of buried vegetation. Climate models foresee heavier, icier precipitation in the Arctic. The suffering falls most heavily on the caribou calves, whose mothers have been unable to feed them. Caribou also have been killed by falling through unusually thin ice.

In the Alaskan Eskimo village of Shishmaref, near Siberia, in the far western reaches of Alaska, 600 people have been watching their village erode into the sea. The permafrost that had reinforced its coast is thawing. "We stand on the island's edge and see the remains of houses falling into the sea," people in the town told Anton Antonowicz of the *London Daily Mirror*. He continued: "They are the homes of poor people. Half-torn rooms with few luxuries. A few photographs, some abandoned cooking pots. Some battered suitcases" (Antonowicz, 2000, 8). Percy Nayokpuk, a village elder, owned a local retail store that by 2001 perched dangerously close to the edge of the advancing sea. "When I was a teenager, the beach stretched at least 50 yards further out," said Percy, who was 48 years of age in the year 2000. "As each year passes, the sea's approach seems faster" (Antonowicz, 2000, 8). Five houses had washed into the sea by the year 2000; the U.S. Army had moved or jacked up others. The villagers also had been told they will soon have to move.

Year by year, the hunting season, which depends on the arrival of the ice, starts later and ends earlier. "Instead of dog mushing, we have dog slushing." said Clifford Weyiouanna, 58, a reindeer herder (Antonowicz, 2000, 8). Villagers have been catching fish, such as flounder, which are usually associated with warmer water. In Alaska as a whole, where 80 percent of glaciers are receding, forests of dead spruce surround Anchorage, a casualty of a spruce-beetle epidemic caused at least in part by rising temperatures, which accelerate the insect's life cycle.

Mosquitoes have come to Banks Island, along with various species of beetles. The insects now arrive earlier and stay longer. "We can't read the weather like we used to," said Rosemary Kuptana, one of about 130 Inuit who live in the Inuvialuit community of Sachs Habour, the only community on Banks Island (Johansen, 2001, 20). Kuptana, 47, grew up in Sachs Harbor and raised three children there. She said that autumn freezes now occur a month later than in previous years; spring thaws come earlier as well. Residents of Sachs Harbour still suffer through winters that most people from lower latitudes would find chilling, with temperatures of minus 40 degrees Fahrenheit or lower. While such temperatures once were commonplace, however, they now are rare. "What happens in the Arctic environment is what is in store for other regions of the world," said Kuptana (Duffy, 2000).

"The permafrost is melting at an alarming rate," said Kuptana (Herbert, 2000). Foundations of homes in Sachs Harbour are cracking and shifting. Melting seaside permafrost, which is carrying Shishmaref to the sea, also threatens Sachs Harbour. Kuptana expressed fear that her community would slide into the Beaufort Sea because once-solid mud is now thawing earlier and freezing later. "What's scary is the uncertainty," she said. "We don't know when to travel on the ice and our food sources are getting farther and farther away" (Herbert, 2000).

At Sachs Harbour, sea ice is thinner and now drifts far away during the summer, taking with it the seals and polar bears upon which the village's Inuit residents rely for food. Young seals are starving to death because melting sea ice separates them from their mothers. In the winter the sea ice often is thin and broken, making travel dangerous for even the most experienced hunters. In the fall, storms have become more frequent and severe, making boating difficult. Thunder and lightning have been seen for the first time, arriving with another type of weather that is new to the area—dousing summer rainstorms. "When I was a child, I never heard thunder or saw lightning, but in the last few years we've had thunder and lightning," Kuptana said. "The animals really don't know what to do because they've never experienced this kind of phenomenon" ("Thunderstorms," 2000).

"We have no other sources of food, the people in my community are completely dependent on hunting, trapping and fishing," said Kuptana. "We have no means of adapting to a different environmental reality, and that is why our situation is so critical" (Johansen, 2001, 20). In 1990, caribou migrating to the coastal plain of northern Alaska found that the earliest spring in nearly 40 years had caused their principal forage to go to seed, depriving them of crucial nourishment (Rauber, 1997). "We don't know when to travel on the ice and our food sources are getting further and further away," said Kuptana. "Our way of life is being permanently altered" (Knight, 2000).

Yupik elders in the small coastal Alaskan village of Kipnuk believe that their village is sinking due to thawing permafrost. Buildings in the village show signs of an unstable ground surface, and also signs of subsurface thawing. In Kotzebue, Alaska, the town hospital was relocated because it was sinking into the ground ("Threat," 1997) Rising seas and coastal erosion directly threaten Tuktoyaktuk, a Dene and Inuit community located at the edge of the Arctic Ocean. Ice that once protected the coast has receded out to sea. Extensive erosion washed away the school, and has forced the village to relocate many other structures.

In Arctic Village, Alaska, home to the Gwichen people, "Summers are hotter, winter days when the temperature plunges far below zero are fewer, and the timberline is creeping higher up the mountainsides" (Knickerbocker, 2001, 9).

Caribous are giving birth before they reach the coastal plain, which reduces survival chances of their young. Lakes in which people recall having hunted ducks into the fall are now gone (dried up) by mid-July. "Everything is out of whack," said one resident who was raised in Arctic Village. "Last year, in the village, it only went below minus 40 degrees [Fahrenheit] once." "I don't remember it being this hot when I was a kid," said another resident, Gideon James, during the last week of July. He was wearing a tank top and shorts as temperatures neared 80 degrees Fahrenheit (Knickerbocker, 2001, 9).

During early December 2001, Sheila Watt-Cloutier e-mailed news that the ice had just formed on Frobisher Bay within sight of her home, several weeks later than usual. Two weeks later, on December 18, she e-mailed again, to say that freezing rain was falling in Iqaluit. Snow had been unusually scarce. She said that now the streets of Iqaluit were streaked with ice, a real novelty in the Arctic during December. "The hunters," she wrote, "will not be pleased. . . . We have been hoping for blizzards and snowfall but certainly not rain. . . . The ice will get messy and more dangerous for the seal hunters and fishers. . . . Indeed we are living climate change already. It is not in the future!" (Watt-Cloutier, 2001).

CLIMATE CHANGE AND SMALL ISLAND NATIONS: WAITING TO DROWN

On many small islands around the world, rising seas are everyday news, and global warming is on everyone's lips. Nicholas Kristof of the *New York Times* described how Teunaia Abeta, a resident of Kiribati Island in the Pacific, watched in horror as

> a high tide came rolling in from the turquoise lagoon and did not stop. There was no typhoon, no rain, no wind, just an eerie rising tide that lapped higher and higher, swallowing up Abeta's thatched-roof home and scores of others in this Pacific Island nation. "This had never happened before," said Abeta, 73, who wore only his colorful lava-lava, a skirt-like garment, as he sat on the raised platform of his home fingering a home-rolled cigarette. "It was never like this when I was a boy." (Kristof, 1997)

Two islands in the Pacific Ocean's Kiribati archipelago have vanished beneath the rising seas, as 16 percent of the world's coral reefs died in just nine months in 1998 when water temperatures in parts of the tropics exceeded all records.

The Alliance of Small Island States (AOSIS), including the Philippines, Jamaica, the Marshall Islands, the Bahamas, Samoa, and others, has been one of few voices at climate talks favoring swift, worldwide reductions in greenhouse-gas emissions. Their self-interest is evident: with warming already forced, but not yet fully worked into the world's temperature equilibrium, many small island states will lose substantial territory and economic base within the next few decades.

For every other country on Earth, competing economic interests drive climate negotiations. For the small island nations, the driving force is survival. "For us, it's a matter of death and life, whereas in terms of the citizens of the industrialized countries, [global warming] will affect their lifestyles basically, but not to the extent they will be disappearing," said Bikenibeu Paeniu, Tuvalu's prime minister (Webb, 1998).

In no other place is global warming more urgent an issue, in a practical sense, than in equatorial Kiribati (pronounced "Kirabas"), a chain of coral atolls in the Pacific, near the Marshall Islands. Kiribati is a republic, a member of the United Nations, and home to about 79,000 people. Its land area, at any point, is no more than two meters above sea level. Additionally, the living corals that have raised the islands above sea level are threatened with demise by warmer ocean waters. Ierimea Tabai, who became president of Kiribati after its independence from Britain in 1979, has said: "If the greenhouse effect raises sea levels by one meter, it will eventually do away with Kiribati. In fifty or sixty years, my country will not be here" (Webb, 1998).

Kiribati's 33 islands, comprising 277 square miles, are scattered across 5.2 million square kilometers (2 million square miles) of ocean, including three groups of islands: 17 Gilbert Islands, eight Line Islands, and eight Phoenix Islands. Kiribati includes Kiritimati Island (formerly called Christmas Island), the world's largest coral atoll (150 square miles). The island of Kiribati, located roughly halfway between California and Australia, more than doubles its surface area at low tide.

During 1997, the area was devastated by El Niño, which brought heavy rainfall, a half-meter rise in sea level, and extensive flooding. About 40 percent of the atolls' coral was killed by overheated water, and nearly all of Kiritimati Island's roughly 14 million birds died or deserted the island.

Sea-level rise provoked by global warming could imperil more than 3,000 small isolated islands, grouped into 24 political entities in the Pacific Ocean, with a population of about 5 million people in 800 distinct cultures. Many coral atolls also contain permanent areas of fresh water (lagoons), which are vulnerable to saltwater intrusion as sea levels rise.

Central lagoons of the small islands are social centers and a source of food. According to Kristof, "Children play in the water from infancy. Many adults fish or sail for a living. The Kiribati men in their loincloths go out in outrigger canoes each day to catch tuna, and the women wade out on the coral reef to dive for shellfish and net smaller fish for dinner" (Kristof, 1997). "People here think of the ocean as their source of livelihood, as their friend," said Ross Terubea, a Kiribati radio reporter. "It's hard to think that it would destroy us" (Kristof, 1997).

By 1998, rising sea levels already were swallowing some small Pacific islands and contaminating drinking water on others. The rising sea has endangered sacred sites and drowned some small islands near Kiribati and Tuvalu, including the islet of Tebua Tarawa, once a landmark for Tuvalu fishermen. Kiribati already has moved some roads inland on its main island as the rising Pacific Ocean eats into its shores. Rising sea levels are seeping into soils on some islands, as water tables rise. Soils in some areas are becoming too salty for growing most vegetables. In Tuvalu, according to a dispatch from Reuters, farmers "are beginning to grow their taro crops in tin containers filled with compost instead of traditional pits" (Webb, 1998). Soil contamination is also an issue in the Bahamas, where fear has been expressed that the limestone that underlies the soil on many islands will absorb saline seawater like sponges.

Given expectations that global temperatures will rise further in coming decades, the sea-level rise of the twentieth century is expected to be dwarfed by the rise during the twenty-first century. Scientists who attend to this issue typically estimate sea-level rise possibilities in wide ranges, because no one knows how much ice will melt given a certain rise in temperatures over the poles. A team led by J. J. Wells Hoffman in 1983, for example, estimated that sea levels in the year 2100 would be between 58 and 368 centimeters higher than they were in the 1980s (Hoffman and Titus, 1983). R. Thomas, in 1986, put the estimated range at 56 to 345 centimeters (Thomas, 1986). The estimates vary by a factor of roughly seven, an indication of the uncertainty that plagues forecasts of how much the sea will rise under various assumptions about global warming.

In Tuvalu, a chain of nine Pacific coral atolls midway between Hawaii and Australia—with 10,600 people living on 16 square miles of dry land, on which the highest point is 15 feet above sea level—bridges and burial grounds have subsided under the advancing ocean. Tuvalu's authorities are seeking international help to move the entire population. Most may move to New Zealand during the next decade. During the year 2000, "Tuvalu's late prime minister, Ionatana Ionatana, told his New Zealand counterpart, Helen Clark, that homes for at least 3,000 people—and possibly the whole population—would be needed in five to 10 years" (Dutter, 2001, 8). Roughly 5,000 Tuvaluans already live in New Zealand. Teleke P. Lauti, Tuvalu's assistant minister of natural resources and environment, said that "when a cyclone hits us, there is no place to escape. We cannot climb any mountains or move away to take refuge" (Dutter, 2001, 8).

About 1,500 of the world's leading marine biologists met during October 2000 at an international symposium about coral reefs in Bali, Indonesia. One assessment said that between 50 and 90 percent of the coral reefs stretching from the east coast of Africa to the west coast of India, including the Maldives and the Seychelles, are now dead or dying. "There is a very good probability that coral reefs as we know them now will be gone in 30 to 50 years," said Ove Hoegh-Guldberg, who calculated that 26 percent of the world's coral reefs already have been destroyed (Connor and Ecott, 2000, 19). The Maldives, an array of 1,200 islands in the Indian Ocean, is home to 270,000 people. Eighty percent of the Maldives' land is less than one meter above sea level.

REFERENCES

Antonowicz, Anton. "Baking Alaska: As World Leaders Bicker, Global Warming Is Killing a Way of Life." *London Daily Mirror,* November 28, 2000, 8–9.

The Bonn Declaration. Third International Forum of Indigenous Peoples and Local Communities on Climate Change, July 14–15, 2001, Bonn, Germany. [http://www.ienearth.org/climate_1-p2.html#bonn2001]

Connor, Steve, and Tim Ecott. "Islanders Threatened by Coral Destruction." *London Independent,* October 26, 2000, 19.

Declaration of the First International Forum of Indigenous Peoples on Climate Change, Lyon, France, September 8, 2000. [http://www.ienearth.org/climate_1-p2.html]

Duffy, Andrew. "Global Warming Causing Arctic Town to Sink, Says Inuit Leader 'Warning Signal.'" *Montreal Gazette,* April 18, 2000. [http://www.climateark.org/articles/2000/2nd/glwatosi.htm]

Dutter, Barbie. "Islanders Plan Their Flight from Rising Sea." *London Daily Telegraph,* July 16, 2001, 8.

Hahn, Tamar. "Inuit Activist Puts Face to Global Warming." Earth Times News Service, November 16, 2000. [http://www.earthtimes.org/nov/profileinuitactivistputsnov16_00.htm]

Herbert, H. Josef. "Inuit Say They Are Witness to Global Warming in the Arctic." Associated Press, *Milwaukee Journal-Sentinel,* November 19, 2000. [http://www.jsonline.com/alive/news/nov00/warming20111900.asp?format=print]

Hoffman, J. J. Wells, and J. Titus. *Projecting Future Sea-level Rise.* Washington, D.C.: Environmental Protection Agency, 1983.

Jaimet, Kate. "Global Warming 'Lethal' to Rare Northern Caribou." *Ottawa Citizen,* November 30, 2000, A-4.

Johansen, Bruce E. "Arctic Heat Wave." *The Progressive,* October 2001, 18–20.

Knickerbocker, Brad. "To Some Here, Global Warming Seems Real Today." *Christian Science Monitor,* August 1, 2001, 9.

Knight, Danielle. "Inuits Tell Negotiators of Climate Change Impact." Interpress Service, November 16, 2000. [http://www.oneworld.org/ips2/nov00/01_44_005.html]

Kristof, Nicholas. "For Pacific Islanders, Global Warming Is No Idle Threat." *New York Times,* March 2, 1997. [http://sierraactivist.org/library/990629/islanders.html]

Lalonde, Michelle. "Natives to Appeal Mining Decision." *Montreal Gazette,* December 17, 2001. [http://www.nationalpost.com/]

Rauber, Paul. "Heat Wave." September, 1997. [http://www.sierraclub.org/sierra/199709/HeatWave.html]

Thomas, R. "Future Sea-level Rise and its Early Detection by Satellite Remote Sensing." In *Effects of Changes in Atmospheric Ozone and Global Climate,* vol. 4. New York: United Nations Environment Programme/United States Environmental Protection Agency, 1986.

"The Threat of Climate Change to Arctic Human Communities." Greenpeace, 1997. [http://xs2.greenpeace.org/~comms/97/arctic/library/region/people.html]

"Thunderstorms Are Latest Evidence of Climate Change." Associated Press Canada, November 15, 2000. [http://abcnews.go.com/sections/science/DailyNews/arctic_thunder001115.html]

Watt-Cloutier, Sheila. Personal communication with author, December 3, 2001.

Webb, Jason. "Small Islands Say Global Warming Hurting Them Now." Reuters News Service, 1998. [http://bonanza.lter.uaf.edu/~davev/nrm304/glbxnews.htm]

COLOMBIA

INTRODUCTION

The roughly 5,000 U'wa, an indigenous people living in the Colombian highlands, have engaged in one of the cardinal conflicts in indigenous environmentalism by threatening to commit mass suicide if Occidental Petroleum exploits
oil on their homelands. The U'was' ultimatum resonated during the 2000 presidential campaign in the United States, during which Al Gore was accused by
environmental activists of betraying his environmentalist credentials by holding a sizable block of Occidental stock inherited from his father. The U'was'
message also hung over Occidental's annual stockholders' meetings in Los
Angeles. As of this writing, however, Occidental has avoided a public-relations
nightmare by failing to find enough oil on U'wa land to drill commercially.
Finally, by May 2002, Occidental withdrew from the area.

In the meantime, Colombia continues to experience environmental conflicts with native peoples on several other fronts, including widespread indigenous opposition to U.S.–financed spraying of cocaine crops with toxic
herbicides, coal strip mining, and construction of hydroelectric dams.

THE U'WA: MASS SUICIDE AS AN ALTERNATIVE TO OIL EXPLORATION

Five thousand U'wa, a seminomadic native tribe in Colombia's mountains,
have stopped exploration for oil in their homeland by threatening to walk, en
masse, off a 1,400-foot cliff to their deaths. Occidental Petroleum seems to
have been stymied, at least for a time, by the ultimatum and also by its inability to find commercially exploitable oil deposits in the area.

The threatened mass suicide became an issue at Occidental's 1997 shareholders' meeting in Santa Monica, California, as a representative of the U'wa
toured the United States in the company of several environmental groups.
"This oil belongs to the land, and cannot be taken from it," Roberto Cobaria
told audiences in the United States (Johansen, 1998, 5). Occidental, which
was using exploration permits granted by Colombia's government, asserted

that the U'wa have been forced to risk their lives by local antigovernment guerrillas who are looking for publicity. Occidental's spin doctors had not studied the U'was' history, however. Four centuries ago, a portion of the tribe jumped off a sacred cliff rather than submit themselves to Spanish colonial rule.

Occidental already operates the Cano oil field in Colombia, which delivers an average of 180,000 barrels of oil a day. This field, which makes Colombia self-sufficient in oil, is being depleted, thus the government's interest in opening a new oil field at Samore, in the region that the U'wa call home. The government also is looking for a way to control the countryside, where its authority is challenged not only by political opposition and environmental activism, but also by drug-inspired criminal activity.

Under Colombian law, the U'wa have no claim to the area in which Occidental wishes to drill. The land lies outside reservation boundaries as defined by the central government, but inside territory utilized by the U'wa as they migrate. A Colombian administrative-law judge dismissed the U'was' claim to the land in 1997 under a legal doctrine that sanctions the government's claim to all mineral rights within its borders. Exploration rights to the U'wa territory once included Shell Oil as a partner with Occidental. In 1998, however, Shell announced the intention to sell its 37.5 percent share in the Samore area.

Shell also sold to Occidental its interests in another Colombian field, the Cano Limon Project, whose pipeline was bombed by guerillas 65 times during 1997, costing the company $85.6 million in lost revenues and spilling more than 200,000 barrels of oil. The number of bombings rose to 99 in 2000 and 170 in 2001. The bombings cut Cano Limon's oil production to 19 million barrels in 2001, down 58 percent from 1999's level. Rebels have stolen employees' cars and murdered two oil-company engineers. Oil workers' travel outside fenced compounds was, by 2001, restricted to helicopters. By 2001, the drilling of new wells in the Cano Limon area was carried out by convoys with Colombian army air support. Occidental has maintained its operations in the area because the oil is relatively inexpensive to extract, at $10 a barrel (Barrionuevo and Herrick, 2002, A-1).

Colombian oil developments are an increasingly popular target of guerrillas, which increases risk to the environment. At the Cano Limon pump station operated nearby in Colombia by Occidental and Shell, roughly 1.5 million barrels of crude oil have been spilled into the rainforest during the last decade. (By comparison, the Exxon Valdez spill along the Alaska coast was 36,000 barrels.) Much of the spillage has been due to sabotage by guerrillas who represent another major threat to the U'wa. In response, Occidental and Shell pay a war tax of $1 per barrel (about $180,000 a day) to the Colombian military in return for its protection of their installations at Cano Limon. The Colombian military is known for its human rights abuses, and militarizing the U'was' territory will introduce organized violence into the area.

One U'wa woman told the *Guardian* on September 20, 1997, "I sing the traditional songs to my children. I teach them that everything is sacred and linked. How can I tell Shell and Oxy that to take petrol is for us worse than killing your own mother? If you kill the earth, then no one will live. I do not want to die. Nobody does" (Johansen, 1998, 5). The U'wa have taken their case to the Organization of American States by petitioning the Inter-American Human Rights Commission. Martin Wagner of the Earth Justice Legal Defense Fund represented the U'wa at the Organization of American States in Washington, D.C., saying: "Whether it's by the pollution of the land they consider sacred, the increased violence this project will inevitably bring, or by their own hand, oil development means the death of the U'wa" (Wagner, 1997).

During November 1999, the U'wa blockaded a road leading to an Occidental drilling site. They were met by hundreds of baton-wielding police mounted on bulldozers, spraying tear gas. Three U'wa children were drowned when police forced them into the rapids of the Cubujon River. A local official, an advocate of drilling, was quoted by the U'wa as telling them that "those animal Indians have to be evicted violently," as military officials declared that "the oil will be extracted even over and above the U'wa people" (Gedicks, 2001, 55). In the meantime, the U'wa watched guerillas attack oil pipelines immediately north of their homelands, spilling 1.7 million barrels of crude oil into local soil and rivers. "The loss of our U'wa children in the violent evictions, the humiliations of the armed forces, the cries of the U'wa children and elders in the peaceful mobilizations, the challenge to resist the aggressions by the Colombian State and Occidental, will not go unpunished. It will be a bittersweet memory that will remain in the minds of those who participated directly and indirectly in the most difficult moments of this process," said Perez ("U'wa Update," 2001).

Perez took his people's appeal against Occidental directly to investment firms in the United States that invest heavily in the company. After Perez visited offices of Fidelity Investments in Boston during September 2000, that firm divested more than $400 million worth of Occidental stock. The divestiture came after several thousand people demonstrated at 75 Fidelity offices around the world (Gedicks, 2001, 2).

Occidental Petroleum announced in late July 2001 that it had been unable to find oil at the Gibraltar 1 well site on the U'was' ancestral land in northeastern Colombia. The company has begun removing equipment from the site, "a positive turn of events for the valorous non-violent resistance campaign waged by the U'wa" ("U'wa Update," 2001). The news came nine years after Occidental was granted drilling rights to the Siriri block (formerly known as Samore). Occidental's announcement arrived in the U'was' highlands as many of them were taking part in a traditional three-month spiritual retreat of fasting, meditation, teaching, singing, and prayer. The U'wa *Werjayas* (spiritual leaders) and *Karekas* (medicine people) had been praying for months and using traditional rituals to "hide the oil" from Occidental's test drills ("U'wa Update," 2001).

While the U'wa called Occidental's inability to find oil a cultural triumph, the tribe's leaders pointed out that their ancestral land is still threatened by oil exploration by the Spanish company Repsol, which began exploratory drilling in the Capachos 1 block during the year 2000. "This is a battle that we have

U'wa Indians rest under a tent after protests and clashes with police in La China, Colombia, about 200 miles northeast of Bogota on Saturday, February 12, 2000. The 8,000-member tribe is trying to stop American oil company Occidental Petroleum from drilling on land sacred to the Indians. (AP Photo/Ariana Cubillos)

won, but the war continues, because the U'wa territory is not only Gibraltar 1," said Roberto Perez, president of the U'wa Traditional Authority, in a communiqué ("U'wa Update," 2001).

Occidental's operations in Colombia suffered significant losses during 2001; the company's Cano Limon field and pipeline were paralyzed by more than 110 guerrilla bombings during the year. In addition, Occidental's private security contractor, AirScan, was recently implicated in one of the Colombian military's worst civilian massacres, forcing Occidental into the center of yet another controversy. AirScan guided the Colombian military's attack on the Santo Domingo village that killed 12 civilians, including 9 children ("U'wa Update," 2001).

Despite its inability to find significant amounts of oil, Occidental and other oil companies continued to explore until early 2002. Opponents of oil exploitation also continued to resist. During February 2002, 30,000 indigenous Colombians joined a mass sit-in in the state of Arauca in northeastern Colombia to protest the destruction of oil development in their region. "The oil in Arauca has been a curse for us. The only thing that it has brought us, and continues to bring us, is death and destruction," said a Colombian woman in a protest at the U.S. embassy in Bogota, an extension of the provincial sit-in ("Bush Gives," 2002). The sit-in also protested brutality of the Colombian military in the Arauca Province, which includes some of the U'wa ancestral territory.

"The United States is also financing Plan Colombia, the struggle against drug trafficking, which signifies the increase of violence in the department of Arauca, Boyaca and North of Santander and Our Ancestral Territory . . . to protect the Cano Limon Pipeline in Covena, solely for having found oil . . . without seeing that what Colombia needs is more investment in social, health, education and employment programs, so that we can live in Peace," said a February 14, 2002 statement from the U'wa Traditional Authorities ("Bush Gives," 2002). "The government and petroleum multinationals are the first responsible for the social and environmental problem in the Arauca and base of the mountain region, and in [the] second place are the actors of the armed conflict, for the dynamiting actions against the Oil pipeline that cause the contamination of water, pastoral areas and watershed basins of the Arauca River. These actions are affecting climatic changes and the basic sustenance of our communities. We have the right to freedom of expression and thought," said the same statement by U'wa elders ("Bush Gives," 2002).

At its annual shareholders' meeting on May 2, 2002, Occidental announced that the company will return to the Colombian government its controversial oil block located adjacent to the traditional territory of the U'wa. The U'wa and their American environmental supporters rejoiced, following a nine-year campaign to halt the oil project. U'wa spokesperson Ebaristo Tegria said, "This is the news we have been waiting for. Sira, the God of the U'wa has accompanied us here in Colombia and our friends around the world who have supported us in this struggle. Now Sira is responding to us. This is the result of the work of the U'wa and our friends around the world" ("Occidental Petroleum," 2002).

Following Occidental's withdrawal, U'wa traditionalists issued a statement which read, in part: "The money king is only an illusion. Capitalism is blind and barbaric. It buys consciences, governments, peoples, and nations. It poisons the water and the air. It destroys everything. And to the U'wa, it says that we are crazy, but we want to continue being crazy if it means we can continue to exist on our dear mother EARTH" (U'wa Traditional," 2002). The statement also demanded compensation for lives lost during the U'was' struggle with the oil industry. In the meantime, Ecopetrol, Colombia's state oil company, said it reserved rights to continue the exploration begun by Occidental.

During March 2003, the U'wa Council received official notification from the president of Ecopetrol, Isaac Yanovich, regarding the discovery of petroleum in Gibralter I of the Samorè block. Once again, the U'wa told the oil interests, "The U'wa will never negotiate or sell our mother earth, nature, the environment, our culture, our history, and our higher laws. For the U'wa, all of this is not to be sold. It represents our right to live, which takes preference over any right, be that economic, social, or political" ("Communique," 2003).

THE UNDERSIDE OF U.S. ANTI-DRUG SPRAYING

During 2000, 2001, and 2002, the U.S. government launched its biggest offensive yet against coca and poppy fields in Colombia with a $1.3 billion aid package. Crop-dusting aircraft and escort helicopters sprayed almost 130,000 acres between December 2000 and June 2001. By the end of the year 2000, the United States had sprayed two-thirds as much land in Colombia as was sprayed with Agent Orange in Vietnam ("Harper's Index," 2001, 13).

Martin Hodgson of the *Guardian* reported from Bogota that "indigenous leaders, environmentalists and small farmers have repeatedly complained that glyphosate is sprayed indiscriminately on both legal and illegal crops, poisoning water supplies and making people sick" (Hodgson, 2001, 13). Government officials dismiss the substance as harmless to humans, but farmers whose crops have been sprayed have reported skin rashes, vomiting, and respiratory disorders (Hodgson, 2001, 13). Peasant groups and indigenous organizations have called for a complete cessation of the aerial-eradication program. They want it to be replaced by local development schemes that would promote rural employment and reward farmers for growing legal crops.

The main fumigant is glyphosate, an herbicide manufactured by Monsanto, which is sold commercially in the United States under the brand name Roundup. Farmers say that the herbicide has contaminated local rivers, affecting their health, and killed livestock. The antinarcotics police have been adding a Colombian-made soapy substance known as CosmoFlux to the spraying mix that weighs down the herbicide and helps it stick to the coca and poppy leaves.

On July 31, 2001, a Colombian judge ordered the government to suspend aerial eradication of drug crops with glyphosate. The ruling provided a serious challenge to a U.S.–sponsored antinarcotics campaign. Judge Gilberto Reyes-Delgado, who ruled in favor of Indian groups that had protested the spraying, asked the government to provide information regarding the weed killer's impact on human health and the environment.

Judge Reyes-Delgado's ruling followed a complaint by a Colombian human rights group, Minga, acting on behalf of 62 indigenous communities in the country's southern Amazon region. "Obviously, the judge found our arguments valid," said Minga attorney Tito Augusto Gaitan, who noted that the judge not only ordered the suspension in the Amazon region but nationwide. A circuit court judge on August 6 ruled that "evidence presented by the plaintiffs and the government did not provide conclusive proof of lasting damage to human health, the local ecosystem, or the cultural integrity of indigenous communities" ("Colombia: Flap," 2001, 25). Judge Reyes-Delgado's ruling came just days after the United Nations called for an independent audit of the spraying. Colombian antinarcotics police said that they would continue to spray heroin poppies and coca bushes.

The same day that Judge Reyes-Delgado issued his ruling, the United States delivered the first three of 16 Blackhawk helicopters, doubling the eradication capacity of the antidrug program, "for use by an elite U.S.-trained army battalion operating in narcotics-producing zones." The United States also was set to deliver 14 new crop-dusters to Colombian forces by September (Hodgson, 2001, 13).

Hugh O'Shaughnessy, a reporter for the *London Observer*, described health effects that some in Colombia attribute to antidrug pesticide spraying in a Kofan village:

> Franci sits on the veranda and whimpers. The little girl is underweight. Her armpits are erupting in boils. Like most of her people, she has suffered from respiratory problems and stomach pains since the aircraft and the helicopter gunships came over at Christmas and again at New Year dropping toxic pesticides on their villages. (O'Shaughnessy, 2001, 20).

Pineapples being grown by the villagers were stunted and shriveled, according to O'Shaughnessy's account: "The once-green banana plants are no more than blackened sticks. The remains of a few maize plants can be seen here and there, but the food crops have been devastated. There is hunger at Santa Rosa. . . . Peasants, already miserably poor, are getting hungrier" (O'Shaughnessy, 2001, 20). Chickens and fish also have been poisoned in farmyards and ponds. The tilapia that had brought a new prosperity to farmers who had built fish ponds are dying by the thousands, along with dogs, pigs, and other livestock. Indigenous people in the area generally survive by fishing and growing bananas, but the area has recently become a corridor for arms and drug smuggling.

Children from local schools showed signs of serious skin infections that heal over but continually recur. Gloria, a teacher at the school at El Placer, reported similar illnesses. "About 230 of the 450 pupils at our school have gone down with diarrhea, and respiratory and constantly recurring skin infections," she said (O'Shaughnessy, 2001, 20). Elsa Nivia, a Colombian agronomist who works with the Pesticide Action Network, said that during the first two months of 2001, local authorities reported that 4,289 people had suffered skin or gastric disorders while 178,377 domesticated animals were killed by the spraying, including cattle, horses, pigs, dogs, ducks, hens, and fish (O'Shaughnessy, 2001, 20).

Ironically, according to O'Shaughnessy, the coca bushes that the spraying was meant to eliminate have generally survived, as farmers chop off poisoned leaves to permit roots to survive. "In the front line of America's war on drugs," he wrote, "it is humans and the environment that have become the victims." While the Colombian and the U.S. governments have asserted that the aerial spraying has not caused any injury or significant damage to the environment, "the reality," wrote O'Shaughnessy, "is that the results on the ground are disastrous. . . . On the hills of Putumayo their lime-green leaves are holding the promise of new thrice-yearly harvests from which the narcotic [cocaine] will be manufactured again: their flourishing presence mocks the politicians and soldiers in Washington and Bogota" (O'Shaughnessy, 2001, 20).

Antolio Quira, a Coconuco Indian who served in Colombia's Senate between 1991 and 1994, said that current drug-eradication efforts are a continuation of fumigation against marijuana crops in the Sierra Nevada of northern

Colombia, home to Cogui, Arwaco, and Arzario Indian communities. As a result of these fumigations, said Quira, "hundreds of Indian children were blinded, their [parents'] cows died, and even now, 15 years after the fact, Indian children are being born with birth defects, and the land is barren and unusable" (Murillo, 1996, 5). Quira and other opponents of spraying criticize the government for ignoring local peoples' need for ways to make alternative livings in areas where coca and poppy cultivation presently are the only cash crops. "For us, coca is something we've used for hundreds of years," said Quira. "Chewing coca is an indigenous tradition that we've been doing since before the Spanish came. For the Indian, coca is life. For the white man, it means death" (Murillo, 1996, 5).

Luis Fernando Arango, a conservative lawyer and university professor who opposes the spraying, said: "Anyone who protests about this is labeled a drug dealer. Years into the future a lot of old men with dandruff will get together in Geneva and talk about it. But by then there will be no countryside left" (O'Shaughnessy, 2001, 20).

In the midst of the Colombian natives' protests of the antidrug war, one of their best-known leaders was kidnapped. Three suspected paramilitary gunmen on motor-scooters allegedly abducted Kim Pernia Domico, a 43-year-old environmental activist from the Embera Katio tribe, on June 2, 2001 (McGirk, 2001, 15). Nearly 1,000 indigenous people, members of 84 tribes, subsequently defied Colombian government warnings to disperse as they maintained a public vigil for Pernia at Tierralta, in Cordoba Province, at an encampment where hammocks were slung between trees, and dozens of cooking fires smoldered (McGirk, 2001, 15). Pernia "led opposition to the multinational-funded Urra Dam and then demanded compensation when his tribal fishing waters were spoilt [and he] was by far the region's most prominent indigenous leader" (McGirk, 2001, 15). Contractors and other businesses had been stopped in their tracks by his international appeals against the dam.

During July 2001, Ecuador asked Colombia to stop aerial crop-spraying near the 389-mile border because, according to Ecuadorian officials, "The chemical applied to eradicate coca . . . could harm Ecuadorians' health" ("Ecuador Asks," 2001). Ecuador has complained that spraying the herbicide glyphosate could make Ecuadorians ill and damage crops in the region's jungle towns. At about the same time, local media reported "an increase in headaches, fever and rashes among some Amazon village residents since the spraying began" ("Ecuador Asks," 2001). Ecuador's Foreign Affairs Minister Heinz Moeller said that the government was worried that uncontrolled aerial spraying could waft over the border.

Government investigators have been inundated with complaints from farmers and other rural people throughout areas of Colombia that have been sprayed. The health department in Putomayo collected affidavits from residents who displayed symptoms of herbicide spraying. Nancy Sanchez, coordinator of the health department's human rights section, said that evidence had been collected of "intoxication, diarrhea, vomiting, skin rashes, red eyes,

A Colombian air force crop-dusting airplane sprays a coca field high in the mountains of Colombia. The spraying program, underwritten in part by the U.S. government, is part of a wider program to wipe out Colombia's cocaine-production capabilities. (NewsCom.com/Zuma Press)

headaches" ("Colombia Faces," 2001, 5). The reports indicated an unusual number of skin diseases in children.

In mid-January 2002, the public interest environmental-law firm Earthjustice called on the United Nations Commission on Human Rights to bring pressure to bear on the United States and Colombia to halt aerial spraying of herbicides used to eradicate coca and poppy plants. Earthjustice advised alternative methods, and it issued its call with support from the Amazon Alliance, a group of Amazonian peoples' organizations. According to these groups, the aerial spraying "deprives the affected residents of Colombia and Ecuador of their rights to a clean and healthy environment, health, life, sustenance, property, inviolability of the home and family, and access to information" ("Aerial Herbicide War," 2002).

According to the Environment News Service, the aerial spraying also has drawn objections from 141 scientific, medical, environmental, and human rights groups from across the United States and around the world. In August, these individuals and groups wrote an Open Letter to the U.S. Senate that said, "From an environmental perspective, applying a concentrated broad-spectrum herbicide over delicate tropical ecosystems is almost certain to cause significant damage. Moreover, human health impacts from a concentrated mixture are obviously more likely" ("Aerial Herbicide War," 2002).

The Earthjustice statement to the United Nations Commission on Human Rights lists health problems from the spraying that include "gastrointestinal disorders (e.g. severe bleeding, nausea, and vomiting), testicular inflammation, high fevers, dizziness, respiratory ailments, skin rashes, and severe eye irritation. The spraying may also have caused birth defects and miscarriages" ("Aerial Herbicide War," 2002). The spraying has destroyed more than 1,500 hectares (3,700 acres) of legal food crops such as yucca, corn, plantains, tomatoes, sugar cane, grass for livestock grazing, and fruit trees, said Earthjustice, and has resulted in the death of cows and chickens, as well as numerous species of wildlife ("Aerial Herbicide War," 2002).

One Colombian people, the Nukak, faced extinction by the 1990s due to the politics of cocaine. The Nukak, whose habitat lies between the Inirida and Guaviare rivers, have watched a growing number of outsiders immigrate to their fertile lands to grow coca. This invasion has been followed by two others: Colombian army troops, spraying the coca under the U.S. aegis, and the guerillas of the Revolutionary Armed Forces of Colombia (FARC) Thus, the Nukak have quickly found themselves caught in Colombia's civil war. Never a numerous tribe, the Nukak numbered fewer than 1,000 in 1990. By the middle 1990s, the outside invasion made them vulnerable to a number of infectious diseases that cut their population in half. Plans to provide the Nukak with doctors and other medical help have foundered as politics hinders funding. A number of Nukak have gone to work on the colonists' coca plantations, which exposes them to herbicide spraying as well as the colonists' imported diseases.

Another problem, according to a report in the hemispheric indigenous newsmagazine *Native Americas*, is "the long-standing custom amongst the colonists of 'adopting' Nukak children and raising them themselves" ("Colombian Government," 1998, 8). Many Nukak children are being raised away from their families and culture. "The cumulative effect of these problems," according to the *Native Americas* report, "is that those Nukak who have survived so far are being sucked into the colonist towns [away from the forest], and their society is disintegrating" ("Colombian Government," 1998, 8).

THE EMBERA: CONFLICTS WITH DAM CONSTRUCTION

Colombian paramilitary forces have killed at least six indigenous people in Colombia's indigenous Embera Katio community who protested construction of a 350-megawatt Urra hydroelectric dam on the Sinu River. Ten others have been abducted, according to Amnesty International. The Embera Katio people have opposed construction of the dam because it imperiled fishing grounds that are basic to their traditional economy. In addition to reducing the Emberas' fishing harvests, changes in water flow wrought by the dam also have flooded indigenous homes and agricultural areas, destroying crops, notably banana harvests. The dam also inundated several sacred sites.

The Embera indigenous people have lived in the Darien Gap region of Panama and northern Colombia for several centuries. Their centuries-old relationship with the land changed on November 18, 2000, when the Urra company began filling a reservoir behind its dam on the Sinu River in the Colombian province of Cordoba, following authorization from the Colombian Environment Ministry.

The Emberas' objections to the dam date to 1996, when the board of directors of Urra refused to abide by an agreement with indigenous groups reached only a few weeks earlier. In that agreement Urra, the Colombian government, and regional indigenous groups made arrangements for compensation from the company that included sharing the income generated by the sale of the electricity produced by the hydroelectric project.

After several decades of study, a corporation that combines Canadian and Swedish interests began building the Urra dam on the Sinu River during 1995. The Embera Katio won a brief injunction suspending construction, but subsequent legal rulings resulted in the 1998 flooding of a fertile valley, destroying the Embera Katio's banana plantations (Wilson, 2001). In addition to Urra's refusal to pay promised compensation, the company also ignored a court injunction obtained by lawyers for the Embera Katio meant to halt construction of the dam. The Colombian Supreme Court's ruling on November 10, 1998 had said that the project could be completed only with the Emberas' consent. The ruling was met with a threat from the paramilitaries that more of the Embera would die in a massacre if such consent was not forthcoming.

Instead of sharing profits, the indigenous people found themselves evicted from their homes by force of arms. The Environment News Service reported: "On January 29 [2001], a paramilitary force entered the Embera Katio community of Kipardo in the Resguardo Karagabi, Karagabi indigenous reserve, forcing families to leave their homes" ("Colombian Indigenous People," 2001). Two days later, the same paramilitary group established a checkpoint on the Sinu River, which runs through the reserve, and abducted 10 members of the community. According to the Environment News Service (citing reports from London-based Survival International), work on the dam has devastated the river's fish population, one of the Emberas' main sources of food. Completion of the dam flooded much of their land and destroyed their livelihood, causing widespread hunger ("Colombian Indigenous People," 2001). One of the most important Embera leaders, 60-year-old Alonso Dominco Jarupia, was fatally shot outside his house on August 25, 1998, probably by assassins hired by the landowners.

During June 2001, several hundred Embera Katio people converged on the town of Tierralta, "arriving on airplanes and in caravans of cramped buses and wooden rafts, filling the central square of this frontier town with garish hammocks, tarps and the acrid smell of camp-fire smoke" (Wilson, 2001). The people rallied to demand release by the paramilitaries of Kim Pernia Domico, another prominent leader, who had been kidnapped in Tierralta on June 2, 2001 by three gunmen presumed to be members of the right-wing paramilitary United Self-Defense Forces of Colombia.

Scott Wilson of the *Washington Post* described "the Indians gathered in the cluttered square—their faces and legs marked with ritual tattoos, walking on bare, broad feet, speaking in languages that predate the Spanish colonization, seek[ing] the return of a man who tried to keep war and economic interests from overwhelming tribal land" (Wilson, 2001). At least 10 leaders of the Embera Katio and Zenu tribes, in Cordoba and neighboring Antioquia and Choco provinces, were killed by paramilitary forces and left-wing guerillas between 1998 and 2001.

THE TABACO: COAL STRIP MINING

In order to expand South America's biggest coal strip mine, Cerrejon Norte, Intercor, a subsidiary of ExxonMobil, evicted residents of the indigenous community of Tabaco (as well as several other nearby villages) in the province of La Guajira, Colombia, during August 2001. The Cerrejon mine is controlled by ExxonMobil in a consortium with three other multinationals: Swiss Glencore (Swiss), Australian Billiton, and the British firm Anglo-American.

Before their eviction, residents had held out for an adequate relocation package that would have allowed the whole community to stay together and move to a new site where they could have continued to practice small-scale agriculture. The relocation arrangements offered by the company broke up the community and left most of the former villagers lacking funds to buy land on which to live and farm. The villagers feared that insufficient compensation would lead to poverty and unemployment in nearby towns, which are already swollen with people displaced by Colombia's civil war (Doran and Solly, "Action Alert," 2001).

Initially, villagers in Tabaco complained that their houses' walls were cracking because of blasting from the mine. They also said that their main water source was polluted with coal dust. Pasture land also was lost as mining operations encroached (Solly, Pérez Araújo, and Moody, 2001). Some villagers departed as soon as the company's intentions became clear, but others stayed. After a villager sold a home, the company constructed large earth banks around the property. These banks collected standing water and became breeding grounds for mosquitoes. The town's school and community center were deliberately destroyed. Meanwhile, mining operations and test drilling moved ever closer to the community (Solly, Pérez Araújo, and Moody, 2001).

Members of the Tabaco community sought refuge in the local Catholic church, whose building they regarded as property of the Tabaco community.

They were told by the church's Italian priest, Marcelo Graciosi, that the building had been sold to Intercor to block the refugees. The church was then leveled (Solly, Pérez Araújo, and Moody, 2001). Graciosi sold the church for 38 million pesos (approximately U.S. $16,550) to be delivered on condition that the building be destroyed ("Tabaco Attacked," 2001). The church was demolished early one morning during Holy Week 2001 by a gang of armed civilians hired for the task. People in Tabaco were especially angered because the church had been built by their own hands with no financial assistance from the Catholic Church.

On August 9, 2001, more than 200 antiriot police arrived in Tabaco, along with several Intercor officials, all of whom were wearing bulletproof vests. According to an account in the newspaper *La Guajira*, "An official of the Fonseca office of the Colombian Institute for Family Welfare (I.C.B.F.), Albin Gomez Pèrez, was given the mission of gathering up the children so that their parents would not use them as human shields" as the village was destroyed ("Tabaco Attacked," 2001). The newspaper account described how people's protests were drowned out by the noise of advancing mining machinery, as their homes were leveled. Later in the day, several hundred more police arrived; by the end of the day, roughly 15 police were standing in the ruins of Tabaco for each of the local dispossessed residents. Later, the homeless families were forced to sign a bond offered by Intercor that legalized the destruction of their homes. Their reward for signing was return of their children, who had been seized earlier in the day.

Other communities were similarly dissembled to make way for the expanding strip mine. According to local observers, nearby Manantial and Carracoli were "simply broken up by violence and dispersed without compensation" (Solly, Pérez Araújo, and Moody, 2001). At Espinal, according to the same observers, "Police trucks arrived one day to remove the villagers to a new site at Rio de Janeiro. Those who co-operated received some funding for new community facilities. Those who resisted were forcibly removed at night to an unproductive, waterless site a few kilometers from Rio de Janeiro" (Solly, Pérez Araújo, and Moody, 2001).

REFERENCES

"Aerial Herbicide War on Drugs Poisons Land, Water." Environment News Service, January 15, 2001. [http://ens-news.com/ens/jan2002/2002L-01-15-04.html]

Barrionuevo, Alexi, and Thaddeus Herrick. "Wages of Terror: For Oil Companies, Defense Abroad Is the Order of the Day." *Wall Street Journal*, February 7, 2002, A-1, A-12.

"Bush Gives Oxy a hand in Escalating Colombia's War." *Drillbits and Tailings* 7:2 (February 28, 2002): n.p. [http://www.moles.org]

"Colombia Faces 'Apocalypse.'" *Native Americas* 18:1 (spring 2001): 4–5.

"Colombia: Flap over Fumigation." *World Press Review* (October 2001): 25.

"Colombian Government Neglect Leaves the Nukak Facing Extinction." *Native Americas* 14:3 (fall 1997): 8.

"Colombian Indigenous People Killed Opposing Dam." Environment News Service, February 10, 2001. [http://ens.lycos.com/ens/feb99/1999L-02-10-01.html]

"Communique to the National and International Community in Response to the Discovery of Petroleum in Our Ancestral Territory Kera Chikara." *Drillbits and Tailings* 8:3 (April 11, 2003): n.p. [project_underground@moles.org]

Doran, Chris, and Richard Solly. "Action Alert: Pressure Point." Colombia Solidarity Campaign, London, August 3, 2001.

"Ecuador Asks Colombia to Halt Aerial Coca Fumigation." Reuters News Service, July 20, 2001. In *Cultural Survival News Notes*, July 26, 2001. [http://www.cs.org/main.htm]

Gedicks, Al. *Resource Rebels: Native Challenges to Mining and Oil Corporations*. Cambridge, Mass.: South End Press, 2001.

"Harper's Index." *Harper's*, October 2001, 13.

Hodgson, Martin. "Colombia Halts Drug Poison Spraying." *The Guardian* (London), July 31, 2001, 13.

"Hotspots: Colombia." *Drillbits and Tailings* 6:7 (August 31, 2001): n.p. [http://www.moles.org]

Johansen, Bruce E. "U'was in Colombia Say They will Commit Suicide if Occidental Drills." *Native Americas* 16:1 (spring 1998): 5.

McGirk, Jan. "Tattooed Tribes Keep a Vigil for Kidnapped Chief." *The Independent* (London), June 26, 2001, 15.

Murillo, Mario. "Colombian Indians Indignant over Fumigations, Children Blinded." *Native Americas* 13:4 (winter 1996): 5.

"Occidental Petroleum Abandons Oil Development on U'wa Land." Environmental News Service, May 3, 2002. [http://ens-news.com/ens/may2002/2002L-05-03-01.html]

O'Shaughnessy, Hugh. "A Deadly Trade: How Global Battle against Drugs Risks Backfiring: The Global Battle Against Narcotics Is Going Awry. Chemical Spraying of Coca Bushes Is Poisoning Colombian Villages." *The Observer* (London), June 17, 2001, 20.

Solly, Richard, Armando Pérez Araújo, and Roger Moody. "Urgent Action on Crisis Involving Exxon in Colombia." August 3, 2001. [http://www.minesandcommunities.org/Action/action9.htm]

"Tabaco Attacked by the Armed Forces and by the Mining Company Intercor." *La Guajira*, August 12, 2001. [http://bargeldsparen.tripod.com/id107.htm]

U'wa Traditional Authorities. "U'wa Statement [Regarding] Oxy Withdrawal from Sacred Territory." May 7, 2002. [http//www.indymedia.org]

"U'wa Update: Colombia's U'wa Tribe and Supporters Celebrate Oxy's Failure to Find Oil." Indigenous Environment Network: Project Underground, August 9, 2001. [http://www.ienearth.org/mining_campaign.html#u-wa-0809]

Wagner, Martin. Statement at Organization of American States, Earth Justice Legal Defense Fund, October 7, 1997. [http://www.moles.org/ProjectUnderground/drillbits/971007/97100703.html]

Wilson, Scott. "Colombian Indians Resist an Encroaching War; Indigenous People Join to Search for Leader." *Washington Post*, June 18, 2001. [http://www.globalexchange.org/colombia/washpost061801.html]

CONGO BASIN

DEFORESTATION AT "ALARMING RATES"

Logging in the Congo Basin has been much less publicized than that of the Amazon Valley, but deforestation there "has reached alarming rates" in the six countries lying within the Basin (Cameroon, Central African Republic, Congo, Republic of Congo, Equatorial Guinea, and Gabon) ("Central Africa," 2001).

A study reveals that more than 11 million hectares are being currently exploited by European companies, most of which are French, notably Bollore, Rougier, Thanry, Interwood, and Pasquet. According to an analysis published in the *World Rainforest Movement Bulletin*, "Those companies usually operate without any forest-management plan and they rarely comply with existing legislation. Tree felling outside the concession area, cutting without respecting the established minimum exploitation diameters and the cutting of endangered species constitute common practices for this group of companies" ("Central Africa," 2001).

Forestry-industry advocates tout the benefits of industrial-scale logging, including employment, infrastructure development, and construction of school buildings, health centers, and churches. In reality, according to the *World Rainforest Movement Bulletin*, while in 1999 U.S. $609 million worth of wood was imported by the European Union from the Congo Basin, local indigenous communities "received little or none of [the] money" generated by logging ("Central Africa," 2001).

REFERENCES

"Central Africa: European Union's Major Responsibility over Deforestation." *World Rainforest Movement Bulletin* 53 (December 2001): n.p. [http://www.wrm.org.uy/deforestation/logging.html]

CONGO REPUBLIC

PYGMIES: LOGGING THREATENS LIVELIHOOD

The Congo Basin, which contains the second largest area of tropical rainforest in the world after the Amazon Basin, has been known for its high biodiversity. This rainforest is also home to several culturally diverse peoples, who depend on the forest to survive. By the turn of the millennium, increased logging was denuding many Pygmies' homelands, as forest workers' demands for bushmeat (country food) was diminishing stocks used by local indigenous peoples.

According to a report by the World Rainforest Movement, "Logging operations will increasingly disrupt the lives of local people, especially Pygmies, who depend on the forest for their livelihoods. As the north is being opened up to logging operations, the demand for bushmeat to supply workers' communities will increase and will contribute to the impoverishment of forest opened up by roads. This will have dire consequences for Pygmy groups who use forest areas for their subsistence activities" ("Congo Republic," 2002). Pygmies who have been forced off the land generally have had difficulty getting jobs with the logging companies due to discrimination from Bantus.

Expansion of the timber industry has been eased by government policies of the Congo Republic. With support of the International Monetary Fund, the Congo (with its capital at Brazzaville) has liberalized its natural-resource sector to increase profits, notably with a new forestry law initiated in 2000. According to the WRM, "Henri Djombo, the forestry minister, has estimated that the log production in Congo Brazzaville will double, or triple within the next two-to-three-year period" ("Congo Republic," 2002).

By the year 2002, according to the WRM, large-scale logging operations were underway in the northern regions of Sangha and Likouala. The German-owned Congolaise Industrielle des Bois is the largest forestry operator in the country, harvesting 1.15 million hectares, as other companies also move in to secure positions. The Congo Republic has emerged generally as an important supplier of tropical timber to Europe.

REFERENCES

"Congo Republic: Increased Logging Activities." *World Rainforest Movement Bulletin* (April 2002): n.p. [http://www.wrm.org.uy]

COSTA RICA

RESISTING CENTRAL AMERICA'S LARGEST DAM, BAUXITE MINING, AND AN ALUMINUM SMELTER

Since the 1970s, the Costa Rican government has been conducting studies aimed at eventual construction of the Boruca Hydroelectric Project on the river Térraba which, with a 1,500-megawatt generating capacity, would be the largest hydroelectric dam in Central America. Plans call for the dam to supply power not only to an aluminum smelter, but also (through regional power grids) to parts of Mexico and several countries in South America.

If the 260-meter dam is built, according to a report in the *World Rainforest Movement Bulletin*, 25,000 hectares would be flooded, including the entire Rey Curré indigenous reserve and parts of territories occupied by two other native peoples, the Térraba and Boruca. At the same time, the Ujarrás, Salitre, and Cabagra indigenous reserves also would be affected by dam-related infrastructure, including roads. For both the indigenous and peasant communities living in the area, the building of the dam would require the probable relocation of several thousand native people ("Costa Rica: Indigenous," 2001). The dam and related infrastructure would require investment of U.S. $3 billion, mostly from Canadian corporations. In addition, significant archaeological sites and important pre-Columbian settlements would be lost to the flooding.

Plans for the Boruca dam have triggered several protests by indigenous Costa Rican citizens who stand to lose their way of life to the rising waters behind it. Potential flooding provoked by the dam is only part of several industrial-scale changes that several indigenous peoples believe will ruin their traditional way of life. The Aluminum Company of America (ALCOA), having found significant bauxite deposits in the subsoil of the El General Valley in Costa Rica in 1970, won a concession from that country's legislature allowing it to mine as much as 120 million tons of bauxite. The company also was granted a license to build a bauxite refinery in the area. Because aluminum plants require large amounts of low-cost electrical energy, ALCOA also expressed a keen interest in construction of a hydroelectric dam on the Rio Grande de Térraba, which would be dammed, forming an artificial lake.

The dam will not provide local residents with electricity, since most of them do not have the infrastructure required to use it. Instead, according to the *World Rainforest Movement Bulletin*, in this case the project is aimed—in the words of President Miguel Angel Rodríguez (quoted in the Costa Rican newspaper *La Extra*, April 4, 2001)—"at providing Mexico and the United States with cheap energy" ("Costa Rica: Indigenous," 2001).

In March 2001, local indigenous peoples signed a manifesto, which concluded as follows:

> Our history, our identity and our cosmovision have since time immemorial been intimately linked to the Earth, the rivers and every expression of nature in our territories. To abandon our territories for us implies death, the end of our history and we declare:
> * Our total opposition to the Boruca Hydroelectric Project;
> * We call on national and international solidarity;
> * We urge international financial institutions to abstain from financing this project. ("Costa Rica: Indigenous," 2001)

Opponents of the Boruca project also have asserted that it "will accelerate deterioration of soils, vegetation and the hydraulic regime, due to the promotion it will give to the building of highways and roads on lands that are not apt for agriculture in general and due to the displacement of the population in the reservoir depression, the stimulation of migration towards the zone, speculation over private land and national reserves and destructive exploitation of forests by logging companies" ("Costa Rica: Opposition," 2001).

The Boruca people have collectively asked:

> Did the emissaries of power think that the "docile Indians" would be willing to leave the bones of our ancestors, our plantations and our humble homes? They underestimated us because they did not know us (and they still do not know us) because the god that inspires them has made them overbearing. The spirit of all our ancestors, the mountains and the river, the air and the landscape have no price. They have not realized yet that there are things that money and manipulation cannot buy. But they live and breathe for the god of money, they cannot understand. That is why they treat us this way. ("Costa Rica: Opposition," 2001)

REFERENCES

"Costa Rica: Indigenous Territory Threatened by Hydroelectric Dam." *World Rainforest Movement Bulletin* 46 (May 2001): n.p. [http://www.wrm.org.uy/bulletin/46/Costa Rica.html]

"Costa Rica: Opposition to Hydroelectric Dam." *World Rainforest Movement Bulletin* (November 2001): n.p. [http://www.wrm.org.uy]

DAM SITES AND INDIGENOUS PEOPLES

A report by the World Commission on Dams estimates that 40 million to 80 million people have been physically displaced by dams worldwide, a disproportionate number of them being indigenous. "Indeed," wrote Balakrishnan Rajagopal in the *Washington Post*, "this 'development cleansing' may well constitute ethnic cleansing in disguise, as the people dislocated so often turn out to be from minority ethnic and racial communities" (Rajagopal, 2001). In the Philippines, almost all the large dam-building projects have been undertaken on lands occupied by the country's six to seven million indigenous peoples. In India, 40 percent to 50 percent of the people displaced by development projects—a total estimated at more than 33 million since 1947—are tribal people, who account, in total, for only 8 percent of the country's population of one billion (Rajagopal, 2001).

According to Rajagopal, international human rights monitors remain oblivious to the violence against indigenous peoples that is implicit in such developments. "A biased focus on international criminal justice—the pursuit of a Milosevic, for example—has blinded the world's conscience to mass crimes that are often as serious as those that occurred in Rwanda and the former Yugoslavia," he wrote (Rajagopal, 2001).

Rajagopal described forcible dislocations that destroy the livelihoods of entire communities as large dams and "inappropriate agricultural projects" alter the land-use patterns that traditionally have supported farming, grazing, and fishing for many generations. "The number of people forcibly dislocated is probably far larger than reported, as the displaced are systematically undercounted—for example, by as much as 47 per cent in the case of the projects funded by the World Bank," he asserted (Rajagopal, 2001).

The James Bay and Pimicikamak Crees who have struggled with Canadian dam builders in Quebec and Manitoba for decades vis-à-vis recognition of their needs, have become accustomed to placing their struggle for survival in the international context of dam building. They have played a leading role in formulating a united indigenous position on the types of hydroelectric development that ignore existing flora and fauna, including indigenous peoples. The

Crees' position, quoted in detail below, helps to place dam building in an international indigenous context:

> On the occasion of the release of the W.C.D. [World Commission on Dams] *Final Report,* indigenous peoples, organizations and individuals around the world endorsed the attached call on financial institutions to immediately implement strict guidelines to prevent and address the adverse impacts of water and energy projects.
>
> We, the James Bay Cree Nation and Pimicikamak Cree Nation, indigenous peoples in the boreal, sub-arctic regions of Canada, have suffered ongoing violations of our fundamental human rights (as was recently recognized by two United Nations human rights monitoring bodies) as a result of massive hydro-electric development in our traditional lands.
>
> As peoples who have been dispossessed and devastated by the adverse biophysical, socioeconomic and cultural affects of water and energy projects, we call upon international financial institutions to refuse funding to all water and energy projects for which the consent of the peoples or communities affected has not been obtained; and endorse the call on all public international financial institutions to immediately implement, in direct cooperation with affected peoples, including indigenous peoples, stricter guidelines for all current and future water and energy projects; to halt all projects which do not comply with these guidelines; and to immediately fund reparations mechanisms and otherwise address the devastating consequences of energy development projects on the peoples and communities affected by them.
>
> Tens of thousands of square kilometers of our traditional hunting grounds and waters have been flooded or rendered inaccessible; our fish and waters have been contaminated with methyl-mercury; and our environments, economies and societies assaulted by rapid and imposed change.
>
> We have been dispossessed, displaced and environmentally, culturally, economically and socially devastated by large hydro-development projects, initiated and built in our traditional lands by the state-owned electricity corporations Hydro-Quebec and Manitoba Hydro respectively, against our wishes and without our consent. The human-rights dimensions of what was done to us as a result of these large dam mega-projects has until now never adequately been understood.
>
> We have long known that these projects and their impacts constituted violations of our human rights. However, this has only recently become adequately acknowledged. In December 1998, the United Nations Committee on Economic, Social and Cultural Rights assessed Canada's compliance with the International Covenant on Economic, Social and Cultural Rights [and found] Canada in violation of its international human-rights obligations.
>
> . . . The governments of Canada and the relevant provinces and the hydro-electric utilities have benefited from over 20 years of multi-billion dollar revenues at our expense. However, they have to date in no way adequately mitigated, remediated or compensated us as peoples for the profound and ongoing injuries and losses we have suffered. Deprived of adequate lands and resources, we now endure mass poverty and unemployment, ill health including epidemics of infectious disease and suicide, and crises of hopelessness and despair. Moreover, promises of economic development assistance, employment, training

and community development, made to us in formal treaties entered into as minimal, after-the-fact dispensations, have never been meaningfully fulfilled. This state of affairs led a June 1999 Inter-Church Inquiry into Northern Flooding to conclude, in the case of the Manitoba project affecting [the] Pimicikamak Cree Nation, that it has been a moral and ecological catastrophe. ("Response to the Final Report," 2000)

REFERENCES

Rajagopal, Balakrishnan. "The Violence of Development." *Washington Post*, August 9, 2001, A-19.
"Response to the Final Report: James Bay Cree Nation and the Pimicikamak Cree Nation." In *Dams and Development: A New Framework for Decision-making: The Report of the World Commission on Dams*. November 16, 2000. [http://www.dams.org/report/reaction_cree.htm]

ECUADOR

INTRODUCTION

Indigenous groups in Ecuador have increasingly opposed oil exploitation on their territories in the Amazon rainforests, sometimes with violence. Along with environmentalists, indigenous leaders warn that a consortium comprised of British Petroleum and Amoco will inherit a nightmare when it takes over operations begun by the Atlantic Richfield Company (ARCO), which has been pushing forward with plans for oil exploration despite local opposition (Knight, 1999). Widespread indigenous opposition also extends to plans for building an oil pipeline across Ecuador's Andes to ports on the Pacific Ocean.

OIL EXPLORATION IN THE ECUADORIAN AMAZON

Several Shuar tribes formed a confederation in 1964, mainly in response to an invasion prompted by a so-called agrarian reform law announced that year by Ecuador's dictator Rodriguez Lara, which "called for massive colonization of the Amazonian rainforest by landless peasants from the Andean region" (Weinberg, "Shuar Federation," 2001, 11). "In the familiar pattern," wrote Weinberg in *Native Americas*, "the peasant colonists were quickly followed by more powerful interests—primarily the cattle and oil interests. The ecological impacts were predictable and devastating. The Shuar's hunting-and-gathering way of life was abruptly interrupted as the forest rapidly disappeared" (Weinberg, "Shuar Federation," 2001, 11).

The Ecuadorian Amazon is home to between 100,000 and 250,000 indigenous people, including the Shuar, Achuar, Quichua, Cofan, Siona, Waorani, and Secoya, all of whom use the rainforest for hunting, subsistence farming, and fishing, as well as small-scale cultivation of cash crops. Many of these native people took refuge in this area to escape slavery by Amazonian rubber barons between 1880 and 1920.

In 1967, Texaco located the first notable oil deposits in the Ecuadorian Amazon area at Lago Agrio, near the homeland of the Huaorani. Indigenous resistance intensified in 1972, as the Ecuadorian Amazon was opened to industrialization, following construction of a 420-kilometer oil pipeline from the Oriente ("the East," as the area is known in Ecuador west of the Andes) oil fields over a 4,300-meter-high pass in the Andes, into the port of Esmeraldas on the Pacific Coast. With the pipeline, the first roads were constructed into Ecuador's northeastern Amazon area, opening the region to colonization, as farmers from the overcrowded coastal and mountain regions of the country swarmed over the spine of the Andes. Within two decades, the population of Napo Province soared from a few thousand to more than 100,000 people ("Yasuni National," n.d.). Petroecuador, the state oil company, cooperated with Texaco in developing oil resources in the area. By 2000, oil production accounted for 70 percent of Ecuador's export earnings (Gedicks, 2001, 68).

During the mid-1980s, newly discovered oil reserves were identified in the Pastaza and Napo river valleys, including 150 million barrels of heavy crude beneath the Waorani Ethnic Reserve and Yasuní National Park. After several years of delays due to environmental protests, construction of a new road and pipeline began in December 1992. The development extended from the Napo River south into the center of the Waorani reserve, amidst continuing opposition from Ecuadorian indigenous peoples and environmental organizations. The new oil fields were developed with promises from oil companies and government agencies to restrict immigration by squatters. Wells also were to be drilled in clusters to reduce deforestation. Both promises quickly were broken.

Following development of oil resources in the Oriente, road building and logging accelerated deforestation to a rate of about a million acres a year, one of the highest rates in Latin America. The indigenous Cofan, comprising 15,000 people in 1970, declined to about 650 by the late 1990s as their forest environment was stripped in what one observer called "an environmental free-fire zone" (Gedicks, 2001, 72). Texaco's oil operations generated an average of 3.2 million gallons of toxic waste daily. Over three decades, accidental oil spills from the Trans-Ecuadorian Pipeline discharged almost 17 million gallons of crude oil into the headwaters of the Amazon river basin.

In protest of this pollution, during 1992, 1,500 indigenous residents of the rainforest walked across the Andes 140 miles to Quito, Ecuador's capital, to negotiate title to 13,000 square miles of ancestral land. The title was negotiated, but oil extraction continued. As ARCO's operations continued, a class-action lawsuit against Texaco was filed in the United States by indigenous communities in the Ecuadorian Amazon, charging that during two decades of oil drilling in the Amazon, the company dumped more than 3,000 gallons of crude oil into the rainforest. The indigenous communities also asserted that Texaco dumped a toxic cocktail of chemicals into unlined pits that reached streams and eventually polluted major waterways, ignoring industry standards for waste disposal. Lawyers for the indigenous communities estimated that Texaco saved U.S. $3 to $4 a barrel—a total of $6 billion during the company's 20 years of operations in Ecuador—by dumping the waste into the water rather than removing it from the ecosystem. The plaintiffs asserted that the oil company's damage to their homelands totaled over U.S. $1 billion. In a statement, the plaintiffs said that, if the case goes to trial, punitive damages against Texaco could increase that amount significantly.

While the indigenous residents gained surface rights, the state reserved the right to allocate subsurface rights (including oil exploitation). In 1998, ARCO Oriente, a subsidiary of California-based ARCO, signed a contract with the government for exploration and exploitation rights in Oil Blocks 10 and 24, comprising several hundred thousand hectares in the eastern provinces of Pastaza and Morona Santiago, located in a primary tropical rainforest.

THE ACHUAR, SHUAR, AND WAORANI RESIST DEVELOPMENT

"We are at war with ARCO. We have seen the pollution and health problems oil companies caused in indigenous territories in the northern part of the Ecuadorian Amazon and we have decided we do not want this in our territory." said Santiago Kawarim, president of the Interprovincial Federation of the Achuar Indigenous Group of Ecuador (known by the Spanish acronym FINAE) (Knight, 1999). Kawarim said that the company also is creating phantom organizations consisting of selected Achuar and Shuar people who were in favor of oil development to give an impression that they represent majority sentiment. "They are trying to divide us in order to gain access to our territory," said Kawarim. "If they continue to push there will be a confrontation which will only endanger our communities" (Knight, 1999). A number of indigenous ARCO employees have been severely beaten by Achuar and Shuar people who consider them to be traitors.

Herb Vickers, a spokesman for ARCO in Quito, said that the company is "focusing . . . on working more at the local level because it believes the larger indigenous organizations do not represent the people as a whole" (Knight, 1999). Vickers maintained that a majority of the indigenous peoples want help with such things as road construction, to which Kawarim replied: "This is a complete lie" (Knight, 1999). Vickers asserted that ARCO was using the latest technology to reduce the impact of oil drilling on the environment, including construction of a 136-kilometer pipeline using helicopters, instead of building a road that could provoke a flood of illegal squatters and logging in the rainforest. Indigenous and environmental groups replied that the use of helicopters was no solution because their noise scared away wildlife on which many people depend for their livelihoods. The local peoples also asserted that construction of the pipeline and a flowline (which some local residents said leaked crude oil) has provoked soil erosion that harmed fish populations in nearby rivers (Knight, 1999).

Approximately 1,600 Waorani in the Ecuadorian Amazon are among the most threatened tribal peoples in the country. The Waorani are hunters, who use blowguns, poisoned darts, and spears to capture birds, monkeys, and wild pigs. Waorani women tend small vegetable gardens and collect wild fruit. Traditionally, the Waorani move from one communal house to another every few months. These dwellings, situated in the forest interior, take only a few days to build.

In some cases, outsiders have invaded the Waorani Ethnic Reserve, established to protect Waorani lands. Oil exploration began in this area during the 1940s. One small group of Waorani, known as the Taga-eri, have resisted all European-American attempts at contact, and have killed some of outsiders who have tried. There are known to be at least 12 Taga-eri, and possibly as many as 40.

Most of the interlopers work for oil companies pushing southward into the territory of the Waorani and other tribal peoples. The tribes further north, such as the Siona and Secoya, already have lost their lands to thousands of formerly landless colonists who used oil roads built to facilitate oil exploration to enter the area and establish homesteads. Large parts of the area also have been contaminated by oil spills, drilling wells, and waste pits.

A NEW TRANS-ANDES OIL PIPELINE

The president of Ecuador, Gustavo Noboa, in mid-2001 approved construction of the new trans-Andes pipeline, extending from the Amazon oil fields to Ecuador's Pacific Coast north of Quito. According to an account in *Drillbits and Tailings*, with this approval, Noboa declared war on environmental groups in Ecuador, in particular Acción Ecológica, an Ecuadorian grassroots organization opposed to the construction of the trans-Andes pipeline. "I'm not going to let anyone screw with the country, I'll give them war!" said Noboa ("Ecuadorian President," 2001). Following Noboa's approval of the pipeline project, more than 2,000 people marched in opposition outside the pipeline consortium's offices. Following this demonstration, more than 100 activists stormed and occupied the office of Pablo Teran, Minister of Energy and Mines.

Ecuador's highest court in 2001 also approved construction of the new pipeline. The decision, which was extensively criticized by environmental and indigenous groups, will allow construction of the pipeline across the 19,200-hectare Mindo-Nambillo Cloud Forest Reserve, which incorporates three distinctive ecosystems: Andean forest, subtropical forest, and cloud forest. The ecological reserve is also home to 450 species of birds and 370 species of wild orchids. The proposed pipeline quickly became an object of major civil unrest in Ecuador.

The consortium of international companies supporting the pipeline includes Alberta Energy, Repsol-YPF, Agip, Perez-Companc, Kerr-McGee, and the Los Angeles–based Occidental Petroleum, whose operations on the U'wa people's land in Colombia have been the subject of widespread controversy (see the entry on Colombia, this volume). Germany's largest public bank, WestLB, was the lead financier behind the project, providing a syndicated loan to the consortium in the amount of U.S. $900 million.

Along most of its proposed route, the proposed 500-kilometer pipeline would run parallel to an existing pipeline, which originates in the Amazon rainforest and traverses the Andes to the Pacific Coast. Following a landslide in May 2001, the existing pipeline ruptured, spilling 7,000 barrels of oil. This accident was the fourteenth major oil spill in the area in three years (since 1998). During 2001, the same pipeline also was ruptured by at least five bombings, an indication that the new pipeline, if constructed, could face similar problems ("Consortium Officials," 2001). During three years (1998–2001), a total of 145,000 barrels of oil were spilled from the older pipeline into the ecosystem, causing soil and groundwater contamination. The proposed pipeline is also in an area prone to landslides and earthquakes ("Ecuadorian President," 2001).

According to *Drillbits and Tailings*, a 157-kilometer section of the proposed pipeline, known as the northern route, would deviate from the path of the existing pipeline and cut through the Mindo-Nambillo Cloud Forest Reserve, which is nestled within a cloud forest in the Andes. The Reserve is rich in biodiversity and home to more than 450 species of birds, 10 percent of which are endangered ("Ecuadorian President," 2001).

The U.S. $1.1 billion pipeline project, which has been delayed for 10 years mostly due to Ecuador's economic and political instability as well as mounting protests from indigenous peoples and environmentalists, was approved by Ecuador's government and was scheduled for completion in mid-2003. The pipeline will open large parts of Ecuador east of the Andes to oil production for the first time; it is expected to double Ecuador's oil production and lure foreign investment (Carrere, 2001).

The pipeline project is the first of its kind under new regulations in Ecuador that allow private ownership of hydrocarbon-producing facilities. This law was implemented in Ecuador as part of an economic bailout package required (as a precondition for loans) by the International Monetary Fund and the World Bank to address the country's nearly U.S. $16 billion external debt, which has lead to the failure of some Ecuadorian banks.

Ricardo Carrere, writing in the *World Rainforest Movement Bulletin*, observed that "much of the heavy crude reserves that would flow through the pipeline starting in Lago Agrio (Amazonia), are likely to be found in pristine and protected areas of high bio-diversity and tourism activity" (Carrere, 2001). The Amazon region of northeastern Ecuador is one of the most biologically diverse regions on Earth; some areas contain 400 different tree species within a single square kilometer. Tropical ecologist Norman Myers has been quoted as saying that this area "is surely the richest biotic zone on Earth, and deserves to rank as a kind of global epicenter of bio-diversity" (Gedicks, 2001, 67).

Critics of the pipeline have denounced it as a violation of the Ecuadorian Constitution, which requires prior consultation with affected communities. On June 6, 2001, the Second Court of the Constitutional Tribunal in Quito ruled against a constitutional injunction sought by Acción Ecológica and other Ecuadorian environmental advocates, which sought to delay construction of the pipeline. The next day, the Ecuadorian government awarded an environmental license to the pipeline consortium and approved its construction by consortium member Techint. "Additionally," according to a statement issued in Quito by Acción Ecológica, "the government declared the project a national interest so as to impede all opposition, whether by social justice groups or affected communities" (Arias, 2001).

Entrance to the Forest Protector of Mindo, 84 kilometers to the north of Quito (the capital of Ecuador), is about to disappear because of construction of an oil pipeline. (NewsCom.com/Zuma Press)

Following failure of its appeals to Ecuadorian courts, Acción Ecológica said that it may appeal to international tribunals "to protect the human rights recognized by the Ecuadorian Constitution and international agreements" (Arias, 2001). Ecuadorian environmental groups maintained that "the decisions regarding the O.C.P. ["Oleoducto de Crudos Presados"] and its route definition are not the result of technical studies nor objective analysis, but rather obey political decisions and corporate calculations that have nothing to do with the best interest of the Ecuadorian people" (Arias, 2001).

The threat of indigenous challenge to the O.C.P. and other energy colonization of the Ecuadorian Amazon has been prompting militarization of the area. Atossa Soltani, executive director of Amazon Watch, was quoted by Bill Weinberg in *Native Americas* as saying that "Ecuador is going the way of Colombia. There were three pipeline bombings in Ecuador in the past year [2000]; ten oil workers kidnapped, and one—an American—killed. This has mostly been in the border region near Colombia, where the new pipeline is planned. Oil companies, including Oxy, have asked for extra military commitments from the Ecuadorian government and are complicit in Ecuadorian government plans for militarization of the border region" (Weinberg, "Amazonia: Planning," 2001, 13).

The Energy Intelligence Group's *Oil Daily* reported in its August, 3, 2001 edition that "in a bid to improve security in areas bordering Colombia, Ecuador's oil companies—16 private companies and the state company Petroecuador—and the country's armed forces signed a framework agreement under which each company can arrange to pay the military for protection" (Weinberg, "Amazonia: Planning," 2001, 13). The same article pointed out that construction of the pipeline was compelling oil companies doing business in Ecuador to rely more often on the military to protect their investments.

Nine women environmentalists were violently assaulted during a sit-in at offices of the O.C.P. Consortium in Quito, Ecuador. The sit-in—organized by Acción Ecológica and Oilwatch International, a Quito-based network of communities in South America, Asia, and Africa affected by the oil industry—had been called to support a general strike in the oil-producing region of Lago Agrio. According to a report in the August 31, 2001 edition of *Drillbits and Tailings*, company security guards destroyed television reporters' cameras and confiscated photographers' film and equipment as they assaulted activists and journalists to prevent coverage of the sit-in. One journalist from *El Universo*, a leading Ecuadorian daily newspaper in Quito, "was reportedly locked in a room and beaten by O.C.P. employees." "The O.C.P. Consortium has clearly demonstrated how it intends to treat Ecuadorians who exercise their legitimate right to protest the impacts of the company's operations," said Alexandra Almeida, speaking for Acción Ecológica ("Consortium Officials," 2001).

All of the civil resistance was organized to protest construction of the new pipeline. On February 28, 2002, several thousand striking O.C.P. construction workers and local residents in the northern Ecuadorian Amazon were attacked by the country's armed forces. Local newspapers reported that three children died by asphyxiation from tear gas and dozens of others were wounded in the attacks. The armed forces were called in to break the general strike ("Ecuador: O.C.P.," 2002).

During the strike, demonstrators erected roadblocks and occupied more than 60 oil wells and 5 refineries. A report by *Drillbits and Tailings* reported that the strikers "halted all construction on the pipeline and [brought] oil production to a near standstill" ("Ecuador: O.C.P.," 2002). Ecuadorian President Noboa declared a state of emergency. Under the decree, the state has prohibited public meetings, restricted the movement of key civic leaders, and shut down a local radio station for broadcasting messages that were deemed "against the state of emergency" ("Ecuador: O.C.P.," 2002).

On October 11, 2001, about 40 women and children arrived in the Mindo-Nambillo Cloud Forest Reserve and began peacefully blockading construction machinery owned by Techint, a member of the O.C.P. Consortium in charge of building the pipeline. The protest sought to stop construction of the pipeline by impeding clearance of its route, which began during the first week of October. Most of the blockaders live in communities near the pipeline's route. "The blockade has virtually stopped the crews from destroying this globally significant cloud forest reserve," according to a statement released by the environmental group Acción Ecológica ("Women and Children," 2001). Police quickly cleared the blockade, however. At the same time, in Germany, activists pressured that country's largest public bank, Westdeustche Landesbank (WestLB), to withdraw from a $900 million financing package it was arranging for the project.

During January 2002, local indigenous people reported uncontrolled spillage of oil into the environment from an abandoned exploratory oil well in the Ecuadorian Amazon, months after government authorities first were notified, according to Patricia Medici, a conservation biologist with the Lowland Tapir Project. Medici said that reports from local indigenous peoples indicated that in October 2001 the Organization of the Shiwiar Nationality of the Ecuadorian Amazon announced over a local radio station, Radio Puyo, that they had notified Petroecuador of the oil spill, but no one from the oil company had come to investigate. The spill, which was discovered by local inhabitants of the community of Chuindia while hunting, was still killing wildlife in the southeast region of Pastaza province in eastern Ecuador three months after the initial discovery. According to Medici, "They found a tapir still alive, but with its whole body shaking, and with the snout, lips and tooth flesh in a state of advanced decay." "Continuing forward," she reported, "they found a huge pool of petrol, and bodies of dead tapirs and peccaries, some of them just skeletons. The oil spill was on a salt lick, such that the animals have consumed oil while licking for salt" ("Oil Spill," 2002).

During January 2002, according to the *World Rainforest Movement Bulletin*, the inhabitants of local communities, students, and environmentalists carried out the "permanent and peaceful occupation of the most fragile zone of the Los Guarumos forest, at the entry of the Mindo-Nambillo Cloud Forest Reserve, to halt the construction of a new [O.C.P.] oil pipeline which will cross the whole of Ecuador" ("Ecuador: Human Shield," 2002). According to World Rainforest Movement (WRM) reports, "A number of activists have climbed the trees and built platforms and others are chained to the trunks, with the purpose of preventing the building teams entering the protected zone. Their intention is to remain in the zone until the building company and the government desist in their intention of destroying this unique ecosystem housing a diversity of fauna and flora species, many of which are in danger of extinction" ("Ecuador: Human Shield," 2002). The so-called human shield began activation as the pipeline consortium faced challenges to its funding from German banks. Observers in Ecuador told the WRM, "It is obvious that the O.C.P. Consortium does not want bulldozers facing the defenders of trees at the very time that the $900 million loan is running a serious risk in Germany" ("Ecuador: Human Shield," 2002).

RESISTANCE TO THE O.C.P. ON THE PACIFIC COAST

Inhabitants of Esmeraldas, on the Pacific Coast, also have resisted the new pipeline because of previous oil spills. One such spill on February 27, 1998 was said to have "left many dead and left irreversible impacts on the environment" (Arias, 2001). Thirty years of oil exploitation in the area have inflicted on local peoples "loss of ancestral territories, traditional practices and even the extinction of entire cultures like those of the Tetetes and the Sansahuari" (Arias, 2001).

A letter to the directors of Chevron-Texaco Corporation, written in November 2001 by Jose Luis Guevara, a leader in the Committee of the Affected, Province of Esmeraldas, Ecuador, summarizes many of the reasons why several indigenous peoples in the area have chosen to adamantly oppose the proposed trans-Andes pipeline:

> We would like to explain to you that our city is a zone that is situated near the north coast of Ecuador on the Pacific Ocean. Esmeraldas, with 500,000 inhabitants, was a tropical paradise of abundance and well-being, until the infrastructure of the petroleum industry arrived in our city in 1970. Along with the construction of the pipeline and the petroleum terminal in Balao, built by Texaco, a refinery was also built to process the petroleum, which the company extracted.
>
> These petroleum facilities have brought destruction, poverty, displacement and malnutrition to the people of Esmeraldas, as well as huge environmental disasters with incalculable consequences, which have submerged our province in terrible misery. The SOTE (Trans-Ecuadorian Oil-Pipeline System) and the petroleum terminal of BALO have been abandoned by Texaco. They are a time bomb—the local population lives permanently with the risk of breaches and the possibility of fires.
>
> The Esmeraldians have paid for the irresponsibility of Texaco with our own lives. Many spills and fires have been produced. People have burned to death and some have disappeared in these tragic events. Esmeraldas is a city of black and mulatto people, which has been impoverished by the petroleum activities of Texaco, because it took away our natural subsistence, the jungle, the clean waters of the rivers and the ocean, which was once full of animals which gave us food. Texaco did not care at all; it cared nothing for our lives, we black Ecuadorians. This is a clear example of environmental racism. Now that Chevron and Texaco are one company, Chevron has inherited the responsibility for the damages and the assaults that our people suffer on a daily basis. (Guevara, 2001)

REFERENCES

Arias, Natalia. "Environmentalists Affirm Position in Defense of Life as Construction for O.C.P. Pipeline is Given Green Light." Amazon Watch and Acción Ecológica, 2001. [http://www.amazonwatch.org/newsroom/newsreleases01/jun0601_ec.html]

Carrere, Ricardo. "Ecuador: Action to Stop the Oil Pipeline Continues." *World Rainforest Movement Bulletin* (September 2001): n.p. [http://www.wrm.org.uy]

"Consortium Officials Violently Remove Journalists and Activists from Company Headquarters." *Drillbits and Tailings* 6:7 (August 31, 2001): n.p. [http://www.moles.org/ProjectUnderground/drillbits.html]

"Ecuador: Human Shield in Defense of the Mindo Forest against Oil Pipeline." *World Rainforest Movement Bulletin* 54 (January 2002): n.p. [http://www.wrm.org.uy]

"Ecuadorian President Approves Pipeline Backed by Occidental." *Drillbits and Tailings* 6:5 (June 30, 2001): n.p. [http://www.moles.org/ProjectUnderground/drillbits/6_05/3.html]

"Ecuador: O.C.P. Protesters in Amazon Attacked by Military." *Drillbits and Tailings* 7:2 (February 28, 2002): n.p. [http://www.moles.org]

"Ecuador's Highest Court Upholds $1.1 Billion Dollar Oil Pipeline." *Cultural Survival News Notes,* August 30, 2001. [http://www.cs.org/main.htm]

Gedicks, Al. *Resource Rebels: Native Challenges to Mining and Oil Corporations.* Boston: South End Press, 2001.

Guevara, Jose Luis. "Committee of the Affected, Province of Esmeraldas. Community Voices: Letter to ChevronTexaco from Communities in Esmeraldas, Ecuador." *Drillbits & Tailings* 6:9 (November 30, 2001): n.p. [www.moles.org]

"Hotspots: Ecuador." *Drillbits and Tailings* 6:7 (August 31, 2001): n.p. [http://www.moles.org]

Kimerling, Judith. "Oil, Lawlessness, and Indigenous Struggles in Ecuador's Oriente." In Helen Collinson, ed., *Green Guerillas: Environmental Conflicts and Initiatives in Latin America and the Caribbean.* London: Latin America Bureau, 1996.

Knight, Danielle. "Indigenous Groups 'at War' with US Oil Giant." Interpress Service: World News, July 1999. [http://www.oneworld.org/ips2/july99/04_04_005.html]

"Oil, Mega-development Plans Would Destroy Amazon." *Indian Country Today,* October 10, 2001, A-4.

"Oil Spill Contaminates Ecuadorian Amazon." Environment News Service, January 10, 2002. [http://ens-news.com/ens/jan2002/2002L-01-10-01.html]

"To the Ends of the Earth: Revenge of the Lost Tribe: the Amazon's Indigenous Peoples." N.d. [http://www.channel4.com/plus/ends/tribe4.html]

Weinberg, Bill. "Amazonia: Planning for the Final Destruction." *Native Americas* 18:3/4 (fall–winter 2001): 8–17.

———. "The Shuar Federation: A Rainforest People Confront Modernity." *Native Americas* 18:3/4 (fall–winter 2001): 11.

"Women and Children Begin Blockades of Pipeline Construction Crews in Threatened Ecuadorian Cloud Forest Reserve." Amazon Watch, October 11, 2001. [http://www.amazonwatch.org/newsroom/newsreleases01/oct1101_ec.html]

Yasuni National Park and Waorani Ethnic Reserve, Ecuador, n.d. [http://www.nmnh.si.edu/botany/projects/cpd/sa/sa8.htm]

ERITREA

THE TIGRE, BENI AMER, HIDAREB, AND KUNAMA TRIBES: DEFORESTATION OF THEIR HOMELANDS

Indigenous ethnic groups in the western Lowlands of Eritrea, at the eastern-most extension of the Sahel—notably the Tigre, Beni Amer Hidareb, and Kunama tribes—are losing their forested homelands to industrial logging. These homelands lie between Eritrea's border with the Sudan and the Eritrean-Ethiopian highlands, an area of hills and plains covered mainly with semidesert scrub and savanna woodlands, laced by three river valleys clothed in dense woodlands, including mixed acacia and dom palm (Carrere, 2001; Connelly and Wilson, 2001).

Several hundred thousand indigenous people depend on these woodlands. Their distinctive survival systems have been tested by numerous natural and human-made stresses during the last half-century. Major droughts and war have led to a collapse of their farming system, provoking many deaths and a mass exodus of environmental refugees. Between 1998 and 2000, the Lowlands also were invaded and laid waste by Ethiopian armies (Carrere, 2001).

All these stresses have destroyed parts of the forests that play a crucial role in the indigenous peoples' livelihoods. Members of affected tribes rely on the forest for housing, tools, and some of their food. The dom palm fiber is the principal source of cash income for the majority of the Lowlands population. During peacetime, and when rainfall levels allow at least some herding and raising of crops, the poorer members of the community or those who cannot farm land themselves (including many women widowed by the war) make a living by cutting, weaving, and selling palm. Dom palm nuts also supply a food of last resort during the hungry weeks before the annual harvest. During drought years, they become a staple food (Carrere, 2001; Connelly and Wilson, 2001).

One ethnic group—the Kunama—has a distinctly different approach to the forest. They cut very little palm for income, but collect food from more than 20 tree species. These include the dom palm and others that they use as food reserves for drought years when their crops fail. For them, riverside forests function as insurance, rather than as a regular income source (Carrere, 2001).

Agricultural extension services of the Eritrean government have created obstacles for the indigenous peoples' traditional way of life because of an unfounded belief that cutting of palm leaves damage the trees, but also (and mainly) because the government has other priorities. The forests occupy fertile land with high water tables, which is ideal for irrigated agriculture of cash crops such as onions and bananas, sought by the government to attract foreign investment and to raise hard currency through exports (Carrere, 2001).

REFERENCES

Carrere, Ricardo. "Eritrea: Sustainable Forest Use Threatened by Government Policies." *World Rainforest Movement Bulletin* (September 2001): n.p. [http://www.wrm.org.uy]

Connelly, Stephen, and Nikky Wilson. "Trees for Semi-nomadic Farmers: A Key to Resilience." *LEISA Magazine*, April 2001. [http://www.ileia.org/2/17-1/10-11.PDF]

FIJI

THE NAMOSI, SERUA, NADROGA, AND REWA: FIGHTING A PROPOSED COPPER MINE

Introduction

During 2001, the Japanese mining company Nittetsu-Nippon proposed to pulverize the copper-rich hills of Fiji, endangering its ecologically fragile Waisoi Valley and the Coral Coast. According to a report in *Drillbits and Tailings*, the ore that the company proposed to mine is only one-half of one percent copper, which would make the proposed mine "among the biggest producers of crushed rock among copper mines worldwide" ("Namosi Copper," 2001).

100,000 Tons of Waste a Day

The account in *Drillbits and Tailings* calculated that the mine, if it is developed at the headwaters of two rivers, as proposed, "would dump 100,000 tons of tailings, equal to the weight of 1 million people, into the waters of the Navua Delta every day. The large open pits would eventually flood after the copper is extracted, forming toxic lakes from the contaminated aquifer" ("Namosi Copper," 2001).

An alternative dumping site for the tailings may be the Beqa Lagoon, which French ocean explorer Jacques Cousteau once called the "soft coral capital of the world." According to *Drillbits and Tailings*, "Sea life thrives in its deep upwelling, and it is one of the most seismically active areas in the world. The instability of the area makes submarine tailings disposal—the dumping of mine waste into bodies of water—a significant threat to the ecosystem" ("Namosi Copper," 2001).

Roughly 40 small indigenous villages have pooled their resources to oppose the proposed mine. The villagers are seeking to persuade local landowners, authorities in the Namosi district, and the Fiji government to oppose the mine. The Fiji Tourism Forum, which has a vested economic interest in protecting these areas, formed an Environment Sub-Committee to educate local communities about the impacts a mine of such scale will have on the environment and traditional lifestyles ("Namosi Copper," 2001). "The [awareness] program aims to educate, hence empower, the communities to understand the forms and characters of development, and how they will be affected, should a mine of this size be allowed into a small, ecologically fragile small island," said Kalaveti Batibasaga, a member of the committee ("Namosi Copper," 2001).

According to *Drillbits and Tailings*, the mine would bring about 10,000 additional people to the region, "an increase in population that is sure to impact these traditional communities" ("Namosi Copper," 2001). In addition, although indigenous leaders in the provinces have sold logging, fishing, and mining rights to outside corporations in the past, rarely have the communities benefited from the jobs these operations generated.

Tourism and Copper Mining Don't Mix

The four provinces that would be affected by the proposed mine—Namosi, Serua, Nadroga, and Rewa—are major sources of tourism income for Fiji. These areas, along with the Coral Coast and Waisoi Valley, are renowned for majestic landscapes, rich biodiversity, and pristine waters, all of which may be threatened by the proposed copper mine. If the Namosi mine is approved, the Nittetsu-Nippon Company plans to use the Waisoi and surrounding valleys for waste storage.

Existing mines on and near Fiji have been beset by labor problems and occasional strikes. The Emperor Mine, operated by Emperor Gold Mining on the Isle of Man, for example, has endured a multiyear strike by the Fiji Mine Workers Union, which began in 1991. Major strike issues have been over pay and working and housing conditions, which are described by the union as "appalling" ("Mining," n.d.). The company refused to recognize the union, fired its 420 members, and hired nonunion labor.

REFERENCES

"Mining in the South Pacific: On the Other Side of the Forests." N.d. [http://parallel.acsu.unsw.edu.au/mpi/docs/niart.html]

"Namosi Copper Mine Proposes to Dump 100,000 Tons of Waste per Day." *Drillbits and Tailings* 6:5 (June 30, 2001): n.p. [http://www.moles.org/ProjectUnderground/drillbits/6_05/4.html]

FOREST STEWARDSHIP COUNCIL

The Forest Stewardship Council (FSC) by late 2001 had certified 59 million acres of forest in 47 countries, about 4 percent of the world's forests presently subject to logging. The FSC maintains 10 principles leading to a product's certification, including: Obey the law, protect the rights of native peoples, limit waste, preserve species, contribute to the economic well-being of nearby communities, and preserve the forest.

The furniture retailer Ikea now sells ecofriendly teak patio tables and chairs. Home Depot stores in Seattle stock so-called certified hardwood flooring (Murphy, 2001, 13). "Lauan doors, lauan plywood, and ramin dowels: These things are among the most egregious products [from illegal logging in Indonesia], but they're also industry standards," said Tim Keating, executive director of Rainforest Relief, a New York-based group (Murphy, 2001, 13). Logging companies pay a fee for FSC certification, a process that may take several years. As a result, very few retailers advertise ecofriendly wood products.

During April 2001, for example, P.T. Diamond Raya became the first Indonesian logging operation granted a certification by the Indonesian Eco-labeling Foundation, which works jointly with the FSC. Later in 2001, two more tracts of Indonesian forest were certified.

The FSC's rigorous certification system involves a chain of custody certification that requires the tracking of specific parcels of harvested wood from the jungle, to the mill, to the container ship, and eventually to retail stores. This detailed process is meant to prevent "greenwashing," whereby documents that appear legitimate are used to cover trade in illegally logged wood.

European retailers make broader use of FSC certification than retailers in the United States. A report in the *Christian Science Monitor* said, "In Britain, for example, consumers can buy everything from a 2-inch by 4-inch stud to a piece of furniture, the handle on a broom, paper, [and] anything sold as a derivative product," according to Hank Cauley, FSC's executive director in the United States (Murphy, 2001, 13). Even in the United Kingdom, however, only 1.5 percent of the retail trade in wood was FSC-certified by mid-2001 (Murphy, 2001, 13).

The World Rainforest Movement (WRM) has asked whether the FSC certification program will benefit rainforests and the Earth in the long run. "The main problem," according to WRM, "appears to be the need perceived by the FSC to supply the world market with as much certified wood as possible" ("Is Certification," 2001). The WRM questions the FSC's efforts to certify as many large-scale logging operations as possible. Too little has been done, according to WRM, to certify small-scale forestry activities carried out by local communities.

"One underlying problem," according to the *World Rainforest Movement Bulletin*, "is that the FSC is focused on how to log. It has proved very difficult to get it to deal with the issue of where—and where *not*—to log. To date, the FSC continues the focus on forests as sources of timber, largely failing to support or promote other revenue streams (fruits, medicines, resins, non-timber fibers, etc.). As such, it has not helped those arguing for less logging and for forest areas to be off limits (permanently or under moratorium) to commercial logging" ("Is Certification," 2001). According to the WRM, the real problem—taking the forests' and indigenous peoples' long-range needs into account—is overconsumption of wood and wood products. The WRM asserts that the social and environmental costs of extraction must be included in the price of timber.

REFERENCES

"Is Certification the Solution?" *World Rainforest Movement Bulletin* 53 (December 2001): n.p. [http://www.wrm.org.uy/deforestation/logging.html]

Murphy, Dan. "The Quest for Certifiably Eco-friendly Lumber." *Christian Science Monitor*, August 23, 2001, 13.

FRENCH POLYNESIA

THE TE AO MAOHI MOOREA: A CANOE BLOCKADE AGAINST DREDGING A LAGOON

Introduction

Indigenous residents and environmental activists on Moorea Island in Te Ao Maohi (French Polynesia) have stopped the dredging of Moorea Lagoon for a tourist hotel. At the end of February 2000, according to a report from the scene, "Moorea residents surrounded the dredge with canoes to halt the dredging operation. With the support of environmental activists from the association Paruru ia Moorea and other groups, the Maohi community in Moorea is seeking to halt the extraction of sand from their lagoon" ("Protect Moorea," 2000).

Indigenous People Oppose Additional Tourism

The indigenous residents took their action contrary to the wishes of the territorial government of President Gaston Flosse, which has been trying to boost the tourism industry to generate revenue to replace French military spending that was reduced after cession of French nuclear testing in 1996. Business interests have been given tax write-offs and incentives to build new tourist hotels, even though existing hotels are underused ("Protect Moorea," 2000).

Local reports describe the activities of one investor in tourism projects, the Wan family, which also dominates the country's black pearl industry. Robert Wan and his brother Louis have diversified their investments from pearls into tourism, and have developed a contract with the Outrigger chain, which owns hotels in Papeete, Bora Bora, and Moorea ("Protect Moorea," 2000).

More specifically, according to these reports, the Société Moorea Lagoon Resort, a company controlled by Louis Wan, "has purchased the Moorea Lagoon hotel at Pihaena on Moorea (an island located near Tahiti). The company is planning a major reconstruction and expansion project, with plans to dredge the Moorea lagoon to obtain more than 10,000 cubic meters of sand for the project . . . an area as big as two soccer fields" ("Protect Moorea," 2000).

Local residents oppose the dredging plan, fearing that it will damage the lagoon's ecology. More than 2,000 people on the island signed petitions opposing dredging. Finding petitions ineffective, residents decided to initiate a canoe blockade to stop the dredging. During late February 2000, about 2,000 indigenous residents surrounded the dredge with canoes to halt any dredging operation in Moorea Lagoon. Wan responded by taking three environmental associations to court, but his initial application was rejected.

Inhabitants Refuse to Obey the Bailiff

Wan's appeal included a request for the government to deploy *gardes mobiles* (riot police) to clear the protesters. Wan sought to invoke a law allowing a fine of U.S. $10,000 for acts of opposition registered with the *huissier* (official bailiff). On February 26–27, 2000, the bailiff arrived at the lagoon to register the names of people using their canoes on the lagoon. Protesters refused to give their names, as one stated he would "rather die than let this sand go cheap for a hotel beach!" ("Protect Moorea," 2000).

This was not the first time that indigenous people on the island had organized to resist commercial exploitation of their island. During the late 1980s, the Japanese company Nishikawa announced plans to build a golf course and hotel at Opunohu on Moorea on land that includes several *marae* (sacred sites). Human rights activists from Tahiti joined with church leaders and local communities to form the association Paruru ia Opunohu, which opposed the transfer of land. After months of organizing, the local community won a referendum against the project.

According to local reports, another Japanese hotel project on Tupai was blocked by a land occupation, "led by community elders and the community organization Paruru in Tupai. Construction of the Rivnac hotel by a subsidiary of Lyonnaise des Eaux and the Meridian group, at Punaauia on the west coast of Tahiti, was delayed by sit-ins and protests between 1996 and 1998" ("Protect Moorea," 2000).

REFERENCES

"Protect Moorea Lagoon!" Pacific Concerns Resource Centre, February 28, 2000. [http://www2.planeta.com/mader/ecotravel/resources/rtp/rtp.html]

GHANA

GOLD-MINING TAILINGS SPILLS

Introduction

Two major cyanide spills during fall 2001 hit villages in western Ghana with such intensity that many indigenous people found themselves facing hunger and disease. According to reports by *Drillbits and Tailings*, within two weeks after a tailings dam broke and spilled gold-mine waste into the river Asuman, another spill occurred near the village of Kubekro at a gold mine operated by Satellite Goldfields.

Spill Ruins Life-Sustaining Marsh

Drillbits and Tailings reported that "the cyanide-laced tailings were spilled into an important marshland that provided the local people with mud fish, local medicines and bamboo for a wide range of uses and that is host to a spectacular array of plant and animal species" ("Ghanaian Communities," 2001). Virtually all life forms in the river and its tributaries were obliterated, and people's livelihoods were endangered. Fear was expressed that the cyanide and heavy-metal toxic residue from the spill could remain in the area for decades, posing a health and environmental threat to people and wildlife in the area ("Hotspots: Ghana," 2001, 3).

Following a spill from the Tarkwa mine in October 16, 2001, disaster faced the village of Abekoase. *Drillbits and Tailings* reported that thousands of cubic meters of mine wastewater flowed into the river Asuman when a tailings dam ruptured, contaminating its watershed with cyanide and heavy metals. Local sources said that hundreds of dead fish, crabs, and birds littered the banks of the river, while others floated on its surface. Reports indicated that "people in the villages of Abekoase and Huni have lost their clean drinking water and their livelihood as they can no longer sell or eat produce from their farms through which the river runs" ("Ghanaian Communities," 2001).

Daniel Owusu Koranteng, executive director of the local mine-monitoring organization, the Wassa Association of Communities Affected by Mining (WACAM), asserted that the Satellite Goldfields Corporation should compensate villagers whose economic well-being had been severely damaged by the spills. He added that compensation should include resettlement on uncontaminated land. Local activists also called for examination of international lenders' and corporations' role in Ghana's continued impoverishment. Ghana was hailed by the World Bank during the 1990s as a showcase, as gold became its largest earner of export income; it endured two decades of financial discipline by the World Bank, only to realize that the economic circumstances of most people have not improved. Government tax holidays lasting as long as 10 years, and lax environmental laws, have increased corporate profits but ruined many indigenous peoples' lives, according to local activists. By 2001, two-thirds of Ghana was under concession to various corporations that were mostly foreign-owned ("Ghanaian Communities," 2001).

People Forced off Farming Lands

"Everywhere you go," reported *Drillbits and Tailings*, "you see huge cavities in the ground, discarded pits where thriving villages once stood and where nothing now grows. People have been forced off their farming land, losing their only source of income" ("Ghanaian Communities," 2001). Describing the disaster at Abekoase as "unprecedentedly serious," Joshua Awuku Appau of the Accra-based Greenearth organization severely criticized continued environmental damage wrought by the use of cyanide to leach gold from ore. "There is the need for a long-term monitoring program along the whole river system, but there is still a risk of another catastrophe as long as cyanide is being kept behind a dam, which is often too weak. This is unacceptable, and the mining industry must learn that clean rivers and healthy ecosystems are more precious than gold," Awuku stressed ("Ghanaian Communities," 2001). Sixty percent of the rainforests in Ghana's Wassa West district already have been destroyed by mining operations, which also have polluted surface and groundwater with cyanide and other chemicals ("Ghana: IMF," 2002).

A report compiled by the World Rainforest Movement describes how gold mining has devastated local communities:

> Nearby villages suffer from contaminated water supplies and cracked buildings from the mines' blasting. In many cases, the land used for mining operations in Ghana has been forcibly acquired from peasant farmers under ambiguous regulations. Sometimes this acquisition occurred with no compensation. In some instances, the mines have been responsible for the dislocation and forced resettlement of communities numbering in the hundreds and even thousands. Numerous violations of human rights, including shootings and beatings, have also been committed in relation to the mines. ("Ghana: IMF," 2002)

The Asuman River area was hit by another cyanide spill early in 2003. Water from a ventilation shaft of an abandoned underground mine within the mining concession of Goldfields Ghana Ltd. in Tarkwa seeped into the Asuman River in the Wassa West District of the Western Region, sparking fears of contamination. Area residents, chiefs, and opinion leaders said contaminated water filled the mine shaft and flooded the river, a source of drinking water for the people of Abekoase and surrounding communities. The October 2001 cyanide spill killed nearly all life in the river and some of its tributaries.

REFERENCES

Anane, Mike. "Ghana: Cyanide Spill Worst Disaster Ever in West African Nation." Environment News Service, October 24, 2001.

Atarah, Linius. "Environment—Ghana: Gold Mine Spills Cyanide in River—Again." InterPress Services, November 1, 2001.

"Ghana: IMF, Mining and Logging." *World Rainforest Movement Bulletin* 54 (January 2002): n.p. [http://www.wrm.org.uy]

"Ghanaian Communities Hit Hard by Two Cyanide Spills." *Drillbits and Tailings* 6:9 (November 30, 2001): n.p. [www.moles.org]

"Ghana: NGOs Criticise Gov't Handling of Cyanide Spillage." *Ghanaian Chronicle*, November 21, 2001.

"Hotspots: Ghana." *Drillbits and Tailings* 6:7 (October 31, 2001): 3. From Environment News Service, October 24, 2001.

GUAM

THE CHAMORRO: MILITARY PCB POLLUTION

One-third of Guam's 216 square miles is occupied by U.S. military installations; the rest of the island is home to roughly 150,000 indigenous Chamorros and mixed-blood Japanese, Filipinos, Chinese, and Koreans. The indigenous population, roughly 80,000 before sustained contact with outsiders, had fallen to fewer than 5,000 people by the year 2000. Alexander Ewen wrote in *Voices of Indigenous Peoples: Native People Address the United Nations* that "the indigenous Chamorros have been protesting to the United Nations and [U.S.] Congress for years about environmental contamination from the military bases, and they have also intensified their campaign to re-take land seized by the United States for military purposes" (Ewen, 1994, 141).

While the U.S. Environmental Protection Agency (EPA) asserts that the level of polychlorinated biphenyl (PCB) contamination in samples of taro and yams tested from Tanapag village in Saipan (an island north of Guam) is only one-tenth the level considered toxic, a citizens' group has urged residents to avoid eating the root crops in the face of conflicting evidence. The group's members believe that PCBs dumped by the U.S. military have filtered into the drinking water of Guam and nearby islands; the drinking water has been tested at levels that are clearly unsafe for human consumption.

The Pacific Insular area program manager of the EPA, Norman L. Lovelace, said differences in results reported by the EPA and independent specialists resulted in confusion that made it appear that the level of concentration was about 1,000 times higher than reality. The Guam attorney general's office, however, citing a report on the results of an investigation made by the Test Law Practice Group and Meridian Hydrotechnical Services, disclosed high concentrations of PCBs in taro and yams tested from Tanapag.

Not all of the pollution on Guam comes from military bases. The Guam Waterworks Authority faced thousands of dollars in local and federal fines during 2001 for discharging raw sewage, which "spewed out of manhole covers along the island's main thoroughfare, Marine Drive, and along scenic Harmon Cliffline situated above the tourist belt Tumon Bay" (Sagapolutele et al.,

2001). Local officials estimate that $200 million will be required to upgrade Guam's aging sewer system.

REFERENCES

Ewen, Alexander. *Voices of Indigenous Peoples: Native People Address the United Nations*. Santa Fe, N.M.: Clear Light, 1994.

Sagapolutele, Fili, Aldwin R. Fajardo, Eric Say, and Radio Australia. "Government Briefs." *Pacific Magazine*, March 2001. [http://cust16530.lava.net/PM/pm32001/pmdefault.cfm?articleid=23]

GUATEMALA

INTRODUCTION

Indigenous peoples face a number of varied problems in Guatemala, including inundation by a hydroelectric dam of land used for centuries by the Achi Maya; spoilation of wetlands used by the Champericos, which have been polluted by commercial shrimp farming for export; and oil exploration in a Mayan biosphere reserve. In each of these cases, ways of life that have been relatively stable for many generations have been imperiled by commercial interests from the outside. In the case of the Achi Maya, several hundred people have died protesting these intrusions.

THE ACHI MAYA: PROTESTING HYDROELECTRIC FLOODING

The Chixoy hydroelectric dam, in Guatemala, was built after at least 376 Maya Achi people (mainly women and children) died protesting its construction. Most were from Río Negro, one of the villages that were flooded by the dam ("Guatemala: A Dam," 2001).

Violence against the indigenous people of the area began in 1980, when military police came to Río Negro and shot seven people. In July of that year, two representatives from the village agreed to go to a meeting requested by the National Institute for Electrification (INDE). The representatives took with them the village's only documentation of resettlement and cash payment agreements. The mutilated bodies of the two men were found a week later, and the resettlement documents were never recovered ("Guatemala: A Dam," 2001).

In February 1982, according to a report in the *World Rainforest Movement Bulletin*, "seventy-three men and women were ordered by the local military commander to report to Xoxoc, a village upstream from the reservoir, which had a history of land conflicts and hostility with Río Negro. Only one woman returned to Río Negro. The rest were raped, tortured, then murdered by the

Xoxoc Civil Defense Patrol, one of the notorious paramilitary units used by the state as death squads" ("Guatemala: A Dam," 2001).

The *World Rainforest Movement Bulletin* reported that

> worse was yet to come. On 13th March [1982], the military rounded up all the women and children, marched them to a hill above the village and proceeded to torture and murder 70 women and 107 children. Witness for Peace produced in 1995 a report based on interviews with survivors, where the terrible way in which these people were murdered is described in detail. Two months later a further 82 people were murdered. ("Guatemala: A Dam," 2001).

The news magazine *Native Americas* reported that in November 1993, forensic specialists began to exhume bodies from the largest massacre. "We believe that these shocking revelations require an independent investigation to discover whether or not [World] Bank project staff knew about the massacres," said Patrick McCully, campaign director of International Rivers Network ("Guatemala Massacre," 1996, 7). The Inter-American Development Bank and the World Bank have provided more than U.S. $300 million in loans for this project. "The Chixoy massacres hold important lessons for the funding [of] forced resettlement in countries with repressive regimes," said McCully. "An investigation into this matter is also extremely important given the tendency to ignore or suppress information on the real impacts of their projects on local people" ("Guatemala Massacre," 1996, 7). In Guatemala as a whole, 72,000 Guatemalans died as a result of government-sanctioned terror during the early 1980s.

THE CHAMPERICO: WETLANDS RUINED BY SHRIMP FARMING

Beginning in May 2001, the Champerico community of Guatemala denounced contamination of its wetlands and the logging of mangroves, as well as restriction of access to fishing grounds in an area where three-quarters of the community's protein comes from fish. The local people also assert that many fish are dying because of pollution by Camarones del Sur (Camarsa), which raises shrimp, mainly for export.

Contamination of wetlands and restriction of access to fishing grounds violate Guatemalan law. Local indigenous people have found themselves fighting for recognition of the law by national authorities. Confrontations between Camarsa and local residents have triggered various demonstrations, which resulted in the death of a young man, Moytin Castellanos, and several injuries ("Guatemala: Security," 2001). The local indigenous community has demanded that Camarsa immediately cease its operations.

During June 2001, local people blockaded the shrimp factory, preventing workers from entering it. This demonstration ended after a confrontation in which Fernando Chiyoc died and seven other people received bullet wounds from the security guards and other Camarsa employees. Following the shootings, Mike Corser, a U.S. citizen and an engineer at Camarsa, was arrested, along with nine of the plant's security guards. All were charged under Guatemalan law with homicide and attempted homicide.

Trópico Verde, a Guatemala-based environmental group, together with the artisan fishermen from Champerico, have carried out research documenting contamination of wetlands and a serious lack of compliance with the environmental laws of the country. A full report on the issue—"The Impact of Shrimp Farming Activities in Champerico, Retalhuleu, Guatemala"—is available in Spanish on the Web (http://www.wrm.org.uy/paises/Guatemala/Champerico.html).

OIL EXPLORATION IN THE MAYA BIOSPHERE RESERVE

The Guatemalan Human Rights Court on January 31, 2000 ruled that the Guatemalan government had violated the fundamental human rights of all Guatemalans by allowing oil exploration in the Maya Biosphere Reserve and Laguna del Tigre National Park. The court found that several government officials from national to local levels had engaged in "injurious behavior, that has harmed Guatemalans," especially those who live in the reserve. "Each time they [petroleum activities] disrupt the right to a clean environment, to the right of individual dignity, to the right of the preservation of the cultural and natural patrimony of the country, and the right to social and economic development . . . [they violate human rights]," said the ruling ("Oil Operations," 2000).

According to a report in *Drillbits and Tailings*, the ruling also pointed out that violation of the Law of Protected Areas constitutes "an administrative tendency detrimental to the citizens of Guatemala, and especially to the communities neighboring the ravaged protected areas" ("Oil Operations," 2000).

The environmental group Trópico Verde said that the offending authorities had allowed approximately 50 percent of the Biosphere's buffer zone and 40 percent of its Multiple Use Zone and Core Zone to be opened or auctioned for oil extraction. The official government reaction to the ruling has been to focus almost exclusively on one of the nine concessions illegally approved in the Maya Biosphere Reserve and has failed to address objections by residents of Carmelita and Uaxactun.

The 8.3-million-acre Maya Biosphere Reserve, which is situated in northeastern Guatemala near the borders with of Mexico and Belize, contains more than 3,000 plant species, as well as about 10,000 Mayan archaeological sites. When it was created in 1989, all oil exploration and development was prohibited, despite the fact that the area contained existing oil-drilling facilities.

On February 15, 2000, roughly 5,500 farmworkers occupied a building at a refinery owned by the U.S. company Basic Resources in La Libertad, 80 miles from the Reserve, shutting down a major pipeline valve. According to *Drillbits and Tailings*, Basic Resources is the primary company extracting oil in the Maya

Biosphere Reserve, with existing facilities in the Laguna del Tigre National Park.

The protesters demanded that the company provide contracts for workers transporting asphalt from that area to other regions. These type of jobs would be threatened upon completion of another pipeline. In addition, the farm workers are requesting some 10 million trees to help reforest certain areas of the Reserve ("Oil Operations," 2000).

REFERENCES

"Guatemala: A Dam and the Massacre of 400 People." *World Rainforest Movement Bulletin* 42 (January 2001): n.p. [http://www.wrm.org.uy/bulletin/42/Guatemala.html]

"Guatemala Massacre Related to World Bank Project Under Scrutiny." *Native Americas* 13:2 (Summer 1996): 7.

"Guatemala: Security for Shrimps; Insecurity for the Local Population." *World Rainforest Movement Bulletin* 48 (July 2001): n.p. [http://www.wrm.org.uy/bulletin/48/Guatemala.html]

"Oil Operations in Guatemala Declared Threat to Human Rights." *Drillbits and Tailings* 5:3 (February 28, 2000): n.p. [http://www.moles.org/ProjectUnderground/drillbits/5_03/2.html]

GUYANA

INTRODUCTION

Logging and gold mining, two staples of industrial intrusion into indigenous lands around the world, are the biggest problems for surviving tribal peoples in Guyana. The Isseneru have been struggling for many years with the toll of mercury poisoning caused by gold mining in their territories, while the people of the Akawaio Nation are working for legal recognition of their rainforest lands before commercial interests log them to the ground. In both cases, the indigenous peoples of Guyana find themselves realizing nearly nothing from commercial exploitation that ruins their lands and their health. Increased gold mining and logging have been promoted by Guyana's government to help reduce its international debt, and to meet the terms of adjustments required by the International Monetary Fund and the World Bank.

THE ISSENERU: MERCURY POISONING FROM GOLD MINING

Gold mining is exposing residents of Isseneru, indigenous peoples who live on the Essequibo River in Guyana's Upper Mazaruni, to unhealthy levels of mercury contamination. "We have obtained evidence that the residents of the village of Isseneru are significantly exposed to mercury and that they are closely associated with gold mining activity," said Institute of Applied Science and Technology Chairman David Singh. "It was important to seek the means by which the mercury loading might be minimized through the use of appropriate technologies," Singh said ("Isseneru Villagers," 2001).

Long-term exposure to mercury vapor can damage the brain, nervous system, and kidneys. Such exposure also may disrupt the functioning of the thyroid gland. Severe brain damage also is possible to children born to mothers who have been exposed. Babies fathered by men who have been exposed to mercury vapor risk miscarriage. Consumption of organic mercury in fish can permanently damage the brain and cause mental and physical defects, especially in children. According to Cultural Survival, short-term exposure to mercury vapor also "can permanently damage the lungs and can lead to death" ("Isseneru Villagers," 2001).

These indigenous peoples have been resisting invasion of their lands by miners since the late 1950s. In 1959, one-third of the Upper Mazaruni indigenous reservation was classified by the government as a mining district. In the beginning, the area was invaded by small numbers of independent miners. A few years later, however, the first of several multinational companies, most of them from Canada, began mining gold on an industrial scale in the area. The Canadian companies Golden Star Resources and Vanessa Ventures have been most notable for their environmental impact.

Large-scale fish kills and deaths of several hogs were reported in August 1995 after a wastewater dam at the Omai gold mine (about 100 miles from the coast) broke and spilled 3.2 million liters of cyanide-laced gold-mining waste into the Essequibo river, "a frightening reminder of the consequences of uncontrolled industrial mining" ("Little Progress," 1997, 12).

Studies by the Pan American Health Organization revealed that all aquatic life in a four-kilometer-long creek that runs from the mine into the Essequibo was killed (Chatterjee, 1998). According to a report in *Native Americas*, "Dead hogs and fish were seen floating down the river" ("World Bank," 1996, 4). The mine was closed following the accident, a blow to the Guyanese economy because it had been providing about 25 percent of the nation's export income ("World Bank," 1996, 4).

The Omai gold mine disaster did not end in August of 1995 for the Essequibans in Guyana. Resumption of operations and the continuous dumping of chemicals into the river was ignored except by the people who have had to live with it. Led by Gustav Jackson, a Guyanese geologist and an environmental scientist, a group of professionals from both Guyana and North America joined efforts to secure clean (i.e., cyanide-free) drinking water for the people of the Essequibo.

Spillage of mining waste occurred in a formerly pristine tropical rainforest. According to a report in the *World Rainforest Movement Bulletin*, the Upper Mazaruni river basin once comprised a luxurious forest that is home to Akawaio (Kapon) and Arenuca (Pemon) indigenous peoples. Their ancestral territory also encompasses parts of the Gran Sabana in Venezuela, as well as northern Roraima State in Brazil. These peoples' traditional economy long has been based on seasonal migrations between the lower and upper reaches of the Mazaruni and Kamarang rivers for hunting, fishing, and farming.

According to the *World Rainforest Movement Bulletin*, the companies use "missile dredging," which it describes as the use of "enormous vacuum cleaners shaped like missiles, that are attached to river dredges to remove alluvial deposits," to extract gold-bearing ore ("Guyana: Transnational Mining," 2001). The missile dredges destroy riverbanks and nearby forests, meanwhile increasing sedimentation and killing large numbers of fish that once fed local people. Miners also use mercury to separate gold from ore.

The *World Rainforest Movement Bulletin* described the environmental impacts of mining in the area: "The water is discolored and heavy with sediments, [and] piles of debris accumulate at the river banks, some of which have disappeared because of missile dredging. Environmental impact assessments required by law exist only on paper" ("Guyana: Transnational Mining," 2001). Noise caused by mechanized mining also chases away game animals, which local indigenous peoples depend upon for food.

The invasion of miners also has brought an increased incidence of malaria to the area, along with alcoholism, prostitution, drug use, and violence. The miners have moved in without consent of local indigenous people, who have had no role in advising the government on the future uses of their lands.

Gold mining in Guyana continues to expand. According to the *World Rainforest Movement Bulletin*, "On November 2, 1998, the government of Guyana and Vancouver-based mining company, Vannessa Ventures, signed an agreement granting Vannessa more than 2 million hectares of land in which to conduct geophysical and geological surveys for gold and primary diamond sources over the next two years. This concession includes the heavily forested Kanuku mountain range, as well as the upper reaches of the Corentyne River on the border with Suriname in the eastern region of Guyana. The area is part of the ancestral territory of the Wai Wai, Wapisiana, and Macusi indigenous peoples. They have vigorously objected to any mining or logging company operating on their lands and are demanding that their rights to their ancestral lands be legally recognized and respected" ("Indigenous Peoples Fight," 1998).

Toward that end, Guyana established a native claims process in 1967, but more than 30 years later only a quarter of the land that native peoples assert they own has been recognized as such. In October 1997, community leaders of the Wai Wai, Wapisiana, and Macusi peoples formed the Touchau's Amerindian Council of Region 9 to defend their ancestral territories from miners and loggers. Six Akawaio and Pemon indigenous leaders from the Upper Mazaruni also, for the first time, filed a lawsuit in the High Court of Guyana asserting their land rights. "Our communities have been requesting title to these lands, which we know to be ours, since the Amerindian Lands Commission visited our communities in 1967. Since then we have attempted to discuss this matter on many occasions without result" said a statement presented to the High Court ("Indigenous Peoples Fight," 1998).

THE AKAWAIO NATION: SEEKING A LAND BASE BEFORE IT IS LOGGED AWAY

In addition to Canadian gold-mining companies, several Asian logging companies also operate on native lands in Guyana. Like many other indigenous peoples the world over, the Akawaio Nation of Guyana (in South America) is working to gain recognition of its traditional land base before a timber frenzy strips it bare.

The Guyanese government in 1997 awarded the Canadian Buchanan Group a 1.5-million-acre exploratory lease in the Middle Mazaruni region of Guyana, the traditional homeland of the Akawaio. Buchanan proposed a complete cultural and economic makeover for the Akawaio—logging of the forest that sustains their traditional economy and putting the native people to work in its mills, until the resource has been exhausted. "When the timber is gone," wrote Mark Westlund in *Native Americas,* "the company will move on, leaving the ecosystem devastated, and the Akawaio without their traditional way of life" (Westlund, 1997, 11). The Guyanese government is accelerating the logging with tax breaks. The government charges royalties, taxes, and fees that are as low as a tenth of what is charged for similar concessions in Africa and Asia.

The Akawaio hope to stall logging by gaining new titles to their lands. Their titles as now written are inadequate because they allow the government to assert ownership of subsurface minerals, a situation that has allowed indigenous lands to be inundated by hundreds of small-scale miners, most of whom are moving from exhausted deposits (usually of gold) in Brazil.

REFERENCES

Chatterjee, Pratap. "Gold, Greed and Genocide in the Americas: California to the Amazon." *Abya Yala News: The Journal of the South and Meso-American Rights Center* (1998): n.p. [http://saiic.nativeweb.org/ayn/goldgreed.html]

"Guyana: Transnational Mining Companies' Impacts on People and the Environment." *World Rainforest Movement Bulletin* 43 (February 2001): n.p. [http://www.wrm.org.uy/bulletin/43/Guyana.html]

"Indigenous Peoples Fight for Territorial Rights in Guyana." *World Rainforest Movement Bulletin* 17 (November 1998): n.p. [http://www.wrm.org.uy/bulletin/17/Guyana.html]

"Isseneru Villagers Face Risk of Mercury Contamination." *Cultural Survival News Notes,* July 19, 2001. [http://www.cs.org/main.htm]

"Little Progress in the Recognition and Demarcation of Indigenous Lands, in Guyana." *Native Americas* 14:1 (spring 1997): 11–12.

"Omai Update: The Disaster Continues; Water Is More Valuable Than Gold." N.d. [http://saxakali.com/CommunityLinkups/omaiupdate799.htm]

Westlund, Mark. "Akawaio Nation in Cross-hairs of Timber Frenzy." *Native Americas* 14:1 (spring 1997): 11.

"World Bank Quietly Insures Major Polluters." *Native Americas* 13:1 (spring 1996): 4–5.

THE HAUDENOSAUNEE (IROQUOIS) ENVIRONMENTAL WORLDVIEW: DIFFERENT TREES IN A DIFFERENT FOREST

> You who are wise must know, that different Nations have different conceptions.
>
> —Cannasatego, an Onondaga Haudenosaunee spokesman,
> addressing English colonial officials, 1744

INTRODUCTION

The Haudenosaunee (pronounced "Ho-dee-no-show-nee") are "the People of the Longhouse." They are also known as the Six Nations or the Iroquois Confederacy. The member nations are the Mohawks, Oneidas, Onondagas, Cayugas, Senecas, and Tuscaroras. Today their reservations are in what is now New York, Quebec, Ontario, Wisconsin, and Oklahoma. At the time of first European contact, the core of their homeland stretched across what is now the state of New York and northward into the St. Lawrence River Valley. During the colonial period of U.S. history, Haudenosaunee political power reached westward to the Mississippi and Lake Superior, southward to the Gulf of Mexico, eastward to the Atlantic, and northward into the northern tributaries of the St. Lawrence River. By 1850, after centuries of pressures by Europeans and Americans of European heritage, they were dispersed to their present locations. Today, the philosophical influence of the Haudenosaunee is global because of the world's growing interest in philosophies that offer alternatives to Western ideas (Akwesasne Notes, 1978, passim).

Oren Lyons, a Faithkeeper of the Onondaga Nation, one of the internationally known Haudenosaunee speakers, notes:

> Today's times are very perilous for all. It's not only Indian children who are endangered. It's not only white people's children who are endangered. It's not only black people's children who are endangered, or any of our brothers around the world's children; it is *all* of *our* children who are endangered. We have to set our houses straight. . . . And, to take on those terrible obligations of survival that we face today, we must support one another.

That's how it's going to be. We're going to have to be united to survive on this great Turtle Island. I'm not just talking about Indians. I'm talking about everybody; we must be united on principles. We must remember the spiritual law that governs everything, and continue thinking, really, truly, about the seventh generation coming. We must continue to have great compassion and love for the unborn in future generations. (Lyons, 2000, 70, 74–75)

The *seventh generation* is a key to Haudenosaunee environmental ethics. The Haudenosaunee believe that individual humans and human communities must be responsible for taking actions that positively affect seven generations hence. Thus they must also avoid actions that might negatively affect the seventh generation.

THE THANKSGIVING ADDRESS

Haudenosaunee environmental ethics are reflected in the Thanksgiving Address, also known as "the Words Before all Else" or "the Opening Address." The Haudenosaunee believe that humans have the spiritual responsibility to give thanks for the entire universe. Thus, whenever the Haudenosaunee gather, someone—an elder if possible—begins the meeting with the Thanksgiving Address. It is not a thanksgiving to the Creator for objects that are inferior, but rather it is an expression of the Haudenosaunee's intense gratitude to the Creator for being a part of a world of equals. The Haudenosaunee realize that, as part of a world of equals, they have responsibilities that go beyond the obligations to their own human community.

The Thanksgiving Address can be lengthy, or it can be brief. The words of the Thanksgiving Address also may vary with the speaker, because the goal is to remind people of a set of broad principles, not a single exact text. Whatever brief precision or lengthy eloquence a speaker may choose, the purpose of the Thanksgiving Address is to remind all who are present that all life is interdependent. The goal of all who listen is to bring their diverse strengths together so that the individuals present can become a community, united in their right thinking, in order that all may focus on being of one mind.

The following points are covered in the Thanksgiving Address. After each point, those who are listening affirm in their minds or orally a concept that translates as, "Now our minds are one."

Greetings to the People
Thanks are extended to Mother Earth
Thanks are extended to the Waters
Thanks are extended to the Fish
Thanks are extended to the Plants
Thanks are extended to Food Plants
Thanks are extended to Medicine Herbs
Thanks are extended to the Animals
Thanks are extended to the Trees
Thanks are extended to the Birds

Thanks are extended to the Four Winds
Thanks are extended to the Thunders [the Thunder Beings of lightning and thun-
    der, "Grandfathers" who bring rain]
Thanks are extended to the Sun
Thanks are extended to Grandmother Moon
Thanks are extended to the Stars
Thanks are extended to all Spiritual Messengers
Thanks are extended to the Creator
Closing Words

As an example of how the Thanksgiving Address begins, an English translation of a Mohawk version declares:

Today we have gathered and we see that the cycles of life continue. We have
been given the duty to live in balance and harmony with each other and all living things. So now, we bring our minds together as one as we give greetings and
thanks to each other as People.

[And the listeners respond:] Now our minds are one. (Swamp and Stokes,
1993, 1)

The same translation ends with the Closing Words:

We have now arrived at the place where we end our words. Of all the things we
have named, it was not our intention to leave anything out. If something was forgotten, we leave it to each individual to send such greetings and thanks in their
own way.

[And the listeners respond:] Now our minds are one. (Swamp and Stokes,
1993, 1)

"TO BE OF ONE MIND"

The Haudenosaunee express their spiritual commitment to the whole community and to the world as "to be of one mind." To be a confident individual
human means to adhere to the Creator's spiritual teachings. These spiritual
instructions are at the foundation of a spiritually created interdependence.
They are a primary goal of one's thinking. "To be of one mind" does not mean
conformity, because each individual human is respected for his or her unique
characteristics and strengths. "To be of one mind" means that each human is
committed to directing their individual strengths so that the spiritual instructions of the Creator are carried out. These spiritual instructions are the standard for all actions. "To be of one mind" is to be committed to this spiritual
standard. (One of the many reasons the introduction of alcohol by the Europeans was so destructive to the Haudenosaunee was that alcohol prevented the
individual from clearly seeing the spiritual instructions common to every
being, humans and nonhumans alike. Alcohol prevented the community from
understanding and respecting the individual's thoughts, because these thoughts

were so clouded by the alcohol. Alcohol severely inhibited the individual and the community from being "of one mind.")

Paul Williams, a Canadian lawyer and historian, describes the importance of the Thanksgiving Address, "the words that come before all else," in its contemporary context:

> When nations bring their minds together, they also bring their most sacred thoughts and ways. There is no higher purpose than peace. What people do to bring their minds together also brings their spirits together to create this peace. Every Council of the Haudenosaunee begins internally and externally with the same words. Even before we put our minds together politically, we put our minds together as human beings with a place in this world, a humble place.
>
> All of these principles still guide the laws and thoughts of the Haudenosaunee in Council today. And the Good-Minded ones, the ones that in English are called Chiefs, they carry certain responsibilities that guide and inform their thinking. In any Council, in any decision, the law requires that they ask themselves: what will this do to the seven generations yet to come? What will this do to the natural world? What will this do to peace? These are three lenses through which the lawmakers must see each question. They are lenses that I think every lawmaker in this world would do well to carry. (Williams, 2000, 36)

A WORLD OF EQUALS

As the Thanksgiving Address indicates, the Haudenosaunee are confident that they are a part of an interrelated structure of living things in which all beings, including humans, are responsible for maintaining a balance with, rather than a dominance over, all others. Thus, the Haudenosaunee do not share the Judeo-Christian belief in a God-given right to dominate the earth and all other beings. This Judeo-Christian concept might be termed *the Genesis pyramid,* deriving from God's command, in the Bible, that humankind should be fruitful, multiply, and subdue the Earth.

The Judeo-Christian ethic proclaims that humans are at the top of an ecological pyramid of living things. Just as there is no Genesis pyramid among the Haudenosaunee, there is no scientific evolutionary pyramid that would create any kind of a hierarchy, human or nonhuman. Nor is there a Marxist belief that humans must work out their destiny without spiritual aid and insight. In fact, the Haudenosaunee environmental ethics cannot be separated from their economics, religion, philosophy, community behavior, or individual behavior. All must be practiced simultaneously, and all must be practiced in a continually conscious cooperation with all creation. The human contribution to this cooperation is based on spiritual obligations carried out through spiritual ceremonies.

In the Haudenosaunee worldview, the entire world is believed to be alive with the spiritual energies of all beings, such as the deer, the eagles, the trout, and the corn. Each species has its own function, assigned by the Creator, and each species has a sense of its own community. Each species has religious instructions that have been provided by the Creator and that each species is obliged to carry out. All beings are equally conscious of other beings. There are no unconscious objects. There are no inferior beings. There are simply beings with different functions.

The Haudenosaunee believe that all beings have a spiritual essence (or soul). For example, the Haudenosaunee conceptualize the trees as having spiritual values that are equal to human spirituality. They believe that the spiritual essence of each and every mortal being, human and nonhuman, is of equal spiritual value. For these reasons, the Haudenosaunee follow what some scholars have termed an "environmental religion" (Vecsey and Venables, 1980, 1–37; Capps, 1976).

THE ENVIRONMENT AND INTERDEPENDENT RESPONSIBILITIES

With regard to the environment, there were significant differences between Haudenosaunee and European worldviews in 1492. Within the Haudenosaunee worldview, it is impossible to conceptualize a single, temporal, sovereign power with the right to determine the fate of an entire environment. As noted above, the Haudenosaunee are confident that all life forms, humans and nonhumans alike, are interdependent. The Haudenosaunee spiritual worldview also maintains the broad concept that humans and nonhumans consciously understand that they share a reciprocal responsibility for the land. A temporal human government that has the right to determine the fate of the environment is therefore not a logical concept.

DIFFERENT TREES IN A DIFFERENT FOREST

The Haudenosaunee have very definite responsibilities to the environment. These responsibilities must be—not *can be*, but *must be*—the focus of their lives, their government, and their ceremonies. Even when faced with what would appear to be the same physical reality or the same moral dilemmas as nonnative neighbors, the Haudenosaunee respond differently because their entire reality is different. As a result, the different questions and solutions posed by the Haudenosaunee are more than mere choices of different paths through the same forest. The Haudenosaunee see different trees in a different forest.

The basic principles of the Haudenosaunee worldview may be summarized as:

> Peace to enable the world to function;
> Power to maintain the peace;
> Right thinking (moral thinking to sustain both peace and power, sometimes
>   defined in English as "righteousness");
> Balance among all life forms;
> Spiritual equality of all life forms;
> Different functions for all life forms, even as each function is spiritually equal in
>   importance to the others;
> Interdependence of all life forms;

Conscious interaction among all life forms;
Reciprocity among all life forms; and
Respect shared mutually by all life forms.

COMMUNAL ETHICS

Within the Haudenosaunee worldview, environmental ethics encourage and depend upon communal ethics. The Haudenosaunee view of the environment always has been based on the premise that both humans and nonhumans are consciously interdependent and interactive. The environment is interdependent spiritually as well as biologically. From this strong sense of interdependence comes a stress on communal ethics. Communal human ethics are thus environmental ethics. They are a logical extension of how the Haudenosaunee define a communal, interdependent world. Because the Creator filled the world with symbiotic, equal souls who nevertheless carry out specific and separate functions, the most logical premise upon which to base an organized human community was/is communal. In this sense, all beings, with their equal souls, are your relations, your relatives.

SPIRITUAL SELF-AWARENESS

The Haudenosaunee believe that each being possesses an individual consciousness that is motivated by spiritual self-awareness. All beings have the constant ability to make conscious decisions. All beings have a sense of spiritual responsibility to all other beings, not just to their own species. That said, the Haudenosaunee do not pretend to be able to define the complexity of their environment. For example, while they are confident that the hawks are carrying out the hawks' spiritual responsibilities, the Haudenosaunee do not know the specifics of the Creator's instructions to the hawks. Since the Haudenosaunee can never know the details of what the Creator has taught other beings, they are content in the confidence that the other beings are following their instructions. In that same light, the Haudenosaunee realize that they cannot interfere with other beings to the extent that an imbalance is created, which might prevent other beings from carrying out their spiritual instructions.

ALL BEINGS ARE CONSCIOUS OF THEIR HISTORY

Within the Haudenosaunee worldview, all beings have a sense of history. Each generation of every life-form understands where their ancestors have come from, and how each generation must take future steps according to spiritual instructions. All history is therefore interdependent, the accumulation of the conscious, and moral actions of all beings. All history is in fact a transformation of the environment carried out by all beings.

THE HAUDENOSAUNEE ARE STILL IN THE GARDEN

The Haudenosaunee are confident that they are exactly where the Creator has always intended them to be. Unlike those influenced by Judeo-Christian thought, the Haudenosaunee have not been cast out of the Garden of Eden. They are still in the Garden. They are not born in sin. They are born as the Creator intended. Therefore the Haudenosaunee are not inevitable sinners who must overcome a corrupt human nature. In their spiritual construct, there are mistakes but there is no blame. Far more importantly, there is no sense of guilt or sin. All the quirks and all the strengths of human nature (including the sex drive of every person) are exactly what the Creator intended. The same standard is applicable to all nonhumans, who are also regarded as having characteristics that are exactly as the Creator intended.

SACRED CIRCLE, SACRED CONSUMPTION

The Haudenosaunee environmental philosophy assumes that every human being is essentially good. Every individual also is assumed to be trying to do his or her best. These assumptions are extended to all beings. All beings are carrying out specific spiritual instructions that the Creator views as necessary, as good. Part of this good is that beings eat other beings. This concept is not unique to the Haudenosaunee (Arnold, 2001). Thus hunting, fishing, and the harvest killing of berries, nuts, and vegetables (such as corn) all take place within this worldview. The predators—humans and wolves, for example—are intended to survive by killing other equal life-forms, such as deer. The deer in turn kill equal life-forms, such as the meadow grasses. When predators die, their bodies replenish the soil that feeds the grass that feeds the deer. The Creator thus intended the world to function as a sacred circle of sacred consumption.

Moral responsibilities, not moral dilemmas, emerge from this circle. Believing, understanding, and following this worldview are the spiritual obligations of all beings in what is perceived to be an interdependent world. Among the humans, each individual contributes his or her individual strengths to the human community, which in turn contributes to the rest of the world, where each component (hawks, trout, corn, etc.) makes its own unique but essential contribution. To be confident as an individual human while contributing to everyone else, human and nonhuman, is a primary goal of one's actions.

REBALANCING THE WORLD

Although the Haudenosaunee lack guilt, they have a pervasive sense of moral responsibility. This moral responsibility lies in rebalancing whatever has been set out of balance, even if temporarily through an action such as hunting. This rebalancing begins even before a hunt, a trek to a fishing station, or a harvest of berries or corn, through prayers, sacred singing, and ceremonies. These

spiritual offerings continue during the taking of other life, and conclude in ceremonies after the taking of a life that bring closure to the entire process. Ceremonies such as Midwinter allow entire Haudenosaunee communities to express
these concepts collectively (Kurath, 1964, passim; Shimony, 1994, 140–64;
Tooker, 1970, 7–37). The West, in contrast, does not believe that a material
loss can be balanced by a spiritual ceremony. The Haudenosaunee remain confident that spirituality does bring balance to the physical world. However, there
are mistakes.

THE FUR TRADE: A POLITICAL NECESSITY AND A HORRENDOUS ENVIRONMENTAL MISTAKE

During the colonial period, fur trade in the North and deerskin trade in the
South was carried out with European settlers by virtually all Indian nations east
of the Mississippi. The trade in fur and skins was an environmental disaster.
The Haudenosaunee faced a moral dilemma, as did the other First Nations,
with regard to this trade. If the Haudenosaunee did not obtain furs, they could
not obtain the guns required to defend their confederacy. However, by integrating Haudenosaunee culture with the market economies of the Dutch,
English, and French, the view that all life was equal was compromised. After
1640, most authors agree (Thomas Eliot Norton takes exception) that the fur-
bearing animals in the Haudenosaunee homeland had been slaughtered to such
a degree that the Haudenosaunee were increasingly forced to look to the
homelands of other Indian nations to obtain the furs their economy required.
This, in turn, initiated wars to control those areas (Hunt, 1940, 34 and passim;
Norton, 1974, 10; Trelease, 1960, 118).

By 1700, the Haudenosaunee realized that they had to stop the cycle of
unending warfare that was causing the English, French, and various Indian
nations to compete in the slaughter of fur-bearing animals. A recognition of
this political reality had to come before a spiritual renaissance of their worldview. Although peace negotiations were carried out in 1700 and 1701 by the
Haudenosaunee with the French and their Indian allies, the peace did not hold
and warfare continued throughout the eighteenth century. Moreover, the
negotiations implied what might be termed an environmental compromise in
the sense that all the First Nations, including the Haudenosaunee, seem to
have believed that there were enough fur-bearing animals in the western Great
Lakes to continue the fur trade unabated (Aquila, 1983, 85–128 and passim;
Dennis, 1993, 229–71; Richter, 1992, 197–235).

Further complicating Haudenosaunee efforts to resolve their dilemma was
the fact that Catholic and Protestant Christian missionaries succeeded in
undermining the Haudenosaunee worldview after 1654 (Huey and Pulis,
1997, 19–20). It is probable that one reason some Haudenosaunee felt com-
fortable in converting to Christianity was the fact that the converts no longer
had to be concerned about an interdependent world of equals but could
instead place themselves, along with their white friends, atop the Genesis

pyramid, where the exploitation of animals and the environment was a mandate from Heaven.

For the traditional Haudenosaunee, the dilemma was "war versus the environment" in both spiritual and political terms. Because the French and English colonists continued their wars, the Haudenosaunee never completely resolved their dilemma. Despite these pressures, there is some evidence that the Haudenosaunee took steps to encourage fur-bearing animals in their territories to recover, and thus to rebalance their worldview. By 1774, there was a sufficient population of fur-bearing animals in the Haudenosaunee homeland to permit limited hunting (Johnson, 1921–1965, 8:683–84).

HANDSOME LAKE

Between 1799 and 1815, new spiritual messages were revealed to a Haudenosaunee teacher, Handsome Lake (Ganio'dai'io'). The spiritual teachings, the Gai'wiio ("Good Message"), which Handsome Lake received from spiritual messengers and then passed on to the Haudenosaunee between 1799 and his death in 1815, did much to ensure the survival of those Haudenosaunee who did not wish to convert to Christianity. The spiritual messages received by Handsome Lake inspired many Haudenosaunee to rise above the social and economic depression that was part of the legacy of the postrevolutionary period, especially as the ever-increasing settlements of non-Indians drove away the deer and other animals the Haudenosaunee had depended upon for much of their food.

Included among the teachings inspired by spiritual messengers were these telling insights into the environmental challenges that faced the Haudenosaunee around 1800: "At some future day the wild animals will become extinct. Now when that comes the people will raise cattle and swine for feast food at the thanksgivings" (Cornplanter and Parker, 1968, 42). The Haudenosaunee could take up white farming methods, but "let them not be proud on that account" (Cornplanter and Parker, 1968, 38). When the Haudenosaunee raised farm animals such as horses and cattle, these animals were not viewed as non-Indians viewed them: "Now all this is right if there is no pride. No evil will follow this practice if the animals are well fed, treated kindly and not overworked" (Cornplanter and Parker, 1968, 38).

THINKING IN A "DIFFERENT BOX"

Haudenosaunee beliefs give them a perspective different from that of the Judeo-Christian tradition, even in the midst of mistakes. The Haudenosaunee experience of the fur trade is a sobering reminder that no philosophy is consistently carried out by any human society, because all human societies make mistakes. But the principles of a philosophy are not revealed simply by how well a society follows them. Perhaps even more importantly, the strengths of a philosophy are revealed by the questions and answers that a particular philosophy poses when a human society inevitably violates that philosophy.

When the Haudenosaunee made mistakes in the past, and when they make contemporary mistakes, their philosophy encourages them to ask questions that are premised on a worldview that is interdependent. This philosophy includes humans within a world of equals rather than the Western worldview that isolates humans from the world. In this sense, the Haudenosaunee are not simply thinking "outside the box," they are thinking in "a different box." As Haudenosaunee history during the fur-trade era demonstrates, this does not necessarily mean that the dilemma will be resolved easily or quickly. But the Haudenosaunee realized *why* their environment was in trouble, with spiritual insights that their European counterparts could only define in terms of using up a physical resource.

DAILY LIFE

Haudenosaunee concepts of the environment are reflected in the practical daily life of Haudenosaunee communities. Rather than compartmentalize aspects of society, the Haudenosaunee integrate the environment with government, religion, and society. This integration can be summarized by the following goals:

Balance: Among women and men; among all human generations; within a nation; within the confederacy; and among all nonhuman beings. Distribution of power, carried out with a sense of responsibility: to achieve balance, there was and is a distribution of political power among women and men; within a nation; within the confederacy; and within the world (environment) among all nonhuman beings.

Consciousness of Purpose: All humans and all other beings act consciously, not instinctively, and these conscious actions are regarded, whether human or nonhuman, as all beings make their best efforts at following the religious instructions of the Creator.

Reciprocity: All humans and other beings owe spiritual obligations to each other so that all may survive on earth.

Rationality of Purpose: All humans and other beings are regarded as believing each is acting rationally; if change is desired, other viewpoints are regarded as rational at least to the being holding them; no viewpoints are regarded as inferior or irrational because that would assume that a being was consciously deciding to act in contradiction to the Creator's instructions. If a person or other being was regarded as acting in ways contradictory to the Creator's instructions, the others seek to heal, not exterminate, that being.

Direction, Not Details: An emphasis on broad, general "directions" and conduct, not on detailed dogmas.

Accountability: Everyone in Haudenosaunee society, all other humans, and all beings in general are believed to be accountable to the Creator and other spiritual forces. This accountability is reciprocal: the Creator and other spiritual forces meet the mortal needs of humans and the rest of the world in exchange for proper behavior by all temporal beings. Proper behavior for humans includes giving thanks to the Creator and other spiritual forces, and spiritually reinforcing the world in general through ceremonies (including, for example, prayers to the deer in exchange for the deer's sacrifice to provide humans with food). Humans are accountable to other humans and to nonhumans. Nonhumans are accountable to all humans. Clans, towns, the five founding nations, other member nations, and the Confederacy are all interdependent in their accountability among each other, their responsibilities to all the nonhuman beings, and especially to the Creator and other spiritual forces.

CONCLUSION

Because the Haudenosaunee worldview perceives the interdependence and communal ethics of all the beings residing within their environment, Western concepts of the environment such as the private ownership of real estate and the exploitation of natural resources cannot be applied to the Haudenosaunee worldview. Of course the Haudenosaunee recognize the physical boundaries where they live and make their living. But they occupy this environment, this land, while sharing it with all other beings. All life has reciprocal responsibilities and obligations. All life is interdependent, and the Haudenosaunee are confident that this interdependence is the intent of the Creator.

REFERENCES

Akwesasne Notes. *Basic Call to Consciousness.* Akwesasne, Mohawk Nation, via Rooseveltown, N.Y.: Akwesasne Notes, 1978.

Aquila, Richard. *The Iroquois Restoration: Iroquois Diplomacy on the Colonial Frontier, 1701–1754.* Detroit: Wayne State University Press, 1983.

Arnold, Philip P. *Eating Landscape: Aztec and European Occupation of Tlalocan.* Boulder, Col.: University Press of Colorado, 2001.

Capps, Walter Holden, ed. *Seeing With A Native Eye.* New York: Harper & Row, 1976.

Cornplanter, Edward, and Arthur C. Parker. "The Code of Handsome Lake." In Arthur C. Parker, *Parker on the Iroquois,* with an introduction by William N. Fenton. Syracuse, N.Y.: Syracuse University Press, 1968.

Dennis, Matthew. *Cultivating a Landscape of Peace: Iroquois-European Encounters in Seventeenth Century America.* Ithaca, N.Y.: Cornell University Press, 1993.

Huey, Lois M., and Bonnie Pulis. *Molly Brant: A Legacy of Her Own.* Youngstown, N.Y.: Old Fort Niagara Association, 1997.

Hunt, George T. *The Wars of the Iroquois: A Study in Intertribal Trade Relations.* Madison, Wis.: University of Wisconsin Press, 1940.

Johnson, Guy. "Proclamation." October 4, 1774. In James Sullivan, ed., *The Papers of Sir William Johnson.* 14 vols. Albany, N.Y.: University of the State of New York, 1921–1965, 8:683–84.

Kurath, Gertrude P. *Iroquois Music and Dance: Ceremonial Arts of Two Seneca Longhouses.* Bulletin 187, Bureau of American Ethnology. Washington, D.C.: Smithsonian Institution, 1964.

Lyons, Oren. "The Canandaigua Treaty: A View from the Six Nations." In G. Peter Jemison and Anna M. Schein, eds., *Treaty of Canandaigua 1794.* Santa Fe, N.M.: Clear Light Publishers, 2000, 67–75.

Norton, Thomas Eliot. *The Fur Trade in Colonial New York, 1686–1776*. Madison, Wis.: University of Wisconsin Press, 1974.

Richter, Daniel K. *The Ordeal of the Longhouse: The Peoples of the Iroquois League in the Era of European Colonization*. Chapel Hill, N.C.: University of North Carolina Press, 1992.

Shimony, Annemarie Anrod. *Conservatism among the Iroquois at the Six Nations Reserve*. 1961. Reprint. Syracuse, N.Y.: Syracuse University Press, 1994.

Swamp, Jake, and John Stokes. *Thanksgiving Address*. Onchiota, N.Y. and Corrales, N.M.: Six Nations Indian Museum and the Tracking Project, 1993.

Tooker, Elisabeth. *The Iroquois Ceremonial of Midwinter*. Syracuse, N.Y.: Syracuse University Press, 1970.

Trelease, Allen W. *Indian Affairs in Colonial New York: The Seventeenth Century*. Ithaca, N.Y.: Cornell University Press, 1960.

Vecsey, Christopher, and Robert W. Venables, eds. *American Indian Environments: Ecological Issues in Native American History*. Syracuse, N.Y.: Syracuse University Press, 1980.

Williams, Paul. "Treaty Making: The Legal Record." In G. Peter Jemison and Anna M. Schein, eds., *Treaty of Canandaigua 1794*. Santa Fe, N.M.: Clear Light, 2000, 35–42.

Robert W. Venables, Ph.D.
American Indian Program
Cornell University

HONDURAS

INTRODUCTION

Protest of hydroelectric construction in Honduras can be hazardous to one's health, as illustrated by the murder of environmental activist Carlos Roberto Flore in 2001. In a pattern of governmental behavior that is very familiar throughout the world, dam building and gold mining are being promoted in Honduras by the World Bank and the International Monetary Fund to earn foreign exchange, pay down debt, and keep pace with the cash demands of a globalized economy.

A MURDER FOLLOWS PROTEST OF DAM CONSTRUCTION

During July 2001, about 100 campesinos from the municipality of Gualaco (department of Olancho) camped in front of the National Congress of Honduras calling for justice regarding the assassination of community member Carlos Roberto Flore, who was killed June 30, following his opposition to the Babilonia River hydroelectric project. Eyewitnesses said that Flore was assassinated by security guards employed by the private company, Energisa, which is building the dam. On July 18, the Gualaco community was joined on the congressional grounds by about 1,000 campesinos from COPINH, the Civic Council of Popular and Indigenous Organizations of Honduras.

On July 20, unknown persons smashed the lights of a vehicle owned by Daniel Graham, a U.S. citizen who was assisting the Gualaco community's protests. Three days later, unknown persons driving a yellow car also fired two shots at the mayor of Gualaco, Rafael de Jesus Ulloa, Father Fredy Cornelio Benitez Alvarez, and Sister Carmelita Luis David Perez (all of Gualaco), when they were driving in Olancho. Each of them had spoken out against the dam ("Indigenous Protest," 2001).

Energisa won a government contract in May 2000 to build and run a hydroelectric dam on the river Babilonia. According to *Cultural Survival*, many local people believe that the dam "will seriously damage the environment and ruin the livelihood of local people, who may be forced off their land" ("Community Leader," 2001).

THE SPREAD OF GOLD-MINING CONCESSIONS

More than 30 percent of Honduras's territory was licensed to foreign mining companies during the late 1990s (Marsh, 2001). The Asociaciòn de Organismos No Gubernametales (ASONOG), based in Santa Rosa de Copán, spent more than a year investigating the massive recent intrusion into Honduras of gold-mining companies from Canada and the United States. The Honduran mining department issued mining concessions totaling 21,000 square miles to foreign companies, mostly from the United States, Canada, and Australia. Much of this mining licensing took place under the aegis of recovery from Hurricane Mitch, which ravaged Honduras and Nicaragua during fall 1998. Four weeks after the hurricane, during December 1998, a new mining law was enacted that reduced taxes and gave foreign companies nearly unlimited power to petition for the removal of traditional communities located near mineral deposits.

The political and economic climate had been changing in Honduras even before Hurricane Mitch. One of the first mining companies to take advantage of the newly favorable business climate was Greenstone Resources Limited, based in Toronto, Canada. Greenstone gained the mining concession for several hundred acres in Copán (in western Honduras) during the mid-1990s, then promptly moved to evict the local residents (Marsh, 2001).

Local residents surrendered to Greenstone after two years of enduring corporate tactics that included shutting off their water and intentionally crushing one local resident with a bulldozer. Greenstone promised local people resettlement and title to new lands, which was not forthcoming before the company went bankrupt, its assets auctioned to another company. Greenstone had been the target of numerous environmental complaints, including illegal discharge of waste into a nearby river, using cyanide within 25 yards of occupied homes, and causing the deaths of farm animals (Marsh, 2001).

Elsewhere in Honduras, Entre Mares, a subsidiary of the U.S. company Glamis Gold, faced civil and criminal charges for taking water needed by nearby communities and for cutting down swaths of a forest without permission. According to one observer, "Entre Mares, which commenced operations without a required environmental license, has been the target of numerous protests by members of the Committee to Protect the Environment in the Valley of Syria" (Marsh, 2001).

Honduras has been granting widespread mining concessions under pressure from the International Monetary Fund (IMF), recovering from Hurricane Mitch by attracting foreign venture capital through offering low taxes, unlimited access to water, legal rights to expropriate campesino and indigenous lands, and relief from the country's few environmental regulations. During

December 2000, Honduras eliminated its export tax on mining products, and it cut land-use fees to as little as $1,500 a year for large mines, with a nominal one percent municipal tax (Marsh, 2001).

Government advocates of the new legal climate for mining assert that the Honduran economy will benefit from rising employment offered by the mines. Opponents, including many indigenous peoples who fear that their lands may be devastated by the new mines, say employment has not materialized. In San Andrés Minas (the community forcibly removed by Greenstone), for example, only 11 local people are employed at the mine. Overall, the mine employs 144 people, fewer than half of the 370 jobs promised by the company when it convinced the government to grant it the concession (Marsh, 2001).

The land taken for mining had been used previously to grow corn and graze cattle, employing many more indigenous people. Taken from the point of view of the community that has watched the mine destroy its traditional economy, the mine is a liability, not an asset. Additionally, indigenous people in eastern Honduras are fighting to protect rivers, which run through their ancient tropical rainforest, from dredging associated with mining. They and other Honduran activists have been seeking to prohibit use of cyanide in mining operations. They also are seeking to prohibit the expropriation of campesino and indigenous lands, and to strengthen mining and environmental regulations.

In El Porvenir, Honduras, near a densely populated region, local groups organized to resist a large gold mine because of toxic releases and its use of scarce fresh water; they also resisted pressure to evict local residents from their homes to make way for expanding mining operations. Subsistence farmers in the area report that areas proposed for their resettlement are unsuited for growing crops or raising animals. One community near the San Andres mine was located only 42 meters from a cyanide leach pad. Homes in another village shake every time a nearby mountain is dynamited to mine gold ("Gold Mining," 2001).

REFERENCES

"Community Leader Protesting Dam in Honduras Murdered." Global Response's Quick Response Network. *Cultural Survival News Notes*, July 19, 2001. [http://www.cs.org/main.htm]

"Gold Mining in Honduras Project." Center for Economic and Social Rights, May 11, 2001. [http://www.cesr.org/PROGRAMS/honduras.htm]

"Indigenous Protest in Honduras." *Cultural Survival News Notes*, July 27, 2001. [http://www.cs.org/main.htm]

Marsh, Michael. "Honduras Is Worth More Than Gold." 2001. [http://www.menominee.com/nomining/honduras.html]

INDIA

INTRODUCTION

India, the world's second most-populous country, still has room for a large number of indigenous peoples. Some of these groups (on the isolated Andaman and Nicobar Islands in the Bay of Bengal) have recently experienced their first cases of sometimes-fatal influenza and measles, caught from tourists intent on having an exotic experience, no matter what the human cost. On the mainland, uranium poisoning has reached a remarkable degree of severity among indigenous peoples in the Jharkand Tribal Belt of Bihar. As many as 40,000 indigenous people have rallied in the Kashipur region of Orissa to blockade representatives of multinational companies exploring the region for bauxite, the basic material used to manufacture aluminum, provoking attacks by police that have killed several people. When Enron went bankrupt in the United States, indigenous peoples in India cheered as they watched the country's largest foreign investment, the Dabhol electricity plant, sink as well. The plant had been proposed by Enron to produce electricity from natural gas at more than six times the previous prevailing rate. Several proposed hydroelectric dams also have incited indigenous protests in India; one of them, a complex of dams along the Narmada River, has provoked tens of thousands of people to occupy dam sites (and engage in pitched battles with police), as others pledge to drown in rising waters rather than abandon their homes.

IMPORTED DISEASES ON THE ANDAMAN AND NICOBAR ISLANDS

Fewer than 400 surviving indigenous Jarawa people on the Middle and South Andaman islands (in the Bay of Bengal, south of Burma) are being threatened by tourists, who are infecting the islanders with their first cases of influenza, measles, and other imported diseases. Survival International has drawn attention to the fact that illegal settlers and tourists are making more frequent incursions into the Jarawas' protected reserve. "The organization, which campaigns for the rights of tribal peoples worldwide, said that police have been routinely bribed to turn a blind eye. Survival international said: "This is extremely dangerous for the tribe as it is likely to result in the spread of contagious diseases which could easily prove fatal" (Barton, 2001, 2).

Officially, travel is severely restricted to the Andamans and Nicobars, an Indian government territory of almost 300 islands (39 of which are inhabited). A permit is required, good for only 30 days in specified areas. Of four indigenous tribes originally living on these islands, only the Jarawa and the Sentinelese remain. Influenza, measles, and other imported illnesses carried by foreign colonists devastated the Andamanese and the Onge almost immediately after first sustained contact with outsiders during 1997. Within two years, measles was epidemic on the islands. In the meantime, the government's permits were not enforced, and its restrictions were generally ignored as tourists seeking an exotic experience carried their germs and viruses into areas set aside as protected tribal reserves.

The next few years of the twenty-first century will be crucial for the Jarawa people of the Andaman Islands in the Indian Ocean. The Jarawa have only recently started coming out of their forest home to meet outsiders, and nobody else speaks their language. They have resisted government plans of removal and forced resettlement that would destroy their livelihood and risk exposure to sometimes-fatal diseases. Forcible settlement has all but wiped out other tribes on the Andaman Islands—the Great Andamanese were reduced from 5,000 to about 40 people in 150 years. In 1999, plans to settle the Jarawa were temporarily shelved following an international outcry.

Survival International has urged the Indian government to close the Andaman trunk road through the Jarawans' territory that exposes them not only to tourists, but also to game poachers. The Jarawa have opposed the road since its construction was first proposed during the 1970s. They attacked road-construction crews and felled trees to block work. In retaliation, construction crews strung up high-voltage lines, which electrocuted an unknown number of Jarawanese ("India: Road Brings Death," 2002).

On May 7, 2002, the Supreme Court of India issued an order that closed the Andaman trunk road and directed removal of settlers from tribal reserves. This order may provide the tribe of 250 to 300 people with a margin of safety from encroachment from outsiders who have been destroying the tribe's forest, poaching their game, and introducing epidemics ("India's Supreme Court," 2002). The question now becomes enforcement, given the fact that many existing laws enacted to protect the indigenous peoples of the Andaman Islands are being regularly violated. By mid-2002, many settlers had moved away from the Jarawa tribe in the Andaman Islands, in accordance with the Indian Supreme Court ruling.

"The government of India is very keen to further promote the Andaman Islands as a tourist destination," said Alka Kohli, director of the India Tourist Office in London. She added, however, that "permission is not given for visiting the Jarawa reserve" (Barton, 2001, 2). The standard bribe at the reserve's border has increased from 500 to 2,000 rupees, according to one report. Illegal tours to the islands have been booked in Madras (Barton, 2001, 2).

URANIUM POISONING IN THE JHARKAND TRIBAL BELT OF BIHAR

Radiation from uranium mines in Bihar state in eastern India has been linked to genetic mutations and slow, often painful deaths among more than 30 indigenous tribal peoples. The radiation originated in three mines that are operated by the Uranium Corporation of India and owned by the country's Atomic Energy Department. The mines are located in or near the villages of Jaduguda, Batin, and Narwarpahar in the tribal belt of India known as Jharkand. Uranium from this area fuels India's entire nuclear-power industry.

Nearly 30,000 people in 15 villages live near the mines and their associated tailings ponds. According to a report in *Drillbits and Tailings,* many of these people assert that they have suffered radiation poisoning. In a survey conducted by the Jharkand Organisation Against Radiation (JOAR) (including residents of 7 villages located within a 1-kilometer [0.6 miles] radius of a radioactive waste pond), 47 percent of the women reported disrupted menstrual cycles; 18 percent said they had suffered miscarriages or had given birth

A truck transports uranium ore with little protective cover from a mine near Jaduguda, in the eastern Indian state of Bihar, July 2, 2000. The dispute over possible dangers of the mine activities could spread to other regions as the government steps up its hunt for new sources of uranium. (AP Photo/Saurabh Das)

to stillborn babies during the preceding 5 years. Nearly all of the women complained of fatigue, weakness, and depression ("Uranium Mine," 1999).

The same survey also described medical pathologies afflicting people in nearby villages, including "skin diseases, cancer, tuberculosis, fertility loss, bone and brain damage, kidney damage, hypertension, disorders of the central nervous system, congenital deformities, insomnia, nausea, dizziness, and pain in the joints and abdomen" ("Uranium Mine," 1999).

Indian authorities have been quite high-handed at times in determining which villages will be displaced for uranium-waste facilities. The *Drillbits and Tailings* report described the fate of Chatijkocha, "a prosperous village whose economy was built on agriculture and forest products and was home to famous theater groups" ("Uranium Mine," 1999). The same village had become a refuge for many people who had been forced out of their homes by nearby uranium mines.

On January 26, 1996, according to the same report, mine management "assisted by the Central Reserve Police Force, the Central Industrial Security Force, the Bihar Police, a set of heavy bulldozers and a high-powered shovel, entered the Chatijkocha village and started bulldozing the houses without any warning" ("Uranium Mine," 1999). In the village of Dungardihi, open tailings ponds slosh onto well-traveled roads during the monsoon season. Tailings effluent mixes freely with drinking water supplies in some villages. Employees in the mines are said to go about their work in cotton uniforms, instead of plastic protective suits, as they freely inhale uranium dust and radon gas. At least 71 miners died between 1994 and 1997, according to *Drillbits and Tailings* ("Uranium Mine," 1999).

DEATHS OF BAUXITE MINING PROTESTERS

Indigenous communities in the Kashipur region of Orissa, India have sought international support following the killings of three community members and the serious maiming of nine others by Indian government police on December 16, 2000, during protests of plans to mine bauxite, the basic raw constituent of aluminum. Protesters also objected strenuously to plans for a large aluminum smelter that they fear will devastate the local environment.

The $1 billion mining project has led to confrontation between the tribal villagers, who stand to lose their lands and forests, and the company's supporters, including the state government of Orissa and the police ("India: Mining Ancestral Lands," 2001). On December 16, 1999, 3 men were killed and 9 others seriously injured by armed police during protests by more than 4,000 people from 15 villages in the area.

The exploring companies, according to a report in *Drillbits and Tailings,* include the Indian companies Hindalco and Utkal Alumina (the latter is 35 percent owned by Alcan, a Canadian company). The other partner is Norsk Hydro, a Norwegian corporation with investments in light metals, oil, petrochemicals, and agriculture. The bauxite in question is located on sacred tribal lands in eastern Orissa. The project, as planned, would process one million tons of bauxite a year ("India: Mining Ancestral Lands," 2001). Following the demonstrations and shootings, Norsk Hydro briefly curtailed project activities. The company has faced tensions over the project since 1993, when the predominantly tribal population of the region first heard of plans by Utkal to mine bauxite there ("India: Mining Ancestral Lands," 2001).

The government of India, which already had transferred ownership on the land in question to the companies, "has asked the people to vacate the land. . . . [and has] met strong resistance from the people of Orissa" ("Indian Police," 2001). Residents of the local communities responded by blocking local access to company surveyors, and by blockading government offices.

On December 16, 2000, according to *Drillbits and Tailings*, police response was swift:

> Two platoons of police descended on the [blockade], and except for a few women, the terrified villagers fled for the hills. According to one of the women, Indumati Jhodia, the police stated that they had orders to fire, and were to spare neither woman nor child if the men were not to appear before them. "When we tried to reason with them, they assaulted Danei Jhodia, a 55-year-old woman, mercilessly. She fainted and we went inside our doors in fear. There was a lathi charge [assaults with bamboo poles], cries and firing." ("Indian Police," 2001)

The mineral-rich area is already studded by 172 different mines. According to *Drillbits and Tailings*, the mines have not brought promised jobs and income, but have burdened much of the area with environmental degradation. A large aluminum smelter in the area (owned by NALCO) has polluted the once-pristine Bramhani River, where a tailings dam collapsed, causing a flash flood of toxins and producing 40,000 environmental refugees, according to the National Center for Advocacy Studies ("Indian Police," 2001). Local people believe that much of the local hilly terrain will be pulverized by strip mining, leading to erosion and toxic overflow during the monsoon season.

According to a *Drillbits and Tailings'* report, Andhra Pradesh contains large deposits of bauxite for which the state government has been issuing leases since 1952, allowing industrial-scale mining operations in the homelands of several indigenous communities, including Bhagata, Khond, Konda Reddi, Samantha, and others. In 1995, a political party called Samata approached the Supreme Court with a "Special Leave Application" and eventually won a historic judgment in favor of the various indigenous communities in July 1997 ("Indian Police," 2001).

The Baphlimali Hills, site of the proposed mine, is a source for 350 streams, including tributaries that feed local rivers. To local indigenous peoples in this region, Baphlimali is a sacred life-giver. Although the region's forest cover has been depleted over the years, there is still enough left to sustain more than 70 villages in the region. Kutrumali, a huge mountain that the companies plan to mine, has forests covering around 10 to 15 percent of the plateau top. Forests that remain after previous mining are crucial for local tribal peoples' food supplies during the dry months. Local people have asserted that Utkal's plans to mine 200 million tons of bauxite from the Baphlimali plateau will destroy this watershed. The number of people to be affected by mining range from 750 (Hydro's estimate), to 3,500 (Utkal's estimate), to 60,000 (Norwegian Agency for Development Cooperation estimate) ("India: Mining Ancestral Lands," 2001).

Critics of the proposed mine assert that Utkal has downplayed the importance of the ecosystem it plans to mine, and has misrepresented sludge-deposition rates in its application for environmental clearance. The same critics accuse Utkal of having presented misleading data about the region's economy and ecological status. According to Norwatch, the deforestation caused by the mines and associated smelter will be aggravated because of the hilly terrain, resulting in more frequent flash floods, landslides, and nutrient depletion of water resources. Simultaneously, forest loss also would aggravate existing losses of habitat for the region's wildlife, including bears, jackals, wolves, sambars (a deerlike animal), spotted deer, leopard cats, and tigers ("India: Mining Ancestral Lands," 2001).

ENRON, VELDUR, AND SKY-HIGH ELECTRIC RATES FROM NATURAL GAS

Just before dawn on June 3, 1997, police officers forcibly entered the homes of several women in Veldur, a fishing village in western India, dragging them into waiting police vans and beating them with sticks. The only so-called crime committed by the women, one of whom was three months pregnant, was having led a peaceful protest against the environmental impact of a massive new natural gas plant being built near Veldur by Enron (Chatterjee, 2000). Upon its anticipated completion, Enron's Dabhol Power Project would have been the world's largest natural-gas-fired electricity power plant. Before Enron went bankrupt late in 2001, the second phase of the project had received a go-ahead from local officials.

Enron had begun work on the 2,015-megawatt Dabhol plant during early 1995 in a joint venture with Bechtel and General Electric at an estimated cost of $2.8 billion. That year the company began clearing ground 100 miles south of Bombay on several hundred acres of porous red rock and shrub-covered hills on a volcanic outcrop overlooking the Arabian Sea. The company also constructed large docking facilities to be used to off-load the plant's fuel—liquefied natural gas purchased from an Enron subsidiary in Qatar.

Enron denied any connection to the arrests and beatings of the women in Veldur. "The Dabhol Power Company does not employ, second or subcontract police officers at the site. By law, we are required to offset the cost of police officers placed near our site if police officials deem it necessary to preserve law and order when protests occur. We have no authority over their actions," the company asserted in a November 17, 1997 statement (Chatterjee, 2000).

Enron's statement contradicted a January 1999 report by Human Rights Watch (HRW) investigators, who found that Enron was directly paying police salaries. Police subinspector P. G. Satoshe, who was in charge of this operation, told HRW that Enron was picking up the tab for policing Dabhol. "I calculate the number of officers there and according to the [government-set] rates, submit a report to the superintendent of police in Ratnagiri. . . . I do not handle any money. The company pays directly to the government," he told HRW (Chatterjee, 2000). Beatings of protesters at company gates had become a routine response to at least 30 separate protests, according to HRW.

Kenneth Roth, HRW's executive director, said that the protests resulted from the Maharashtra State Electricity Board's decision to add the rising costs of the Dabhol plant's construction onto the bills of electricity consumers. During 1999, the local monopoly utility bought a quarter of its power from the newly completed first phase of the Dabhol power project at about 4.14 rupees (roughly U.S. 10 cents) per kilowatt hour, nearly double the average previous cost of producing electricity from existing sources. Enron itself projected that electricity generated from its natural gas would rise to 33 cents per kilowatt hour by 2017, more than 6 times the previous cost (Chatterjee, 2000).

During November 2001, the *Wall Street Journal* reported that activists in India were "rejoicing" at the news of Enron's collapse on United States financial markets (Raghavan, McKay, and Zuckerman, 2001, A-6). The company's stock, which had traded as high as $80 to $90 a share two years earlier, ended the month at 70 cents a share after investors learned that executives had lied massively about profit levels and outstanding debt. The collapse came at a crucial time for the Dabhol plant—days before a meeting of the Maharashtra board at which an activists' motion to cancel Enron's contract was to be heard. Dabhol's lenders canceled a meeting planned for early December in London, signaling probable withdrawal from financing the controversial power plant. "We've been following it on the Internet," said Girish Sant, an Indian activist who has been highly critical of Enron's role in the Dabhol plant for years (Raghavan, McKay, and Zuckerman, 2001, A-6).

Enron's collapse sent "shivers" through India's banking system, according to a report in the *Wall Street Journal* (Bailey, 2001, A-8). The $2.9 billion Dabhol project was India's largest single foreign investment. Some of the $1.4 billion in loans for the plant were backed by its future generating capacity. Nearly a quarter of the loans issued by India's banks already were nonperforming, according to a Standard and Poors report. During 2000, the Maharashtra Electric Company, the plant's only customer, stopped buying its electricity, asserting that its rates were too high (Bailey, 2001, A-8).

SAYING "NO" TO THE NARMADA DAM COMPLEX

The indigenous organization Narmada Bachao Andolan (NBA) has led a 15-year struggle against a series of large dams being built or proposed for the Narmada River. The dams eventually could uproot hundreds of thousands of people, mainly in tribal villages along the river's course, destroying their ways of life and cultural heritage. The requirements of large-scale industry and urban electricity consumers have come into direct conflict with the traditional lives and economies of many indigenous peoples along the banks of the river. Constructed in its entirety, the Narmada Valley dam project will be the second largest hydroelectric complex in the world, after the Three Gorges dam project in China (Rajagopal, 2001).

During the 15 years that indigenous peoples have resisted the imposition of new dams along the Narmada River, tens of thousands of protesters have occupied dam sites to prevent construction, provoking intervention by police. A number of residents have obstructed construction by refusing to move, risking drowning as waters rise behind proposed dams during the monsoon season. According to outside observers, "The governments of India and the states of Gujarat, Maharashtra and Madya Pradesh have all failed to provide basic compensation or rehabilitation for 'oustees,' and have engaged in violent repression of peaceful protesters" (Tobin, 2001).

During the fall of 2000, India's Supreme Court allowed one of the dams, the Sardar Sarovar Project (SSP), to be raised to 90 meters, 10 meters above its present height of 80 meters. Advocates of the dam want it eventually to reach 145 meters in height. Many people who are threatened with inundation say they will refuse to leave their homes if monsoon rains raise water levels behind the dam.

Given the enormous scope of protests, the World Bank, which originally planned to fund the Narmada project, withdrew in 1993 after it was criticized for violation of its own internal regulations on resettlement, rehabilitation, and environment. Every funding source since then, Japanese and German included, has withdrawn after encountering intense local, indigenous criticism. The project continued with funding from Indian state governments which redirected scarce funds from health and education projects (Rajagopal, 2001).

India's government has promoted industrialization via dam building since its independence from Great Britain in 1947. Jawaharlal Nehru referred to large dams as India's "secular temples" (Tobin, 2001). The government planned to supply irrigation and electric power with the dams. Dam building enjoyed widespread political support in India across the political spectrum until the NBA began to speak up for the people who were losing their homes and traditional economies in 1985.

The Narmada projects include several dams, some of which have been completed. Two projects, the SSP and the Maheshwar Hydro-electric Project, have provoked the most vehement opposition. Most of the people displaced by the SSP will be adivasi (tribal) villages of small farmers. These two dams are being built to provide electricity and irrigation for parts of three Indian states: Gujarat, Maharashtra, and Madya Pradesh.

According to official surveys, the SSP will displace about 40,000 people. The government says that the Maheshwar dam will submerge between 9 and 13 villages. The NBA, by contrast, asserts that about 200,000 people will lose their homes to the SSP and at least 50,000 will lose their homes to the Maheshwar, including 61 villages (Tobin, 2001).

Observers from Ohio's University of Toledo reported that "people have repeatedly told us they are willing to die, but they will not leave this land. Several years ago, a similar tactic led to a temporary freeze of the SSP dam height at 80 meters. Six people, including NBA leader Medha Patkar, had made the decision to allow themselves to be drowned. But after days in neck-level, rising water, government ordered the Indian police to forcibly drag out the protesters and institute a temporary hold on construction" (Tobin, 2001). Opponents of the dams also have used sit-ins, hunger strikes, and other forms of protest. With support from around the world, the NBA has pressured international banks and companies to withdraw their support of the Narmada projects.

During January 1998, 20,000 villagers sneaked through the predawn darkness to initiate a three-week occupation of the Maheshwar dam site as construction was beginning. The occupation took place shortly before elections, so the government promised to convene an independent inquiry and halt construction until the inquiry was completed. Two months later, after the election, the government reneged and resumed construction (Tobin, 2001). On April 22, 1998, several thousand people were arrested as they again stormed the dam site. The next day, more than a thousand peaceful protesters "were met with fierce, violent repression from the Special Task Force, known as the 'black cats,' for their distinctive uniforms. The STF officers used clubs and horses to beat protesters, sending about 50 to the hospital" (Tobin, 2001).

By law, all project-affected persons are to receive land in compensation for territory submerged by rising waters behind dams, an important matter because most of them are farmers. In the bureaucratic language of the project, they are supposed to be "rehabilitated." The government asserts that all affected people have received fair compensation but, according to outside observers, "We spoke to persons [who] claim they have never been offered any compensation, others that were given useless 'titles' to land already committed to other oustees and still others that were promised new house plots but farming land that [was] already underwater due to pre-monsoon rains" (Tobin, 2001).

REFERENCES

Bailey, Rasul. "India's Banking Sector Feels Enron's Aftershocks." *Wall Street Journal*, December 6, 2001, A-8.

Barton, Robin. "Tourism Threatens Health and Welfare of India's Island Tribes." *London Independent*, April 22, 2001, n.p.

Chatterjee, Pratap. "Enron In India: The Dabhol Disaster." CorpWatch: Holding Corporations Accountable. July 20, 2000. [http://www.corpwatch.org/issues/politics/featured/2000/enronindia.html]

"India: Mining Ancestral Lands for Corporate Profits." *World Rainforest Movement Bulletin* (November 2001): n.p. [http://www.wrm.org.uy]

"Indian Police Kill Three Protesting Mining by Canadian Alcan Inc." *Drillbits and Tailings* 6:5 (January 31, 2001): n.p. [http://www.moles.org/ProjectUnderground/drillbits/6_01/4.html]

"India: Road Brings Death to Isolated Tribe." Survival International, February, 2002. [http://www.survival-international.org/jarawauab0202.htm]

"India's Supreme Court Closes Isolated Jarawa Tribe's 'Road of Death.'" Survival International, May 27, 2002. [http://www.survival-international.org/enews.htm]

Jayaraman, Nityanand. "Norsk Hydro: Global Compact Violator." *Corporate Watch*, 2001. [http://www.igc.org/trac/un/updates/2001/norskhydro.html]

People's Commission on Environment and Development in India. [http://www.pced india.com/peoplescomm/advocacy]

Raghavan, Anita, Peter McKay, and Gregory Zuckerman. "Enron's Woes Touch Firms around the World; Scandinavian Paper Makers; Tokyo Traders Feel Pain; Some Rejoice in India." *Wall Street Journal*, November 30, 2001, A-6.

Rajagopal, Balakrishnan. "The Violence of Development." *Washington Post*, August 9, 2001, A-19.

Tobin, Francis X. "Save the Narmada." University of Toledo, June 28, 2001. [http://comm-org.utoledo.edu/pipermail/announce/2001-June.txt]

"Uranium Mine Poisons Indigenous Communities in India." *Drillbits and Tailings* 4:7 (June 1, 1999): n.p. [http://www.moles.org/ProjectUnderground/drillbits/990601/99060101.html]

INDIGENOUS ENVIRONMENTALISM AND ECONOMIC DEVELOPMENT

One of the most important—and often the most vexing—questions debated among North American indigenous peoples today concerns the creation of reservation economic bases that produce necessary cash income while being culturally appropriate and sustainable. Casinos, the reservation cash-cow du jour, sometimes produce mountains of money as they transform parts of reservations into annexes of the non-Indian economy, with all of their imported artifices and vices.

One question that concerns many responsible students of Native American economic infrastructure is: What is going to be left behind that is sustainable and culturally appropriate after gambling loses its luster, and after non-Indian interests find ways to cash in on the bonanza?

Modern Tribal Development by Dean Howard Smith (2000) is one of the first books to confront this question directly. It is an extension of the author's involvement in programs such as the National Executive Education Program for Native American Leadership (NEEPNAL), which aims to increase "the cultural integrity and sovereignty of the Native American nations. . . . leading to cultural integrity, self-determination, and self-sufficiency" (Smith, 2000, x).

In this context, Smith, an associate professor of economics at Northern Arizona University as well as a Grand River Mohawk, raises some very important questions:

- How do native peoples maintain their cultural individuality and secure their cultural integrity?
- How do tribes develop their sovereign rights as stipulated in the Constitution and the laws of the United States and the treaties? and, most importantly,
- How do tribes become fully self-determined and self-sufficient, thereby securing their rights and cultures? (Smith, 2000, 3)

Development can be beguiling but costly in terms of cultural integrity and self-sufficiency. Early in the history of European colonization in the Americas, Smith points out, missionaries and merchants introduced guns, after which many Native American hunters lost their traditional tracking and bow-hunting skills. "When the shotgun shells ran out," Smith observes, "they became beggars" (Smith, 2000, 8). One recalls the debates engaged by Canassatego, Haudenosaunee Grand Council *Tadadaho* (speaker) during the 1740s, regarding whether European artifices should be adopted at all.

Even today, some Native American nations have made conscious decisions to forgo the types of economic development that would make money and build infrastructure on a so-called mainstream capitalistic model. The Hopis, for example, have passed up substantial potential income from tourism so that they can maintain their religious traditions. The Hualapai have closed large parts of their lands to outsiders so that they can live unmolested in a traditional manner. The Havasupai have refused uranium mining; they believe that mining uranium desecrates Mother Earth.

Often, debates regarding the appropriateness of development have split native nations, sometimes with deadly results. The Akwesasne Mohawk civil war during 1990, which degenerated to gunfights (and two deaths), was essentially a debate over which path economic development should take. One faction interpreted sovereignty as the freedom to establish businesses, including gaming, and to smuggle goods across the international border. Another faction opposes the transformation of the Akwesasne into a one-stop cash transfusion from non-Indians looking for discount smokes, cheap gas, or a lucky strike on the video slots. The same debate has split the Senecas, with a seriousness that has cost several lives.

Smith points out, briefly but cogently, the fact that "Native Americans had extensive and vibrant economic systems of production and trade during the centuries of pre-contact" (Smith, 2000, 23). Extensive trade networks laced the hemisphere long before the arrival of Columbus; some of today's interstate highways follow these original routes. The Senecas, for example, maintained extensive, lush cornfields, which General John Sullivan's soldiers marveled at before they laid them to waste. The contributions of Native American cultures and peoples to our staple diet (not to mention our political institutions) are well known to serious students of Western Hemispheric history.

Smith also points out the growing role of tourism on many reservations, meanwhile stressing the need to develop tourism industries that do not disrupt traditional life. Dances, rodeos, and powwows are among the draws. In the meantime, some Native American nations face difficulties keeping tourists away from religious ceremonies that are meant to be private. Without cultural education, many non-Indians often cannot tell the difference between public and private activities.

Instead, writes Smith, instead of being assimilated into an industrial capitalistic system, traditional Native American regard for the environment can be used to design "a new type of system that incorporates competitive behavior, social compatibility and adaptation, and environmental concerns" (Smith, 2000, 77). Smith has a handle on the conceptual framework required to develop reservation economies that will enhance social values and culture rather than simply make money for off-reservation interests. Here, we have a truly exciting beginning. Next, Smith and others will have to face the hard part of this chore: How do we bring these thoughts to street-level reality in some of the poorest patches of earth on Turtle Island?

REFERENCES

Smith, Dean Howard. *Modern Tribal Development: Paths to Self-sufficiency and Cultural Integrity in Indian Country.* Walnut Creek, Calif.: AltaMira Press, 2000.

INDONESIA

INTRODUCTION

For comparative purposes, Indonesia, the fourth most populous nation on Earth, probably hosts more specific instances of indigenous conflict with industrial development than any other place which functions (increasingly nominally in the first years of the twenty-first century) under a single national government. Cobbled together by Dutch administrators from thousands of islands (and nearly as many individual indigenous traditions), Indonesian peoples present a vast array of responses to the intrusions of industrialism. Some peoples on East Kalimantan (Borneo) have chosen to accommodate—in one case, former headhunters have turned their traditional longhouses into bed-and-breakfast establishments for Japanese tourists. Others resist intrusion actively, such as the Dayaks, also of East Kalimantan, who have made life dangerous for corporate gold-miners who disregarded their traditional, small-scale mining customs. Local people in Sumatra have taken to the streets in opposition to pollution caused by a large pulp-and-paper mill, and have shut it down, after several people died at the hands of police and troops.

Indonesia has become a case study in just how quickly the resources of the Earth are being consumed by the industrial engines of capitalism. On Borneo, the devastation of natural habitats is reflected in heretofore unknown attacks on humans by sun bears which, like the polar bears of the Arctic, no longer find the natural environment supplying them adequate food. Deforestation rates in this island nation are among the highest in the world.

RAPID DEFORESTATION

Decades of uncontrolled exploitation have resulted in massive deforestation across Indonesia. The deforestation rate between 1984 and 1998 was around 1.6 million hectares each year, and recent deforestation has been closer to 2.0 to 2.4 million hectares each year, one of the highest deforestation rates in the world. By 1997 Indonesia had lost 72 percent of its original forest cover; more than half of Indonesia's remaining forests were threatened (Bryant, Nielsen, and Tangley, 1997).

Indonesia is home to one-tenth of the world's tropical forests, which are quickly being drawn into the global economy as merchandise. The last trees of central Java's once-great teak forests have ended up as lawn chairs in the United States, Europe, Japan, Australia, China, and other countries. Dozens of workshops around the forest are turning the teak logs into chairs, mainly for export. Those chairs, and other manufactured products, spell the end of the majestic forests that once blanketed large parts of Indonesia. Their disappearance also has hastened the extinction of innumerable animal and plant species indigenous to this country. "We are facing a cataclysm," said Togu Manurung, the director of Forest Watch Indonesia, an environmental organization that documents the destruction of the country's forests (Gargan, 2001, A-7).

Citing rapid deforestation, the Indonesian Forum on the Environment in 2001 called for a moratorium on industrial logging for two to three years, or until the forest industry and forest management can be reformed. The group proposed that "the logging moratorium would be phased in . . . with the objective of stopping the illegal and destructive logging, and promoting sustainable forest management. Illegal logging can most easily be stopped during a period when no industrial logging is allowed" ("Portrait of Indonesian Forestry," 2001). In the short term, according to this proposal, forest-based industries would be forced to rely on plantation-produced or imported wood, "or else close their operations" ("Portrait of Indonesian Forestry," 2001).

By way of precedent, Indonesian environmental activists cited similar moratoria in other Asian nations. In 1988, for example, after 15 years of logging blockades by rural communities, and following devastating floods and landslides linked to logging, the government of Thailand banned all logging throughout the country. The ban was still in place in 2001. A decade later, facing similar circumstances, China banned logging in its remaining native forests. The decision followed major flooding in lowland areas near the Yellow and Yang Tse rivers in 1998. The floods killed thousands of people, while millions were evacuated from their homes. Economic damage stemming from the floods in China was estimated to have been more than $10 billion. The Chinese logging ban affected one million forestry workers. Half were employed in plantation projects (mainly using native species and designed to restore degraded lands). One quarter became forest guards, involved in protection and restoration of the native forest areas. Closer to Indonesia, during 1999, the government of Papua New Guinea placed a moratorium on the granting of new logging concessions.

In an unpublished report on the state of Indonesia's forests, the World Bank predicted that all of Sumatra's lowland forests will be logged before 2005. Kalimantan's (Borneo's) lowland forests are expected to be gone by 2010. Swamp forests, according to the report, will disappear within the next five years. During the 1990s, the rate of Indonesia's deforestation accelerated from 2.47 million acres to 4.2 million acres annually (Gargan, 2001, A-7). Freeboot capitalism has been even more rapacious in the forests than the former so-called New Order of the Suharto dictatorship.

Borneo's Tanjung Puting National Park, designated by the United Nations as a biosphere reserve, is being systematically and illegally logged, according to Forest Watch, Telepak Indonesia, and Indonesia's Ministry of Forestry and Estate Crops. Suripto, the secretary general of the forestry ministry (like many Indonesians, he goes by one name), charged during the year 2000 that lumber companies and sawmills owned by Abdul Raysid, a member of Indonesia's Parliament, were illegally processing ramin logs. The ramin is the most valuable tree in the national park, whose blond, straight-grained wood is used extensively in furniture, wood moldings, window blinds, and pool cues. Suripto's findings were ignored by authorities, as the logging continued (Gargan, 2001, A-7).

Indonesian forestry minister M. Prakosa and trade and industry minister Rini M. S. Suwandi announced a moratorium on the export of logs and wood chips during October 2001 to "safeguard the conservation of Indonesian forests" ("Indonesia: Low Expectations," 2001). The main result of the ban may not be the conservation of forests, however, but to ensure a supply of raw material for domestic wood-processing industries. According to the *World Rainforest Movement Bulletin*, "The new ban will be a boost to the country's highly-indebted wood industry and the timber tycoons whose businesses have suffered from a lack of raw materials. A total of 128 companies are under the control of the Indonesian Bank Restructuring Agency (IBRA). A previous commitment to close half of these companies has not been fulfilled" ("Indonesia: Low Expectations," 2001). Additionally, according to the *World Rainforest Movement Bulletin*, the export ban is not likely to significantly reduce timber smuggling, "while Indonesia's notoriously corrupt police force, government apparatus and courts continue as before" ("Indonesia: Low Expectations," 2001).

During late May 2002, Indonesia's president, Megawati Soekarnoputri, ordered a temporary halt to logging in an attempt to save the country's remaining forests. The president's office cited a World Bank study warning that Indonesia will lose all of its forests in the next 15 years if, according to an Environment News Service report, "the government does not act quickly and strongly against deforestation activities" ("Indonesian President," 2002). Deforestation is threatening extinction of several animal species, including endangered tigers, elephants, rhinos, and orangutans. The orangutan population declined by 50 percent during the 1990s, due primarily to destruction of its forest habitat, up to 80 percent of which has been lost during the last 20 years ("Indonesian President," 2002).

Now the question for Indonesian forests becomes one of enforcement, since much present logging already is illegal, and the power of the central government over the country's many islands has been waning. Illegal logging activity has been increasing in many Indonesian forests despite governmental edicts. According to the Environment News Service, "The Environmental Investigation Agency reported in 1999 that Tanjung Puting National Park was full of logging camps and an extensive network of wooden rails had been built to drag

A logging operation in the West Kalimantan region of Borneo. (Painet)

out the timber. In the east of the park, a logging road was built to truck out the illegal timber. Steel barges were observed loaded with illegal wood, and investigators tracked the timber to local sawmills and factories" ("Indonesian President," 2002). In Gunung Leuser National Park, according to the same report, investigators witnessed loggers with chainsaws operating in the Suaq Balimbing research area, which provides prime orangutan habitat and is the only place where these apes have been observed using tools ("Indonesian President," 2002).

THE PENAN: OBSTRUCTING LOGGING

The Penan are one of the few surviving nomadic peoples of the rainforest in Sarawak (the northern portion of Borneo, controlled by Malaysia). Of roughly 9,000 Penan alive today, only about 300 continue to live as nomads, in the traditional way. They live in a forest laced by rivers, as well as one of the world's most extensive network of caves and underground passages. The Penan are experiencing one of the highest rates of logging on earth. Some Penan also are threatened by a massive dam project. The proposed Bakun dam will flood 70,000 hectares of land, displacing indigenous peoples, wildlife, and large areas of rainforest.

The Penan are one of several indigenous tribes known collectively as the Dayak. For centuries before the arrival of European colonizers, the Dayak lived in communal longhouses, surviving by hunting and slash-and-burn agriculture in the lush rainforest. Aside from occasional raids by notorious headhunters, life was generally peaceful.

The Penan live amidst awe-inspiring scenery. The peaks of Batu Lawi, a Penan sacred site, are said to be places of spirit and mystery, described by Richard Lloyd Parry of the *London Independent*:

> [T]wo immense stone pinnacles . . . thrust up out of the jungle in a remote part of Malaysian Borneo. One peak is broad and squat, draped with clinging trees; the taller is thin and sharp with bare, vertical sides rising to its 6,700-foot peak. The Penan people, whose homeland this is, speak of them as a husband and wife. From where I stood, rinsed in sweat from the climb, they looked like the bunched knuckles of a fist, its index finger pointing into the sky. (Parry, 2001, 18)

Rapid deforestation began in the Penan's homeland during the 1960s. An Internet Web page of the Sarawak Peoples Campaign described a "frenzy of logging that has gripped Malaysia. . . . a rate of forest destruction twice that of the Amazon and by far the highest in the world" ("Sarawak Peoples Campaign," n.d.). In the Baram River drainage alone, more than 30 logging companies, equipped with as many as 1,200 bulldozers, worked 1 million acres of forest on lands traditionally belonging to the Kayan, Kenyah, and Penan. More than 98 percent of Sarawak's timber is exported in the form of raw logs, virtually all of which is destined for Asian markets ("Sarawak Peoples Campaign," n.d.).

Japanese companies supply the bulldozers and heavy equipment that is utilized to extract the logs. Japanese interests provide the insurance and financing for the ships that carry the raw logs to be processed in Japanese mills and sold as lumber to construction firms often owned by subsidiaries of the same companies that first secured the wood in Sarawak. Once milled in Japan, the wood produced by the oldest and perhaps richest tropical rainforest on Earth is used principally for packaging material, storage crates, and furniture. Roughly half of it is used in construction, mostly as plywood cement forms which are used once or twice and then discarded ("Sarawak Peoples Campaign," n.d.).

During 1987, having appealed in vain to the government for more than seven years to end the destruction of their traditional homelands, the Penan declared:

> We, the Penan people of the Tutoh, Limbang, and Patah Rivers regions, declare: stop destroying the forest or we will be forced to protect it. The forest is our livelihood. We have lived here before any of you outsiders came. We fished in clean rivers and hunted in the jungle. We made our sago meat and ate the fruit of the trees. Our life was not easy but we lived it contentedly. Now the logging companies turn rivers to muddy streams and the jungle into devastation. Fish cannot survive in dirty rivers and wild animals will not live in devastated forest. You took advantage of our trusting nature and cheated us into unfair deals. By your doings you take away our livelihood and threaten our very lives. You make our people discontent. We want our ancestral land, the land we live off, back. We can use it in a wiser way. ("Sarawak Peoples Campaign," n.d.).

The statement concluded: "We are a peace-loving people, but when our very lives are in danger, we will fight back. This is our message" ("Sarawak Peoples Campaign," n.d.).

On March 31, 1987, armed with blowpipes, a group of Penans erected a blockade across a logging road in the Tutoh river basin. In April, about 100 people blockaded a road though their territory at Uma Bawang. By October, Penans from 26 settlements had joined the protest. Blockades have continued since that time, as "whole villages moved onto logging roads, building makeshift shelters directly on the right-of-way. Often the protests lasted for months, and when they finally were suppressed by government forces, new ones sprang up in other areas. At their peak, the blockades halted logging in half of Sarawak ("Sarawak Peoples Campaign," n.d.).

The government and the logging companies have reacted to these peaceful protests by changing the law to make blockading of roads a criminal offense. Hundreds of indigenous people were harassed, arrested, and imprisoned as a result. Environmental activists have been accused of undermining the nation's economic security and have had their passports confiscated. In the meantime, loggers continued to strip the forests.

The Sojourn of Bruno Manser

Bruno Manser, a European environmentalist who became a legend among the Penan, was seen for the last time in their homeland during May 2000. Manser, a former Swiss cowherder, moved to the Penans' homeland, where for six years, according to an English account, "he hunted with a blowpipe, lived off snake and monkey meat and helped the Penan in their struggle against the logging companies which were stripping the rainforests where the nomads roam" (Parry, 2001, 18). While he lived, Manser (who was described as the "Swiss Tarzan" and the "Wild Man of Borneo") made the Penans' causes fashionable in Europe. Prince Charles gave a speech in which he described the Penan as victims of genocide. The European Parliament adopted a resolution calling for the suspension of Malaysian timber imports.

Born in 1954 in Basel, son of a factory worker, at age 19 Manser served three months in prison for refusing to serve in the military as required of all Swiss men. He later lived 12 years in the Alps, working in the mountain pastures, learning to lay bricks, carve leather, and husband bees. He wove, dyed, and cut his own clothes, and he manufactured his own shoes.

Manser first visited Sarawak in 1984 as a tourist, then returned a decade later to hike far inland, searching for the nomadic Penan. He found them, was adopted, and became a Penan spokesman to the outside world. He spoke the Penan language fluently, quizzing indigenous people on pronunciation, as he wrote phrases in a notebook. He took detailed notes and many photographs. Manser also nearly died of malaria, was bitten several times by poisonous snakes, and fell from trees several times while hunting.

Manser's activism did little to forestall industrial pressure to log the Penans' forests and force them into resettlement camps. Manser watched the forests of the Penan shrink. During the past 40 years, 90 percent of the virgin jungle that housed the Penan has been logged to make plywood for construction in Taiwan, South Korea, and Japan. Beginning in 1987, the Penan and Dayak organized blockades of logging roads. One of the blockades lasted eight months before police demolished it. Penan communities opposing the logging were harassed by hired thugs; in 1994, the mutilated body of a blockade organizer named Abung Ipui, who earlier had disappeared, was found in a river (Parry, 2001, 18).

The logging companies suffered large losses from stalled production, and hundreds of native people were arrested. Manser drafted petitions on the Penans' behalf, made translations, and discussed tactics with Penan leaders. He also overstayed his visa at about the same time that the Malaysian government began to regard him as seriously subversive. Malaysian commandos hunted Manser and captured him twice, only to see him escape. At one point, he was smuggled out of Sarawak and returned home to Switzerland for a time.

During the spring of 2000, Manser returned to the rainforests of Sarawak. Shortly after that, on or about May 25, he disappeared. The Penan mounted extensive searches of the rugged countryside, but were unable to find any trace of him, his clothes, or his equipment. Many of the Penan believe Manser was murdered for voicing their opposition to the reduction of their nourishing rainforest.

The nomadic Penans' desire to remain in the forest was not shared by Taib Mahmud, the chief minister of the state of Sarawak, whose point of view is shared by Mahathir Mohamad, Malaysia's president. "There is nothing romantic about these helpless, half-starved and disease-ridden people and we will make no apologies for endeavoring to uplift their living conditions," Taib has been quoted as saying. Mahathir said that "we don't mind preserving the Sumatra rhino in the jungle, but not the Penan" (Parry, 2001, 18).

These points of view contrasted with those of the Penan people quoted in Manser's book *Voices from the Rainforest*. In the book, Manser quoted an elderly Penan man who had seen enough of the world to compare his home with life outside. "Our land provides us with food for free, and so, without a *sen* in our pocket, we have enough," he said. "What is it about the people in the town in their shops? Why do they have to install fans and air-conditioners in their apartments? They live in the heat because they have destroyed their forest. Here, under the big trees, is cool shadow. We don't want to change places with them" (Parry, 2001, 18).

THE ATTACK OF THE SUN BEARS

Andrea Johnson had little warning before the attack. A recent Harvard graduate, she had spent several months in Borneo tracking orangutans. Fighting boredom, she watched one of the rare orange-furred apes in the rainforest canopy above, stuffing herself with fruit. Then, Johnson heard heavy breathing from a dead tree trunk, and looked down in surprise. "At first, I thought the noise was bees," she said. That's when a furry black missile landed on her backpack and "knocked me to the ground as I turned to run," Ms. Johnson said. "It gashed my leg with its claws, and after I shouted and kicked, it took off" (Murphy, 2001).

Johnson had been attacked by a sun bear, a cuddly-looking little creature, the smallest member of the ursine family, which heretofore had not been known for attacking humans. The attack on Johnson was the beginning of a trend; after she was attacked, four other people were assaulted by sun bears in Gunung Palung National Park after 16 years without a single such incident (Murphy, 2001). The bears have been attacking humans because their habitat has been reduced and their traditional food supplies diminished, a study, in microcosm, of problems facing the environment and indigenous peoples of Borneo. "The bears are stressed, the whole system is stressed," said Johnson (Murphy, 2001).

The largest trees in Kalimantan, Indonesia's two-thirds of the Pacific island of Borneo, have stopped reproducing in response to climate change and canopy fragmentation caused by massive logging in surrounding areas. "We're coming to the end of the line," warned Lisa Curran, an ecologist with Yale University, who led a 15-year study of the forests in Gunung Palung, a national park (Murphy, 2001). Higher temperatures are helping to kill the forest, and dwindling forests are among the factors contributing to global warming. "It's heartbreaking," says Curran. "You walk in parts of the forest that used to be these beautiful cathedrals. Now you want to cry." Researchers say logging already affects more than 50 percent of the park (Murphy, 2001).

Curran found that the trees utilize a rare cooperative reproductive strategy that makes them unusually vulnerable to habitat change:

> Almost all *dipterocarps* fruit in a simultaneous, six-week flurry every few years, in an event spanning hundreds of square miles. The fruiting starts rush-hour in the rainforest. Long-tailed parakeets migrate inland, and macaques, or leaf monkeys, and orangutans converge to eat. They are followed by bearded pigs, which live for years in solitude before forming huge herds that sweep across the landscape during the fruiting. No matter how [much] the animals gorge, they are overwhelmed by the abundance, which is the point of the event, called a "mast" after an Old English word for nuts. The leftovers have time to germinate. . . . But the mast can only work on a huge scale. If too few trees are involved, or if the trees are so weakened by climate change that they put out only small amounts of fruit, animals like the bearded pig don't sweep across the landscape. They circle, and every seed is eaten. (Murphy, 2001)

For eons, until 1991, nature had replenished itself this way. After that, every mast event in Gunung Palung failed. No new seedlings have taken hold, as older trees have been removed by loggers.

The sun bears have endured habitat destruction that forces more of them to compete for less space. Beginning during the 1980s, nearly every acre of Borneo not legally protected was intensely logged. During the 1990s, as Indonesia's political infrastructure crumbled, even areas that had been legally off-limits, such as national parks, were logged. In 15 years, an area of forest roughly the size of Ohio has been felled (Murphy, 2001). The climate change that is killing the trees is affected by local deforestation, which has made Borneo hotter and drier, as well as a global trend toward warmer temperatures. The Earth as a whole has been warming and intensifying the erratic El Niño weather pattern, which often causes droughts across Southeast Asia.

THE IMPACT OF TOURISM: EX-HEADHUNTERS' BED-AND-BREAKFAST

To see just how far, and how quickly, the tendrils of industrial capitalism have spread, come to today's Kalimantan. Tours to the area (with native guides) are popular enough to have been written up as a tourist destination in Tokyo's *Daily Yomiuri*. Direct, six-hour flights are available from Tokyo to Kota Kinabalu, one of the larger cities in the area. Agop Batu Tulug, one of several tourist destinations, is located 800 meters from the Kinabantangan River in Kampung Batu Putih, about 40 kilometers from Kota Kinabatangan township. The journey to Agop Batu Tulug takes about two hours by car from Sandakan, the second-largest city in Sabah.

Tourists may arrange a guided tour to a community of Batu Tulug longhouses in the jungle, where they lodge in traditional dwellings and socialize for a few days with extended families of Borneo tribespeople. The jungle setting may seem old-fashioned, but the author of one Japanese newspaper account observed that young people from the longhouses travel to a town 20 miles away that offers an Internet café. At the longhouse, electricity was less than a year old in April 2001, when the author of the account visited. At that time, some of the longhouse residents were discussing acquisition of a telephone (Fujii, 2001, 10). The villagers are quoted in this account as trading on the imminent demise of their tribal lifestyle. Soon, they told the tourists, this type of experience will not be available at *any* price.

The residents of this jungle bed-and-breakfast fatalistically share bouts of ritual alcoholism with the tourists:

> Then came show time, led by children who performed a quick traditional bamboo stick dance. Lucky tourists got to try the dance—jumping up and down without getting their legs caught in the bamboo—before the huge drinking session that followed. . . . The highlight was a merciless drinking party with the villagers featuring *mentako*, a transparent liquor similar to Japanese sake, and *lihing*, a sweet rice liquor. "Bottoms up," the [indigenous] Rungus men cheered as a glass filled to the top with either of the liquors were passed from one person to

another. Not surprisingly, the locals far outpaced the Japanese tourists. (Fujii, 2001, 10)

The local people were doing their best to accommodate change. Asked about the uncertain future of their traditional lifestyle, longhouse parents, who rely on their children's employment in town, said calmly, "We can't do anything about it" (Fujii, 2001, 10). The largest ethnic group in Sabah is the Kadazan-Dusun tribe, which makes up one-fourth of Sabah State's population. They originally were farmers, but these days, many of them hold important urban jobs.

> A visit to Monsopiad Cultural Village, a 20-minute drive from Kota Kinabalu airport, offers an insight into the traditional lives of the Kadazan people. The site is a living museum dedicated to the famous headhunter Monsopiad. Forty-two skulls of plunderers killed by this great warrior 200 years ago are still worshiped in one of the wood-and-bamboo-stilt houses. (Fujii, 2001, 10)

Herman Scholz, a German who lives in the village and oversees its operation, said that little effort has been directed at preserving the skills and beliefs of earlier generations. Skills are not being passed on to younger generations, he said. "Kadazans are nice, friendly people, but the sad thing about them is that they are not proud of their culture," Scholz said. "Their attitude is that of the oft-heard phrase, 'Apa buleh buat,' meaning 'What can we do about it?' and they generally do not try to maintain their culture" (Fujii, 2001, 10).

ARMED RESISTANCE TO DEVELOPMENT IN EAST KALIMANTAN

While some native people of Borneo welcome tourism, others have been rebelling against increasing immigration and development of industrial infrastructure in their homelands. At least 400 people died during February 2001 in bloody fighting there. Madurese migrants were attacked by the indigenous Dayak people of Central Kalimantan Province as "scapegoats for the failed economic and social policies that Indonesia, with American support, has followed for far too long" (Abrash, 2001, A-21).

"For decades," wrote Abigail Abrash, an advisor to Indonesia's National Commission on Human Rights, "the Dayaks—like Papuans, Acehnese, Moluccans and other indigenous communities throughout the Indonesian archipelago— have been on the receiving end of destructive policies imposed by the central government in Jakarta and backed by repressive political and military force" (Abrash, 2001, A-21). Indigenous lands have been seized for mining and logging, single-crop plantations, factory fishing, and building of roads and other infrastructure that will tie the area into the worldwide web of industrial capitalism.

Until chaos in Jakarta stopped government transmigration policies, the government was relocating Indonesians, including the former Madura Islanders (who were attacked in Borneo), from populous islands, such as Java, to Sumatra, Borneo, and others. Abrash described "boomtowns attracting thousands of other 'spontaneous' migrants who further rend a social fabric woven together over hundreds, sometimes thousands, of years in the fragile ecosystems of Indonesia's more than 10,000 islands" (Abrash, 2001, A-21). Native peoples on these islands face a centralized legal system which, according to Abrash, "offers no adequate respect for community land rights and no effective protection for traditional livelihoods" (Abrash, 2001, A-21). Complaints brought to Indonesia's National Commission on Human Rights nearly always concern taking of land and resources, usually from indigenous islanders.

A U.S. State Department report on human rights, released early in 2001, charged that the Indonesian government had "used its authority, and at times intimidation, to appropriate land for development projects, particularly in areas claimed by indigenous people, often without fair compensation." It further stated, "When indigenous people clash with private-sector development projects, the developers almost always prevail" (Abrash, 2001, A-21). Roughly two-thirds of Indonesia's military budget comes from private-business income, so the armed forces have a vested interest in divesting indigenous islanders of their land and resources.

Soldiers guard Madurese refugees on an overcrowded truck evacuating them from violence-wrecked Sampit, Central Kalimantan on February 28, 2001. The soldiers were deployed to help contain violence by indigenous Dayaks against Madurese settlers in Central Kalimantan, which has claimed at least 270 lives. (NewsCom.com/iPhoto Inc.)

A lieutenant commander in Indonesia's elite American-trained special forces described for a human rights investigator in 1998 the role that Indonesia's military played in establishing and protecting natural-resource operations and other development projects: "The military is here to make sure that investors can come in," he said (Abrash, 2001, A-21). This arrangement was introduced under former president and army general Suharto, whose government, which ruled from 1966 to 1998, based its so-called New Order on removal of nonrenewable resources and low-wage labor. These economic policies also were tacitly supported by the International Monetary Fund and the World Bank because they increased Indonesia's earnings of foreign exchange, regardless of their impact on indigenous peoples.

International corporations supplied financial resources and expertise with the army's protection until Indonesian central authority began to crumble during the 1990s. At this time, according to Abrash's analysis, the impact of these economic strategies on local peoples such as the Dayaks (and on Indonesia's environment) has worsened, in large part due to the International Monetary Fund's structural-adjustment policies and to increased investment from abroad in ecologically destructive projects. "The palm-oil, timber, and mining businesses, all benefiting from multinationals' investment, have been particularly harmful to the Indonesian environment—and, often, to the indigenous people whose lands they use" (Abrash, 2001, A-21).

AN OIL BLOCKADE IN THE RAINFOREST

A violent clash occurred on October 8, 2000 between protesting community members in East Kalimantan, Indonesia and the Indonesian Mobile Police Brigade (BRIMOB), leaving 7 protesters dead, 16 seriously wounded, and 2 others missing. The violence ended a 14-day blockade of Unocal's Tanjung Santan oil refinery, at which local community people were protesting pervasive air and water pollution. According to an account in *Drillbits and Tailings*, local residents blockaded Unocal's transportation routes, demanding compensation for the company's environmental degradation of nearby waters and farmlands.

Unocal first responded to the blockade by stating that the Tanjung Santan Terminal would temporarily stop production. Instead, company spokesperson M. Ramli asked 60 police and BRIMOB officials to break up the blockade. According to *Drillbits and Tailings*, Unocal asked the mobile police to intervene to break the blockade. Responding to the request, according to the account, "The troops brutally attacked the protesters by firing shots, kicking, and beating them with rattan sticks" ("Protesters Shot," 2000). The company justified the police intervention on grounds that the blockade was impeding oil and natural gas production.

Drillbits and Tailings reported that local discontent regarding Unocal operations in East Kalimantan has grown significantly during the last two years,

as rice fields and shrimp stocks have been repeatedly devastated by faulty waste-disposal systems. A water sample taken earlier this year after the break of a Unocal pipeline showed levels of heavy metals high above standard limits; rice fields flooded by such waters in 1998 remain so contaminated that they still lie barren today. ("Protesters Shot," 2000)

Local residents also reported that Unocal's operations are responsible for unacceptably high levels of sulfide and ammonia in the air around its facilities.

GOLD MINING AND WATER SUPPLIES IN EAST KALIMANTAN

According to a *Drillbits and Tailings'* report, since 1992, Rio Tinto's Kelian Gold Mine in East Kalimantan has produced more than 400,000 ounces of gold per year using the cyanide heap-leaching process, producing cyanide-laced tailings. The tailings are held in a dam and treated in a polishing pond near the Kelian River. Water from the polishing pond pours into the river through an outlet. The company claims that the water is clean, while the community says that people cannot drink or bathe in the water because it causes skin lesions and stomachaches ("Rio Tinto Kelian Mine," 2000).

This mine pollutes a tributary of the Mahakham River on which thousands of indigenous people rely for water. "The Kelian mine has consistently manipulated environmental reports, and has wiped out without recognition the local community's traditional mining rights," asserted Mohammed Ramli, speaking for the Indonesian Mining Advocacy Network (JATAM), a group that represents affected communities, according to an analysis by the Mineral Policy Institute ("Rio Tinto's Shame," 2000).

The East Kalimantan Kelian gold mine, which is 80 percent owned by international mining giant Rio Tinto, has been causing serious pollution problems which endanger the local community's health, according to accounts from the scene. "Locals suffer from skin rashes when they bathe in the river. They can no longer catch the fish they rely upon as a protein source, and the water is so contaminated with insufficiently treated mine wastes that it's too dangerous to drink," said Ramli. In addition, the first five kilometers of the river near the mine have been artificially diverted without taking into consideration biological effects, leaving the previous watercourse devoid of life ("Rio Tinto's Shame," 2000). Community protests of the mine's pollution have been met with fire from police.

Rio Tinto's 1997 Health, Safety and Environment Report described a massive acid-drainage problem at the Kelian mine. Levels of manganese in water discharged from the mine during 1997 averaged 800 micrograms per liter. This level would not be legal in European or North American drinking water; it also exceeds the World Health Organization's recommended limits of 100 to 500 micrograms per liter. On 9 occasions in 1997 (and 105 occasions in 1996), manganese levels were more than 200 times the amount permitted in drinking water under European Union directives (50 micrograms per liter). Rio Tinto's research unit pledged to investigate methods of reducing manganese pollution, but seemed to be in no hurry. The company said that a new method currently under trial would not be implemented at Kelian until 1999 at the earliest ("Rio Tinto's Environmental Record," 1999).

Wastewater from the same mine also contained more than 500 kilograms of cyanide in 1997. While this level was about half the cyanide emissions for the previous year, the Kelian mine still was responsible for the worst cyanide pollution levels of any Rio Tinto gold or copper mine. (Cyanide compounds are used to extract gold from ore.) Rio Tinto implies that high cyanide levels are not a problem because "any residual-free cyanide breaks down rapidly in the presence of sunlight and does not persist in the environment" ("Rio Tinto's Environmental Record," 1999).

The Kelian mine also releases large amounts of suspended solids into the river. These are fine particles of soil and rock produced during the processing of ore. At 1,600 tons, the amount of suspended solids in the water, discharged by P.T. Kelian Equatorial Mining (PT KEM) (the Indonesian subsidiary of Rio Tinto), is the second highest of all Rio Tinto's operations worldwide. In 1996, levels of suspended solids were even higher, at 4,700 tons, as PT KEM diverted part of the Kelian River. Nevertheless, Rio Tinto puts much of the blame for the high turbidity of the river water on the operations of small-scale community miners ("Rio Tinto's Environmental Record," 1999).

For its mining operations, PT KEM takes more than 6 million cubic meters of fresh water per year from the Kelian River. Only about 4 million cubic meters is recycled within the mine. Wastewater containing high levels of manganese, cyanide compounds, and mud has been discharged into the Kelian River.

Because it has polluted local drinking-water supplies, PT KEM has been required to provide drinking water for the local population since the start of its operations in the early 1990s. However, not all of the indigenous people or more recent settlers have access to piped drinking water. Water from the Kelian River is used for all other household and agricultural needs, including bathing, laundry, and preparing food, regardless of the levels of pollution ("Rio Tinto's Environmental Record," 1999).

Local indigenous peoples also have questioned Rio Tinto's assertions that it has rehabilitated more than 500 hectares of forest that had been cleared in connection with mine operations. Instead, say local peoples, rehabilitation has consisted largely of "planting non-native tree species several hundred kilometers away in the Bukit Suharto Park—much of which went up in flames in . . . forest fires [during 1999]" ("Rio Tinto's Environmental Record," 1999).

Several thousand Dayak people blockaded operations at the Kelian mine in East Kalimantan, Indonesia after a breakdown in negotiations between them and Rio Tinto. The blockade began on April 29, 2000 and lasted 10 days, forcing the company to maintain only minimal operations as its stockholders met in London. *Drillbits and Tailings* reported that local police arrested a dozen community representatives and held them overnight in an attempt to force them to lift the blockade. Rio Tinto, which owns 90 percent of its Indonesian subsidiary PT KEM and the mine, has stalled negotiations with the community meant to address concerns over compensation for land despoiled by mining ("Rio Tinto Kelian Mine," 2000).

"In the name of the Kelian community of West Kutai district, East Kalimantan, Indonesia, we state that PT Kelian Equatorial Mining (PT KEM) has not been genuinely committed to settling the issues and demands raised by the people. The company has only paid lip service to various activities—community development projects, recruitment of local workers, environmental management and mine closure plans—as a form of propaganda," said a statement from the community read at Rio Tinto's Annual General Meeting in Britain ("Rio Tinto Kelian Mine," 2000).

Rio Tinto and PT KEM signed a contract with the Indonesian government in 1985 and began exploration for gold in 1987. Since that time, the community has alleged numerous human rights violations, destruction of property, and pollution of their lands and rivers. The community began the blockade in frustration over RioTinto's continued efforts to stall negotiations, as well as its violations of conditions of the negotiations.

During 1998, heavy monsoon rains caused toxic tailings to overflow in Rapak Lama village, Marangkayu subdistrict. The flood mixed toxic tailing and chemicals from Unocal's processing plant. Another, similar flood occurred on February 11, 2000. The floods caused the company's tailing pipe to overflow, washing its toxic contents into local rice fields, which contaminated more than 400 hectares of indigenous peoples' rice fields, wiping out many villages' rice crops. According to an on-the-scene report, local people "noticed milky-white water with foams and brownish blobs on the surface around the plantation, in its irrigation gutters, and the mouth of the pipe close to [the] UNOCAL factory fence" ("Unocal Tailing," 2000). Company officials took water samples but did little else to stem the flow of putrid water that was destroying the fields

"In the evening," according to a report by local people in *Minergy News*, "UNOCAL workers removed the brownish blobs on the surface of the water, yet the water remained milky white with foams" ("Unocal Tailing," 2000). Later, the head of the Marangkayu District Police (Kapolsek Marangkayu) and one of Unocal's security personnel visited the mouth of the pipe where the tailing-carrying water emerged. They were heard to have said, "If only the security had covered the . . . pipe-mouth with sand earlier in the morning, none of the locals would have been able to discover them" ("Unocal Tailing," 2000). Samples were again taken, but local people were not informed of the results.

A report in *Minergy News* described the damage to the local environment:

[Many] hectares of rice fields, fish/shrimp embankments, and plantations were severely contaminated by the waste of the oil-processing plants. Tailings were dumped in a 100-metre distance from the beach causing the local fishing people to lose their livelihood since their daily catch smell of oil and [are] unfit for human consumption. Their previously white-sand beach is now brown and the air is filled with the suffocating stench of crude oil. ("Unocal Tailing," 2000)

The Indonesian government's National Human Rights Commission investigated allegations of human rights abuses by Rio Tinto, publishing a report during 2000 that documented "egregious violations" ("Rio Tinto Kelian Mine," 2000). According to *Drillbits and Tailings*,

> The report revealed that the Indonesian military and company security forcibly evicted traditional gold miners, burned down their villages, and arrested and detained protestors. PT KEM employees have also been named in a number of incidents of sexual harassment, rape and violence against women between 1987 and 1997. Local people have lost homes, lands, gardens, fruit trees, forest resources, family graves and the right to mine for gold in the river. ("Rio Tinto Kelian Mine," 2000)

In a statement addressed to Rio Tinto shareholders at their 2000 annual meeting, the Kelian community demanded fair compensation for land, crops, and property; restoration of small-scale mining rights; remediation of contamination of the rivers by cyanide, metals, and dust; legal action against PT KEM staff and local authorities for human rights violations; and genuine community development.

THE DAYAK: TRADITIONAL GOLD-MINING PRACTICES

The Muluy Dayak villagers of East Kalimantan depend on the forest and river resources for their livelihood. They live in small houses (as large as four by eight meters) built of bark, *sungkai*, and *ulin* timber as well as corrugated iron. They gather gaharu resin, and hunt birds and game, including mouse deer and porcupine. As agriculturalists, they cultivate mountain rice, fruit, honey, rattan, and coffee. River fish provide them with a major source of protein.

Small-scale gold mining has long been used by the Muluy Dayak community as part of its traditional way of life. *Adat* (customary law) governs traditional gold-panning activities, which utilize simple equipment made from materials collected in the surrounding forests. Industrial-scale mining violates all of these rules. For this reason, the Muluy Dayak vehemently oppose large-scale mining that pollutes the earth. A report in the Indonesian-language newsletter *Gaharu*, published by the East Kalimantan nongovernmental organization, Plasma, offered an unusually detailed account of traditional small-scale mining practices. This account was translated by *Down to Earth* from "Bila Orang Muluy Menambang Emas," a statement by the Muluy Community composed during June 2001.

The Muluy villagers own gold-mining areas communally under *adat* agreements handed down from generation to generation. A person who discovers gold shares the discovery with the community so that labor and profits may be shared. The gold is sold only when members of the community have pressing needs that require cash (for food, electronic appliances, or other goods not produced locally). The gold is taken to markets at Long Ikis, Banjarmasin, and Balikpapan.

Traditional gold-mining tools include panning vessels, made from local wood and coconut shells, which are used to separate gold from river-bottom stones and mud. Unlike industrial-mining operations, the villagers use no chemicals. Several rituals must be performed before panning starts. According to an account in *Down to Earth*, community members must go to the river and state their intention to pan for gold in the local language (*bemamang*), to secure the approval of the water spirits (*tondoi danum*) and the sacred figure, *Nabi Haidir*. The Muluy then make an offering from red-colored rice, which "is placed on a palm leaf with an egg and is taken to the water by four or five elders, whilst chanting *bemamang*" ("Indigenous Small-scale Mining," 2001). This offering is then released into the river, and watched. If the offering sinks after floating four to five meters, the water spirits are believed to have granted the people's request to pan for gold. If the offering does not sink, permission has not been granted. Anyone who attempts to take gold under these conditions will not be successful, according to ritual belief.

Several taboos restrict certain behaviors while mining under penalty of bad luck for the entire community. Cursing while panning is prohibited, as are "promiscuity, stealing from another group's area, saying the word 'gold' or 'stone,' killing animals that come to the panning location and going to the toilet in the panning area" ("Indigenous Small-scale Mining," 2001). Panning is performed during a six-month season, which is concluded with a traditional *adat* ceremony.

Anyone who wishes to mine gold in the Muluy Dayaks' territory is expected to abide by *adat* traditional law by making an application to the *adat* leader. The decision is made by community consensus. If an applicant is denied, he is expected to leave. A miner who is accepted must mine within defined boundaries, using traditional mining equipment. The guest miner also must also describe to the *adat* leader the results of their mining efforts. Half of the gold mined by an outsider is to be paid back to the Muluy community.

According to the report in *Down to Earth*, a similar traditional mining system prevailed in the Kelian area of East Kalimantan when industrial-scale gold miners arrived during the 1980s. Local Dayaks reached an *adat* agreement with outsiders during the 1980s gold rush, every part of which was disregarded by Rio Tinto. The company was supported by Indonesian authorities who do not respect the traditional system with legal authority ("Indigenous Small-scale Mining," 2001).

During the 1990s, the mining company P.T. Aneka Tambang, which is partially owned by the Indonesian government, surveyed gold-mining potential in the Muluy Dayaks' traditional territory. Other freelance entrepreneurs arrived, bearing modern equipment, such as diesel pumps, plastic hoses, lights, trucks, mercury, and other chemicals used to leach gold from ore. The Muluy Dayaks routinely refused to allow gold mining by such methods after a community meeting called by the *adat* leader. The community was concerned about degradation of local water supplies by large-scale mining methods.

Local indigenous people soon found themselves barred by force of arms from traditional mining areas that had been seized by colonists. The Dayak miners were earning income in what they believed to be a legal manner on traditional lands. The government and the company assert that they are illegal squatters. On January 18, 2002, for example, a 20-year-old Dayak man was shot and seriously injured at an Australian gold mine in Central Kalimantan. At Aurora Gold's Indo Muro Kencana mine, a security officer shot the local man in the head at close range. The shooter was a BRIMOB officer, a member of the Indonesian government's military-style police force that provides security at many industrial-scale mine sites.

The next day, roughly 100 angry Dayak briefly closed Aurora Gold's roads and processing plant, demanding that the company take responsibility for the shooting and cease to allow the posting of BRIMOB security personnel at its mine site ("Indonesian Man Shot," 2002). The mine company's secretary, Michael Baud, said the shooting was "part of a continuing series of incidents which have occurred, with illegal miners entering the mining site, being warned off by the Indonesian security forces and not responding" ("Indonesian Man Shot," 2002). The miners that the company maintains are illegal are carrying on traditional small-scale mining practices.

Aside from outsiders' control of their traditional mining sites, local people objected to the intruding mine's toxic emissions, which were killing fish on which the indigenous peoples depend for sustenance. Clean water also became scarce for local Dayak and their livestock, for the same reason. Local people asserted that "BRIMOB police have bulldozed and burned to the ground whole villages, harassed people at gunpoint, and arrested and detained without valid charges community members including women and children" ("Indonesian Man Shot," 2002).

Shootings at Aurora Gold during 2001 resulted in two deaths of small-scale miners, according to Geoff Evans, director of the nonprofit Mineral Policy Institute in Sydney, an advocacy group that monitors Australian-based mining companies. "On June 5, 2001," said Evans,

> the same security force shot at community small-scale miners. Running for their lives, six people fled into a flooded mine pit. From the edge of the deep pit, the police fired gunshots and threw stones at the men. A 28-year-old and an 18-year-old were killed, and three others were wounded. Two months later on August 27, a boy in his early teens was shot and seriously wounded. ("Indonesian Man Shot," 2002)

Baud repudiated Evans' account. He said that these deaths "were drownings by people who were operating illegally in the mine and fell into the water at the bottom of a disused pit and drowned" ("Indonesian Man Shot," 2002).

SUBMARINE TAILINGS DISPOSAL

Indigenous peoples on the northeast tip of Celebes Island (south of the Philippines) have rallied against submarine tailings disposal (STD), by which the effluent from local strip mines is pumped into nearby ocean waters. Its advocates tout STD as environmentally friendly because it removes tailings from the land. They rarely investigate what it does to the fish on which many native peoples depend for protein. The indigenous point of view on STD has been relayed by the Web sites of the Indonesian Forum on the Environment, as well as in the online editions of *Kerebok* (*kerebok* is the indigenous Dayaks' term for gold pawning). *Kerebok* is a monthly online bulletin published by the Secretariat of Mining Advocacy Network (Jaringan Advokasi Tambang [JATAM]) to provide information on mining issues.

Meares Soputan Mining (MSM) has tried to convince people in six villages in Likupang Minahasa Regency to allow a pipe to be placed for STD on Paser Beach. These people have reacted negatively, based on the experiences of the Buyat Bay people, whose traditional fishing economy was devastated by an earlier STD project. The local people were doubly irritated at the company's representatives who insisted that the tailings would be harmless in the local aquatic environment. "This operation area will change the beautiful panorama and water flow as well as have drastic impacts on the surrounding fish population. Moreover, it will jeopardize the life source of 3000 fishermen in Rinondoran village," according to one local source ("Dig In!" 2001).

Local people have asserted that they knew the sea better than experts contracted by mining companies who argued that no fish live below 200 meters, where they proposed to dump their tailings. A piece in *Kerebok* exclaimed: "How could they even say that there are no fish at a 200 meter depth? That is a lie, man" ("Dig In!" 2001).

The *Kerebok* commentary continued:

> People participating in the S.T.D. International Conference visited Rinondoran village and Barnabas, a prominent leader in the village, testified to the trauma experienced by the Buyat Bay people. This is understandable since the distance between Likupang and Buyat is not that far. They are linked in terms of their culture and come from the same regions of Sanger and Minahasa that can be reached through land transportation. One must not forget that the ability of Newmont Minahasa Raya to circulate the information they want known about S.T.D. is a simple task for them. The people of Rinondoran are determined not to allow the same mistake that happened in Buyat Bay from occurring in their village. ("Dig In!" 2001)

"The people, in fact, do not want to lose their natural maritime richness," said *Kerebok*. "If somehow MSM manages to get a license from the district government to dispose tailings to the bottom of the sea, we, all the people from Rinondoran, are prepared to fight a hard and long struggle. We want to live like we live right now, with no interruptions to our life now" ("Dig In!" 2001).

Local accounts asserted that, failing to convince the people of Rinondoran that STD would not harm the fish that provide their main source of protein, MSM representatives moved on to a neighboring village, Kalinaung, also in the regency of Likupang, where people's attention was drawn with liberal amounts of money. "As a result," said *Kerebok,* "Not only will the [STD] pipe be placed, but the local land will also be declared M.S.M. property. . . . Some sources state that thousands of hectares will be taken" ("Dig In!" 2001).

Elsewhere, the Misima Placer Nuigini mine (majority-owned by Placer Dome Corporation of Canada) has been criticized by the Papua New Guinea government for dumping gold-mining tailings into the ocean. The Papua New Guinea government has noted that "soft wastes pollute the sea for nine kilometers along the coast." Reports also note impacts on drinking water and fishing ("Mining in the South Pacific," n.d.).

FOREST MANAGEMENT AND INDIGENOUS PEOPLES ON JAVA

Debate has broken out on Java, Indonesia's most populous island, regarding whether indigenous people have a role in ownership and management of forests. Indigenous peoples occupied a legislative chamber during October 2001, asserting that they had not been included in forest management for decades. Most of West Java's forests have been controlled by a state corporation, P.T. Perhutani. Indigenous people have traditionally found themselves restricted to five percent of the forests.

Yudi Widhiana, a member of Commission B of the provincial legislative council, said 80 percent of the West Java forests had been controlled by P.T. Perhutani, while most of the remaining land was controlled by the state's Natural Resources Conservation Agency and public agencies ("Who Really Owns," 2001). In the meantime, expansion of industry and the spread of human settlement have contributed to deforestation across Java, sparking conflict over control of remaining forests. In West Java, 70 percent of 770,000 hectares of forest have been destroyed ("Who Really Owns," 2001).

According to an account in the *Jakarta Post,*

> Representatives of indigenous people swarmed the provincial legislature . . . to protest the interference of the legislature and N.G.O.s [nongovernmental organizations] in forest management. Dressed all in black and wearing green headbands, locals rallied passersby in front of the legislative building. It was the first time in history that locals from various areas of West Java [had] marched to the legislature to protest the alleged mismanagement of forests. ("Who Really Owns," 2001).

Once inside the legislative chambers, the indigenous peoples demanded a role in forest management, saying that they no longer trusted the legislators. "We no longer trust NGOs and legislators who always claim to represent the public. They are all nonsense," said Akbar Saputra, a member of the indigenous delegates, while pounding the table in the meeting room, bursting into tears ("Who Really Owns," 2001).

Indigenous representatives from Kampung Naga Tasikmalaya, Kampung Dukuh Banjar, Ciamis, Gunung Paok, and Sumedang also attended the same meeting. They said that they had been gradually expelled by outsiders, who had created problems with their forest-related policies. A representative of Kampung Naga and Dukuh said that water had become scarce in their villages following the destruction of forests by outsiders. A representative from Gunung Paok and Sumedang told those gathered at the legislative council that he was arrested and beaten by local police for reporting illegal logging in the village. "The police told me to pay 5 million [rupiahs] if I wanted to be released. How could I afford that much money?" said one man. "I won't do anything if [I] see any more illegal logging in the future" ("Who Really Owns," 2001).

THE BATAK: SHUTTING DOWN PULP-AND-PAPER MANUFACTURING IN NORTH SUMATRA

Following a popular revolt against pervasive pollution, the paper-pulp and rayon-fiber plant owned by P.T. Indorayon Inti Utama (Indorayon) in North Sumatra was shut down by thousands of local protesters. Indorayon's financial backers then withdrew, tired of waiting for the company to break a deadlock with the Batak community in the Porsea district that compelled the company to lose two years' worth of production, as it compiled U.S. $400 million in debts. During 27 months of local conflict near the plant that at times pitted its workers against opponents of the plant, at least a dozen people were killed and several hundred injured.

Foreign banks and bondholders that owned 86 percent of Indorayon stopped making monthly U.S. $1 million payments on September 1, 2000. Shortly thereafter, the company laid off its 7,000 workers and closed its doors. Indorayon was brought down by resistance of indigenous peoples near its plant because, according to its opponents, it flouted environmental regulations and the rights of the community.

According to a report in *Down to Earth*, "Long-standing grievances against Indorayon over environmental and health issues erupted soon after the downfall of the Suharto government. Production virtually came to a halt in mid-1998 when thousands of local residents prevented trucks from bringing raw materials to the mill for four months. Months of violent confrontations between local people and the security forces resulted" (Carr, 2001). A review by Indonesia's government revealed that the company had violated pollution and toxic-waste regulations.

After a short-term shutdown during 1998, environmentalists and local indigenous people argued that Indorayon should become the first industrial company to be closed for environmental reasons by Indonesia's central government. In May 2000, the government decided that the paper-pulp side of Indorayon's operations could start up again, but the production of dissolving pulp (the raw material for rayon fiber) should not be resumed. The decision provoked appeals from all directions. Environmentalists "argued that the company's past pollution and community record justified a complete shutdown"

(Carr, 2001). While the government stalled, community pressure and with-drawal of investors doomed the plant.

The Indonesian government was reluctant to let a company once listed on the Jakarta and New York stock exchanges go bankrupt, at a time when it was trying to attract foreign investment. The investment could have increased Indonesia's tax revenues and income from exports as required by an International Monetary Fund plan for Indonesia's economic recovery. According to the *Jakarta Post* (December 2, 1999), closure of Indorayon was projected to cost Indonesia at least $50 million in lost tax revenues and other fees (Carr, 2001). Mas Achmad Santosa, executive director of the Indonesian Centre for Environmental Law, was quoted in the *Jakarta Post* on May 15, 2000, saying, "Unfortunately, what the government cares about now is getting as many investments as possible. The preservation of the environment has taken a back seat" (Carr, 2001).

Indorayon's closure became a landmark case for the Indonesian environmental movement. Indonesia's largest environmental group, WALHI (Indonesian Forum for the Environment), filed a lawsuit against the company's owners and five government departments, accusing all of failing to comply with the country's environmental laws. The case was thrown out on technicalities. In the meantime, according to an account in *Down to Earth*:

> Inhabitants of Sodorladang and other villages near the Indorayon plant suffered a decade of polluted air and water. The acrid fumes which poured out of the smokestacks day and night could be smelled several kilometers away. Local people blamed the high incidence of asthma, chest infections and other respiratory ailments on the factory, but health-care facilities are so poor that there is no proof. The evidence of acid rain is obvious: corrugated iron roofs of houses and churches used to last two generations; since Indorayon, they corrode away within five years. (Carr, 2001)

On a day-to-day basis, residents of the area found their once-fresh water fouled by the plant's chemical effluent. Two years after the effective closure of the plant, according to local accounts, "Trucks no longer thunder through Batak villages every minute day and night, destroying roads and bridges. The air is refreshingly clear, as elsewhere in the Lake Toba region, and local people are again able to drink the water and to fish in the River Asahan" (Carr, 2001).

Elsewhere in Sumatra, a Tanjung Enim Lestari (PT TEL) plant in South Sumatra started production late in 1999. The pulp-and-paper mill, one of the largest in Indonesia with production of as much as 1 million tons of pulp a year, was a target of criticism from people in the communities in the Muara Enim district, who complained about a wretched stench that tainted water supplies within weeks of its first production. The company's environmental impact statement revealed that even when its waste-treatment units were working optimally, more than 18 tons of sulphurous gases were released each working day. According to *Down to Earth*'s account, "Giant pipes, over two meters wide, pour 80,000 cubic meters of waste per day into the River Lematang—the main source of water for drinking and all other domestic needs for the tens of thousands of people whose homes live along its banks" (Carr, 2001). These discharges deplete oxygen levels in the river and make the water murkier, affecting the aquatic ecosystems on which local fishing people depend for a living. This was the day-to-day routine. When waste-treatment plants failed, as happened at Indorayon on several occasions, the pollution was even worse, resulting in massive fish mortality in local rivers.

Development of the pulp-and-paper industry in Indonesia has been aided by the country's large areas of fast-growing tropical trees, cheap labor, lax environmental standards, and access to expanding Asian markets. Paper pulp can be produced in Indonesia, for example, for a third as much as it costs in Sweden (Carr, 2001). The growth of the pulp-and-paper industry in Indonesia is also linked to rapid deforestation there. In 1998, Indonesia exported 6.7 million tons of paper and pulp, three times the level in 1997, even as domestic demand fell by 50 percent to 1.3 million tons (Carr, 2001). During the 1960s, 1970s, and 1980s, traditional lands of several indigenous peoples were converted to production of commercial timber, mainly for export in the form of pulp and paper. According to *Down to Earth*, "Deprived of their agro-forestry systems or subsistence agriculture, local communities were dispossessed and destitute. Local people are not allowed to cultivate the land, cut timber for firewood or to build homes" (Carr, 2001). As a result, Indorayon and other pulp-and-paper factories came to symbolize oppression, destitution, and poverty to many indigenous peoples who live on the land.

Disregarding prior protests, the former Indorayon plant in early 2001 reorganized and proposed to resume production under a new name, P.T. Toba Pulp Lestari. The company proposed to produce pulp and paper only, leaving aside the rayon operation that had been shut down at Indorayon by the Indonesian government. The new management asserted that it would establish an independent commission and an audit committee before December 2001 to review social and environmental issues related to its operation. The company also promised to upgrade its technology and improve its treatment of the community in which it operates. According to *Down to Earth*, a new lineup of foreign investors had stepped in with U.S. $4 million on top of the $25 million already invested in the project. The consortium included the Bank of Boston, Bank of New York, Bank Namura, ABN Amro Bank, and Credit Lyonnaise (Carr, 2001). Local indigenous protests resumed near the plant, and the provincial governor, Rizal Nurdin, ordered military personnel to guard the plant against protesters.

ALLEGATIONS OF CORPORATE TORTURE IN ACEH, SUMATRA

The Indonesia Human Rights Network expressed support for a lawsuit filed in Washington, D.C. that named ExxonMobil as responsible for murder, torture, kidnapping, and 12 other charges at its liquefied natural gas operations in

Aceh, on the northern tip of Sumatra. The suit, filed on behalf of eleven Acehnese villagers, asserted that "ExxonMobil hired the Indonesian military to provide security for the corporation's facilities in Aceh" ("ExxonMobil Sued," 2001). According to the suit, these troops, under the employ of Exxon-Mobil, committed human rights abuses against the local population. The suit alleges that "troops stationed at the ExxonMobil facilities tortured villagers in buildings on ExxonMobil property, according to the filing by the International Labor Rights Fund" ("ExxonMobil Sued," 2001).

"Heavy corporate activity in the Lhokseumawe Industrial Zone (ZILS) has led to pollution that has degraded the quality of life of local inhabitants. Both PT Iskandar Muda fertilizer factory and Mobil Oil Indonesia (now known as ExxonMobil Indonesia) have been discovered dumping industrial waste into local waters. This has led to decreased fish harvests by local fishermen. Air pollution by fertilizer factories within the ZILS has created health concerns and reports indicate that the level of reported respiratory problems, particularly in children, has risen dramatically as a result" ("Building Human Security," 2001).

Islamic leaders in Aceh have supported an armed insurrection known as Darul Islam (House of Islam) that aimed to establish an Islamic state in Indonesia. The modern independence movement began in 1976 when Hasan di Tiro, leader of the rebel movement GAM, unilaterally declared Aceh an independent state. The Indonesian military has been used to crush insurrection in Aceh on several occasions. Between 1990 and 1998, an estimated 1,000 to 3,000 people were killed, and an additional 900 to 1,400 disappeared and were presumed dead ("Building Human Security," 2001). According to one local report, since 1998 the rebel movement, now known as the Free Aceh Movement (Gerakan Aceh Merdeka, or GAM), "has gained momentum and popular support. It is now estimated that two-thirds of all villages in Aceh are under rebel control" ("Building Human Security," 2001). "Approximately 1,000 people have been killed between 1998 and 2001 by Indonesian forces. Large numbers of Acehnese have been displaced from their homes" ("Building Human Security," 2001).

LOGGING, PEARL HARVESTING, AND TOURISM ON TOGEAN

On September 15, 2001, more than 500 people from several neighboring indigenous communities on Togean (also known as Togian), a small island which is enclosed by Sulawesi (Celebes) in central Indonesia, set fire to logging equipment owned by the timber company Argo Nusa, a subsidiary of the timber conglomerate Jayanti Group. Ethnic Togeans from Bungayo (a group of villages on the island of Talatakoh) descended on the logging camp, demanding that the company stop operations and leave the island immediately. Enraged over the exploitation of their lands, they burned two logging trucks and a bulldozer. Community members later assembled outside the local parliament, demanding an explanation for the government's misappropriation of their traditional lands and forests (Faisal, 2001). As local authorities arrested six youths for taking part in the riot, villagers converged on the police station, prepared to force their release. Two days later, the company's representatives departed the island and the youths were released.

On September 18, 2001, the Alliance of Indigenous Togean Peoples and the Togean Women's Solidarity held demonstrations in the Kayome Forest, on the island of Batudako. For generations the Bobongko people have relied on the forest, taking only what they needed for food and housing. They are opposed to any project that would create plantations or otherwise destroy the forest of which they are a part (Faisal, 2001).

During 1998, the governor of Central Sulawesi granted a concession to Cahaya Flora Perkasa Kencana, a logging syndicate from Kalimantan, to operate in the island without consulting village elders. The company began industrial-scale logging, "depriving the Bobongko of their traditional lands," and laying the basis for an uprising (Faisal, 2001). A large group of people of Batudako gathered at the logging camp September 18 and seized two bulldozers, a large generator, and a television set (Faisal, 2001).

According to a report in the November 2001 edition of the *World Rainforest Movement Bulletin*, "People from nearby islands have also come to the Kayome forest, [expressing] solidarity among the indigenous community to drive corporations from their homes. . . . Several hundred people, representing four different ethnic groups (the Bobongko, Togean, Saluan and Bajau Indigenous peoples) and Solidaritas Perempuan Togean got together at Kayome" (Faisal, 2001). In addition to protesting the stripping of their forests, local islanders also took issue with large pearl-harvesting industries, Tamatsu and Cahaya Cemerlang, based in Japan and Australia respectively, that have set up operations in the islands. Foreign tour operators operating on the islands also have drawn their ire.

REFERENCES

Abrash, Abigail. "The Amungme, Kamoro, and Freeport: How Indigenous Papuans have Resisted the World's Largest Gold and Copper Mine." *Cultural Survival Quarterly* 25:1 (spring 2001): 38–43.

———. "The Victims in Indonesia's Pursuit of Progress." *New York Times*, March 6, 2001, A-21.

Bryant, D., D. Nielsen, and L. Tangley. *The Last Frontier Forests: Ecosystem on the Edge*. Washington, D.C.: World Resources Institute, 1997.

"Building Human Security in Indonesia." Harvard College, September, 2001. [http://www.preventconflict.org/portal/main/maps_sumatra_impacts.php]

Carr, Francis. "Indorayon's Last Gasp?" *Down to Earth* (January 2001): n.p. [http://www.gn.apc.org/dte/CInd.htm]

"The Damming of the Mamberamo River in West Papua Threatens Villages." *Cultural Survival News Notes*, August 17, 2001. [http://www.cs.org/main.htm]

"Dayak People of Indonesia Stake Claim on Toorak Home of Rio Tinto Board Member." Oxfam Community Aid Abroad, January 29, 1998. [http://www.caa.org.au/pr/1998/indon2.html]

"Dig In!" Waste Flow, Blood Flow!" *Kerebok* 2:10 (May 2001): n.p. [http://www.jatam.org/xnewsletter/kk10.html]

"ExxonMobil Sued For Atrocities in Indonesia." *Cultural Survival News Notes*, June 21, 2001. [http://www.cs.org/main.htm]

Faisal, Agus. "Indonesia: Togean People Defend Their Forests, Lands, and Ocean." *World Rainforest Movement Bulletin* (November 2001) n.p. [http://www.wrm.org.uy]

"Freeport-McMoran Admits Funding Millions to Indonesian Military." *Drillbits and Tailings* 8:3 (April 11, 2003): n.p. [project_underground@moles.org]

Fujii, Miki. "Life Among Borneo's Ethnic Tribes." *Daily Yomiuri (Tokyo)*, April 7, 2001, 10.

Gargan, Edward A. "Lust for Teak Takes Grim Toll; Illegal Logging Decimating Indonesia's Majestic Forests." *Newsday*, June 25, 2001, A-7.

"Indigenous Small-scale Mining under Threat." *Down to Earth* 48 (February 2001): n.p. [http://www.gn.apc.org/dte/48ssm.htm]

"Indonesia: Low Expectations on Log Export Ban." *World Rainforest Movement Bulletin* 53 (December 2001): n.p. [http://www.wrm.org.uy/deforestation/logging.html]

"Indonesian Man Shot at Australian Gold Mine." Environment News Service, January 22, 2002. [http://ens-news.com/ens/jan2002/2002L-01-23-01.html]

"Indonesian President Calls for Logging Halt." Environment News Service, May 27, 2002. [http://ens-news.com/ens/may2002/2002-05-27-19.asp#anchor2]

Kaltim, Jatam. "Kronologis Jebol dan Meluapnya Air Limbah di Rapak Lama, Marangkayu." *Kompas Daily*, February 16, 2000.

"Mining in the South Pacific: On the Other Side of the Forests." N.d. [http://parallel.acsu.unsw.edu.au/mpi/docs/niart.html]

Murphy, Dan. "Why Borneo's Sun Bears Now Attack." *Christian Science Monitor*, August 27, 2001, 8.

"New Digging: Placer Dome, Inc. Returns to Exploit Another Area of Indonesia." *Kerebok* 2:10 (May 2001): n.p. [http://www.jatam.org/xnewsletter/kk10.html]

Parry, Richard Lloyd. "The Hunt for Bruno Manser; In 1984, Bruno Manser went to Live with a Tribe of Rainforest Nomads. He Dressed as a Native, Hunted with Darts, and Became the Most Famous Green Activist of His Generation. Then, Last Year, He Vanished. A Tragic Accident? Or was the 'Wild Man of Borneo' Murdered?" *London Independent*, September 23, 2001, 18–22.

"The People's Refusal over the Gold Mining Plan in Tahura Poboya-Paneki." *Kerebok* 2:10 (May 2001): n.p. [http://www.jatam.org/xnewsletter/kk10.html]

"Petaka Pembuangan Tailing ke Laut." Published by JATAM, April 2001. MinergyNews.com, April 28, 2001.

"Portrait of Indonesian Forestry: Supply, Demand and Debt: A Call for Moratorium on Industrial Logging." Indonesian Forum on the Environment, April 24, 2001. [http://www.walhi.or.id/KAMPANYE/Moratorium.htm]

"Protesters Shot at Unocal Refinery in Indonesia." *Drillbits and Tailings* 5:17 (October 20, 2000): n.p. [http://www.moles.org/ProjectUnderground/drillbits/5_17/1.html]

"Resource Boom or Grand Theft?" Australia West Papua Association, Sydney. N.d. [http://www.cs.utexas.edu/users/cline/papua/deforestation.htm]

"Rio Tinto Kelian Mine Shut Down by Community Blockade." *Drillbits and Tailings* 5:8 (May 16, 2000): n.p. [http://www.moles.org/ProjectUnderground/drillbits/5_08/1.html]

Rio Tinto's Environmental Record in East Kalimantan." *Down to Earth* (September 1999): n.p. [http://www.gn.apc.org/dte/Cklpl.htm]

"Rio Tinto's Shame File: Indonesian Landowners' Discontent Represented at Rio Tinto AGM." Mineral Policy Institute, May 22, 2000. [http://www.mpi.org.au/releases/rio_agm.html]

"Sarawak Peoples Campaign: The Penan of Sarawak." N.d. [http://www.rimba.com/spc/spcpenanmain1.html]

"Unocal Tailing Pipe Flooded; Rice Fields in Maraangkayu Contaminated." *Minergy News*, Indonesia. 2000. [http://www.minergynews.com/ngovoice/voice3.shtml]

"Who Really Owns West Java Forests?" *Jakarta Post*, October 22, 2001, n.p. (in LEXIS)

IRAQ

THE KURDS: POISON GAS

Introduction

More than 200 attacks using poison gases have been attributed to the Iraqi armed forces as retaliation for the indigenous Kurds' support of Iran during a war with Iraq during the 1980s. In addition to killing between 50,000 and 200,000 indigenous people in northern Iraq, the attacks have caused deformations in many survivors at the DNA level (ensuring that the defects will be passed on generation to generation). In addition, formerly fertile farm fields and water sources have been poisoned (Goldberg, 2002, 61).

Iraq's Military Offensive

In August 1988, shortly after the cease-fire that ended the Iran-Iraq war, the Iraqi government led by Saddam Hussein launched a major military offensive against the Kurds in northern Iraq, using mustard gas and other nerve-destroying agents that long have been illegal under several international protocols. A Physicians for Human Rights (PHR) team concluded that bombs containing mustard gas and at least one unidentified nerve agent had been dropped on Kurdish villages in northern Iraq ("Nerve Gas," 1993). "These chemical weapons attacks were part of a genocidal campaign carried out against Kurdish civilians," said Kenneth Anderson, director of the Arms Project of Human Rights Watch and a member of a forensic team that visited Iraqi Kurdistan in June 1992 ("Nerve Gas," 1993).

The Kurdish city of Halabja, 15 miles from Iran's border, suffered widespread (and, by the year 2002, well-documented) attacks that killed many thousands of people, left others debilitated for life, and ruined what was once a fertile agricultural area. On the late morning of March 16, 1988, reported Jeffrey Goldberg in the *New Yorker*, an Iraqi air force helicopter flew over the city of 80,000 people, taking pictures with still and video cameras. A half-hour later, Iraqi

A Kurdish refugee camp on the Iraqi-Turkish border, 1988. (NewsCom.com/Zuma Press)

ground troops fired artillery shells filled with poison gases into the city. People in the city recalled smelling garlic, sweet apples, and garbage. Soon, sheep, goats, cows, and birds fell onto their sides, dying (Goldberg 2002, 53).

Shelters Became Gas Chambers

Residents of Halabja hid in shelters, prepared for a bombardment, but soon they became very ill. Some felt painful sensations in their eyes, "like stabbing needles" (Goldberg 2002, 53). Children began to vomit. The people realized that their shelters were becoming gas chambers. The gas was heavier than the ambient air, so it followed people who hid below ground level. People emerged from their shelters, running, weakened by nausea, trying to avoid the spreading clouds of gas. Leaves began to fall from the trees, even though it was early spring.

People fleeing from Halabja began to die as they ran—children and old people first. "They were running, then they would stop breathing and die," said one eyewitness (Goldberg 2002). Some people stripped off their clothes and died laughing madly (Goldberg 2002, 53). Others became disoriented and died where they stood. Many children went blind as they ran: "The children were crying: 'We can't see. My eyes are bleeding!'" (Goldberg 2002, 54). Several people tried to avoid the gas by jumping into ponds; they died, and their decomposing remains later poisoned the water table, making the land useless for farming.

The nerve gases were dropped on small villages as well as larger cities. Eyewitnesses have said that Iraqi warplanes dropped three clusters of four bombs each on the village of Birjinni on August 25, 1988. In Birjinni, a small village of about 30 houses, a mosque, and a school, observers recalled "seeing a plume of black, then yellowish smoke, followed by a not-unpleasant odor similar to fertilizer, and also a smell like rotten garlic. Shortly breathing, their eyes watered, their skin blistered, and many vomited—some of whom died. All of these symptoms are consistent with a poison-gas attack" ("Nerve Gas," 1993). At least four people were killed during the attack on Birjinni, two in an orchard and two brothers in a cave where they sought refuge. The remaining villagers fled.

The gas attacks in Birjinni and Halanja were similar to many others in Kurdistan during the ensuing several months. While the Iraqi government later denied the attacks, several international eyewitnesses confirmed them with photographic evidence of the dead and deformations among the living. Physicians for Human Rights also detected trace elements of the gases in the area years after the attacks ("Nerve Gas," 1993). Alastair Hay, a consultant to PHR and a senior lecturer in chemical pathology at the University of Leeds, said, "this discovery not only confirms eyewitness accounts and medical examinations of Kurdish people that nerve gas as well as mustard gas were used against them" ("Nerve Gas," 1993).

Genocide via Infertility

In addition to deaths, deformations, and ecological damage, the genocidal effects of the nerve gases used against the Kurds took another form: infertility. By the year 2000, miscarriages were outnumbering live births in the Kurdish territories, and many children born alive were deformed, with harelips, cleft palates, and other afflictions (Goldberg, 62). Residents of cities attacked by the nerve agents also were afflicted by cancer rates (notably colon cancer) 5 to 10 times those of unaffected regions. Local hospitals lacked radiation or chemotherapy capabilities, so cancers often were cut out by surgeons.

After the attacks, surviving men often were taken away by the Iraqi army and probably killed, leaving behind legions of widows doing their best to look after sick, deformed children. Goldberg reported that "most of the Kurds who were murdered . . . were not killed by poison gas; rather, the genocide was carried out, in large part, in the traditional manner, with roundups at night, mass executions, and anonymous burials" (Goldberg 2002, 62).

REFERENCES

Goldberg, Jeffrey. "The Great Terror: In Northern Iraq, There Is New Evidence of Saddam Hussein's Genocidal War on the Kurds, and of its Possible Ties to Al Quaeda." *New Yorker* (March 25, 2002): 52–73.

"Nerve Gas Used in Northern Iraq on Kurds: Medical Group Proves Use of Chemical Weapons through Forensic Analysis." Physicians for Human Rights, April 29, 1993. [http://www.phrusa.org/research/chemical_weapons/chemiraqgas2.html]

IRIAN JAYA/PAPUA NEW GUINEA

INTRODUCTION

Irian Jaya (Papua New Guinea)—a province of Indonesia on the west; formerly known as Australian New Guinea on the east—is one of the most environmentally exploited large islands in the world. In West Papua, one of the world's largest gold and copper mines continues to grow, spewing waste that turns forests into moonscapes. Several hundred local indigenous people have died following protests of the mine's environmental and safety record during a quarter-century of strenuous opposition to the destruction, by strip mining, of mountains they regard as sacred.

If the results had not been so tragic, some of the environmental history of Papua New Guinea could provide the makings of an environmental soap opera. What other reaction could a serious student have when the chief executive officer of Freeport McMoRan, owner of one of the largest gold and copper mines in the world (which is sawing off the top of one of the tropics' few snow-capped mountains), compares the mine's environmental impact to his urinating in the ocean? In addition to opposing Freeport's Grasberg mine, indigenous peoples in Papua New Guinea have also resisted dam construction, logging, and several other types of mining.

FREEPORT'S GRASBERG MINE: TIDAL WAVES OF WASTE

New Orleans–based Freeport McMoRan's Grasberg mine, in West Papua's Jayawijaya district, operates in conjunction with Rio Tinto (formerly Rio Tinto Zinc). Grasberg, the largest gold mine (and the third-largest copper mine) in the world, is situated on the 16,500-foot-high, snow-capped Mount Jaya, a few hundred miles south of the equator in Papua New Guinea, in an area considered sacred by indigenous people in the area. The Grasberg mine contains gold, silver, and copper valued at $50 billion (Bryce, 1996). According to one observer, "Freeport's Grasberg mine is essentially grinding the Indonesian mountain into dust, skimming off the precious metals, and dumping the remainder into the Ajkwa River" (Bryce, 1996).

The Mineral Policy Institute has called for an end to Rio Tinto's environmentally destructive mining activities at the Freeport Mine, "a mine described as having them world's worst record of human rights violations and environmental destruction" ("Rio Tinto's Shame," 2000). The Freeport mine uses Lake Wanagon, an alpine lake that also is considered sacred by the indigenous Amungme people, to dispose of waste rock from its massive gold- and copper-mining operation near the Grasberg gold mine.

Freeport mines 78,000 tons of ore a day, as well as additional overburden, nearly all of which is dumped as mine waste and tailings into the rivers surrounding the mine, and others in the area, "making the water toxic and thick with silt, smothering and killing all plant life along the previously fertile river banks," and contaminating drinking water supplies ("Resource Boom," n.d.).

In Freeport's five-square-mile strip mine, between 80 and 100 giant trucks haul 600,000 tons of rock daily from a pit almost 3,000-feet deep. Twelve miles of conveyor belts carry ore to a milling plant that uses more than a billion gallons of water a month. Most of the machinery in the mine was dismantled and hauled up rock walls in pieces on an aerial tramway (Roberts, 1996). Mount Jaya, 16,500-feet high, towers above the mine and contains three of the world's eight remaining equatorial glaciers. Gold production from the mine averages between 1 million and 1.5 million ounces a year; copper production averages 1 billion pounds a year. The mine employs about 17,000 people, 89 percent of them non-Papuans (Roberts, 1996, 14).

Freeport has not paid any mining royalties (or any other compensation) to the roughly 4,000 Amungme indigenous people displaced by the growing mine's concession area of 9,266 square miles since strip mining began there in 1972. Many of the displaced people have moved to the lowlands, where malaria and other diseases have killed several hundred of them (Roberts, 1996, 15).

Mine tailings are dumped into a tributary of the Ajikwa River, after which they flow down steep mountainsides into rainforests at lower elevations, producing a desolate landscape. The scene was described by one observer: "Dead and dying trees are everywhere, their broken branches protruding from tracts of gray sludge. . . . Vegetation is being smothered by accumulated sludge that is several yards deep in places. . . . By the company's own calculations, 51 square miles of rainforest is expected to be destroyed before the century is out" (Roberts, 1996, 16).

An estimated 3 billion tons of rock will have been processed by the time the mine is exhausted in about 2040. According to the Mineral Policy Institute, "This waste is acidic and contains heavy metals. The water from Lake Wanagon flows into the Ajkwa River system that flows down to the Arafura Sea. In addition the mine dumps 300,000 tons of waste tailings into the Ajkwa River every day" ("Rio Tinto's Shame," 2000).

In 1977, local indigenous peoples affiliated with the Free Papua Movement issued their own critique of its environmental record by blowing up one of its ore pipelines. According to Al Gedicks, writing in *Resource Rebels* (2001), reaction of the Indonesian military was swift and emphatic:

The Indonesian military responded by sending United States–supplied OV-10 Bronco attack jets to strafe and bomb villagers. The retaliation was code-named Operation Tumpas ["annihilation"]. Papuans claim that thousands of men, women, and children were killed in this action; the government admits to 900. Reports of the use of these counterinsurgency aircraft did not appear in the world press until a year later. (Gedicks, 2001, 95).

Local protests of the Grasberg mine have continued for many years. In 1996, after an indigenous man was hit and injured by a car driven by a Freeport employee, 6,000 tribal people laid siege to the mine's offices. When Freeport Chief Executive Officer Jim Bob Moffett arrived at a local airstrip on March 12, 1996, a group of similar size gathered at the airport to demand that the mine be shut down. Moffett was quoted in the September–October 1996 issue of *Mother Jones* as saying that the environmental impact of the mine is the equivalent of "me pissing in the Arafura Sea" (Bryce, 1996).

Moffett earned $83 million (salary, bonuses, and stock options) during 1995 and 1996; according to *Business Week,* he was the tenth-highest compensated CEO in the United States. "Looking at it another way," reported the *Austin (Texas) Chronicle* in April 1997, "Moffett's pay was nearly three times the total amount that [Freeport] has agreed to pay several thousand Amungme tribal members who have been displaced by the company's mining projects in Indonesia" (Ziman, 1998). Meanwhile, members of the indigenous Amungme tribe have literally watched their mountain disappear. Mining has removed enough earth to lower the mountain by 400 feet in seven years, and now the Ajkwa river is so badly polluted from the mine that Kwamki-lama residents have been warned by Freeport's own employees not to drink the water or eat plants that grow near the water (Ziman, 1998). During 2002, Moffett's executive compensation (salary plus bonus plus exercised stock options) as CEO of Freeport-McMoRan was reported by the Wall Street Journal to be $9.11 million. He also held stock options valued at an additional $10 million ("The Boss's Pay," 2003, R6).

During March 1997, several thousand villagers rioted in the towns of Timika and Tembagapura, located near the mine. Four people were killed and more than a dozen injured as protesters damaged Freeport's equipment. The Australian Council for Overseas Aid (ACFOA) and the Catholic Church of Jayapura reported that Freeport turned a blind eye while the Indonesian military killed and tortured dozens of native people in the area surrounding the mining concession. "Villagers were beaten with rattan, sticks, and rifle butts, and kicked with boots," one tribal leader told Catholic Church officials. "Some were tortured until they died" (Ziman, 1998).

Even as Freeport adamantly denied responsibility for alleged human rights violations, the company and the Indonesian military responded to local indigenous protests by spending $35 million to assemble barracks and other facilities to house and support 6,000 troops, "more than one soldier for each adult Amungme" (Gedicks, 2001, 106–107). The company asserted that the

ACFOA had backtracked on its original claim that Freeport was involved in the killings (Bryce, 1996). Indonesian military troops routinely guarded the area around the mine, and Freeport provided them with food, shelter, and transportation (Bryce, 1996). The Indonesian government maintained a nine percent share in the mine, enough to earn several hundreds of millions of dollars a year in royalties, taxes, and benefits, making Freeport Papua New Guinea's largest single taxpayer.

During May 2000, the Grasberg mine's waste-rock disposal dam collapsed, killing four workers, and, according to one account,

> sending several 40-meter-high "tidal waves" of waste roaring down the Wanagon river towards Banti village. Incredibly, there was no loss of life at Banti despite most people being asleep when the waves arrived, passing just meters below homes, killing livestock and destroying the village graveyard. Adding insult to injury, 30 minutes after the flood reached Banti, an early-warning system installed by Freeport rang the alarm. ("Rio Tinto's Shame," 2000)

One witness at the site reported in the *Jakarta Post* that a "150-foot-high wave had . . . destroyed pig sties, vegetable gardens, and a burial ground . . . about seven miles downstream of the basin" (Gedicks, 2001, 30). One report described the resulting tidal wave of waste as "a mini-tsunami" ("Freeport Faces," 2000). The spill occurred, coincidentally, one day before the annual shareholders' meeting for Freeport McMoRan Copper and Gold.

Within days of the spill, on May 8 and again on May 18, protests against Freeport shut down the company's offices in Jakarta and prevented about 1,000 Freeport employees from entering their workplaces. In addition to protesting the environmental devastation and deaths caused by the spill, the protesters demanded that Freeport Indonesia provide a larger proportion of its earnings to support local people in the impoverished province surrounding the mine.

Soon after the accident, about 600 Amungme people from Banti, Tsinga, and Arwanop blockaded the Freeport mine's access road, preventing workers' buses from entering the mine. Roughly 100 police confronted the blockade but failed to break it until representatives were allowed to meet personally with Hermani Soeprapto, Freeport's general manager, and addressed their grievances to the company.

Indonesian environmental officials later told Freeport that it must submit a comprehensive new plan and obtain government approval before opening a replacement dump for its waste rock. Freeport also was instructed to clean up all destruction and pollution caused by the waste released during the accident. In addition, Freeport was told to allow a criminal investigation by the police and government officials into the four men's deaths caused by the collapse. The company also was ordered by the government to compensate losses suffered by residents of Banti.

The dam at Lake Wanagon has failed three times (June 20, 1998; March, 20, 2000; and May 2000) due to the company's dumping of overburden. After the third breach, dumping was halted pending an investigation. The investigation, conducted by Freeport and the Institute of Technology of Bandung (Indonesia), cleared the company to continue operations in January 2001.

Construction of a dormitory town at Tembagapura in association with Freeport Indonesia's mining operation at Mount Carstenz led to eviction of indigenous Amungme, who were barred from entering the town, which houses as many as 20,000 workers and family members ("Resource Boom," n.d.). Freeport moved the 1,000 inhabitants of a village, Lower-Waa, to the coastal lowlands, where, in one month, 88 of them died of malaria ("Resource Boom," n.d.).

"Freeport has taken over and occupied our land," said Tom Beanal, leader of LEMASA, an acronym for the Amungme Tribal Council, the community organization of the indigenous Amungme people. "Even the sacred mountains we think of as our mother have been arbitrarily torn up by them, and they have not felt the least bit guilty. Our environment has been ruined, and our forests and rivers polluted by waste" (Ziman, 1998). "They take our land and our grandparents' land," said Beanal. "They ruined the mountains. They ruined our environment. . . . We can't drink our water anymore" (Bryce, 1996).

Indigenous peoples living in the area that is being mined by Freeport have no legal title to their lands under Indonesian law. Their land is classified as *tanah negara* (state-owned land) under the terms of the Indonesian Constitution. The same central government has granted Freeport a legal right to use the land largely as it sees fit, with only the lightest of environmental oversight.

During October 1995, after a lengthy investigation, the Overseas Private Investment Corporation (OPIC), a federal agency that supports American companies doing business overseas, canceled Freeport's $100 million political-risk insurance policy, citing environmental problems at the mine. In a letter dated October 10,1995, OPIC told Freeport the mine had "created and continues to pose unreasonable or major environmental, health, or safety hazards with respect to the rivers that are being impacted by the tailings, the surrounding terrestrial ecosystem, and the local inhabitants" (Bryce, 1996).

Freeport's Grasberg mine is only the best known and largest of several mineral-extraction projects on Papua New Guinea that have stirred protests by the island's indigenous peoples. In Papua New Guinea, Rio Tinto's Panguna mine, which was one of the world's largest open-pit copper mines before it was closed, dumped more than one billion tons of mine waste into the Pangana, Jaba, and Kawerong rivers, killing all aquatic life in the 480-kilometer river system. The waste formed a deposit approximately 20 kilometers long, as much as a kilometer wide, and several meters deep along these rivers, with a copper-contaminated outwash fan in Empress Augusta Bay covering roughly nine square kilometers ("Mining in the South Pacific," n.d.).

The U.S. State Department at one point attempted to defeat a lawsuit alleging genocide and environmental damage that had been filed by Bougainville landowners against Rio Tinto's operation of the Panguna mine. State Department officials wrote to the judge hearing the case, saying that airing of the class action suit would affect U.S. relations with Papua New Guinea. The government of Papua New Guinea also tried to block the lawsuit, according to a report November 30, 2001 by the Australian Broadcasting Corporation ("Hotspots: Bougainville," 2001).

Rio Tinto, a mining partner of Freeport in West Papua, is the parent company of C.R.A., an Australian mining company that operated the huge Bougainville copper mine, which was established by Rio Tinto during the 1970s while Papua New Guinea was still an Australian protectorate. A guerrilla movement has been campaigning since the 1980s for compensation of Bougainville's traditional landowners, who had been dispossessed by the company's operations. An estimated 5,000 civilians were killed (or died of imported diseases) in the Bougainville area during the 1990s.

The Kutubu project, operated by the U.S.-based Chevron Oil Company, was Papua New Guinea's first oil-extraction project; critics alleged that its environmental and social impacts were much greater than reported, "including significant impacts on biodiversity, and the risk of [oil] spills, with the benefits to landowners being relatively low. One academic expert on the area described the impact as "devastating" ("Mining in the South Pacific," n.d.). Oil is piped 176 kilometers to the Kikori estuary. In 1996, nonessential staff had to be evacuated after threats from landowners dissatisfied with royalty payments.

The Porgera gold mine, maintained by Placer Pacific of Australia (majority shareholder for Placer Dome of Canada), has been one of the dirtiest operations on Papua New Guinea. It dumps 40,000 tons of tailings and waste rock daily into the Strickland/Fly River catchment basin. Environmental sampling has indicated levels of metals as much as 3,000 times levels permitted by government regulations, which are not enforced.

The Ok Tedi mine has been a subject of controversy because its tailings killed all aquatic life along 70 kilometers of the Ok Tedi River of Papua New Guinea, following pollution by tailings and waste rock. Legal suits by indigenous landowners forced B.H.P. Billiton of Australia, owner of the mine, to reach an out-of-court settlement for $550 million (Australian). Members of communities affected by the mine's pollution blockaded it and shut it down on November 25, 2001, costing the company close to U.S. $1 million in lost production.

Late in 2001, B.H.P. Billiton convinced the Papua New Guinea government to endorse three acts of legislation affecting the Ok Tedi mine, which "the environmental and human rights communities [say will allow] the company [to] escape responsibility for the damage it has caused to the environment and communities living near the Ok Tedi in the western province of Papua New Guinea" ("B.H.P. Billiton," 2000). According to a summary provided by Slater and Gordon, an Australian law firm that has filed a lawsuit against the company, the agreements will give B.H.P. and Ok Tedi Mining "unrestricted legal indemnity for the pollution and destruction caused now and into the future by the operations of the Ok Tedi mine. The mine's owners will have no obligation to stop tailings entering the river system in future, and will be permitted to increase the amount of copper it is currently permitted to dump into the river system" ("B.H.P. Billiton," 2000). The agreements release Billiton from any liability under an ongoing suit in Australia's Victorian Supreme Court; under the same agreements, landowners also lose their common-law rights to enforce a 1996 settlement as well as future legal rights to sue the mine for any environmental damage.

At about the same time, representatives from four indigenous communities living near the mine presented a petition demanding compensation and a share of B.H.P. Billiton's 52 percent stake in Ok Tedi. The local people also demanded compensation for environmental damage from the date the mine opened (1981) to the present. The government of Papua New Guinea denied a request from local landowners to grant them 12 percent of the benefits from the Program Trust Company, to which B.H.P. Billiton's interest in the mine was sold in 2002 ("Hotspots: Papua," 2001).

Papua New Guinea's Prime Minister, Sir Mekere Morauta, has called the Ok Tedi mine a national asset. Morauta argues that closure of the mine could devastate the nation's economy and cause the ruin of communities that depend on it. The mine accounts for 10 percent of Papua New Guinea's gross national product and 20 percent of its export income.

As the mine's owners were sealing their deal with the Papua New Guinea government, the Australian Conservation Foundation issued a report saying that nearly 70 kilometers of the Ok Tedi River has become "almost biologically dead," and 130 kilometers of riverbank have been "severely degraded." Fish stocks have declined between 50 percent and 80 percent, according to the mine's own internal report. Roughly 30,000 downstream landowners have lost their ability to live off their own land. A scientific peer review group employed by the mine's management identified potential for a total collapse of the fishery ("B.H.P. Billiton," 2001).

In addition to widespread logging on traditional lands in West Papua, some of the world's largest transnational mining corporations have been active in exploiting the island's oil and minerals, including Union Oil, Amoco, Agip, Conoco, Phillips, Esso, Texaco, Mobil, Shell, Petromer Trend Exploration, Atlantic Richfield, Sun Oil, and Freeport (United States of America); Oppenheimer (South Africa); Total (France); Ingold (Canada); Marathon Oil and Kepala Burung (United Kingdom); and Dominion Mining, Aneka Tambang, B.H.P. Billiton, Cudgen R.Z., and C.R.A. (Australia). Petromer Trend and Conoco have produced 300 million barrels of oil from the field at Sele near Sorong, valued at an estimated $4.5 billion ("Resource Boom," n.d.). Mining concessions in the Ertsberg and Grasberg mountains, the Paniai and Wissel lakes region, Fak Fak, the Baliem Valley, the "Bird's Head" western tip, and the Papua New Guinea border area have involved dislocation and suppression of Papuan peoples, sparking popular uprisings followed by military reprisals.

By early 2002, B.H.P. Billiton had officially exited Papua New Guinea, leaving behind operations at the Ok Tedi gold mine, leaving 30,000 people displaced by pollution. The company admitted in 1999 that the sediment load (a buildup of waste rock or tailings) has caused a 90 percent fish kill in the lower Ok Tedi River ("Hotspots: Papua," 2002).

According to a report in *Drillbits and Tailings*, "Part of B.H.P. [Billiton's] plans to exit Ok Tedi included a transfer of its 52 percent equity interest to the P.N.G. Sustainable Development Program. S.D.P.L. is a special-purpose development fund that will channel dividends to sustainable development projects, mainly in the Western province. According to the P.N.G.-based Environment Watch Group, the deal is 'a slap in the face' especially because B.H.P. has destroyed the Fly River" ("Hotspots: Papua," 2002).

Freeport-McMoRan later disclosed in a report to the Security Exchange Commission that it paid the Indonesian national military, Tentara Nasional Indonesia (TNI) an estimated U.S. $5.6 million in 2002 for security purposes, most of it associated with the Grasberg mining operation.

According to a report from *Drillbits and Tailings*, "In [its] 2002 annual report, entitled *Real Assets, Real Value,* Freeport boasts that the Grasberg mining complex is the flagship of its operations. The Freeport Indonesia unit produced and sold 1.5 billion pounds of copper and 2.3 million ounces of gold during 2002. The report heralded the world's lowest copper production cost at less than US $0.08 per pound." The number of soldiers deployed at such vital Freeport locations has increased from 200 soldiers to over 2,300 soldiers since 1996 ("Freeport-McMoran Admits," 2003).

DAM DEVELOPMENT IN PAPUA NEW GUINEA

A proposed $6 billion dam on the Mamberamo River in West Papua (New Guinea) would flood one of the richest biological areas of the world. Not only would this project reduce biodiversity in the area, but it also could devastate the cultures and traditions of 35 nomadic tribes in the area. Wimpie Dilasi, a leader of the regional tribes' federal council, said that these plans, especially the 10,000-megawatt dam, will create widespread misery. Local indigenous people have only rarely been consulted regarding their homelands' future.

In 1997, an Indonesian government official arrived at Lau, West Papua, a group of palm-thatched houses along a remote tributary of the Mamberamo River. The envoy said that everyone in Lau would soon be forced to move into the surrounding mountains. Their land, an area larger than England, was slated to be flooded. On April 2, 2001, West Papua's governor, J. P. Salossa, announced that the Mamberamo megaproject was proceeding with "renewed momentum" with funding from the World Bank and the Asian Development Bank ("Damming," 2001). A Lau village chief told the visiting coordinator of the World Wildlife Fund, "I would rather be shot in the head than be resettled" (Carrere, 2001).

During 1999, a South Korean firm, P.T. Kodeco Mamberamo Plywood, opened a sawmill and planted an oil-palm plantation in the area. Extensive, industrial-scale logging of old-growth rainforests in the company's 691,700-hectare concession already was threatening populations of endangered green turtles and rare birds. According to an analysis by the World Rainforest Movement, this is just the beginning of an industrial future for the area: "Land that has been cleared by P.T. Kodeco will serve as a site for a major industrial estate with metal-smelting works, sawmills, agribusiness plantations, and petrochemical processing factories to be powered by the dam" (Carrere, 2001).

Areas within some mining concessions have been questionably designated earthquake zones, requiring mass resettlement of tribespeoples such as the Hupla of the central highlands, an unnecessary and destructive practice according to the World Rainforest Movement. Communities are often coerced into moving to sites at lower elevations, where they are more prone to diseases such as malaria and where traditional mountain foods do not grow.

THE MOI: LOGGING AND MINING IN WEST PAPUA

Under Indonesian national law, all land not being actively used for agriculture, housing, or industry is deemed state property. The government therefore often concludes that indigenous peoples are not using their communal homelands according to a body of law that recognizes no form of indigenous land rights. Any protest of this situation has led to accusations that the Moi are "security disturbers . . . the official term . . . used to silence any form of indigenous protest" ("Resource Boom," n.d.) Like other indigenous communities, the Moi people have come into conflict with the Intimpura Timber Company.

The livelihoods of the Moi and other indigenous peoples have been threatened, as vast tracts of land have been granted as concessions to timber companies. These are some of the most biologically diverse forests in the world. Numerous species that are unique to the area have been threatened by logging and other development projects.

Indonesia has encouraged the development of a large timber-processing industry by banning the export of raw logs; the country has thereby become one of the world's largest exporters of plywood. As Indonesia's remaining forest resources decline in Sumatra and Kalimantan (Borneo), the forestry industry has targeted West Papua. Four Jakarta-based timber tycoons have divided West Papua between them; their domination has been achieved with support of the government. To exploit the country's resources fully, the government has given the construction of roads in previously inaccessible areas a high priority. "Logging roads are carelessly constructed," according to one observer,

leading to substantial soil erosion and consequent silting of rivers and irregularity of river flow. Roads are routinely built over minor streams; the result is a roadside string of standing pools, which produce unusually high concentrations of mosquitoes and present the threat of malaria and other diseases. . . . The heavy machinery destroys trees used by local people for food sources and traditional medicines. ("Resource Boom," n.d.).

Changes in the land worry Damien Arabagali, a community leader in Toroba, West Papua:

I think nature will pay back the disrespect shown to her. Look around you! It's becoming hotter, drier, more eroded. Here, in Huli country, people are hiding from the sun. They never used to do that before. The land is more barren than ever before. That's why people have to work much harder than ever before. Look at how skinny they are! And because of all this clearing of our forests we have floods, which we never had before. We can see it here. Our fertile swamps, where we plant *kaukau* and our sweet potatoes . . . are drying out. Soon there will be no more swamps. What then? What's the cause of all this? I think it's greed. Something in humanity must be evil in itself. . . . The big companies have become dehumanized. Profits at all costs, no matter what that leaves behind. (Interpress Service, 1993, 89)

REFERENCES

"B.H.P. Billiton Runs from Responsibilities in Papua New Guinea." *Drillbits and Tailings* 6:10 (December 30, 2001): n.p. [www.moles.org]

"The Boss's Pay: *The Wall Street Journal*/Mercer 2002 Compensation Survey." *Wall Street Journal*, April 14, 2003, R6-R10.

Bryce, Robert. "Spinning Gold." *Mother Jones* (September–October 1996): n.p. [http://www.etan.org/news/kissinger/spinning.htm]

Carrere, Ricardo. "Indonesia: Mamberamo Dam Threatens Nomadic Tribes." *World Rainforest Movement Bulletin* (August 2001): n.p. [http://www.wrm.org.uy]

"Freeport Faces Investigation Due to Recent Disaster and Past Mismanagement." *Drillbits and Tailings* 5:9 (May 31, 2000): n.p.

Gedicks, Al. *Resource Rebels: Native Challenges to Mining and Oil Corporations*. Boston: South End Press, 2001.

"Hotspots: Bougainville." *Drillbits and Tailings* 6:9 (November 30, 2001): n.p. [www.moles.org]

"Hotspots: Papua New Guinea." *Drillbits and Tailings* 6:9 (November 30, 2001): n.p. [www.moles.org]

"Hotspots: Papua New Guinea." *Drillbits and Tailings* 7:2 (February 28, 2002): n.p. [http://www.moles.org]

Interpress Service. *Story Earth: Native Voices on the Environment*. San Francisco: Mercury House, 1993.

"Mining in the South Pacific: On the Other Side of the Forests." N.d. [http://parallel.acsu.unsw.edu.au/mpi/docs/niart.html]

"Resource Boom or Grand Theft?" Australia West Papua Association, Sydney. N.d. [http://www.cs.utexas.edu/users/cline/papua/deforestation.htm]

"Rio Tinto's Shame File: Indonesian Landowners' Discontent Represented at Rio Tinto AGM." Mineral Policy Institute, May 22, 2000. [http://www.mpi.org.au/releases/rio_agm.html]

Roberts, Greg. "Mining Big Money in Irian Jaya." *Sydney Morning Herald*, April 6, 1996. In *World Press Review* (July 1996): 14–16.

Ziman, Jenna E. "Freeport McMoRan: Mining Corporate Greed." *Z Magazine*, January 1998. [http://www.zmag.org/zmag/articles/jan98ziman.htm]

KENYA

INTRODUCTION

In Eastern Africa, Kenya has become a key site for indigenous peoples' environmental conflicts. For example, the Kwale, who live on Kenya's Indian Ocean coast, have raised objections to a titanium-mining project planned by Canadian industrialists. The Kwale have proposed that they should become co-owners if titanium on their land is mined at all. The Maasai, a pastoral people who dwell in the Mau Forest during the dry season, have been fighting expropriation of their lands. Since Kenya's independence from Britain during the 1960s, Maasai land has been taken for private farms and ranches, government projects, and wildlife parks. Most recently, the Maasai, whose land base once extended across much of present-day Kenya and Tanzania, now compete for living space with military testing grounds, on which unexpected explosions have killed and maimed a number of people. In the meantime, the Ogiek of Kenya's Rift Valley, known locally as "honey hunters," are being squeezed out of their longtime homeland by logging interests allied with the country's political elite, who have deeded substantial portions of the Ogieks' homeland to themselves while bidding for the votes of landless squatters who also are moving into the Mau Forest.

THE KWALE: OBJECTING TO TITANIUM MINING

Kenyans, notably the indigenous Kwale, have raised objections to a titanium-mining project proposed by a Canadian company with Kenyan affiliates that will strip indigenous people of their land and resources. The mine has been proposed for the homeland of the Kwale, on Kenya's Indian Ocean coast, by three companies: Canada-based Tiomin Resources, its subsidiary Tiomin Kenya, and locally owned Titanium Minerals. Local indigenous advocates have proposed a joint venture with the Kwale, if the titanium is to be mined at all. The government initially approved the mining project without joint participation during September 2000, but it suspended work a month later, following pressure from local indigenous people and the Kenyan media.

According to a report from Nairobi by the Interpress Service, "The project has sparked protests from human-rights groups and environmental organizations, [which] cite the company's poor record in its treatment of indigenous groups elsewhere" (Achieng, 2000). Grievances against the project's sponsors, according to this report, range from poor environmental impact assessments to inadequate compensation for indigenous peoples who have, for generations, occupied the land the companies want to mine. Local economic experts also have accused the company of deliberately overestimating the market value of Kwale's titanium, and of asserting that the minerals are "worth trillions of dollars, enough to wipe out the East African country's [Kenya's] foreign debt and raise the standard of living for its impoverished population" (Achieng, 2000).

"Arabs are rich because of their oil, so why can't we benefit from our titanium? The titanium at the coast is enough to liberate Kenyans from the jaws of poverty once and for all," argues Gideon Hanjari, an economic expert from the region, who supports the mining proposal (Achieng, 2000). While the mining companies have waxed eloquent about the value of titanium under the Kwales' land, they have been criticized for offering compensation far below market value to indigenous people who will be displaced by the proposed mine.

THE MAASAI: FIGHTING LAND EXPROPRIATION FOR MILITARY TESTING

The Maasai, a pastoral people who dwell in Kenya's Mau Forest, once occupied territory extending from Lake Victoria almost to the Indian Ocean. Disease and European colonization during the late nineteenth and early twentieth centuries reduced the Maasai land base substantially. At the beginning of the twentieth century, the Maasais' herding range included large parts of present-day Kenya and Tanzania. With the advent of European colonization, the extent of Maasai lands steadily shrank, until they were moved to a reserve in southern Kenya.

Following Kenyan independence from Britain during the 1960s, Maasai land has been taken regularly for private farms and ranches, government projects, and wildlife parks. According to Survival International, six of Kenya's and Tanzania's national parks alone cover more than 13,000 square kilometers (5,000 square miles) of what was once Maasailand ("Ogiek.org," 2001).

Landless peasants have taken refuge with the Maasai for many years. At first, their numbers were small and they were easily absorbed into Maasai villages. The Maasais' land was held in common at that time, so they experienced no pressure to subdivide and sell it. Following independence from Britain in 1963, however, the communal nature of Maasai land lost its legal sanction as pressure from immigrants intensified to sell individual parcels of what had been common estate. Coming from a migratory society, few Maasai attached monetary value to their land. Some of them found that they had signed their land away without realizing what they were doing. Within a century, the landholdings of the 300,000 Maasai shrank by three-quarters (Interpress Service, 1993, 173).

In addition to pressure from intruders, the Maasai soon faced disregard from Kenya's government, which designated Mukogodo, in the Laikipia district, as a combined reserve for the Laikipia-Maasai and a military training zone, sparking conflict. Between 1995 and 2001, according to one observer, "At least 12 people died and 332 were maimed by explosives left behind by Kenyan, British, and U.S. military personnel" (Stahl, 2001).

The Maasai contend that use of their homeland for military exercises violates Article 28 of the United Nations Draft Declaration on the Rights of Indigenous Peoples, which reads, in part: "Army activities shall not take place on the land of indigenous peoples without their consent. Hazardous material shall not be stored or disposed of on the land of indigenous peoples. Governments shall take measures to assist indigenous peoples whose health has been affected by such material" (Stahl, 2001).

The area around Mukogodo was first set aside for military training during the 1950s under British colonial rule and later, after Kenya's independence in 1963, by the Kenyan government. Since then, according to reports by Cultural Survival observers, "Various Kenyan police and army units, and lately the British Army and the U.S. Marines, have undertaken their training maneuvers on the land where the Maasai herd their cattle. They left behind military training gadgets, bullets, and explosives that have on several occasions killed or severely injured Maasai herdsman or their cattle" (Stahl, 2001).

In these arid and semiarid lands, where water is a precious resource and droughts are common, the Maasai complain that

> the exercises sometimes engage up to 3,000 soldiers whose water demand add to the already severe water crisis in the Laikipia district. It is estimated that 60 per cent of the water supply in Dol-Dol, the main trading center of Mukogodo, is used by the military. Moreover, the soil around the military camps is being compacted to such an extent that vegetation cannot easily regenerate. Trees, often with medical or spiritual value to the local people, have been cut down. Rubber bullets and other military waste left behind pollutes Maasai pastures, and during military maneuvers the pastoralists are forced to move to other rangelands and redefine traditional grazing patterns in order to graze their livestock without disturbance. (Stahl, 2001)

THE OGIEK: "HONEY HUNTERS" FORCED FROM THE RIFT VALLEY

The Ogiek, an indigenous hunting and honey-gathering people who live in Kenya's Mau Forest, are being forced out of a rainforest in the Rift Valley as the country's elite, including its president, usurps the land. The land at issue comprises about 900 square kilometers (about 550 square miles), 200 kilometers (125 miles) northwest of Nairobi, the capital of Kenya.

This hunter-gatherer people, who number about 20,000, is known for collecting honey from beehives, which they place in the high branches of trees in their home forest. The Ogiek have lived since time immemorial in the Mau Forest overlooking Kenya's Rift Valley. According to Kenyan law, this area is protected under Kenya's Forest Act. The Kenyan government, however, plans to open nearly 60,000 hectares of it to developers such as tea planters and loggers, along with landless settlers from elsewhere in the country. The Ogiek and their supporters also assert that international law supports the right of all tribal peoples to ownership of their native lands. The Kenya High Court has issued an order halting the opening of 35,000 hectares in East Mau, but the government has ignored it.

For decades, the Ogiek fought with the first British colonists, then with the Kenyan government, in defense of their right to inhabit parts of the East Mau and South West Mau Forests, seeking recognition of the area as ancestral land. After years of dispute, authorities have refused to recognize the Ogieks' heritage, ordering them to leave the forest, contending that they were allocated separate land years ago and then abandoned it. The Ogiek believe that they have a right to live in their ancestral home. They assert that the government wants to give their land to private individuals rather than conserve it for the benefit of the Ogiek and the entire nation.

The Ogiek selectively hunt animals in the Mau Forest that are not endangered, such as warthogs and tree hyraxes. These animals are hunted for food, not for sport. Recently, the Ogiek have reduced their hunting, focusing more on rearing domesticated animals, such as sheep and goats, as well as growing foods including beans, potatoes, and cabbage.

During March 2000, Nairobi High Court judges Samuel Oguk and Richard Kuloba dismissed a case contending that roughly 14,000 surviving Ogiek people have an ancestral right to live in the Tinet Forest (part of the broader Mau Forest). The judges spoke from behind rows of lawyers in wigs and gowns as representatives of the Ogiek stood before the court, formally dressed "in fur cloaks and head-dresses decorated with cowrie shells" ("Green Smokescreen," 2000, 1). According to an observer affiliated with Survival International, "The tone of the judgement was harsh and contemptuous, as though the judges were determined not only to deny the title of the Tinet Ogiek, but to frighten off any other group of people who might try to make a similar claim" ("Green Smokescreen," 2000, 1). The judges accused the Ogiek of defiling a national conservation area by constructing one-room wooden schoolhouses and small stores. The Ogieks' representatives complained that the judges had ignored other threats to the forest, including logging operations that defied an official ban, a tea plantation owned by a company belonging to Kenya's President Daniel Arap Moi, and farms cultivating flowers for export. The same forest also contained a "handsome house and large estate owned by Zakayo Cheruiyot, permanent secretary for provincial administration and internal security in the office of the president" ("Green Smokescreen," 2000, 1).

The Kenyan government has displayed some contradictions on the issue of Ogiek land ownership. On January 13, 2001, for example, the local district commissioner, Ole Serian, reportedly told the Ogiek that the forest belonged to them. Serian was said to have told the Ogiek that they could arrest anyone doing survey work. The next day, 200 Ogiek, taking Serian at his word, confronted surveyors and confiscated their equipment. Following this confrontation, police arrested one of the Ogiek, Dominic Maritim Wilson Monoso, as the others disappeared with the surveying equipment. Armed police then moved into the forest to protect the surveyors. Monoso was charged with "robbery with violence," an offense punishable by death. Lawyers for the Ogiek retaliated by threatening to arrest the Kenyan minister for the environment.

On February 16, 2001, the Kenyan government issued an official notice announcing its plan to degazette 147,000 acres of the Mau Forest, opening the land to outside settlers. This move effectively would remove approximately 70 percent of the Mau Forest from the legal control of the Forest Act and leave the Ogieks' homeland vulnerable to invasion by land speculators.

The government was racing to complete this legal change as opposition leaders pushed through the country's legislature a new Forest Bill, which could give the Ogiek some legal protection for their forest homelands. Three days after it was issued, the Ogieks' attorneys challenged the degazette order as a violation of the High Court's prior injunction, placing the country's minister of environment, Francis M. Nyenze, in contempt of court, and subject to arrest. The attorneys said that if the minister did not withdraw the notice, they would "move to the High Court for the necessary orders of prohibition and [his] committal to jail for contempt of court" (Kenyans, n.d.). During March 2001, police arrested the chairman of the Sagana Wildlife Protection Self-Help Group, Gerald Ngatia, for leading the local community in uprooting beacons erected by surveyors.

On July 6, 2001, the Ogiek were given a month to leave their homes, according to an order issued by the new Rift Valley provincial commissioner, Peter Raburu. "We are going to flush out everybody residing and cultivating in the forests irrespective of who allowed them in," Raburu said ("Ogiek Given," 2001). The Ogiek then challenged the removal order in the Nairobi High Court. "The administration will be in contempt of court if they dare try to evict us," said Ogiek spokesman Joseph Towett ("Ogiek Given," 2001).

The Ogiek also are concerned about ongoing logging in the Mau Forest that has been destroying their hunting grounds. While the Kenyan environmental ministry complained that the tribe's people were defiling the Mau Forest with their homes, shops, and schools, it was allowing logging companies to cut trees there. The government imposed a partial logging ban; the ban, however, exempted three large logging companies: Pan African Paper Mills, Raiply Timber, and Timsales. According to the government, the three firms were exempted because Raiply and Timsales employ more than 30,000 Kenyans, while Pan African was exempted because it is partially owned by the government.

The government of Kenya during the fall of 2001 announced its intention to "excise" more than 67,000 hectares of forests, part of the less than 2 percent of Kenya's land area that remains forested, including the Ogieks' homeland. If implemented, the announced excisions would open to outsiders the two largest of the five main water-catchment areas of the country: Mount Kenya and the Mau Forest. According to a report in the *Nation*, a newspaper published in Nairobi, "The government's plans to excise the forests provoked public outcry. Thousands of signatures were collected in a petition" (Chesos, 2001). Spearheaded by Professor Wangari Maathai and other environmental activists, the petition urged the government to reconsider its decision. The government refused to relent.

Defenders of the Ogieks' right to live unmolested in their homelands asserted that the Kenyan government was defying international law because the High Court ruling opened a formerly protected area of forest for development. Despite threats from local authorities, the Ogiek again announced that they would take responsible government officials to court. "No amount of intimidation will deter us from demanding our God-given right within the Constitution," Ogiek elders said ("Kenyan Government," 2001).

The Kenyan government asserted that the forests required protection from these hunter-gatherers, who have managed them in a sustainable fashion for thousands of years, as it planned to allow logging companies to operate within the same area. Attorneys for the Ogiek asserted that Kenya's most important rivers flow from the Mau Forest. Loss of tree cover could cause the rivers' volume to diminish dramatically. Popular tourist destinations (including the Maasai Mara and Lake Nakuru, famous for its flamingos) also could be negatively affected, according to the Ogieks' attorneys. A decrease in river flows already has caused power rationing because Kenya derives much of its electricity generation from hydropower.

In late December 2001, Kenya's government issued yet another ultimatum, demanding that the Ogiek leave the Rift Valley's Mau Forest by January 15, 2002. Announcing the order, the Rift Valley provincial commissioner, Peter Raburu, said that " the more than 10,000 illegal settlers in government forests should move out peacefully or else we will forcefully evict them" (Wanjiru, 2001). Raburu said that the destruction of the Mau Forest "could not be condoned," but he did not address the question of timber companies that have been logging the Mau Forest (Wanjiru, 2001).

By February 2002, the Ogiek refused to move as President Moi was implicated as one of the beneficiaries of long-standing plans to cut about 70,000 hectares (170,000 acres) of the Mau Forest. Survival International reported that while the forest clearance was being promoted by Kenyan officials as a program benefiting Kenya's landless poor, documents leaked to the *Nairobi Daily Nation* strongly suggested that large tracts of the forest had been secretly deeded to Moi, the present environment minister Joseph Kamotho, and former first lady Mama Ngina Kenyatta. Survival International's director Stephen Corry said: "The Ogiek, who never damaged the forest, have been evicted from it time and again, while the powerful are allowed to take it over and destroy the natural heritage of these people" ("Kenyan President," 2002). "The tragedy," according to supporters of the Ogiek, "is that if the government's scheme goes ahead, the Ogiek will simply join the numbers of Kenya's dispossessed and die out as a people" ("Kenyan Government," 2001).

The Kenyan government, to win votes of the landless poor, promised them land occupied by native peoples. On paper, some of the land already had been transferred to the landless poor. At ground level, however, the situation was different:

In Kitale, hundreds of squatters vainly wait for the promised 7,234 acres of land hived off Kitalale, Kapolet and Sikhendu forests. In official documents, the squatters are already resettled, having benefited from a process sanctioned by President Moi in a public directive in 1999. But instead of hundreds of huts, magnificent residential homes and well-tended plots dot the once public resource. Among beneficiaries are the chief of a paramilitary unit, a Cabinet minister, several Members of Parliament and members of an Ugandan repatriated clique. ("Kenya: Forest Destruction," 2002)

The scene is similar, according to the World Rainforest Movement, in the Nandi District, another part of the Mau Forest. About 2,891 hectares of that forest were officially sequestered to resettle 200 families that were forced to move from the neighboring Koibem forest by an excision sanctioned in 1999. Instead of resettling them on the entire area, the displaced were distributed on small parcels while the rest of the land was grabbed by well-connected individuals. "Most of the land meant for squatters went to powerful people. The landless, who initially supported the excision, have now realized they were cheated," said environmental lawyer Nixon Sifuna. ("Kenya: Forest Destruction," 2002)

Even as the Ogiek tried to protect their forest homeland in the Kenyan courts, trees were falling to illegal logging. The *World Rainforest Movement Bulletin* described how the Kaptagat Forest, in Keiyo, was irreversibly destroyed. The canopy of trees that frames the Eldama Ravine Road is a mere facade. Behind the thin line of trees, illegal sawmills had been erected along major roads, operating in broad daylight. President Moi was removed from office in 2002 following disclosure that his regime had been responsible for at least $800 million worth of corruption.

REFERENCES

Achieng, Judith. "Kenya: Civic Group to Sue over Titanium Project." Interpress Service, October 12, 2000. [http://www.afrika.no/index/update/archives/2000October13.shtml]

Chesos, Richard. "Carve-Up of Forests Can Go Ahead, Court Rules." *The Nation* (Nairobi, Kenya), October 5, 2001. [AllAfrica.com: http://allafrica.com/stories/200110040519.html]

"Green Smokescreen for Eviction of Forest People." *Survival Newsletter* 42 (2000): 1.

Interpress Service. *Story Earth: Native Voices on the Environment.* San Francisco: Mercury House, 1993.

"Kenya: Forest Destruction for the Benefit of Government Cronies." *World Rainforest Bulletin* 55 (February 2002): n.p. [wrm@wrm.org.uy]

"Kenyan Government Set to Destroy Honey-hunting Tribe." Survival International, December 5, 2001. [is@survival-international.org]

"Kenyan Honey-hunters' Land to be Carved Up?" Survival International, n.d. [http://www.survival-international.org/index2.htm]

"Kenyan President Implicated in Tribal Forest Land Grab." Survival International, February 11, 2002. [http://www.survival-international.org]

"Ogiek Given a Month to Quit Forest." Survival International, July, 2001. [http://www.survival-international.org/index2.htm]

"Ogiek.org: Supporting the Rights of a Kenyan Indigenous Group." Ogiek Welfare Council, 2001. [http://www.ogiek.org/]

Stahl, Johannes. "Indigenous Land Rights and the Military; Army Exercises on Maasai Land." *Cultural Survival News Notes*, 2001. [http://www.cs.org/main.htm]

Wanjiru, Jennifer. "Another Ultimatum." Rights Features Service, December 27, 2001. [URL: dfn.org/focus/kenya/ultimatum.htm]

MALAYSIA (SARAWAK)

ROADBLOCKS GREET LOGGING COMPANIES

In the Malaysian province of Sarawak, on the northwestern coast of Borneo, the Penan, Kayan, and Kenyah built roadblocks during the first half of 2002 that the *World Rainforest Movement Bulletin* described as "human barricades and wooden structures put up at strategic points across access roads to prevent the movement of logging and plantation companies' vehicles . . . into disputed logging sites" ("Malaysia: Self-defense," 2002). At least five blockades were erected and one protest was staged by several native communities during March and April 2002. This was, according to the World Rainforest Movement (WRM), the first time in more than 10 years that several Sarawak native communities had organized blockades simultaneously in various locations to draw the attention of the Malaysian authorities.

"Their rights to traditional native lands have been ignored by the logging companies and state government, thus forcing them into lives of difficulty because of deprivation of forest and river resources," according to the *World Rainforest Movement Bulletin*. "All promises of the Sarawak government concerning biosphere conservation, payment of fair compensation, assistance funds and infrastructure have not been fulfilled," said Sahabat Alam, a Friends of the Earth liaison in Malaysia ("Malaysia: Self-defense," 2002).

The WRM reported that local activists believed the situation of the Penan, who were originally nomadic hunter-gatherers, was most perilous. The Penan said that their rights to their customary lands have long been ignored by the logging companies and the state government, causing a severe decline of their living conditions due to depletion of forests and rivers that contribute to their livelihood. Most of the Penan were living, according to the *World Rainforest Movement Bulletin*, "in dire conditions without adequate food supply, proper housing facilities, accessible health care and education services and other basic necessities like clean water and electricity" ("Malaysia: Self-defense," 2002).

The native peoples behind the blockades were seeking a halt of all logging operations and plantation activities on their customary lands. They maintained that in the future they should be consulted before any economic activities commence on their native lands. In addition, the indigenous peoples demanded that the Malaysian government recognize that local peoples have a right to choose development models that best suit them, chosen with local consent.

REFERENCES

"Malaysia: Self-defense Blockades from Sarawak Indigenous Peoples." *World Rainforest Movement Bulletin*, April 2002. [http://www.wrm.org.uy]

MARIANAS ISLANDS

PCB CONTAMINATION

Introduction

Villagers in the Commonwealth of the Northern Marianas in the Pacific Ocean have demanded that the United States clean up its toxic legacy. Local authorities accused U.S. agencies of "gross negligence and criminality" (Williams, 2000). The islanders were joined in protest by Greenpeace activists aboard the organization's flagship, Rainbow Warrior. Protesters posted signs on toxic dumps and erected a fence around piles of contaminated soil, which were left behind by the U.S. Department of Defense in piecemeal attempts to clean up the village of Tanapag.

Transformers and Contaminated Soil

Electrical transformers containing PCB wastes were dumped on the Mariana Islands during the 1960s, contaminating soil and leaking into the groundwater, where toxicity has been detected since 1992. The islanders were never told that the transformers contained highly toxic chemicals, so they were not sequestered in any way. Within two to three decades, high incidence rates of various cancers (including leukemia), chromosomal changes, and reproductive disorders were reported in the community (Williams, 2000).

In a petition directed at the U.S. Department of Defense and the Justice Department, the islanders' delegates said, in part, that "the U.S. has demonstrated a total disregard for the environment and the health of people both here in Tanapag and at its other dump sites across Asia and the Pacific" (Williams, 2000). Maureen Penjueli of Greenpeace, speaking from Tanapag, said that the islanders "have been exposed to poisonous soils contaminated with PCBs for over a decade due to the negligence of the U.S. agencies" (Williams, 2000). According to Greenpeace, PCB-laced military wastes and other contamination have been left behind by U.S. military operations in many other areas of Asia and the Pacific, including Japan, the Philippines, and Guam.

"We were not told that the waste left by the U.S. was dangerous," said a local resident, Mike Evangelista. "For years we have used it in our communities to make roadblocks and as windbreakers. We even used some of it to make headstones in our cemetery. Now we know why so many of us get cancers and so many of us and our children are sick" (Williams, 2000).

"The global community has accepted the urgent need to rid the world of these toxic chemicals but the U.S. is demonstrating the same contempt for the health and safety of the peoples of the world in these negotiations as it is to the peoples of Tanapag. It clearly regards the rest of the planet as a toxic dustbin at its disposal, an attitude that must no longer be tolerated," said Greenpeace International toxics campaigner Darryl Luscombe (Williams, 2000).

REFERENCES

Williams, James. "Environmental Injustice in the Pacific Islands: United States Accused of Environmental Crimes Over Failure to Clean Up Toxic Dump Sites." Greenpeace, August 4, 2000. [http://headlines.igc.apc.org:8080/enheadlines/965700046/index_html]

MARSHALL ISLANDS

NUCLEAR TESTING

Introduction

The Marshall Islands bore the brunt of open-air nuclear testing by the United States until such tests were banned in 1963. At least 66 nuclear devices were set off over the islands. Some blasts obliterated entire atolls, reducing them to radioactive cinders and forcing the population to relocate.

Nuclear Guinea Pigs

At a meeting of the Atomic Energy Commission held in 1956, Marshall Islanders were assigned a role as nuclear guinea pigs. Merrill Eisenbud, an Atomic Energy Commission official, was quoted as saying:

> Utirik [one of the islands] is . . . by far the most contaminated place in the world, and it will be very interesting to go back and get good environmental data. Data of this type has never been available. While it is true these people do not live the way westerners do, civilized people, it is nevertheless also true that these people are more like us than mice. (Ewen, 1994, 140)

Before their island was bombed during the 1940s, the people of Bikini Atoll were removed to Rongerik Island, where, after the atomic bomb tests, the coconut trees stopped bearing fruit, no small thing in an area where the local traditional economy was based on coconuts. During December 1947, U.S. Navy doctors who visited the refugees reported finding "a starving people" (Wypijewski, 2001, 44). A few months after that, the Bikini residents were moved again, to Kwajalein, where they were hired out as menial labor at a U.S. base. After 1952, downwind of Bikini, stillbirths and miscarriages rose by a factor of 11 times (Wypijewski, 2001, 44).

Taken on July 25, 1946, this photo shows the second U.S. atomic bomb test detonated in the Pacific at Bikini Atoll in the Marshall Islands. The world's first underwater atomic blast created an enormous column of radioactive water that sank nine ships. (U.S. Dept. of Energy)

"Fallout Fell Like Snow"

When hydrogen bombs were tested during the 1950s, the wind blew from the testing grounds to nearby islands, where "native children whirled with delight as fallout fell like snow" (Wypijewski, 2001, 42). After the hydrogen bomb Bravo was detonated March 1, 1954, three inches of radioactive ash accumulated on Rongelap, 120 miles downwind of Bikini. During October and November 1962, nine atmospheric nuclear tests were conducted at Johnston Atoll, including four tests at high altitude. Four decades later, the atoll still was polluted by plutonium from those tests. Johnston Atoll also was used to store hundreds of barrels of Agent Orange after the Vietnam War. According to a report from the Pacific Concerns Resource Center, which is based in New Zealand, "Many of these drums have leaked, polluting the environment with dioxin" ("Johnston Atoll," 2000).

Kwajalein, about 100 miles from Bikini, is the world's largest atoll, with a land area of 5.6 square miles and a lagoon area of 1,100 square miles. Most of the island is occupied by a U.S. base that resembles a Midwestern small town (except for its missile-launching facilities), "the picture of 1950s suburbia recreated on the unsubmerged rim of a sunken volcano" (Wypijewski, 2001,

43). The base is run by U.S. governmental contractors, including Boeing, TRW, Raytheon, and Lockheed.

To make way for this base, the native people who once lived on Kwajalein were removed to a small adjacent island, Ebeye, whose land area is about one-sixth of a square mile. By the late 1990s, the population of this sliver of land was estimated at more than 13,000—a density resembling that of Manhattan Island, except that everyone on Ebeye lives on one level (Wypijewski, 2001, 44–45). The island, which receives more than 100 inches of rain a year, is short of clean water, and residents suffer from cholera outbreaks, tuberculosis, and venereal disease. Alcoholism is endemic. More than half the population is under the age of 17. Ten percent of the deaths on Ebeye are suicides, most of them young boys.

REFERENCES

Ewen, Alexander. *Voices of Indigenous Peoples: Native People Address the United Nations*. Santa Fe, N.M.: Clear Light, 1994.

"Johnston Atoll to be Used as Site for Contaminated U.S. Military Waste." Pacific Concerns Resource Center, New Zealand, May 5, 2000. [http://www.converge.org.nz/pma/jmil.htm]

Wypijewski, JoAnn. "This Is Only a Test: Missile Defense Makes its Mark on the Marshall Islands." *Harpers Magazine*, December 2001, 41–51.

MEXICO

INTRODUCTION

Mexico's indigenous peoples contend with a wide variety of environmental issues. In the South, the Maya of Chiapas (under the Zapatista banner) protest the intrusion of roads, logging, and hydropower dams). Meanwhile, the Huichole Indians of Sinaloa, in Baja, California, suffer from harvesting herbicide-laced tobacco. The workers and their families not only experience toxicity on the job; they also live in the fields, so their bodies are poisoned around the clock. Elsewhere in Mexico, the world's largest producer of silver has been ordered to cut production because children near its facilities were found to have unhealthy levels of lead in their bodies. Near Mexico City, in the state of Morelos (the birthplace of the Mexica (Aztec) god Quetzalcoatl as well as Emiliano Zapata), local indigenous peoples took to the streets to resist construction of a golf course, which would have used much of their water in a historically dry region. The image of Zapata was invoked by the protesters, who eventually killed the proposed golf course.

THE MAYA: OIL EXPLOITATION AND DEFORESTATION IN CHIAPAS

The uprising in Chiapas by the Maya under the Zapatista banner has notable environmental roots, including protests of road and dam construction. The Maya have also resisted exploitation of oil resources, which has opened their homelands to intrusion by outsiders, including the Mexican national state and its armed forces.

Several thousand indigenous Maya and environmental activists forced the Mexican government to stop construction of a highway in Chiapas during mid-2000, asserting that it would be used to militarize the area and ease exploitation of oil resources under the aegis of improving infrastructure. Mexican and U.S. oil interests have long known of significant oil reserves in the Lacandon jungle, at the center of the Zapatista rebellion.

The Zapatista Army of National Liberation asserted that Mexico (with U.S. assistance) was increasing its military presence in an area believed to be rich in oil. "With the highways have come the war tanks, the cannons, soldiers, prostitution, venereal diseases, alcoholism, rapes of indigenous women and children, death and misery," said subcommandante Marcos, speaking for the Zapatista movement ("Militarization and Oil," 1999).

Highway construction was halted when 2,000 Mayas marched into Altamirano, and another 4,000 protested in Ocosingo. According to a report in *Drillbits and Tailings,* protesters and Mexican soldiers skirmished in the remote village of San Jose La Esperanza. About 5,000 people also rallied against road building and militarization in San Cristobal de las Casas. The proposed highway would have wound through the highlands downhill into the jungle lowlands near La Realidad, the capital of the Zapatistas' symbolic homeland, 125 miles south of San Cristobal. One of the reasons for road building has been access to oil resources in the Mayas' homelands. According to *Drillbits and Tailings,* "Various reports in the [Mexican] national press, the *Oil and Gas Journal,* the U.S. Geological Survey, and the U.S. General Accounting Office have reported important petroleum reserves in the Lacandon Jungle" ("Militarization and Oil," 1999).

The Zapatistas oppose oil development in the Lacandon area. They point to problems afflicting the activities of Pemex (Mexico's national oil company) in nearby Tabasco, where 20 years of oil production has caused "abnormal population growth, badly skewed income distribution, tremendous escalation of the cost of living, forced relocations, and environmental destruction and extremely hazardous living conditions for people who reside in petroleum-producing areas" ("Militarization and Oil," 1999).

The Maya also are resisting intrusion of refugees from other parts of Mexico. The Lacandon rainforest of Chiapas has become a dumping ground, used by Mexican authorities for refugees who have been relocated from other areas, where formerly self-sufficient, indigenous *campesinos* have been assigned to prefabricated housing, "fighting off malaria and the jungle, coaxing maize from the thin soil" (Weinberg, 1994, 43). Since the 1940s, expansion of the cattle-raising industry throughout Los Altos ("the highlands") has forced thousands of indigenous *campesinos* from their homelands in Los Altos.

Yet another major taproot of rebellion in Chiapas has been expansion of hydroelectric power, which has created a new class of *expulsados* (refugees). Most of the power is not used in the local area, where only a third of homes are hooked into the electrical grid. The power is exported northward, to urban Mexico City and *malquilodora* factories located on the Mexican side of the U.S. border to take advantage of low wages. "The massive expansion of hydroelectricity in Chiapas during the 1980s also sent waves of Maya refugees into *la selva* [the rainforest]" wrote Bill Weinberg in *Native Americas* magazine. "The flooding of fertile Indian farmland in highland valleys was paid for with the slash-and-burn colonization of the Lacandon rainforest" (Weinberg, 1994, 43). The dams flooded 500,000 acres of the most fertile farmland in Chiapas. With projects often underwritten by the World Bank, Mexico is poised to develop even more hydropower, including the Usumacinta river complex, near the border of Guatemala, which would enable the central government to sell power to Central America.

Although it is dwarfed in size by forests in the Amazon Basin and parts of Africa and Asia, deforestation in the *Selva Lacandona* (Lacandon rainforest) is proportionally among the most severe in North and Central America. Experts estimate that Mexico annually loses a forested area of about 2,300 square miles, an area roughly the size of Delaware. In 1900, forests covered an estimated 77,000 square miles of the country. Today, the figure is one-tenth that amount (Althaus, 2001, 6).

The Mexican Environmental Enforcement Agency during 2001 warned of a forestry collapse across the country, and pointed to the *Selva Lacandona* as one of the most threatened areas. Deforestation has stripped the countryside of some of the most biodiverse habitations on Earth (Althaus, 2001, 6). The *Selva Lacandona* alone is home to nearly 43,000 distinct species of plants, animals, and insects. Some scientists estimate that a single 2.5-acre patch of the forest holds 20 species of animals, 40 of birds, 30 of trees, and 5,000 of insects (Althaus, 2001, 6). Animals threatened by the deforestation of the *Selva Lacandona* include the jaguar, ocelot, howler monkey, spider monkey, and various types of tapir.

Furthermore, wrote Dudley Althaus in the *Houston Chronicle*, "At the same time, soil erosion produced by clearing the forest may accelerate life-choking sedimentation levels in the Usumacinta River system, the world's seventh-largest, into which the *Selva Lacandona*'s rivers and streams feed. The estuaries where the Usumacinta flows into the Gulf of Mexico serve as breeding grounds for economically important fish, shrimp and other sea life" (Althaus, 2001, 6).

THE HUICHOLE: LIVING WITH PESTICIDES AROUND THE CLOCK

Exposure to toxic pesticides is one of the greatest risks faced by indigenous migrant workers in Mexico, where tobacco growers and other agricultural companies employ them as industrial workers. Many of the workers are denied safety equipment and access to showers and facilities to wash their clothes after coming into contact with pesticides. In addition, many of the workers live in the same pesticide-laced fields that they harvest, exposing them to contamination around the clock.

During 1993, for example, an estimated 170,000 field workers arrived in the valleys of Sinaloa during each planting season. Among these workers, roughly 5,000 have been found to suffer from toxic contamination as a result of the handling of, or prolonged exposure to, pesticides that are used in cultivation (Diaz-Romo and Salinas-Alvarez, n.d.).

Of the 35,000 agricultural laborers who worked in the San Quintin Valley of Baja California during 1996, 70 percent were indigenous people. The majority of the indigenous migrant workers who work in the agroindustrial fields of northern Mexico are Mixtecos, Triquis, and Zapotecs from Oaxaca; Nahuas, Mixtecos, and Tlapenecos from Guerrero; and Purhepechas from Michoacan. According to Estela Guzmán Ayala, women (34 percent) and children under 12 years of age (32 percent) constitute two-thirds of the indigenous labor force in the agricultural regions in northern Mexico (Diaz-Romo and Salinas-Alvarez, n.d.).

Ruth Franco, a doctor specializing in work-related health and coordinator of the Program for Day Laborers in Sinaloa, estimated that 25 percent of the 200,000 workers in the Sinaloa valleys during the 1995–1996 cycle were children between the ages of 5 and 14. In the fields where these children and their families work, observers asserted that "thousands of used containers and toxic residues that are generated by the annual use of upwards of 8 million tons of pesticides are criminally disposed of in *ad hoc* trash bins, channels, drains, incinerators, and recycled to storing drinking water" (Diaz-Romo and Salinas-Alvarez, n.d.).

One segment of the migrants includes between 15,000 and 20,000 Huicholes, who inhabit the mountains of the Sierra Madre Occidental. Approximately 40 percent of all the Huichole families leave their communities in the dry season to find employment in the tobacco fields of the Nayarit coast. During the rainy season, the Huichole traditionally cultivate a combination of corn, chile, beans, squash, and amaranth. The Mexican government has made the traditional cycle difficult by promoting monocultural planting, distributing hybrid seeds of corn that require the use of pesticides and artificial fertilizers, and replacing the mixed seeds that traditionally were used by the Huichole and other indigenous agricultural peoples. The use of industrial-scale monocultural agriculture breaks down the indigenous traditions of cooperation while, at the same time, increasing malnutrition and alcoholism. The introduction of herbicides such as Paraquat and 2,4-D gradually destroys communal work, placing the health of cultivators and their families in danger (Diaz-Romo and Salinas-Alvarez, n.d.).

A report on the travails of the Huicholes said: "To arrive at the tobacco fields the Huicholes make a journey from the sierras under subhuman conditions, arriving hungry and thirsty. The 'valuable and appreciated' human merchandise includes pregnant women, babies incapable of crying, mute from pain, who have recently been born to malnourished mothers or mothers with tuberculosis. Vulnerable elders and even the 'strong' men arrive at these centers in weak condition" (Diaz-Romo and Salinas-Alvarez, n.d.).

Favored workers are given purified water, while the remainder are forced to drink water from irrigation channels that draw from the pesticide-laced Santiago River, or local wells that also are contaminated with the chemical cocktail used in the tobacco fields. As they toil in the heat, the workers become drenched with sweat, allowing their bodies to absorb pesticide residues more easily. Nicotine in the tobacco also causes skin irritation and hives, called Green Tobacco Sickness. Child laborers are particularly susceptible to the effects of pesticides and the Green Tobacco Sickness (Diaz-Romo and Salinas-Alvarez, n.d.).

The harvesting families often spend the entire day and night in the fields, living and sleeping in boxes, or under blankets or sheets of plastic, beneath strings of drying tobacco leaves, further exposing themselves to toxic chemicals and tobacco residues. Most have no potable water, drainage, or latrines. Occasionally, "the Huicholes use the empty pesticide containers to carry their drinking water, without paying notice to the grave dangers that this represents, since the majority cannot read the instructions on the labels which may be written in English" (Diaz-Romo and Salinas-Alvarez, n.d.).

SILVER MINING AND LEAD POISONING OF CHILDREN

The Mexican silver-mining firm Industriales Penoles was ordered to halve production at one of its plants as officials tightened emergency measures to combat a lead-poisoning epidemic among children in the northern desert town of Torreon, where 1,166 children were found to have unacceptably high levels of lead in their bloodstreams ("Polluting Mexican Refinery," 1999).

Mexican authorities declared an environmental emergency at the metals refinery on May 5, 1999 and ordered Industriales Penoles, the world's biggest silver producer, to remove hundreds of tons of lead-contaminated dust from Torreon's Luis Echeverria district. "We have declared the firm Met-Mex Penoles in a phase two emergency, under which . . . it has to reduce operations at plant number nine by 50 percent," the government of northern Cohahuila State, where Torreon is located, said in a statement issued jointly with Mexico's Federal Attorney for Environmental Protection ("Polluting Mexican Refinery," 1999). Those steps are designed to reduce gas and dust emissions by about 50 percent. "The symptoms detected in children with high levels [of lead poisoning] are neurological and gastrointestinal," the statement said. "The levels registered represent a significant health risk" ("Polluting Mexican Refinery," 1999).

THE TEPOZTLAN GOLF COURSE "WATER WAR"

The state of Morelos (south of Mexico City) is the legendary birthplace of the Mexica (Aztec) god Quetzalcoatl, as well as Emiliano Zapata, the Mexican revolutionary. More than seven decades after his assassination, Zapata's image led indigenous residents of Tepoztlan, a village of 28,000, to rise in rebellion against a proposal to construct a golf course that would have used more water than everyone else in the town combined. Given the intensity of local indigenous opposition, the project eventually was suspended.

Morelos is a generally dry place, so the main environmental conflict involved usage of water. An environmental assessment disclosed that the golf course would have required 5 million gallons of water per day, while the entire town consumed 3 million gallons daily (Weinberg, 1996, 37). The 18-hole golf course was to have been developed as a centerpiece for 800 luxury homes, a heliport, and a "data center and business park" (Weinberg, 1996, 34). Plans called for golf notable Jack Nicklaus to design the course.

Morelos is regarded by many Mexicans as their nation's spiritual center because it is the ancient home of the Tlahuicas, who "were instrumental in the rise of the high culture of the valley of Mexico just north over the Ajusco mountains" (Weinberg, 1996, 37). The Tlahuicas called their home Cuauhnahuac ("Land of Trees"). Hispanicized, the name became that of the state capital, Cuernavaca.

The area also is known for its agricultural fertility but, like the valley of Mexico, it has become progressively drier over the years, at least in part because of deforestation caused by the expansion of urban areas and sugar plantations. Streams that once ran year-round now often cascade down from the hills only during the summer rainy season. "If we don't protect nature, we will die," said Lazaro Rodriguez, a Tepozteco. "If there are no trees there is no rain, and the land dies" (Weinberg, 1996, 41).

When Mexico City elites targeted their town for development, the Tepoztecos rose en masse against authorities, costing one man his life, and prompting the imprisonment of several other people. The encroachment of farms and ranches upon the communal lands of the village's native people played a role in sparking the rebellion of Zapata in 1910. When the golf course was first proposed in 1995, local people also expressed concern that the heavy use of pesticides required to maintain the greens and fairways would pollute the local water table. Golf courses are heavy consumers of pesticides and herbicides, averaging 5 to 10 times the amount, per acre, used in agriculture (Weinberg, 1996, 35). At one point, local people occupied a luxury home built for Guillermo Occelli, a brother-in-law of former Mexican president Carlos Salinas. They hung a banner on the house reading (in Spanish): "Tepoztlan: Communal land. House of the People" (Weinberg, 1996, 39).

REFERENCES

Althaus, Dudley. "The Fated Forest: Nature's Way; Upsetting the Balance; Deforestation Threatens Thousands of Plant and Animal Species as Well as the Earth's Atmosphere. Studies Show Rain Forests are Declining at Alarming Rates." *Houston Chronicle*, September 30, 2001, 6.

Barreda, Andres. "Militarization and Oil in Chiapas." *La Jornada*, August 17, 1999, n.p.

Diaz-Romo, Patricia, and Samuel Salinas-Alvarez. "Migrant Workers and Pesticides. A Poisoned Culture: the Case of the Indigenous Huicholes Farm Workers." *Abya Yala News: The Journal of the South and Meso-American Rights Center*. N.d. [http://saiic.nativeweb.org/ayn/huichol.html]

"Militarization and Oil in Chiapas, Mexico." *Drillbits and Tailings* 4:16 (October 8, 1999): n.p. [http://www.moles.org/ProjectUnderground/drillbits/4_16/3.html]

"Polluting Mexican Refinery Told to Halve Output." Reuters News Service, May 25, 1999. [http://www.tigerherbs.com/eclectica/earthcrash/subject/mining.html]

Ross, John. "Is Zapatista Rebellion Rooted in Oil?" *Earth Island Journal* 11:2 (1996): 20.

Weinberg, Bill. "Flooding the Jungle." *Native Americas* 11:2 (summer 1994): 43.

———. "The Golf War of Tepoztlan." *Native Americas* 13:3 (fall 1996): 32–42.

MOTHER EARTH, AS ECOLOGICAL METAPHOR

Before sustained contact with Europeans, Native Americans of North America lived in roughly 2,000 distinct societies and spoke several hundred mutually unintelligible languages. These diverse cultures all shared ways of life that involved symbiosis with the natural world. Vine Deloria Jr. has suggested that the natural metaphor of earth as mother can be documented as far back as 1776. On June 21 of that year, at a conference in Pittsburgh during the Revolutionary War, Cornstalk, who was trying to convince the Mingos (Iroquois) to ally with the Americans, said:

> You have heard the good Talks which our Brother [George Morgan] Weepemachukthe [the White Deer] has delivered to us from the Great Council at Philadelphia representing all our white brethren who have grown out of this same ground with ourselves[,] for this Big [Turtle] Island being our common Mother, we and they are like one Flesh and Blood. (Grinde and Johansen, 1995, 26, 28)

William A. Starna, a former professor of anthropology at the State University of New York (Oneonta campus), has called the argument that Native Americans had an environmental ethic "pan-Indian mythology," saying it was invented by non-Indian environmentalists intent on mimicking the hippie movement during the 1960s (Grinde and Johansen, 1995, 30). The historical record, however, is rife with references to *mother earth* by native peoples. To cite one of many examples, the Sauk-Fox chief Black Hawk, who was exiled to a reservation near Fort Madison, Iowa after the three-month war that bears his name, opened a Fourth of July address to a mainly non-Indian audience in the late 1830s by observing, "The Earth is our mother; we are on it, with the Great Spirit above us" (Grinde and Johansen, 1995, 31).

In 1877 the Nez Percé leader Chief Joseph replied to an Indian agent's proposal that he and his people move to a reservation and become farmers. This statement was made a few months before Joseph and his band fled 1,700 miles across some of the most rugged land in North America to avoid subjugation.

Chief Joseph said, "The land is our mother. . . . She should not be disturbed by hoe or plow. We want only to subsist on what she freely gives us" (Grinde and Johansen, 1995, 31).

Smohalla, a religious leader of the Nez Percé , said at the same meeting:

You ask me to plow the ground? I should take a knife and tear my mother's bosom? Then when I die, she will not take me to her bosom to rest. . . . You ask me to dig for stone! Shall I dig under her skin for her bones? Then when I die I cannot enter her body to be born again. You ask me to cut grass and make hay and sell it, to be rich like white men! But how dare I cut off my mother's hair? (Grinde and Johansen, 1995, 31)

A third Nez Percé chief, Tuhulkutsut, joined in: "The Earth is part of my body. I belong to the land out of which I came. The earth is my mother" (Grinde and Johansen, 1995, 32). United States negotiator General Oliver O. Howard then is recorded to have protested, "Twenty times over [you] repeat that the earth is your mother. . . . Let us hear it no more, but come to business" (Grinde and Johansen, 1995, 31). The Lakota holy man Black Elk told John Neihardt, "Every step that we take upon You [the Earth] should be done in a sacred manner; every step should be taken as a prayer" (Grinde and Johansen, 1995, 32).

Ecological metaphors also were woven into the languages of many Native American cultures. "Who cuts the trees as he pleases cuts short his own life," said the Maya. The Maya word for "tree sap" is the same as the word for "blood" (Grinde and Johansen, 1995, 32). Christopher Vecsey and Robert W. Venables, in *American Indian Environments: Ecological Issues in Native American History*, make a case that concepts characterizing the Earth as sustainer, or "mother," the sky as "father," and the realm of the Creator as "Great Mystery" are shared by many native cultures across the continent. They also trace the use of the concept of the "sacred circle" (or "hoop") as a metaphor for physical and spiritual unity (Grinde and Johansen, 1995, 33).

The American Indians' concept of a sacred circle expresses a physical and spiritual unity. This circle of life is interpreted according to the particular beliefs of each Indian nation, but it is broadly symbolic of an encompassing creation. While non-Indians quite willingly admit to the complexity of the circle of things around them, what has been left behind by the scientific, post-Renaissance, non-Indian world is the universal sacredness—the living mystery—of creation's circle.

REFERENCES

Grinde, Donald A., Jr., and Bruce E. Johansen. *Ecocide of Native America: Environmental Destruction of Indian Lands and Peoples*. Santa Fe, N.M.: Clear Light, 1995.

NATIVE AMERICAN CONCEPTIONS OF ECOLOGY

Some Anglo-American scholars believe that Native Americans had no general ecological ethic; as proof, they point to native complicity in the slaughter of beavers and other fur-bearing animals for trade. European or Euro-American traders often contracted with Indians for such large numbers of pelts that the animals almost disappeared. Calvin Martin, for example, has speculated that some native peoples held the beaver responsible for the incursions of immigrants, and so took revenge on the animals. In *Keepers of the Game* (1978), Martin contends that the image of Native American as conservationist is nothing but an example of a non-Indian stereotype:

> Late in the 1960s, the North American Indian acquired yet another stereotypic image in the popular mind: the erstwhile "savage," the "drunken" Indian, the "vanishing" Indian was conferred the title of "ecological" (i.e., conservationist-minded) Indian. Propped up for everything that was environmentally sound, the Indian was introduced to the American public as the great high priest of the Ecology Cult. (Martin, 1978, 157)

In Martin's view, millions of dead beavers rather effectively destroy the veracity of any argument that native peoples generally held nature to be sacred, and that most native peoples took from nature only what they needed. As long as commercially viable numbers of beavers remained, native hunters teamed up with Europeans and European-Americans to take as many as possible within the shortest possible time, motivated by a market economy, not a conservation ethic. Shepard Krech III makes a similar argument in *The Ecological Indian* (Krech, 1999). In his view, Native Americans have been as callous to the environment as Western industrial man.

The native peoples did not initiate the commercial fur trade on their own, however. They had lived in natural symbiosis with beavers and other fur-bearing animals for thousands of years before Europeans imposed mercantile capitalism on them. Immigrants from Europe employed the native peoples in this endeavor, not vice versa. By the time of the fur trade, the market economy was destroying more than the beaver populations. The native societies themselves were being destroyed through the spread of trade goods, liquor, and disease, as well as by the losses of hunting animals and land base.

Native peoples who took part in the fur trade often did so to acquire trade goods that created dependencies of their own, and caused them to abandon traditional beliefs and modes of economy. Within their councils, beginning as early as 1700, native peoples realized what was happening to them, and debated whether European trade goods should be accepted at all. The fur trade was clearly a postcontact phenomenon. Had a trading industry not existed, beavers would not have been hunted to near extinction by native peoples.

Luther Standing Bear, who watched the early years of settlement on the Great Plains, contrasted the European-American and Native American conceptions of the natural world of North America:

> We did not think of the great open plains, the beautiful rolling hills, and winding streams with tangled brush, as "wild." Only to the white man was nature "a wilderness" and only to him was the land "infested" with "wild" animals and "savage" people. To us it was tame. Earth was bountiful, and we are surrounded with the blessings of the Great Mystery. (Standing Bear, 1978, 98)

Most native peoples incorporated nature into their rituals and customs because their lives depended on the bounty of the land. Where a single animal comprised the basis of a native economy (such as the salmon of the Pacific Northwest or the buffalo on the plains), strict cultural sanctions came into play against killing such animals in numbers that would exceed their natural replacement rate. On the Great Plains, the military societies of the Cheyenne, Lakota, and other peoples enforced rules against hunting buffalo out of season and taking more animals than necessary for survival. To do otherwise threatened the viability of the resources that sustained native cultures.

The Sun Dance ceremonies of the Plains reflect celebration of the cycle of life. Like the Christian Easter, the Sun Dances (which the Cheyenne call the "New Life Lodge") are associated with the return of green vegetation and the increase in animal populations (especially the buffalo) during the spring and early summer. The ritual is communal, expressing unity with the earth and knowledge of a nation's dependence on it for sustenance.

The parts of the Sun Dance during which the skin is pierced are not required of anyone not wishing to participate. Some native peoples do not even practice skin piercing during their Sun Dances. In some ways, the central purpose of the Sun Dance is similar to that of the First Salmon Ceremony in the Pacific Northwest: each expresses respect for a natural cycle of life and the people's main food source. The Sun Dance pole is said to unite sky and earth; the four sacred directions also are part of the ceremonial design.

Corn, the major food source for several agricultural peoples across the continent, enjoys a special spiritual significance.

Corn and beans (which grow well together because the beans, a legume, fix nitrogen in their roots) often were said to maintain a spiritual union. Some peoples, such as the U'ma'ha (Omaha) of the eastern Great Plains, "sang up" their corn through special rituals. In addition to "singing up the corn," the Pueblos cleaned their storage bins before the harvest, "so the corn will be happy when we bring it in." The Pawnee grew 10 varieties of corn, including one (called "holy" or "wonderful" corn) that was used only for religious purposes, and never eaten. The Mandan had a Corn Priest who officiated at rites during the growing season. Each stage of the corn's growth was associated with particular songs and rituals, and spiritual attention was said to be as important to the corn as proper water, sun, and fertilizer. Among the Zuni, a newborn child was given an ear of corn at birth, and endowed with a "corn name." An ear of maize was put in the place of death as the "heart of the deceased," and later used as seed corn to begin the cycle of life anew. To Navajos, corn was as sacred as human life.

REFERENCES

Grinde, Donald A., Jr., and Bruce E. Johansen. *Ecocide of Native America: Environmental Destruction of Indian Lands and Peoples*. Santa Fe, N.M.: Clear Light, 1995.

Krech, Shepard, III. *The Ecological Indian: Myth and History*. New York: W.W. Norton, 1999.

Martin, Calvin. *Keepers of the Game: Indian-Animal Relationships and the Fur Trade*. Los Angeles: University of California Press, 1978.

Standing Bear, Luther. *Land of the Spotted Eagle*. Lincoln: University of Nebraska Press, 1978.

NEW ZEALAND

THE MAORI AND THE WESTERN WORLDVIEW

Writing in the *Earth Island Journal*, Brian Tokar described New Zealand as follows:

> The Land of the Long White Cloud is a land of mystery and wonder, where glacial mountain peaks tower over vast coastal rainforests; where there are countless *genera* of trees and birds that cannot be found anywhere else on Earth; where the indigenous Maori population has made its mark on the majority colonial population's language, cultural norms and legal institutions to an extent rarely found in the English-speaking world. (Tokar, 2001)

In 1975, New Zealand established a Commission of Inquiry to address outstanding Maori claims under the 1840 Treaty of Waitangi, which granted the Maori "full, exclusive and undisturbed possession" of their lands, forests, fisheries, and other treasured possessions (Tokar, 2001). The commission has heard about 800 claims, mostly regarding indigenous lands, fishing and mineral rights, language issues, and broadcasting rights. In 1991, a group of Maori claimants raised the stakes by filing Wai 262, the Indigenous Flora and Fauna Claim, which seeks a much broader authority over the uses of native plants, animals, genes, and cultural knowledge.

Under this claim, the patenting of living organisms, the granting of rights to commercial breeders, the export of biological samples, and the misappropriation of Maori cultural symbols are all termed violations of the Waitangi treaty. The claim was conceived as a direct challenge to the Western model of intellectual property rights, as well as to the increasing influence by transnational companies on New Zealand government policies. This claim is seen as part of a long-range educative process, through which it is hoped that Maori cultural principles, including the inseparability of the people, their heritage (*whakepapa*, a concept that situates family genealogies and ties to ancestral lands firmly within their traditional creation stories), and the integrity of the land, will become more meaningfully encoded in the country's legal practices.

Maori elders have sought to challenge the Western view that separates an abstracted nature from the consciousness and evolution of the people and their ways. Maori solicitor Tania Tetitaha explain how her people "draw no distinction between visible and invisible worlds or spiritual and physical beings, and have a landscape where 'culture' is the indivisible sum of all things" (Tokar, 2001).

REFERENCES

Tokar, Brian. "Resisting Global Logging: Forest and Trade Activists Meet in New Zealand." *Earth Island Journal* 16:2 (summer 2001): n.p. [http://www.earthisland.org/eijournal/new_articles.cfm?articleID=167&journalID=46]

NICARAGUA

THE MAYAGNA (SUMO): BATTLING ILLEGAL LOGGING AND CATALOGING ENDANGERED SPECIES

Introduction

The Mayagna (Sumo) Indian Community of Awas Tingni during 2001 won a major legal battle requiring the government of Nicaragua to protect local indigenous rights to land and resources. At the same time, the Mayagna have been cataloging endangered wildlife in their homelands with the help of the St. Louis Zoo.

Nicaragua's Violation of Human Rights

On September 17, 2001, the Inter-American Court of Human Rights decided that Nicaragua violated the human rights of the Awas Tingni Community. It ordered the government to recognize and protect the community's legal rights to its traditional lands, natural resources, and environment. The ruling asserted that the Nicaraguan government wrongly granted foreign companies licenses to log much of the tropical forest in which the community resides. The logging concessions were " smack dab in the middle of their lands," said one observer (Weinberg, 1998, 25).

The court's decision, the first of its kind to uphold international law respecting indigenous land rights, "is precedent-setting internationally," said James Anaya, special counsel to the Indian Law Resource Center, which represented the Awas Tingni Community in the case. "Members of the community have fought for decades to protect their land and resources and against government neglect and encroachment by loggers. This decision vindicates the rights they have struggled so long to protect," he said (Carrere, 2001). The people of Awas Tingni pursued their case in Nicaragua's courts for several years, to no avail.

During 1995, the Indian Law Resource Center filed a petition on the Mayag-nas' behalf before the Inter-American Commission on Human Rights, an independent body of the Organization of American States. According to Ricardo Carrere, writing in the *World Rainforest Movement Bulletin,* "The commission found in favor of the community, but the government ignored the commission's requests for remedial action. In June of 1998, the commission brought the case before the Inter-American Court" (Carrere, 2001).

The commission's decision found that Nicaragua violated international human rights law, according to Carrere, "by denying the community its rights to property, adequate judicial protection, and equal protection under the law. The court ruled that Nicaragua's legal protections for indigenous lands were 'illusory and ineffective'" (Carrere, 2001; "Landmark Victory," 2001). The ruling ordered the government to demarcate traditional Awas Tingni Community lands, as well as those of other indigenous peoples in Nicaragua.

Cataloging Wildlife

As the Sumo fought for their rights in international courts, they also invited the St. Louis Zoo to help them catalog threatened wildlife in their homeland, which they call Mayagna Sauni Bu. The Mayagnans want the zoo's help in learning the abundance and location of different animal species, because they believe that the region is home to many animals that are endangered or extinct in other parts of Central America. According to a report in the *St. Louis Post-Dispatch,* "The Indians also want to know whether their current rate of hunting these animals for food can be sustained without driving the creatures to extinction" (Allen, 2001, A-1). The indigenous people of this area hunt tapirs, monkeys, armadillos, and other animals to supplement a diet of domesticated pigs, chickens, and cattle.

The Mayagna live in the Bosawas Biosphere Reserve, Nicaragua's largest protected area and one of the cornerstones of an international conservation effort called the Mesoamerican Biological Corridor. The Bosawas Reserve includes roughly 2,900 square miles, slightly less area as Yellowstone National Park. With an adjacent reserve across the Honduran border, this forest is the Western Hemisphere's largest area of rainforest outside of the Amazon. The reserve is under pressure by expanding agriculture, illegal logging, and land-tenancy disputes.

Native leaders believe that such knowledge will help them make the case for protecting their homeland as the agricultural frontier—the push of nonnative settlers—moves into their territory. "The indigenous groups want to conserve the environment in order to continue traditional hunting, harvesting and religious practices," said a biologist working with the project, Paul Garber, of the University of Illinois at Champaign-Urbana. "They want to be in a position to make these decisions, based on scientific evidence, rather than have external groups (such as the government of Nicaragua or conservation agencies) make these decisions for them" (Allen, 2001, A-1). Said Anuar Murrar Garay, director of Alistar, a Nicaraguan conservation agency, "The main strategy is to empower them. What we're doing is giving them the tools to manage their lands, if they so choose, in their traditional ways" (Allen, 2001, A-1).

REFERENCES

Allen, William. "Preserving, Exploring 'Central America's Amazon'; St. Louis Zoo Biologist's Mission Is to Save the World's Endangered Wildlife." *St. Louis Post-Dispatch*, January 28, 2001, A-1.

Carrere, Ricardo. "Nicaragua: Indigenous People Win Major Legal Battle." *World Rainforest Movement Bulletin* (September 2001): n.p. [http://www.wrm.org.uy]

"Landmark Victory for Indians in International Human Rights Case Against Nicaragua." Press Release, Indian Law Resource Center, September 21, 2001. [http://www.indianlaw.org]

Weinberg, Bill. "La Miskita Rears Up: Industrial Recolonization Threatens Nicaraguan Rainforests: An Indigenous Response." *Native Americas* 15:2 (summer 1998): 22–32.

NIGERIA

THE OGONI: OIL, BLOOD, AND THE DEATH OF A HOMELAND

Introduction

The 500,000 indigenous Ogoni of the Niger Delta in southern Nigeria have watched as their traditional fishing and farming livelihood has been laid waste by several multinational companies' extraction of oil, with the complicity of the national government, which has allowed large parts of the Ogonis' homeland to be ruined.

The Ogonis' land has been contaminated not only by oil wells and pipelines, but also by gas flares that burn 24 hours a day, producing intense heat and chemical gas fogs that pollute nearby homes and render farm fields barren and unproductive. The constant flaring of natural gas also contributes measurably to global warming. Several Ogoni who protested the ruination of their homeland and the impoverishment of their people have been convicted of false charges and executed.

The constant flaring of oil and gas—24 hours a day, 7 days a week, for as many as 40 years—leaves many people suffering from "blindness, skin problems, and asthma among the young and the elderly headaches and more," according to a *Drillbits and Tailings* observer on the scene (Dayaneni, 2002). In some villages, where flares are maintained in the midst of settlements, children never see darkness. Acid rain provoked by the flares is so strong that it sometimes disintegrates tin roofs, as it reduces crop yields and contaminates the water and soil.

Oil Development in Nigeria

Shell Oil has extracted oil from the Niger Delta since 1958, as part of a joint venture with the Nigerian National Petroleum Corporation, Elf, and Agip. Shell is by far the largest foreign oil company in Nigeria, accounting for 50 per-

The president of the Movement for the Survival of the Ogoni Town (MOSOP), Ledum Anazor Mitee, heads a protest before oil facilities of Shell in Valencia. They accuse this company of causing innumerable environmental and social problems in Nigeria. (NewsCom.com/EFE Photos)

cent of Nigeria's oil production. Nigeria generated roughly 12 percent of Shell's oil production worldwide in the late 1990s ("Shell: 100 Years," 1997).

High-pressure pipelines have been laid aboveground through villages and farmlands, a major reason why the area suffered an average of 190 oil spills per year between 1989 and 1996, involving on average 319,200 gallons of oil per spill ("Shell: 100 Years," 1997). According to one observer on the scene, "Rivers, lakes and ponds are polluted with oil, and much of the land is now impossible to farm. Canals, or 'slots,' have permanently damaged fragile ecosystems and led to polluted drinking water and deaths from cholera. Gas flaring and the construction of flow stations near communities have led to severe respiratory and other health problems" ("Shell: 100 Years," 1997).

By 2000, oil accounted for more than 90 percent of Nigeria's export earnings and roughly 80 percent of government revenue, totaling roughly $20 million per day ("Nigeria: Godforsaken," 2002). More than 90 percent of Nigeria's oil is extracted from the Niger Delta. During the last 40 years, oil worth $30 billion has been extracted from the Ogonis' homeland (Wiwa, 2000). By 2000, the Ogonis' homeland was home to 100 oil wells, two refineries, a petrochemical complex, and a fertilizer complex, while most of the Ogoni people do not have electricity or running water. It is a land where 5 physicians serve 500,000 people (Wiwa, 2000).

According to a local observer, "ChevronTexaco extracts hundreds of thousands of barrels of oil from the Niger Delta in Nigeria every year despite decades of resistance by the people of the Delta. In Opia, one community of the Delta, ChevronTexaco has destroyed the traditional local economy, run pipelines through gardens and villages and leased helicopters to the military to attack local demonstrators" ("Environmental Injustice," 2001).

Local opponents of its activities assert that ChevronTexaco profits heavily from its operations in the Niger Delta and has provided both dollars and infrastructure to the Nigerian military (which uses those resources to suppress resistance and kill activists). For its part, ChevronTexaco accepts no responsibility for environmental problems or human rights abuses. While all meaningful quality-of-life measures indicate that the lives of indigenous people in the Niger Delta continue to decline, the corporation continues to assert that its operations "promote democracy and development" ("Environmental Injustice," 2001).

Obsolete, leaking, rusty oil pipelines have become a major source of contaminating oil spills for the Ogoni. In 1992, a major oil blowout in the village of Botem lasted a week before it was stopped, creating a biological dead zone in the watercourses that supplied drinking water for local residents. Oil spills caused by obsolete pipelines are routinely blamed on sabotage, which allows companies to ignore repairs under Nigerian law. During October 1998, an explosion and leak flooded a large part of the village of Jesse, killing more than 700 people; two years later, two pipeline explosions in southern Nigeria killed 300 people (Gedicks, 2001, 45).

Repression by the "Kill and Go"

The Ogonis' protests of such conditions have been met with brutal repression by Nigerian police. During 1990, people in the village of Umuechem protested oil pollution of their homeland, after which they were attacked by the notorious Mobile Police (known locally as the "Kill and Go") who bombarded the village, killing more than 100 people, as they looted many homes. Survivors were forced to leave their homes (Gedicks, 2001, 46). The Movement for the Survival of the Ogoni People (MOSOP), organized during 1990, adopted an Ogoni Bill of Rights demanding local control of political and environmental affairs, blaming Shell Oil for "full responsibility for the genocide of the Ogoni" (Gedicks, 2001, 46).

Following this declaration, in 1992, the rights of the Ogoni were discussed before the United Nations Commission on Human Rights in Geneva, Switzerland. In 1993, roughly 300,000 people gathered in protest in and around the Ogoni village of Bori to declare Shell Oil *persona non grata* on their land. Shell was forced to suspend oil production for a time due to protests at its facilities. Response to these protests by the Nigerian police and military was quick and forceful. Within the next eight years, according to an account by Al Gedicks (in *Resource Rebels*, 2001), more than 2,000 people were killed and 37 villages were substantially destroyed. About 30,000 other Ogoni were displaced from their homes, as troops took up long-term residence in the area to protect Shell's assets (Gedicks, 2001, 47).

On May 21, 1994, four Ogoni leaders were murdered in Gokana Kingdom, reportedly by angry youths. Ken Saro-Wiwa (the most notable leader of MOSOP) and eight other MOSOP leaders were arrested and accused of involvement in the murders. The day after the crimes, military police (the Internal Security Task Force) stormed into Ogoniland, raiding, burning, and looting villages. While thousands of Ogoni villagers took refuge in the bush, hundreds who did not escape were detained and tortured. On October 31, 1995, Ken Saro-Wiwa and eight other Ogonis were sentenced to death by the Special Tribunal. They were executed on November 10, 1995.

Oil Spills and Wastelands

Large oil spills have turned vast areas of the Ogonis' homeland into wastelands. In mid-2001, for example, a United Nations Internet page described Yaata, an Ogoni village, where "dying vegetation in various shades of ochre stretch as far as the eye can see, poisoned by soil turned soggy and a dark, greasy hue since crude oil began seeping through over a month ago." On April 29, 2001, at the Royal Dutch Shell Yorla oil field, a "quake-like tremor sent shockwaves onto Yaata and surrounding villages" ("Nigeria: Focus," 2001). Within minutes, before people could guess the cause, jets of crude oil were already shooting up 100 meters, raining on the surroundings. The oil plume was quickly followed by strong fumes of natural gas, as the people of the village ran for their lives.

John Nwikine, a student from Yaata, told the United Nations' Integrated Regional Information Network (IRIN) that the villagers "knew from experience that any accidental fire was going to light up the area and spread as fast and as far as the fumes were going" ("Nigeria: Focus," 2001). For nine days, according to an IRIN account, "the shower of crude oil and gas poured on Yaata unabated. The rapidly resulting streams of crude oil swamped neighboring farmlands, forests, streams and rivers" ("Nigeria: Focus," 2001).

Although Shell Oil was quickly alerted to the disaster, its employees were described by indigenous people on the scene as appearing helpless, with teams of Shell workers circling the area in helicopters, without landing. In the meantime, local people organized to alert inhabitants to the danger of lighting fires, limiting the damage as best they could. "Nevertheless," according to the IRIN account, "people in the areas pervaded by the fumes complained of breathing difficulties, in a number of cases combined with cough and runny noses" until experts from Boots and Coots International Well Control of Houston, Texas capped the broken well-head on the ninth day ("Nigeria: Focus," 2001).

When Yaata's residents returned after the spill was stopped,

> They found their village was uninhabitable. Their maize, cassava and yam crops were stained with crude oil, wilted and dying. Much of their livestock had either died or were dying from eating polluted vegetation and drinking contaminated water. Dead fish rose to the surface of creeks and ponds. ("Nigeria: Focus," 2001)

Two months later, no attempt had been made to clean up the spill's damage, aside from the digging of a few trenches to divert some of the oil spill. Residents of Yaata had been forced to move into other villages, giving up their lives' work.

In the meantime, the spilled oil seeped further into the earth, contaminating underground water for miles around, as Shell blamed sabotage and refused to pay compensation. Shell complained that protests by local people had inhibited its ability to properly shut down the well that blew, in effect blaming the Ogoni for the explosive spill in the Yorla field ("Shell Says," 2001). Ledum Mitee, president of MOSOP in 2001, accused Shell of seeking scapegoats for the spill instead of taking prompt steps to contain and clean it. "We are shocked that Shell is already leveling accusations against local people who have risked their lives and health to prevent a fire for the last three days," Mitee said ("Shell Says," 2001).

The Death of Friday Nwiido

The tensions between the oil companies and people in the Niger Delta are a regular feature of daily life that often results in deaths, as illustrated by the demise of one young man, Friday Nwiido, which was described by an Englishman, Nick Ashton-Jones.

On Sunday April 29, 2001, at about 9:30 A.M., a huge explosion took place at the Well 10 facility (Yorla oil field). Then for days crude oil rained sporadically onto adjacent farmlands, settlements, streams, swamps, lakes, and rivers. Jollyboy Olole, an eyewitness, who was inspecting his cassava crops when the explosion occurred, said that the crude "rained into his eyes" (Ashton-Jones, 2001).

On the same afternoon, Shell organized a press conference at its staff club in nearby Port Harcourt. At the conference, Shell staff insisted that the blowout had been caused by sabotage. Three Texan oil engineers who helped to deal with the problem later were quoted as saying that the blowout was caused by "mechanical failure due to gas pressure coupled with corrosion of the facility" (Ashton-Jones, 2001).

In the meantime, Friday Nwiido celebrated his thirtieth birthday that June. He worked as a welder in Port Harcourt, commuting every day on his motorbike. He also was a farmer and fisherman—"a true Ogoni Man," he said (Ashton-Jones, 2001). Seventeen people depended on his earnings, including his mother, his wife, and three children.

Following the blowout at Yorla, a group of local youths, including Nwiido, who had been accepted as a leader, went to see Shell officials in Port Harcourt about clearing up the mess. It was agreed that Shell staff would meet the youths at the site to discuss the situation. However, other Shell workers arrived on the scene, contrary to the perceived agreement, with equipment and contractors. Nonetheless, Nwiido and the other youths began the cleanup. When the operation was not completed the same day, the youths told Shell workers they would return the next day. Due to a communications mix-up, another group was employed. By the time Friday and his cohorts learned what had happened, the work had been completed. By coincidence, Nwiido met the Shell team returning to Port Harcourt as he was returning home on his motorcycle (Ashton-Jones, 2001).

Nwiido stopped the Shell vehicles and asked what was going on. He was told to see the Shell community director in Port Harcourt. Nwiido replied that he could not do that because he was afraid for his safety; he also knew from experience that he would not be allowed inside the gate. Instead, Nwiido said, he wanted to keep one of the vehicles (probably as payment for the work) and, according to local accounts, he was allowed to keep a truck at his house (Ashton-Jones, 2001).

According to Ashton-Jones's account, on June 10 or 11, the police came to Nwiido's house, saying they wanted to remove the truck to Bori, the district center. Friday refused, after which the police returned with the local government chairman, who said that Nwiido's possession of the truck was affecting his job as chairman and that he was willing to pay each of Nwiido's men. The chairman pleaded with Friday, but he refused, saying he would only talk to Shell and not to middlemen. On June 15, the police came to the village with tear gas and guns. The villagers ran to Nwiido, according to an observer on the scene, and he "immediately surrendered himself to the police. With his hands up and outside his own house, he was shot in the legs and put into the boot [trunk] of the police car by the Divisional Police Officer and the Area Commander. That was the last time that Friday was seen" (Ashton-Jones, 2001).

For five days, Nwiido's mother searched for her son. She was not allowed into the Shell compound in Port Harcourt, but was told to go to the nearby Rumuibekwe police station from where she was sent to the Shell clinic. There, an Ogoni nurse told her that her son had died. His body apparently had been deposited in the teaching hospital where she was told that she could only inspect the name in the mortuary register; the body could only be seen with police permission (Ashton-Jones, 2001).

Another account of the incident, from Nwiido's mother, is consistent:

> I don't know who called my son. He suddenly reappeared from his workplace and walked into the invading force with his hands raised in surrender. As he came he was shouting, "I am the one, I didn't hijack any vehicle, Shell is owing me and I want my money" [he cried] as the police fired live bullets at him at close range. He was hit on his thighs several times. He fell down, bleeding profusely. He was carried from the ground by one of the police offices and dumped in the boot. I hired a car immediately and followed the police who were retreating after killing my child. When they noticed that we were following them, they stopped us. We diverted and monitored them up to the police station at Bori, Ogoni. I saw when Shell vehicles stopped and entered the police station. The police held brief talks with Shell and Khana local government officials. (Ashton-Jones, 2001)

Nwiido's mother then undertook a search for her son's body, traveling to Port Harcourt. She received conflicting accounts. "I went to the military hospital," she recalled, "where they told me that there was no body like that. I visited all the government and Shell hospitals in Port Harcourt but I could not find my son. It was at the Shell hospital [that] somebody[,] I will not mention his name, told me that really the boy was brought there alive and after several secret talks with the medical personnel, they transferred him to the Mini-Okoro police station" (Ashton-Jones, 2001). Other people told her that Friday was dead: "At Mini-Okoro, another reliable source told me that the boy was executed on Saturday and that the people who were present during the shooting were the Divisional Crime Officer, Divisional Police Officer, Area Commander, one man nicknamed Ahoada and two others" (Ashton-Jones, 2001).

"To confirm the source," she continued, "I was told that the corpse was deposited at the mortuary of the University of Port Harcourt Teaching Hospital. I went there and saw the corpse of my son." The hospital would not release the body for burial, however. "Up till now," his mother said, "I am still waiting for the corpse of my son. I want to bury him . . . the police, Shell council officials, and the village council are all responsible for my child's death. I will say these things anywhere, any day. Please quote me anywhere, I have read what Shell and the police are saying in the Nigerian media, they are all liars" (Ashton-Jones, 2001).

The Travail of Ogoniland Continues

As the third millennium dawned, an old story continued in Ogoniland: poorly maintained oil-drilling infrastructure continued to leak and blow up, and anyone who spoke out against the devastation risked death. The August 31, 2001 issue of *Drillbits and Tailings* was dedicated

> to the memory of Mr. Vincent Ifelodun Bolarin Oyinbo (also known as Bola), his family and friends, the Ilaje people, all people of the Niger Delta, and the staff and volunteers of Environmental Rights Action/Friends of the Earth, Nigeria. Environmental Rights Action/ Friends of the Earth, Nigeria notified friends and allies that Bola passed on to the great beyond on July 19, 2001 in Lagos, Nigeria at the age of 36 while preparing for a trip to the United States. ERA/FoEN said a death certificate issued by a private medical clinic in Lagos put the cause of his death as cardiac arrest. But relatives and friends say he never had a history of heart problems. Bola was one of the 100 peaceful protesters on the Parabe offshore platform who were attacked by the Nigerian military in helicopters operated by Chevron personnel. He was held for 12 days and tortured by the Nigerian military personnel. Bola witnessed bribes given by Chevron employees to the military personnel who attacked and arrested him. His trip to the United States was to receive medical treatment from injuries he sustained during this attack and to work on a court case filed against Chevron by communities in the Niger Delta. (*Drillbits*, 2001)

Within a week of Bola's death, the Niger Delta was afflicted with yet another massive oil spill. People living in the community of Ogbodo, on the banks of the Miniamu River, "were engulfed with irritating odor and itching every morning." Isiri Alison, an observer, related that "we no longer drink from the rivers. As an emergency measure, Shell supplied [a] few liters of water to the 15 families that make us Ogbodo. Apart from the one I saw with my eyes, everyone here complained that the water Shell supplied is dirty and smells. Many people simply threw theirs away" (*Drillbits*, 2001).

As of August 21, 2001, community leaders from Gokana, Ogoniland reported that fires caused by ruptured pipelines owned and operated by Shell Oil had been burning for two months with no response from authorities. "Agency reports yesterday [August 20, 2001] that the community faced being ravaged by 'devastation of unimaginable proportion' unless urgent steps were taken to put out the 'scores of fires ignited by pipeline excavators,'" reported the Guardian news agency in Lagos ("Shell Oil," 2001). Other news correspondents reported that farmland has been lost to the fires.

In June 1997, when Shell refused to pay compensation for a 1982 oil spill (defying a local court order), members of four affected Ijaw communities gave Shell an ultimatum: to leave the oil producing area by July 8 or to be forced out. Hours before the deadline expired, the leader of the community protest was arrested by the State Security Service. A local observer wrote: "Worried that the said payments will encourage other legitimate compensation demands, Shell has alerted the security forces and this morning Mr. Matthew Eregbene has been whisked away," said a spokesman for the Niger Delta Oil-Producing Communities Development Organization ("Shell: 100 Years," 1997).

In Bayelsa State, the Ijaw Youth Council (IYC) called on the Shell Petroleum Development Company, the Nigerian subsidiary of Shell Oil, to cease all operations at the Nun River flow station in response to the murder of a 22-year-old Ijaw man by the mobile police. At about the same time, the Lagos newspaper *This Day* reported that, by August 2001, 4,835 oil spills totaling almost 2.4 million barrels of crude oil had afflicted the Niger delta (Lwori, 2001).

During November 2001, *Drillbits and Tailings* published details indicating that another oil-related accident in Ogoniland had claimed several more lives: An oil pipeline owned by the Royal Dutch Shell exploded, killing as many as 15 people and injuring 14 others. The explosion occurred at Umidike in Imo State on November 5, 2001. Ironically, *Drillbits and Tailings* also reported that just a month earlier the Shell Petroleum Development Company (SPDC) had called for the introduction of a national environmental and safety standard in the Nigerian oil and gas industry that would have matched international standards ("Hotspots: Nigeria," 2001).

Meanwhile, to quell any popular expression of belief that something might be dreadfully wrong in the oil fields, the Nigerian government set up a special committee to ensure total security for oil-producing areas. The authorities demanded that "a recent siege" in the oil producing areas by "restive youths, communal agitators and economic saboteurs" must end. The new committee signaled increasing vigilance on the part of the military against any sign of unhappiness among the Ogoni. Chief Ekaette explained that the recent "terrorism" had made the assured security of oil installations an urgent imperative. Felix Ekure, the Delta's state chairman of the National Youth Council of Nigeria (NYCN), warned that unless the youths of the Niger Delta were included in the development of the region, "the country may know no peace" ("Hotspots: Nigeria," 2001).

Oil-related disasters in Ogoniland have assumed a near-daily regularity. They are reported on the back pages of local newspapers. For example, the *Lagos Vanguard*, on July 18, 2001, reported how three children had died in Akwa Ibom by drowning in uncapped oil wells belonging to Shell Producing Nigeria. Addressing the World Conference of Mayors in Eket, Governor Victor Attah said that "Shell callously left uncapped wells in which three young children have so far drowned." Narrating the "evil side" of oil exploitation in the area by the ExxonMobil, Addax, and Elf oil companies, Attah said that "pollution, environmental degradation, terminal diseases and birth defects had affected many people in oil producing areas" (Ashton-Jones, 2001).

Also in the July 18, 2001 issue of the *Lagos Vanguard*, on the same page as the Akwa Ibom story, a headline read: "Oil Spill: Strange Illness Hits Rivers Community." In the Ogbodo Isioko community in the Ikwere area of the Rivers region, where the June 25 oil spill of the Shell Petroleum Development Company occurred, residents were said to have reported strange ailments that had claimed four lives. The community said that the spill had "spread quite extensively on the only stream that provided [the] source of drinking water for the area" (Ashton-Jones, 2001).

In the midst of all this, Royal Dutch Shell told a reporter for the *Wall Street Journal* that its "more urgent concern is to protect Ogoni lives and avert disaster" (Moore, 2002, A-10). The company also said it plans to spend $7.5 billion to extract 300 million barrels of remaining oil reserves in the region. This particular account portrayed the oil company as a victim of "local hostility," as well as enterprising thieves who could sell a purloined 25-foot section of oil pipeline for $87, or more than an average Nigerian construction worker earns in a month (Moore, 2002, A-10). Children were said in this account to "flock to the theft sites, collecting leftover oil with plastic bottles to sell to those who use it as medicine or to frighten away evil spirits" (Moore, 2002, A-10). Shell executives are portrayed here as lamenting local hostility that keeps them from helping clean up the mess. "We have pleaded with the Ogoni people to let us come and make those wells safe," said Hubert Nwokolo, Shell's general manager of community development in Nigeria. "What worries me [is that] one day we'll have a blowout and then they'll say, 'Shell, they planned it, they want to kill us all'" (Moore, 2002, A-10).

Shell replies to such criticism by emphasizing that it spends several million U.S. dollars a year on social infrastructure in Nigeria, including building of schools, hospitals, roads, bridges, electrical generation capacity, and water systems in areas that "the government effectively abandoned in the early 1980s" (Hertz, 2001, 170). Shell was trying to answer to demands created by widespread

Women from the Ijaw community sing and dance as they protest against ChevronTexaco inside a fuel station belonging to ChevronTexaco, near Abiteye village in Nigeria, on July 17, 2002. The women refused to leave until they met with senior company executives to press their demands for jobs and community improvements. (AP Photo/Saurabh Das)

social unrest by people who say it is extracting wealth from the country and putting nothing back. Pipelines have been blown up, oil installations invaded, and equipment seized. During the autumn of 1998, civil unrest cut Nigeria's oil production by one-third (Hertz, 2001, 170). During 1999, Shell faced 45 instances of hostage-taking in Nigeria, involving more than 200 employees; during June of that year, 50 young Nigerian men seized a Shell facility and shut it down for five days, costing the company $2.4 million (Hertz, 2002, 170).

Native Women Occupy Shell Facilities

Despite Shell's attempts to emphasize the benefits it bestows on Nigerians, during July 2002, Ijaw and Ogoni and other Nigerian tribal women occupied pumping stations and successfully halted ChevronTexaco operations in the Niger Delta for nearly two weeks. A single action by about 150 women from villages surrounding ChevronTexaco's Escravos oil terminal grew to more than 3,000 women who joined in a nonviolent direct occupation of four other ChevronTexaco facilities in the Delta. According to one account from the scene, "For 10 days, Itsekiri women from Ugborodo in the Warri area of Delta state took over ChevronTexaco's multimillion-dollar tank farm and terminal in Escravos. . . . While it lasted, the blockade disrupted the production of an estimated 500,000 barrels of oil per day. Some 800 workers were trapped in the terminal after 400 of their colleagues were released by the protesting women" (Osadolor, 2002, 47).

The women demanded jobs, education services, health services, and economic investment in their communities. The company agreed to the demands but delivered no follow-through after the women ended their protests. The protests had begun peacefully June 10 after company officials refused to meet with the women to discuss their alleged neglect of community concerns. They had three main demands: "employment for their children, greater economic empowerment, and an enhanced infrastructure" (Osadolor, 2002, 47).

According to an account in *Drillbits and Tailings,*

> Armed with only food and their voices, village women proceeded to occupy the Escravos facility on July 8. They occupied the terminal for eleven days, barricading a storage depot, and blocking docks, helicopter pads and an airstrip, which provide the only entry points to the facility. Their presence prevented 700 workers from working and leaving until the company agreed to certain conditions. On July 18, the company agreed to hire 25 villagers over five years and to help build clinics, schools, and fish and chicken farms. . . . ChevronTexaco officials and representatives of the Ijaw women met at the Abiteye flowstation. They signed a memorandum of understanding, outlining various conditions the oil company must meet in order to continue production. Demands were made on a compensatory impetus for the negative impacts and environmental degradation due to ChevronTexaco's oil activities. ("Women Occupy," 2002)

REFERENCES

Ashton-Jones, Nick. "Causes of Terrorism? Shell Oil in Nigeria, 1993 to 2001." October, 2001. [http://www.shell-terror.net/]

Dayaneni, Gopal. "Field Notes and Reflections from a Project Underground Staffer's Trip to the Niger Delta, March 22 to April 10, 2002." *Drillbits and Tailings* 7:4 (April–May 2002): n.p. [http://www.moles.org]

Drillbits and Tailings 6:7 (August 31, 2001): n.p. [http://groups.yahoo.com/group/protecting_knowledge/message/1770] [http://www.moles.org/ProjectUnderground/motherlode/chevron/chevinfo.html.]

"Environmental Injustice." *Drillbits and Tailings* 6:9 (November 30, 2001): n.p. [http://www.moles.org/ProjectUnderground/motherlode/chevron/wto2001_nov.html]

Gedicks, Al. *Resource Rebels: Native Challenges to Mining and Oil Corporations.* Boston: South End Press, 2001.

"Hotspots: Nigeria." *Drillbits and Tailings* 6:9 (November 30, 2001): n.p. [www.moles.org]

Lwori, John. "Niger Delta Records 4,835 Oil Spills in 20 Years." *This Day* (Lagos, Nigeria), August 3, 2001." In *Drillbits and Tailings* 6:7(August 31, 2001): n.p.

Moore, Sarah. "For Shell, Nigerian Debacle Isn't the End of the Line: Danger Lurks in Ogoniland for People and Firm, but the Place Beckons." *Wall Street Journal,* January 10, 2002, A-10.

"Nigeria: Focus on Ogoni Oil Spill." Integrated Regional Information Networks, United Nations Office for the Co-ordination of Humanitarian Affairs, June 12, 2001. [http://www.reliefweb.int/IRIN/wa/countrystories/niger/20010612.phtml]

"Nigeria: Godforsaken by Oil." *World Rainforest Movement Bulletin* (March 2002) n.p. [www.wrm.org.uy]

Osadolor, Kingsley. "The Rise of the Women of the Niger Delta." *London Guardian*, July 24, 2002, in *World Press Review*, October, 2002, 47–48.

Project Underground. "More Blood Is Spilled for Oil in the Niger Delta." *Drillbits and Tailings* 4:20 (December 11, 1999): n.p.

Robinson, Deborah. *Ogoni: The Struggle Continues*. 2d ed. Geneva, Switzerland: World Council of Churches, 1996.

Saro-Wiwa, Ken. *Genocide in Nigeria: The Ogoni Tragedy*. Port Harcourt, Nigeria: Saros International Publishers, 1992.

"Shell: 100 Years is Enough!" October 1997. [http://www.kemptown.org/shell/rest.html]

"Shell Oil Spills Continue to Ravage Communities and the Environment in Nigeria." *Drillbits and Tailings* 6:7 (August 31, 2001): n.p. [http://groups.yahoo.com/group/protecting_knowledge/message/1770]

"Shell Says Ogoni Oil Blow-out Now under Control." May 7, 2001. Integrated Regional Information Networks, United Nations Office for the Co-ordination of Humanitarian Affairs, June 12, 2001. [http://www.reliefweb.int/IRIN/wa/country stories/nigeria/20010507.phtml]

Wiwa, Owens. "Like Oil and Water: The Ogoni in Nigeria." August, 2000. [http://www.dghonline.org/nl7/owens.html]

"Women Occupy Chevron/Texaco Facilities in the Niger Delta." *Drillbits and Tailings* 7:6 (July 31, 2002).

NOBLE SAVAGE, THE "ECOLOGICAL INDIAN"

During the Age of Discovery, European people were opening their eyes to new ways of thinking, as well seeking land and material riches. Because of these intellectual imports, the Old World also changed. Created of European wish-fulfillment, the image of the Noble Savage was created from this imagery, fashioned by European philosophers. Native societies, especially in America, reminded Europeans of imagined golden worlds known to them only in folk history.

The way that European thinking was shaped by its discovery of the New World (as well as discovery of Africa and parts of Asia) also outlines some of its confusing contradictions. From the beginning, the Noble Savage was idealized in philosophy, as his real-life counterparts were slaughtered to make way for so-called progress. The Noble Savage was an apparition of European imagination, but, like any racial stereotype, it said as much about the very real drives, perceptions, dreams, and desires of its creators as about the newly discovered Americans themselves. The Noble Savage may have been a creature of imagination, but the influence of the concept on European thought was very real, especially during the Enlightenment years, which culminated with the American and French revolutions.

Like most stereotypes, the Noble Savage simplified a complex reality. It also created an image that was, paradoxically, both more and less than reality. More, because it ascribed to the natives more life, liberty, and happiness than many of them actually possessed, creating a myth that imagined an autonomous wild man of the woods and ignoring the very real social conventions and traditions by which Native Americans ordered their lives; less, because the image of the Noble Savage combined many dozens of peoples and belief systems into one generic whole.

As with most stereotypes, distance distorted the reality of the image. Thus, the image of the Indian created by Rousseau or Locke seems utterly more fantastical than that of Franklin, Jefferson, and other influential founders of the United States, who did diplomatic business with American native people in the course of their daily lives. In our time, the Noble Savage is usually dismissed merely as a figment of imagination, ignoring the power the image held in the Enlightenment mind and the impact of its appeal to influential thinkers of the time.

The image of the Noble Savage (like most stereotypes) also engendered its opposite, the so-called bad Indian, who stood in the way of the Europeans' destruction of the naturalness its philosophers so admired. They built churches in a place that was described to them as the Garden of Eden.

Memory outlasted image, and image outlived changing American realities, which themselves had been reworked by desirous imagination, taking what had been real to its most logical (and often patently absurd) extreme. As a vehicle of dreams, the Noble Savage helped reawaken in Europeans a passionate desire for the liberty and happiness that so suffused Enlightenment thought, helping to ignite revolutions on both sides of the Atlantic.

REFERENCES

Grinde, Donald A., Jr., and Bruce E. Johansen. *Ecocide of Native America: Environmental Destruction of Indian Lands and Peoples*. Santa Fe, N.M.: Clear Light, 1994.

Grinde, Donald A., Jr., and Bruce E. Johansen. *Exemplar of Liberty: Native America and the Evolution of Democracy*. Los Angeles: American Indian Studies Center, 1991.

Hertz, Noreena. *The Silent Takeover: Global Capitalism and the Death of Democracy*. New York: The Free Press, 2001.

PAKISTAN

THE KAFIR-KALASH: LAND SULLIED BY TOURISM

One of the major tourist attractions of Pakistan's Chitral region are the Kalash valleys, home of the Kafir-Kalash, "Wearers of the Black Robe," a remote indigenous tribe. A local legend holds that the Kafir-Kalash originated from marriages of indigenous people with soldiers of the legions of Alexander of Macedonia, who settled in Chitral during 326 B.C. Roughly 4,000 Kafir-Kalash live in the valleys of Birir, Bumburet, and Rambur in the south of Pakistan.

Due to construction of roads that can be traversed by jeeps, this once-isolated area has been inundated by foreign tourists. According to one source in Pakistan, "These growing tourism activities have not only affected [the] cultural fabric of [the] Kalash people but the use of . . . food products . . . wrapped in synthetic-based packing material [e.g., plastic] has ruined the valley. You can find wrappers and empty packs of biscuits, pasteurized milk, instant juices, and other commodities everywhere in the Valley of Bumburet" (Iqrar Haroon, n.d.). The same observer reported condom wrappers floating in the water channels of Bumburet." The Kafir-Kalashes' valley now hosts three major motels that channel their revenue to owners in larger cities, such as Islamabad, Lahore, and Peshawar.

Local indigenous peoples have complained that they see little income or other benefits from foreign tourists who disrupt their religious ceremonies and spread their trash around what once was a pristine landscape in one of the most remote areas of Pakistan. The motels shun local produce and import items such as "jams, marmalades, butter, [and] dairy products" from the outside (Iqrar Haroon, n.d.). Even beef is imported, as local people tell observers that "tourism is providing only one thing to Bumburet—left-overs" (Iqrar Haroon, n.d.). Local people are appealing for a garbage dump (something they had never needed for themselves). "It was observed," according to one report, "that there was no garbage dump provided to people by the government or non-government organizations which are champions of environments in Pakistan. This problem needs attention of people involved in the field of sustainable tourism and environments and they should come out of their drawing rooms and offices and should do something" (Iqrar Haroon, n.d.).

REFERENCES

Iqrar Haroon, Agha. "Save Kalash Valley Before It Is Too Late." Ecotourism Society Pakistan, n.d. [http://www.ecoclub.com/news/19.html]

PANAMA

THE NGOBE-BUGLE: WINNING LAND AND RESISTING MINING

Introduction

Despite resistance by non-Indian landholders and miners, Panama's indigenous Ngobe-Bugle by the mid-1990s had won recognition of their title to 11,000 square kilometers in the western part of the country. The Ngobe-Bugle also have expressed their opposition to mining on their lands. One of their leaders, Marcelino Montezuma, explained that the Ngobe-Bugles' rejection of mining projects was related to environmental consequences: "The air, the trees and the river are our lifeline," he said (Hernandez, 1995). Roughly 35,000 Ngobe-Bugle live on lands to which mining rights already had been granted before their title was recognized (Hernandez, 1995).

Like the Ngobe-Bugle, other native peoples in Panama are struggling to attain a measure of sovereignty over their ancestral homelands. The Emberá-Wounan and the Kuna indigenous peoples live in the Darién, San Blás, and Panamá Oriente provinces of Panama, while the Teribe occupy the Bocas del Toro area. On the border with Costa Rica, the Ngobe-Bugle inhabit parts of the provinces of Veraguas and Chiriquí Oriente. According to the 1990 Panamanian National Census, the indigenous population of Panama comprises 180,700 people, 7.8 percent of the country's population.

Defining Land Tenure, Stalling Logging

The 150,000 Ngobe-Bugle of Panama refused to allow engineers, surveyors, and geologists onto their traditional lands until they are legally defined and guaranteed. Marcelino Montezuma, president of the Ngobe-Bugle Congress, said that "without defining our reserve, we will not allow a single mining project to enter our areas because we have no laws to protect us and we will not be allowed to share in any benefits" ("Indians and Police," 1995, 7). The Ngobe-

Bugle and other indigenous peoples of Panama are facing several forms of intrusion into their traditional lands, including industrial-scale logging, road construction, cocoa production, and mining, mainly for copper and gold.

Indigenous peoples in Panama have been struggling for recognition of legal title to their lands as a means of stalling or stopping logging of their forests. In 1850, forests covered an estimated 90 percent of Panama, even after partial deforestation by Spanish colonists since the sixteenth century. Between 1950 and 1960, forest cover diminished from 68 percent to 58 percent. According to official estimates, in 1992 the forest area had been reduced to 3,358,304 hectares, or about 44 percent of Panama's surface area. Since then, forest cover in Panama has been decreasing at a rate of about 75,000 hectares per year due to increased industrial-scale logging, road construction, cocoa production, and mining ("Panama: Mining," 2001).

The first recognition of an indigenous territory by the Panamanian government—the Comarca San Blás—came in 1938. In 1983, the Comarca Emberá-Wounan in the Darién Province was recognized, followed by the Comarca Kuna of Madungandi in the Panamá Province during 1996 and the Ngobe-Bugle a year later. Once a people's territory is officially recognized, its lands must be defined and demarcated under Panamanian law. Each native people has faced a long struggle against the vested interests of miners, loggers, and cattle ranchers who oppose demarcation and recognition of the indigenous territories that may impede their extractive activities.

According to an analysis by the World Rainforest Movement, "mining is resulting in disastrous effects in several areas of the country, generating at the same time conflicts with the indigenous communities that live there" ("Panama: Mining," 2001). By 1994, 25 percent of Panama's land area was covered by mining concessions; another quarter could be affected by pending concessions.

Three-quarters of Panama's mining concessions have been granted on indigenous lands at San Blás, Boca del Toro, Veraguas, and Chiriquí, according to the Panamanian Natural Resources Directorate ("Panama: Mining," 2001). The government has approved extensive copper- and gold-mining concessions within the Ngobe-Bugle and Kuna territories. In the case of Kuna Yala, the concession granted to the Canadian company Western Keltic Mines extends over more than 50 percent of the Kuna territory.

"We first suggested the autonomous territory to [former President Guillermo] Endara, but rather than deal with us he sent in the security forces to mistreat us as though we were about to rob him of the land, which belongs to us," said a statement by Ngobe-Bugle elders (Hernandez, 1995). The Ngobe-Bugle demand for autonomy followed a conference attended by about 5,000 members of the group. The Ngobe-Bugle had to overcome the interests of miners seeking an easy route to extensive copper, gold, and silver deposits in the area, along with pressure from nonnative livestock raisers.

Minamerica Corporation representative Jose Montenegro said that in a few years mining will bring Panama more than U.S. $200 million worth of income, replacing bananas as its main export. The Ngobe-Bugle said, however, that present living conditions of their people demonstrate "quite the contrary" to these assertions. "When they talk about benefits, like work and social progress, the community knows that the qualified workers will be brought from outside and will not include them, and they also run the risk of losing their land," the Ngobe-Bugle elders said (Hernandez, 1995).

Miners' Withdrawal Demanded

"We have issued the government an ultimatum—withdraw the illegal miners from the Calobebora region, or we will throw them out," said Ngobe-Bugle Congress leader Victor Guerra (Hernandez, 1995). The strength of the Ngobe-Bugles' assertions of autonomy has caused some commentators in Panama to compare them to the Zapatistas of Mexico's Chiapas. Ngobe-Bugle elders recalled the story of Chief Urraca, who fought off Spanish colonists for 15 years during the sixteenth century. While they disdain violence as a route to autonomy, the Ngobe-Bugle say they "are losing patience" and are tired of being seen as "animals and not human beings," when they resist eviction from their ancestral homelands by mining corporations (Hernandez, 1995).

On April 11, 1995, the Ngobe-Bugles detained two geologists working for Geo Tec, a mining company, who were exploring their lands. The next day, 600 indigenous people gathered in front of Geo Tec offices in Veraguas. In short order, they were attacked by several hundred police in riot gear who broke up the Indians' assembly with guns, nightsticks, and tear gas. Police shot and wounded one Indian, while two of the police suffered wounds at the hands of machete-wielding Ngobe-Bugles. Eight Ngobe-Bugles were taken into custody, one of them a pregnant woman. From that day forward, Geo Tec posted as many as 50 armed police around its offices in Veraguas.

REFERENCES

Hernandez, Silvio. "Ngobe-Bugle Want Land and Independence." Interpress Service, March 28, 1995. [http://www.stile.lboro.ac.uk/~gyedb/STILE/Email0002069/m21.html]

"Indians and Police Clash in Panama over Mining." *Native Americas* 12:3 (fall 1995): 7–8.

"Panama: Mining, Forests and Indigenous Peoples' Rights." *World Rainforest Movement Bulletin* 46 (May 2001): n.p. [http://www.wrm.org.uy/bulletin/46/Panama.html]

PERU

INTRODUCTION

Opposition to natural resource development (principally oil extraction and gold mining) has sparked major civil unrest in Peru by indigenous peoples who have experienced previous environmental devastation, notably from Yanacocha, Latin America's largest gold mine. Major problems have included mercury poisoning of local water supplies. Elsewhere in Peru, indigenous peoples have organized to clear the air around a lead smelter, and have resisted the extraction of oil and natural gas in the Peruvian Amazon, where disease importation and drug-taking tourists also have posed problems.

INDIGENOUS PEOPLES, GOLD MINING, AND MERCURY POISONING

Hundreds of protesters during 2001 blocked a major highway in northern Peru, alleging that local water supplies in the province of Cajamarca, 530 miles (850 kilometers) northeast of Lima, had been contaminated by toxic mercury from Yanacocha, Latin America's largest gold mine. Denver-based Newmont Mining Corporation, North America's biggest gold miner, holds majority control of Yanacocha. Along with Newmont, the Minera Yanacocha joint venture includes Condesa, a subsidiary of Peruvian Minas Buenaventura, funded by the International Finance Corporation of the World Bank Group.

The protesters demanded that mining be halted pending investigation of water purity. "We reject the environmental contamination from Yanacocha," student leader Jorge Malca told Canal N cable television ("Water Supplies," 2001). Energy and Mines Minister Jaime Quijandria told RPP radio that it was "simply and totally impossible" for the water to have been contaminated with mercury ("Water Supplies," 2001). Suspicious fish, cattle, and human deaths also have been reported among the people of Cajamarca, where Newmont is using cyanide to extract gold from ore (Chatterjee, 1998).

In less than 10 years, a rural agricultural and dairy-producing region in northern Peru has been overwhelmed by a multinational mining operation whose four open-pit gold mines are the most profitable in all of South America. Spread across 25,000 hectares (63,000 acres) of mountaintops, Yanacocha is already the world's second-largest gold mine, and has been growing steadily. The joint-venture company owns mineral rights to an additional 125,000 hectares, including Mount Quilish, the main source of potable and agricultural water for the city of Cajamarca's 130,000 people, as well as for 300,000 more people in nearby areas ("Global Response," 2001).

At the mine sites, huge piles of low-grade ore are soaked in a toxic cyanide solution that leaches out gold and silver. Although Yanacocha managers claim cyanide and other toxic metals cannot escape from the mine site into the local watershed, mining expert Robert Moran said, "All the sites I've ever worked at experience some degree of leakage" ("Global Response," 2001). Mine contamination already has resulted in three major fish kills in area rivers and trout farms.

Local residents have demanded reparations from the World Bank for mercury poisoning that affected as many as 300 villagers 375 miles north of Lima on June 2, 2000. The contamination occurred after a truck carrying 330 pounds of mercury to Lima spilled it over a 27-kilometer (16-mile) portion of a road near Choropampa, 53 miles southwest of the mine. Mercury, a byproduct of the gold mine, is routinely trucked to Lima for use in medical instrumentation and other applications ("Mercury Spill," 2000).

According to an account in the *Denver Post*, about 330 pounds of the poisonous liquid leaked from the truck. Villagers gathered up the poisonous substance. Some believed it still might contain gold, or have value for other reasons. Others thought the mercury might have medicinal uses. Others were simply curious. Many kept some of the mercury in their homes until it made many children ill ("Mercury Spill," 2000). The spill eventually sickened more than 400 people, many of them children, contaminated 80 homes, and cost the company $12 to $14 million in a yearlong cleanup effort ("Peru Villagers," 2000). A 38-year-old woman was flown to Lima in a coma and placed in intensive care. She was examined by a critical-care specialist flown in by Newmont. According to the International Finance Corporation, Minera Yanacocha and Newmont sent medical and toxicology experts to the scene and provided mercury testing for local residents in their homes. The World Bank undertook an investigation of the incident, and issued a report accusing the company of transporting the mercury without appropriate safety measures, in violation of international standards governing the handling of hazardous substances ("Mining in Peru," n.d.).

"This spill is just one more disaster brought by the mine. Many people are sick because they weren't told what the mercury was after it spilled. There were no safety precautions in place," said Segunda Castrejon, president of the Rondas Campesinas Femeninas del Norte del Peru ("Mercury Spill," 2000). On

June 23, 2000, Indigenous Peoples' Day in Peru, hundreds of people marched in the city of Cajamarca protesting the mine and its impacts on the indigenous peoples who live near it. Local people demanded compensation for the families of those affected, as well as closure of the mine. "The people are demanding compensation for this. The company is getting rich off this mine while the local people suffer the impacts," said Julio Marin, speaking for Rondas Campesinas ("Mercury Spill," 2000).

A LEAD SMELTER FOULS THE AIR AT LA OROYA

People living near the Peruvian village of La Oroya face environmental problems similar to those living near the Yanacocha gold mine: pollution of the water supply from cyanide, air pollution, and noisy truck traffic near the mine. The pollution destroys pasturelands and sickens livestock, debilitating local agriculture.

Ore smelters and refineries foul the air with sulfur dioxide and destroy pasturelands needed for livestock. The effects of a lead smelter in La Oroya owned by the U.S.–based company Doe Run have been endured by the community for generations. Nearby rivers contain levels of lead, iron, zinc, copper, and arsenic that exceed the limits for environmental health set by Peru's governmental agencies, according to a September 2000 study by the environmental organization CooperAccion, which also studied air quality in the area, and found concentrations of lead in the air were 800 percent above acceptable levels. The effect on people in the area is serious—local residents have high levels of lead in their blood. Jose de Echave, the deputy director of CooperAccion, noted that its recent study of lead poisoning near La Oroya found that 90 percent of the people tested had levels far above the acceptable lead levels set by the U.S. federal government ("Mining in Peru," n.d.).

"Mining and Communities: Oral and Written Testimonies," published by CooperAccion, describes some of the environmental damage. One local farmer, who grew up only 300 meters from the smelter at La Oroya, said that the fumes routinely burned his throat and nose. Another local inhabitant, who said that the last good harvest he could remember was in 1919 or 1920, blamed the smelter for ruining agriculture in the area. "The smoke fell like a snowfall of arsenic dust on the land, rocks, and pasture. The animals got sick, it was a disaster. How could we live there? There was no harvest and the animals died, like that, in groups. It was as though they were poisoned" ("Mining in Peru," n.d.).

INDIGENOUS PROTESTS OF OIL EXPLORATION

Indigenous peoples from the Peruvian State of Madre de Dios marched in the city of Puerto Maldonado July 18, 2000 to demand that the government deny further mining and logging concessions within indigenous territories. Meanwhile, ExxonMobil, which staked a claim in the nearby Candamo valley in 1996, announced that it would continue exploration for oil and gas in the area. The valley is a complex ecosystem that is home to at least 20 isolated indigenous communities, including the Yoro, Ese'eje, Mascho-Piro, and Amahuaca. Scientists consider the area to be "a complete Amazon in miniature" because of its biodiversity and abundant plant and animal life ("Indigenous Peruvians," 2000).

"By all Peruvians our state is called 'the bio-diversity capital of Peru,' for its biological richness. It is also well-known that this is a land of nearly 20 indigenous peoples, a cultural diversity that has a hand in the biological diversity. What is more, we believe that it is this cultural diversity that guarantees the biological diversity," said a manifesto written in defense of the indigenous peoples of Madre de Dios ("Indigenous Peruvians," 2000).

The indigenous peoples of Peru are facing increasing pressure from international investment in mining, oil, and natural gas, industries that impinge on their traditional lands, cultures, and ways of life. Mobil, which has since merged with Exxon, leased 3.7 million acres (1.5 million hectares) of rainforest in the Peruvian State of Madre de Dios. The area, which is known in the industry as Block 78, includes the 350,000-acre (141,600-hectare) Candamo Valley. In April, ExxonMobil began a second stage of exploration in the northwestern portion of the block at the Karene 3X well once owned by Occidental Petroleum. The study would take six months to complete, during which time the company would decide whether to continue exploration ("Indigenous Peruvians," 2000).

"One question that we must ask the people of Madre de Dios, independent of whether they are indigenous river dwellers, colonists who come here to live and co-exist in this rich biological resource, or the people and enterprises that have adapted their economic activities to the conditions of bio-diversity, is this: looking at sustainable development, is it possible to stop the current reality, where only the interests of the loggers decide the future of our state, people who only see the trees, but not the population that lives with a wealth of natural offerings such as ours," said the manifesto ("Indigenous Peruvians," 2000).

THE URARINA (KACHÁ): OIL DEVELOPMENT, IMPORTED DISEASES, AND "HIPPIE" TOURISTS

The Urarina (Kachá, meaning "the people," in their own language) are a seminomadic Amazonian people who have inhabited the Chambira and Urituyacu river basins north of the Marañon River in Peru for at least 500 years. The roughly 4,000 Urarina, along with the Mayorunas (Matses), are the largest Amazonian indigenous groups in Peru who have no official recognition of their land tenure. Their traditional isolation ended late in the twentieth century at the hands of land-grabbing colonists, loggers, river traders, and drug-taking so-called hippie tourists. Many of the intruders have been importing diseases to which the Urarina and Mayorunas had not heretofore been exposed.

During late January 1997, workers from three multinational oil companies arrived in the territory of the Urarina and Mayorunas to build an oil-drilling site for petroleum extraction. The oil companies began drilling from the Chambira oil field at Santa Martha, an Urarina community on the Chambira River. The Chambira oil fields are owned by Petroperu, the Peruvian national oil company, but the rights to drill directly on the Chambira River were transferred to Enterprise Oil Company, a British company, in 1996, because Petroperu lacked the capital to initiate drilling.

Even before they found significant amounts of oil, the companies found themselves involved in ecological mishaps that offended local residents. On April 30, 1997, the steel bottom of a barge transporting supply oil was punctured by a large capirona tree 10 kilometers from the oil well, causing oil contamination of the entire Chambira River downstream. The barge was then towed upstream, moored at Santa Martha, and surrounded by pylons in an attempt to control the oil leak. Fish in the area were contaminated, and the numbers of Amazonian river dolphins, usually numerous in the Chambira River basin, declined sharply (Witzig and Ascencios, n.d.).

In addition to oil spills, oil-drilling teams brought new diseases into the Urarina lands, compounding diseases imported earlier by loggers, tourists, religious missionaries, and other intruders. Because the drilling teams come from several countries, they are efficient disease vectors for peoples who have been isolated until recent times, as they import various strains of malaria, measles, and whooping cough. At least seven indigenous people died of whooping cough shortly after the arrival of exploration teams. The DTP vaccine that protects against whooping cough has been unavailable in the area. More than 60 percent of the malaria strains appearing in the area have been resistant to chloroquine and pyrimethamine/sulfadoxine, the two cheapest and most easily available drugs used to treat malaria. The Peruvian government has no health clinic in the entire Chambira river basin. The nearest health clinic is in Maypuco, more than a week by canoe from the Urarina territory. The Urarina own no motorized watercraft, meaning that medical care in distant towns is out of reach (Witzig and Ascencios, n.d.). According to one local observer, "To illustrate the seriousness of the epidemic, the Urarina village of Tagual had six people (five children and one pregnant woman) of 80 die of the new strain the week before the last medical survey and supply trip arrived. All other Urarina communities experienced mortality from the new strain, although at a lower rate" (Witzig, n.d.).

During the 1990s, North Americans arranged so-called jungle ecology tours that included two-week trips up the Amazon and Marañon rivers, as well as the lower Chambira River. During the river tour, a shaman from Iquitos manufactured the sacred hallucinogenic *ayahuasca* (*Banisteriopsis caapi*) for the tourists to drink and "experience the jungle like the natives." Finally, they arrived in Urarina villages to look at the Indians and take pictures. Immediately after a tour in the spring of 1995, most of the children in one village that had been visited came down with a respiratory ailment requiring antibiotic treatment (Witzig, n.d.).

While they claim to be selling an authentic Indian experience, the tour operators, who are armed with weapons, actually ridicule genuine Urarina religious ceremonies. Residents of affected villages complained to the Peruvian Ministries of the Interior and Tourism in Iquitos, and the American embassy in Lima, demanding that the individuals responsible be barred from their lands.

In addition to imported diseases, cholera has been known in the area for many years, even before it emerged during the early 1990s in deadlier, more virulent forms. Some communities reported population losses of up to 20 percent, "an incredibly high population mortality rate even from this well-known disease that can kill in less than 12 hours" (Witzig, n.d.).

THE CAMISEA NATURAL GAS PROJECT

The Camisea project, planned for the Peruvian Amazon, would extract natural gas from one of the world's most biologically diverse regions and transport it to Lima and Callao, Peru's capital city and main seaport, through two pipelines that would pass through dense jungles and mountains. The Camisea project would involve land that is legally recognized as a sanctuary for several indigenous peoples. Due to cost overruns and opposition by many environmental and other activist groups, the original consortium, led by Shell Oil, pulled out of the project during 1998. The project was therefore stalled until 2000, when another group, led by the U.S. company Hunt Oil, decided to pursue it.

The gas field is located in one of the world's most biologically diverse regions, on land claimed by several indigenous peoples who have chosen to remain outside the influence of modern industrial civilization and culture. A study by the Smithsonian Institution found, per hectare: 152 different plant species, almost 800 species of birds, 120 species of fish, 86 species of reptiles, as many as 69 species of medium-sized and large mammals, more than 300 species of bats, and more than 600 species of invertebrates. The study said that many other species remained undocumented ("Camisea Natural Gas Project," 2001).

The Camisea project, which is projected to cost $2.7 billion, will develop a massive gas field in the Urubamba (which means "Sacred Valley" in Quechua) river basin of the remote Peruvian Amazon, to exploit the first major natural gas deposit in Peru, which is said to contain an estimated 11 trillion cubic feet of the fuel. Several indigenous peoples living in voluntary isolation inhabit the Camisea area, a pristine area of exceptional biodiversity. Pipelines planned to serve the field will cut through the remote Vilcabamba range, which contains two indigenous communal reserves and a national park, utilizing wells, flow lines, a gas plant, and two pipelines from the interior to the Pacific coast ("Camisea Natural Gas Project," 2001).

The Camisea is home to the Nahua (sometimes also known as the Yora) and Kugapakori Indians. The Nahua live at the confluence of the Mishagua and Serjali rivers, within the Nahua-Kugapakori State Reserve, near the western border of Manu National Park. According to a historical study by Amazonwatch ("Camisea Natural Gas Project," 2001), these bands are so isolated that when Shell Oil explored the area in the mid-1980s, "The Nahua were exposed to whooping cough, smallpox and influenza [for the first time]. An estimated 50 percent of the population died" ("Camisea Natural Gas Project," 2001). The Nahua had lived without outside intrusion until 1984. In the year following first contact, as many as half of the Nahua died from respiratory diseases ("Peru: Illegal Loggers' Invasion," 2001). It is likely that 15,000 surviving native people in the area would suffer diseases, loss of food supply, ruination of archeological sites, and contamination of water if the gas field is further developed.

Developers of the field (and their respective stakes in the development) include gas producers Pluspetrol (an Argentine company, with a 40 percent stake), Hunt Oil (a U.S. company, 40 percent), and S.K. Corporation (a South Korean company, 20 percent). Downstream operators (transporters of gas) include Techint (30 percent), Pluspetrol (an Argentine company, 19.2 percent), Hunt Oil (a U.S. company, 19.2 percent), S.K. Corporation (a South Korean company, 9.6 percent), Grana y Montero (12 percent), and Sonatrach (10 percent) The project is being financed by Pluspetrol and Citigroup (Citibank). The completion date for the project was expected (as of October 2001) to be December 2003.

The project could be stalled or stopped for several reasons, in addition to resistance by local indigenous peoples and the biological sensitivity of the area, including, according to one source: "The lack of Peruvian demand for gas; recent Peruvian political and economic instability; Camisea's history of significant cost overruns; problematic contract negotiations and delays in the bidding process" ("Camisea Natural Gas Project," 2001).

Opponents of the proposed gas-development project who object on environmental grounds have pointed to a Pluspetrol oil spill during October 2000, on the Marañon River, which contaminated Peru's largest protected area, the Pacaya Samiria Reserve. According to Amazonwatch, the protected area's 20,000 inhabitants, many from the Cocamas-Cocamillas people, "suffered diarrhea and skin diseases and saw their food and water supply decimated by toxic pollution" ("Camisea Natural Gas Project," 2001). During November 1999, a rupture in a Pluspetrol oil pipeline in the Pucayacu ravine contaminated the Chambira River. The Urarina suffered severe health problems, as an undetermined number of them died.

As they resist pipeline construction, the Nahua also are afflicted by loss of habitat, as the fingers of industrial-scale logging continue to spread in the Peruvian Amazon. According to Shinai Serjali, a nonprofit volunteer group that works to support the Nahua people in the Peruvian Amazon, between May 2001 and November 2001, 16 logging companies invaded their territory.

The loggers, working illegally, have extracted more than 600,000 board feet, mainly mahogany and cedar, as they have depleted wildlife upon which the Nahua depend. Following the loggers' invasion, several Nahua women have been abducted. A Nahua delegation traveled for two weeks by balsa raft to Puerto Maldonado to seek aid from FENAMAD (a native federation known in Spanish as "Madre de Dios") to protest these abductions to officials in Lima. When the officials did not respond, and illegal logging continued, the Nahua walked to Lima themselves. Following this visit, an official delegation visited the Nahua community, investigated the illegal extraction, confirmed the charges, "and ordered the immediate installation of two guard posts in the area in order to effectively control and immobilize the illegally extracted timber" ("Peru: Illegal Loggers' Invasion," 2001).

On August 28, 2003, the U.S. Export-Import Bank voted to deny financing for the Camisea Gas Project, denying an application for U.S. $213 million in U.S. government loans and loan guarantees to the project. The Export-Import Bank thus joined other major investors—Citigroup and the Overseas Private Investment Corporation, as well as oil giant Royal Dutch Shell—in decisions to pull out of the Camisea project. Less than a month later, however, the Inter-American Development Bank decided to extend credit to the project.

THE AGUARUNA: TAKING THEIR LAND BY FORCE

The Aguaruna people are taking control of their land by force in the face of plans, in some Peruvian political circles, for accelerated colonization to install colonies of landless people from other regions, to create "living frontiers" (Lama, "Indigenous Peoples," 1995). The indigenous Aguaruna attacked landless squatters in a remote jungle area in a clash over land, leaving at least 14 people dead, police in San Ignacio said on January 19, 2002. Two hundred Aguaruna, armed with shotguns, raided the squatters' settlement, roughly 500 miles northeast of Lima near the border with Ecuador, according to survivors of the predawn attack. The Indian villagers were apparently trying to expel landless peasants who began settling the area during 1989, according to *El Comercio,* Peru's largest newspaper ("Settlers Killed," 2002).

San Ignacio Mayor Carlos Martinez said that the Indians had also taken an unknown number of hostages, among them children as young as two years of age. Police who arrived by helicopter to free the hostages were forced to retreat after the Indians attacked them with clubs and other weapons, he said. Luis Aguilar, a regional health official, told a local radio station that more settlers also were killed as they fled along jungle trails ("Settlers Killed," 2002). Survivors walked several hours to the village of Nuevo Trujillo to report the attack.

Amelia Iparraguire, a wounded settler, told Radioprogramas (the local radio station) from a hospital in Jaen that the Indians were armed with shotguns. She said 200 settlers were living in the area at the time of the attack. A judge had ruled in favor of the Indians' rights to occupy their homelands on January 12, 2002. Martinez noted that the Indians had won several legal proceedings in the past that were ignored by the intruders when the edicts were not enforced. The Indians' lack of legal recourse provoked the attacks, in which settlers' homes were burned as they were killed or routed.

In addition to problems with landless peasants looking for homesteads in the jungle, some indigenous peoples in northern Peru and southern Ecuador have been afflicted by border conflicts. Roughly 45,000 Aguaruna and Huambisa who inhabit the Upper Cenepa Valley, site of a Peru-Ecuador border conflict in

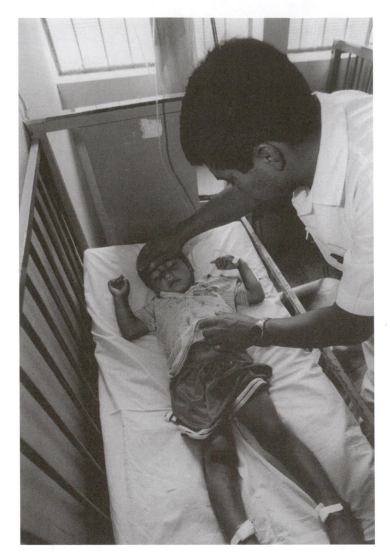

Esteban Alberca Soriano, 5, lays wounded as a doctor examines him, following clashes in Cajamarca, Peru between Indians and settlers on Friday, January 18, 2002. Two hundred Indians from the Aguaruna tribe armed with shotguns attacked the settlement, some 500 miles northwest of Lima near the border with Ecuador, leaving at least 14 people dead. (AP Photo/Domingo Giribaldi–El Comercio)

1995, were attacked by both sides. According to U.S. State Department reports, Aguaruna-Huambisa leaders have long complained about poor living conditions and lack of consultation by the government on matters affecting their land tenure. They have appealed to the courts many times, only to watch illegal squatters continue to usurp their lands.

The war has had a serious impact on the local communities. Some families were divided by the border for 50 years before they were reunited in 1998. The president of the Aguaruna-Huambisa Council, Evaristo Nukuang, reported that 28 local people were killed by landmines during the conflict (Lama, 1995). "No one has included them on the list of casualties, and their families will not receive compensation like the families of the soldiers and other dead," said Nukuang. "There is no peace in our communities, mothers are crying for their children, the children for fear of the bombing raids, and the animals we hunt have all run away. . . . But we are Peruvian and we don't want to leave this territory because we've been here since the time of the Incas" (Lama, 1995).

REFERENCES

"Camisea Gas Field and Pipeline Project." Oxfam America, n.d. [http://www.oxfamamerica.org/advocacy/camisea.html]

"Camisea Natural Gas Project, Peru." Amazonwatch, October 8, 2001. [http://www.amazonwatch.org/campaigns.html]

Chatterjee, Pratap. "Gold, Greed and Genocide in the Americas: California to the Amazon." *Abya Yala News: The Journal of the South and Meso-American Rights Center* (1998): n.p. [http://saiic.nativeweb.org/ayn/goldgreed.html]

Elton, Catherine. "Pristine Amazon Jungle Threatened by Big Oil Firm This Month; Peru Granted Mobil a Deadline Extension on a Decision." *Christian Science Monitor,* September 30, 1999.

"Global Response: Environmental Education and Action Network." October 2001. [http://www.globalresponse.org/gra/current.html]

"Indigenous Peruvians Mobilize While ExxonMobil Further Explores Rainforest." *Drillbits and Tailings* 5:12 (July 20, 2000): n.p. [http://www.moles.org/ProjectUnderground/drillbits/5_12/2.html]

Lama, Abraham. "Indigenous Peoples, the Invisible Victims of War." NativeNet, March 26, 1995. [http://nativenet.uthscsa.edu/archive/nl/9503/0346.html]

"Mercury Spill Poisons Villagers Near the Yanacocha Mine in Peru." *Drillbits and Tailings* 5:11 (June 30, 2000): n.p. [http://www.moles.org/ProjectUnderground/drillbits/5_11/1.html]

"Mining in Peru." Oxfam America. Global Programs: South America. N.d. [http://www.oxfamamerica.org/advocacy/mining.html]

"Mobil Evaluates Next Step in Madre de Dios Basin." *Oil and Gas Journal* (April 24, 2000): n.p.

"Peru: Illegal Loggers' Invasion of Indigenous Community's Territory." *World Rainforest Movement Bulletin* 53 (December 2001): n.p. [http://www.wrm.org.uy/deforestation/logging.html]

"Peru Villagers Poisoned after Truck Spills Mercury." Reuters News Service, June 14, 2000.

"Settlers Killed in Peru Jungle Clash." Associated Press, January 18, 2002.

"Water Supplies Alleged Contaminated by Peru Gold Mine." Reuters News Service, September 28, 2001. [PlanetArk: http://www.planetark.org/dailynewsstory.cfm/newsid/12577/story.htm]

Witzig, Ritchie. "New and Old Disease Threats in the Peruvian Amazon: The Case of the Urarina." *Abya Yala News: The Journal of the South and Meso-American Rights Center* (n.d.): n.p. [http://saiic.nativeweb.org/ayn/urarina.html]

Witzig, Ritchie, and Massiel Ascencios. "Urarina Survival Update: Continued Resource Exportation and Disease Importation by Foreigners and Newly Initiated by Multinational Oil Companies." *Abya Yala News: The Journal of the South and Meso-American Rights Center* (n.d.): n.p. [http://saiic.nativeweb.org/ayn/urupdate.html\]

PHILIPPINES

INTRODUCTION

The Philippines, a chain of 7,000 islands, hosts a wide variety of indigenous peoples, many of whose traditional ways of life are being forced to the margins of survival by intrusions of various extractive enterprises. On Luzon, the largest of the islands, toxic tailings spills from copper mines have caused floods that extensively damaged several villages. Hydroelectric dams have threatened to flood native peoples' lands on Luzon and other islands, as growing urban areas seek new sources of power. The Cordillera region of Luzon Island contains some of the world's largest gold reserves, with a familiar range of environmental problems afflicting native peoples, who, as in other areas, find themselves suffering environmental problems but enjoying none of the wealth that is generated by mining of their lands. "We are sitting on gold, but where is this gold?" asked an Ibaloi elder in Benguet, as he surveyed poverty in a land of abundance ("New Gold Rush," 1996).

THE MARINDUQUE ISLANDERS: COPPER-MINING TAILING SPILLS

During March 1996, the Philippines experienced its worst industrial accident, as toxic spills from a copper mine caused widespread floods and extensive damage to a number of villages on Luzon Island. Mining has been controversial on Luzon since the early 1990s, when the people of Itogon, in the Cordillera region of the island, first protested the open-pit mining of the Benguet Corporation.

On March 24, 1996, Marcopper Mining Corporation's copper mine in Boac, Marinduque, in the southern Tagalog region of Luzon (about 100 miles south of Manila), released toxic mine tailings from the Tapian Pit (holding 23 million metric tons of mine wastes) at the rate of 5 to 10 cubic meters per second into the Makulapnit and Boac rivers. Before the leak was staunched, roughly 1.5 million cubic meters of mining sludge had coursed into the two rivers ("Marcopper," 1996). The site was mined between 1969 and 1996 by Placer Dome (then known as Placer Development) and the Marcopper Mining Corporation. In 1997, Placer Dome abandoned the mining project, selling its 39 percent share in Marcopper to a local mining company, holding it responsible for the unfinished cleanup.

The waste spill quickly caused flash floods that isolated five villages, including 4,400 people, along the Boac River. According to first-person accounts, one village, Barangay Hinapula, "was buried under six feet of muddy floodwater and 400 families had to flee to higher ground. Their sources of drinking water were contaminated while fish, freshwater shrimp, and pigs were killed" ("Marcopper," 1996). Twenty other villages were evacuated. A 27-kilometer stretch of the Boac River was declared dead by government officials. Within three weeks, the Philippine Department of Health issued a report that said some area residents had unhealthy levels of zinc and copper in their blood. According to a report from the scene, indigenous residents of the affected area "also complained of skin irritations and respiratory problems which could have been caused by the poisonous vapors emitted by hydrogen sulfide and nitrous oxide from the mine wastes" ("Marcopper," 1996).

The threat of a major toxic spill from the former Marcopper mine has not ended. According to a report in *Drillbits and Tailings* (April–May 2002), Placer Dome commissioned reports to assess the safety of the mine's infrastructure; the reports said that the Tapian Pit, which leaked in 1996, and the Maguilaguila siltation dam, which burst in 1993, "will both cave in and spill tons of toxic wastes in already devastated Marinduque" ("A Disaster Looms," 2002).

A U.S. Geological Survey report, "An Overview of Mining-Related Environmental and Human Health Issues, Marinduque Island, Philippines, May 12–19, 2000," said:

> [the] high rate of sediment transport from Marcopper will continue to have adverse effects on the aquatic ecosystem, and on the ability of the river system to handle large flood events . . . the fine-grained, metal-rich, and potentially acid-generating nature of the sediments from Marcopper is likely to have been a substantial change from the natural condition of the Mogpog prior to mining. For example, fine-grained sediment from the mine site may fill in the pore spaces of the originally coarser river-bed sediments, thereby adversely affecting the habitat fish and aquatic invertebrates living on the river bottom. (U.S. Geological Survey, 2000)

The U.S. Geological Survey report also criticized a plan to use submarine tailings disposal (STD) to remove waste by flushing it into the ocean. The Geological Survey said that STD would create "a highly acidic, metal-enriched and environmentally detrimental plume" ("Philippine Province," 2002).

The same report stated that "little or no fish and invertebrate aquatic life was visible in the Boac River in its middle stretches . . . local residents also told us that there are periodic fish kills in the lower (presumably estuarine) portions of the Boac River" ("New Digging," 2001). The toll of the spill was still evident four years later.

The 1996 spill could repeat itself, given conditions in the area. The people of Marinduque, who have opposed the mine for 30 years, find themselves in the cross hairs of a disaster waiting to happen. According to Placer Dome's own experts, only the timing remains an issue. In addition, during 2001, engineers for the U.S. Geological Survey said that the waste dams were "virtually certain" to collapse again, placing 100,000 people downstream at risk ("Philippine Province," 2002).

"In December 2001, without warning or consultation, Placer Dome Technical Services pulled out of the Philippines, leaving behind toxic mine tailings in the Boac River, the threat of five dangerously unstable mine structures, and the incomplete compensation of Marinduqueos affected by the 1996 spill," said Ted Alcuitas of MACEC, a Philippines-based organization, and MiningWatch Canada's Research Coordinator Catherine Coumans at Placer Dome's annual shareholders' meeting in 2001 ("A Disaster Looms," 2002). The company contends that the fact that it has departed the site means it is no longer liable for past or potential environmental damage to indigenous lands and peoples.

During January 2002, Philippine President Gloria Macapagal-Arroyo led an official delegation to address Canada's prime minister with a demand that Placer Dome, a Canadian company, clean up the damaged river, repair leaking dams at the mountainous mine site, and provide compensation to people in Marinduque Province whose livelihoods were nearly obliterated by the Marcopper disaster (Jaimet, 2002; NikiForuk, 2002).

THE IBALOI: DAMMING A SACRED RIVER

Elsewhere on Luzon Island, the Agno River has long been regarded as sacred by the indigenous Ibaloi people of Benguet. Known as their cultural heartland, the river valley has hosted the small mines, farms, and homes of the Ibaloi for at least five centuries. The Ibaloi tend rice terraces and orchards. The people also pan for gold in the river and raise cattle in the mountains. Their way of life will come to an end if plans are realized to build the San Roque dam, which, if constructed, will connect two mountains across a deep gorge traversed by the Agno River. The dam, under construction in Central Luzon about 400 kilometers north of Manila, will permanently change the Agno River. The 200-meter-high dam is expected to generate 345 megawatts of power, making it the tallest dam and one of the largest private hydropower projects in Asia. The dam project eventually could significantly affect the lives of more than 20,000 indigenous people.

Plans for the dam state that it will irrigate 87,000 hectares, control floods, and improve water quality downstream. Electricity generated by the dam would be primarily used to power industrial activity and the burgeoning mining industry in northern Luzon. The price for this progress will be paid by the 35,000 indigenous Ibaloi, Kankanaey, and Kalanguya peoples living upstream, who are fiercely opposed to the dam. They believe it will destroy their communities and livelihoods. The indigenous peoples "are concerned that high rates of sedimentation from mining and other land use in the watershed area will lead to increased flooding upstream of the reservoir, inundating their homes and burial sites and negatively impacting water quality" ("San Roque Dam," 2000).

A Philippine legislative delegation visited Tokyo during November 2000 in a bid to stop Japanese funding for the San Roque Dam. Loans for the $1.19 billion project have been provided by a consortium of Japanese and U.S. banks which own the San Roque Power Corporation. "Our last resort to protect our livelihood from the dam is to tell the Japanese people about our fears," Philippine Congressman Ronald Casalan told the press in Tokyo. "I am asking whether the Japanese public is satisfied that their own tax money is used this way—to destroy the environment and cause harm to indigenous people," he said (Kakuchi, 2000). The dam was about 40 percent finished by the end of 2000.

Construction for the project started in May 1998 and (as of 2002) was scheduled for commercial operation by the end of 2004. In the meantime, 1,800 Ibaloi people displaced by two existing dams upstream on the Agno River in the mountainous Cordillera region of Luzon Island—the Ambuklao and the Binga (built between 1954 and 1961)—still were awaiting promised compensation. Some of them had been waiting for more than 40 years. According to observers, "Many have witnessed their lands being consumed by increasing sediment loads backed up behind the dams. They fear the same will happen if the San Roque Dam is built" ("San Roque Dam," 2000). The manager of the Upper Agno Hydroplant Complex, Rene Rivera, asserted that some of the people displaced by the two older dams had been adequately compensated. Rivera told a reporter for the *Baguio (Luzon) Sun-Star* "that there were payments already made to about 240 affected families during the 1950s and 1960s with regards [to] the implementation of the projects" (Cruz, 2001). Many others remain uncompensated, however.

By 2000, more than 600 indigenous families had been evicted to make way for the San Roque dam. Many were struggling to survive in cramped quarters at a resettlement site with no land to sustain themselves, despite guarantees by the Japan Bank for International Cooperation, a primary financier of the project. The lives of another 300 or more soon-to-be-evicted families were being disrupted by excavation for the dam. According to local media accounts, about a quarter of the evicted families were forcibly displaced during early 1998. They lived for almost a year in desperate conditions at a temporary site. They were promised land, houses, alternative livelihood sources, and social services, but instead they were paid only 10,000 Philippine pesos (about U.S. $200 in 2002) per family as compensation. Only in late January 1999 were 147 houses in the new resettlement site handed over to the displaced families. Another 402 families in Pangasinan were required to relocate before the project's completion ("Philippines: Local People," 2001).

A local Internet Web site asserted that the Agno River watershed was being seriously damaged not only by dam construction, but also from erosion provoked by commercial vegetable farming and mining operations. According to advocates for the Ibaloi, "The authorities . . . choose to blame Ibaloi farmers for the erosion. They are planning tree plantations on Ibaloi farms to protect the dam while allowing the more destructive commercial gold and copper mining to continue. Ironically, the San Roque dam will only replace the power lost from the Ambuklao dam which is already useless, having become choked with silt and cracked by earthquakes. The expected life of the new dam may be as little as 28 years" ("Valley of the Dammed," n.d.).

Approximately 4,000 residents, municipal and local officials, and members of the Central Luzon and the Cordillera Peoples Alliance assembled on September 30, 2002 in San Nicholas, Pangasinan for a rally calling for the stoppage of the San Roque dam project ("Philippines: Local People," 2001). Opponents of the dam asserted that "besides being a burden to the Filipino taxpayers, the $1.2 billion dam will only serve the energy needs of the foreign mining companies [that] are out to exploit their natural resources. The project also violates the indigenous peoples' and farmers' rights over their lands" ("Philippines: Local People," 2001). One activist said at the rally that the San Roque dam "will bring hydropower to big business but misery to the people of Dalupirip" (Sarfati, 1998).

Local people said at the rally that the dams are eradicating a gentle way of life. "The Ibalois are peace–loving people. They lived on rice and tapioca, kamote [sweet potato], and the various creatures they caught from the Agno River. They used to hold a *caniao* [ritual] each March to celebrate a bountiful harvest. "We live this way," an elder said. "As long as one made use of the land and the river, without abusing it, one would eat. But more than that, one would live" (Sarfati, 1998). Elders described traditional Ibaloi society, where food is exchanged frequently between households. No child grows hungry in Dalupirip, they said. Every child is the son or daughter of the whole village. And the village is self-sufficient. During a *caniao* or any other ritual, the whole village participates. Everyone helps gather firewood, cook, and clean. In times of need, like when one is sick or a member of the family dies, the whole village helps (Sarfati, 1998).

"Dalupirip is incomparable," said Nanang Patricia, a local resident. "Here, no one steals, no one fights, no one is killed on the streets. It's a very safe place to live in. The food we need, we get from the land and the river. The help we need, we get from each other. In the city, life is different. People do not care enough for each other. There is no room for our ways, our culture and beliefs there. We want our children and grandchildren to grow up here" (Sarfati, 1998).

Aside from those areas to be directly submerged, other households in higher-lying areas could find their traditional economy ruined because of silt accumulation from the damming of the Agno River. The river already is badly silted and eroded due to mining in Itogon and Tuba. Dam construction will supply power to the lowlands, as well as to several new mining ventures expected to move into the Cordillera region. One of the mining ventures was being initiated by Newmont, the biggest gold mining company in the United States, which has filed an application to explore the area around the San Roque dam-site. Sixteen other mining firms also have also filed applications in the Cordillera region. Mining, within a few decades, will replace the traditional peoples and their way of life: the rice paddies, swidden (slash-and-burn) fields, fruit trees, pastures, forest wood (for fuel), gold panning, and fishing sites along the Agno River.

Opponents of the San Roque dam worry because the Philippines, and Luzon Island in particular, lie on the seismically active Pacific Rim, with potential for major earthquakes. Hydrologist Peter Wiling has said that "either an earth-quake or a flood could totally destroy the dam" (Bengwayan, 1999). Wiling said the reservoir was designed to contain a only relatively small flood expected to occur every five years. "It cannot contain larger floods which are expected to be devastating" (Bengwayan, 1999). Any breach of the dam could send cascades of water laced with heavy metals and other toxic materials through down-stream communities. The critics assert that earthquakes could endanger the lives not only of roughly 20,000 tribal Ibaloi people but also those of some 1.5 million residents in 78 towns of the provinces of Tarlac, Pangasinan, and Nueva Ecija located below the dam (Bengwayan, 1999).

In a report made public July 24, 1998, an International Rivers Network (IRN) panel, which had lobbied against the dam, concluded that "irresponsible releases of flood waters [under monsoon flooding downstream of the dam] will result in a severe flooding of the whole Pangasinan and most of Tarlac plains. Inattention by operators on the rising reservoir water level could also result in the breaching of the dam, its eventual loss, and the consequent catastrophic flooding downstream" (Bengwayan, 1999). Northern Luzon averages one or two typhoons a year, with torrential rains and winds.

Geochemist Sergio Feld, a member of the IRN panel, warned that, given increasing mining in the area, "the dam has the potential of becoming a tail-ings dam of silt and sedimentation laced with deadly dissolved chemical con-stituents like mercury, lead, selenium, cadmium, molybdenum, zinc, arsenic, copper, cyanide and even uranium which comes from several mines operating above the damsite" (Bengwayan, 1999). Another member of the IRN panel, seismologist Tizanio Grifoni, contended that existing environmental assess-ments of the dam do not examine damage that could be caused by earthquakes. "The dam has not been designed to withstand earthquakes," he said. "An earthquake could induce seiche waves that could erode the banks, induce land-slides and cause the dam to break, leading to the loss of the dam altogether" (Bengwayan, 1999).

GOLD MINING AMIDST POVERTY IN LUZON'S CORDILLERA

"We are sitting on gold, but where is this gold?" asked an Ibaloi elder in Benguet, as he surveyed poverty in a land of mining abundance ("New Gold

Rush," 1996). The Cordillera region of Luzon Island in the Philippines is endowed with rich mineral resources. The Cordillera (particularly the area near Benguet, which contains much of the Philippines' mineral wealth) is one of the richest mining lodes in the world. Measured in the amount of minerals taken from a given piece of land, the Philippines is the world's second-richest producer of gold (404 pounds per square kilometer), the world's third-richest producer of copper (0.75 pounds per square kilometer), and the world's sixth-richest producer of chromate (0.57 pounds per square kilometer) ("New Gold Rush," 1996). Despite the Cordillera's mineral wealth, however, most of the area's indigenous people continue to endure intense poverty.

Following approval of the Mining Act of 1995, foreign investment in Philippine mining surged. The Cordillera Peoples Alliance opposed the new act, arguing that it served only to surrender the Philippines' sovereignty to foreign corporations. "The new Act underlines the [Fidel] Ramos government's brazen disregard for indigenous peoples' rights and welfare, as the mineral resources being offered to foreign corporations are mostly located in the territories of indigenous peoples," said the Alliance ("New Gold Rush," 1996).

The new mining law modified requirements that had required 60 percent Philippine ownership of mining operations. Under the new law 100 percent foreign ownership was allowed for tracts up to 81,000 hectares after 25 years of operation, providing that at least U.S. $50 million was invested ("New Gold Rush," 1996). The new law also doubled depreciation rates and granted a five-year tax holiday that could be extended for another five years. The new law also allowed repatriation of profit and capital in U.S. dollars, tax-free treatment of capital investments, and assignment of rights that guaranteed unhampered mining operations. Under this law, foreign investors also were granted control over water and timber rights, as well as mineral rights.

In addition to local indigenous peoples, the National Council of Churches of the Philippines also condemned the new law, asserting that a large amount of new mining activity would provoke thousands of indigenous families to be evicted from mining areas. In the meantime, applications for new mine sites burgeoned, with Australian companies in the lead. Indigenous peoples on Luzon complained that the central government, in its rush to expand the Philippines' industrial base, was running roughshod over their inherent rights to ancestral lands, as well as their future prospects for economic prosperity and social development. Thus, they argued, the new mining law disrespected indigenous cultures, indigenous peoples' rights to pursue their cultural development, and political integrity.

Indigenous peoples in the Cordillera reacted skeptically to the new mining law because they already had endured a century during which their lands and rights had been bargained away by the central government in favor of development by foreign companies. In Itagon, for example, nearly all of the rivers had dried up or been contaminated by the Benguet Corporation's open-pit mining operations. Clean drinking water had become scarce, and many rice fields had been abandoned because irrigation water was no longer available. Open-pit mining was encouraged by the new law because it speeded the pace of mining and subsequent profits, while paying little regard to environmental consequences. The new law also broadened companies' easement powers, speeding eviction of indigenous peoples whose homes stood in the way of expanding open-pit mines and infrastructure, such as roads, which are required for industrial-scale mining.

According to indigenous opponents of increased mining on Luzon, the Mining Act of 1995 "threatens to wipe out the very existence of the Cordillera indigenous peoples. For the Cordillera indigenous peoples, land is the source of life—from the land comes the materials needed for their production, and thus, their very sustenance. From the land emanates, and revolves, their collective culture and spiritual life. To destroy their land is to wipe out this web of life, and the very peoples themselves" ("New Gold Rush," 1996).

MINDANAO'S LUMADS: LOGGING, MINING WASTES, AND EVICTIONS

On Mindanao, the southernmost large island of the Philippines, widespread industrial-scale logging and mining have provoked opposition from indigenous peoples who are collectively called Lumads. "Lumad" is a Bisayan term meaning "born of the earth," and thus native or indigenous ("Situation," n.d.). The Lumad population of Mindanao, which is estimated at 2.5 million, comprises about 20 percent of Mindanao's total population.

Philex Corporation security forces reportedly have evicted indigenous peoples from the chromate mines on Mindanao. The Blue Guards of Philex have harassed small-scale miners and environmentalists opposed to its entry with strafing; one man died in a strafing incident. The strafing occurred during Philex's forcible takeover of mine tunnels operated by small-scale miners. In other instances, indigenous peoples have been forced off their lands following construction of roads and importation of grazing animals that ravage their crops. When their land became worthless for agriculture, indigenous occupants were forced to sell it at a low price ("Situation," n.d.). Mine wastes from Philex operations in western Mindanao also have caused the deaths of fish and shellfish in Morsellagos Bay.

In southern Mindanao, indigenous peoples in Pantukan and adjacent towns in Davao del Norte have been faced with relocation due to the Kingking Project, a joint mining venture of Canadian-owned Echo Bay Mines, Toronto Ventures (Pacific) Incorporated (TVI), and Benguet Mining Corporation ("Situation," n.d.). Elsewhere in Mindanao, Echo Bay/TVI and Benguet undertook a feasibility study on the Kingking's Pantukan gold-copper concession and began exploration, with plans to extract 3.8 million ounces of gold and 2.1 billion pounds of copper. Even at the exploration stage, the Kingking Project caused silt buildup in the Kingking River, as it triggered mudslides and flash floods. In June 1996, coconut trees, houses, and a church were buried by flash floods laced with mining wastes. A group of local government officials discovered that mining wastes had been dumped into ravines and gullies, then carried into rivers and the sea by seasonal monsoon rains. Road construction required by mining also caused deforestation in the area, aggravating the effects of flooding rains ("Situation," n.d.).

Flooding in the same area has been aggravated by deforestation, as well as open-pit mining. At the beginning of the 1950s, the Philippine Islands contained about 100,000 square kilometers of primary forests, an area that had shrunk by 90 percent, to 10,000 square kilometers, by 1988. By the year 2001, logging companies had invaded some of the last semiautonomous tribal areas of the islands, in some instances clear-cutting what remained of once-lush forests.

Industrial-scale logging in several areas of Mindanao also has met resistance from indigenous peoples. The Yagatibo Lumads, for example, have warned that should corporate logging operations continue unabated, the fight against the loggers "will become bloody" ("Situation," n.d.). As logging increases, many Subanen Lumad people have been restricted from using the resources of the forest such as rattan, firewood, and lumber for their dwellings. They also have been constrained from *pangayam* (hunting activities). Tree farms' neat ranks of softwood and hardwood trees have replaced indigenous fruit-tree orchards in some areas.

Sometimes establishment of tree farms has been carried out under the guise of biodiversity projects. Lumad leaders have been lured with offers of attractive salaries to become forest rangers before the land is logged ("Situation," n.d.).

In southern Mindanao, the Umayamnon and Ata-Manobo Lumad peoples of Kapalong, Davao Oriental, and Loreto in Agusan del Sur have declared a total ban on logging in their ancestral domains. While they have enforced the ban among members of their tribes, two giant logging companies have entered their territories with intentions to take lumber anyway. Eight native people who actively opposed logging operations have been killed by company guards since the Santa Ines Melale Corporation (SIMCOR) commenced operations in 1985. A local observer commented that "having organized themselves in 1994, the Umayamnon/Ata-Manobo are "mustering the courage to speak out against these atrocities." These Lumads have been said to be "seeking justice and indemnification for the victims of gruesome murders inflicted on their tribespeople" ("Situation," n.d.).

In some areas of Mindanao, forests have been replaced by cattle ranches. For example, during the late 1990s, the Western Mindanao Livestock Development Center in Bayog, Zamboanga del Sur imported about 1,240 cattle from the United States. The firm then established a large, modern stock farm. This project was developed on the lands of the Subanen Lumads, roughly 1,900 hectares in Barangay Bantal, Pulang Bato, Liba, Cagayan, and Bayog in Zamboanga del Sur. Some Lumad leaders in these areas were paid monthly salaries in exchange for their support of the project. A majority of the village residents, however, remained opposed to the project. Many of them were forced from their lands by the stock farm ("Situation," n.d.).

The Philippines' central government has been promoting logging with a so-called industrial tree development plan that allots 500,000 hectares of open and forest lands to local and foreign private companies, in the national interest of alleviating the country's dependence on imported timber. Affected areas include the Cagayan Valley, the Cordilleras, northern Mindanao, and Caraga, all of which are home to considerable tribal populations. According to the *World Rainforest Movement Bulletin*, "This kind of development projects follow a cycle of dispossession and violence, with terrible costs for local people. The pattern is always the same: the soldiers come in first, sow terror in order to displace the locals or discourage opposition, then the project gets implemented" (Carrere, 2001).

The Philippines' Legal Rights and Natural Resources Center described how an entire Banwaon Lumad village was forced from its lands and homes. In other instances, Filipino army soldiers have seized and tortured indigenous leaders who are known for their outspoken defense of indigenous land rights, under a pretext of searches for separatist guerillas. For example:

> On July 21 [2001], in San Luis, soldiers tortured Lolong Badbaran and Eddie Badbaran, both Banwaon rattan cutters. They also detained Dino Rueda, a motorcycle driver; and 60-year-old Linda Loyola, who had witnessed the torture of the Badbarans. The four were tied to the posts of a hut and were not released until the next day. "The government knows that the Banwaon people will strongly oppose the loss of more tribal lands," said Otto Precioso, a Banwaon leader. "That is why they are now terrorizing the communities." (Carrere, 2001)

Similar experiences have been reported by members of the Ata-Manobo Lumad tribe in Talaingod, Davao del Norte, where thousands of hectares of ancestral lands have been usurped by the plywood firm C. Alcantara and Sons, and converted into tree plantations (Carrere, 2001). The Lumads decided to fight back, declaring a *pangayaw* (war of vengeance) in 1994 after hundreds of native peoples were forced from their homes. The government responded by moving troops to the area to act as de facto security guards for the plywood company. Local people believe that the new tree plantations will aggravate the situation.

MINDORO ISLAND'S MANGYAN, ALANGAN, AND TADYAWAN PEOPLES: NICKEL AND COBALT MINING

Mindex, a Norwegian mining company, has confronted indigenous communities on the Philippine island of Mindoro with a proposed nickel and cobalt mine. According to a report in *Drillbits and Tailings*, the Mindoro Nickel Project plans to strip mine 24,000 acres (9,700 hectares) in the central mountains of Mindaro, southwest of Luzon Island. Ore from the mine will be transported 26 miles (43 kilometers) by pipeline to Pili on the eastern side of the island, where Mindex plans to build a plant that will process 40,000 tons of nickel and 3,000 tons of cobalt each year ("Mindex Moves," 1999).

Following indigenous protests on the island, Philippine President Gloria Macapagal-Arroyo in 2001 denied a $30 billion peso (U.S. $600 million) nickel-mining project in Mindoro by Crew Development, citing environmental concerns. Environment Secretary Heherson Alvarez said that Macapagal-Arroyo concurred with the Department of Environment and Natural Resources' assessment that the mining operations of the Canada-based Crew Development Corporation would inflict "irreparable damage to the environment" ("Hotspots: Philippines," 2001).

According to the same report in *Drillbits and Tailings*, "The President sustained our observation that the Aglubang mining project of Crew transcends the threshold of sustainable development because of technical and social evidence." According to Catherine Coumans of MiningWatch Canada, the Department of Environment and Natural Resources in July canceled a permit given to the mine by the previous administration, a first in the Philippines ("Hotspots: Philippines," 2001).

Indigenous peoples of the area, including the Mangyan, the Alangan, and the Tadyawan, joined with Catholic Church groups, human rights and environmental activists, and farmers to form an alliance called ALAMIN, which played a key role in stopping the mining project. According to *Drillbits and Tailings*, "They have organized demonstrations, submitted formal protests to the authorities, and have collected nearly 25,000 signatures against the project" ("Mindex Moves," 1999). Contrary to the rather obvious protests of the local people, Mindex asserted that it faces no social, environmental, or technical challenges to the project and that the local populations welcome it. "We refute the categorical statement of Mindex that the local populations of Oriental Mindoro welcomes the mining project unconditionally. Thus, we present our policy considerations and moral positions in our unified opposition against Mindex mining operations," said ALAMIN in a letter ("Mindex Moves," 1999).

The ALAMIN group expressed concerned that Mindex's proposed mine would remove the remaining forests on the island, causing flooding and erosion, as well as pollution. The group also asserted that the pipeline, which is proposed to run underground except at river crossings, will be constructed in an earthquake-sensitive area. Any leaks or spillage from the pipeline will impact Naujan Lake downstream.

Norwatch, a Norwegian nongovernmental organization that monitors Norwegian companies, reported that the company had planned to discharge more than 4 million tons of waste from ore processing into the ocean each year. This process, known as submarine tailings disposal (STD), is illegal in the United States and Europe. The waste will contain metals and sulfuric acid from the so-called high-pressure acid leach method that the company plans to use to extract the nickel and cobalt from the ore. Norwatch asserted that the analysis of the ore from Mindoro indicates high levels of other metals (including chrome and smaller amounts of copper and zinc) in the ore. "I was born and raised here, and I've made a living from fishing since I was a young boy. Here we also have deep-water fish that is valuable for us. If they discharge millions of tons of waste every year, we are bound to lose our way of income," said Janito Palermo, a 75-year-old fisherman from Barangay Pili ("Mindex Moves," 1999).

The company said that at least 20 families would have been relocated to accommodate the mine. According to ALAMIN, the Mangyan, Alangan, and Tadyawan tribal land claims have been registered as Certificate of Ancestral Domain Claims (CADC). Norwatch reports that a section of the CADC states that the Mangyan have "priority rights in the harvesting, extraction, development or exploitation of any natural resources contained within their ancestral domain" ("Mindex Moves," 1999).

A MINING BAN AT ORIENTAL MINDORO

The Philippine Province of Oriental Mindoro during early 2002 placed a 25-year moratorium on all major mining projects in the region. The ordinance declared that it is "unlawful for any person or entity to engage in land clearing, prospecting, exploration, drilling, excavation, mining, or transport of mineral ores preparatory to all forms of mining operations for a period of 25 years" ("Philippine Province," 2002).

The new law blocked the Crew Development Corporation, a Canadian mining firm, from developing a large-scale nickel- and cobalt-mining project in the area. In July 2001 the central government canceled Crew Development's permit to develop the mine. Crew appealed the decision in November 2001, but the government upheld its decision. The Catholic clergy and local environmental, fishing, and indigenous peoples' organizations asserted that Crew's plan to build the mine and dispose of its wastes would endanger important watersheds, agricultural lands, and fishing grounds, as well as the welfare of local communities ("Philippine Province," 2002). Collapse of a copper-mine tailings dam at Placer Dome's Marcopper mine on the island of Marinduque in 1996 (see above) was cited as a major reason for the mining ban.

REFERENCES

Bengwayan, Michael. "Luzon Dam Plan Riddled with Flaws, Fault Lines." Environment News Service, October 5, 1999. [http://ens.lycos.com/ens/oct99/1999L-10-05-01.html]

Caluza, Desiree. "Ibaloi Leader of Opposition to San Roque Dam Dies." *Philippine Daily Inquirer* (Manila), October 5, 2001. [http://www.inq7.net/reg/2001/oct/05/reg_8-1.htm]

Carrere, Ricardo. "Philippines: Planting Trees and Terror." *World Rainforest Movement Bulletin* (September 2001): n.p. [http://www.wrm.org.uy]

Cruz, Cheryl G. "Runaround Delays Ambuklao, Binga Dams Damages Claims." *Baguio (Luzon, Philippines) Sun-Star*, August 10, 2001. [http://www.sunstar.com.ph/baguio/08-10-2001/topstories2.html]

"A Disaster Looms for Communities in Marinduque, Philippines." *Drillbits and Tailings* 7:4 (April–May, 2002): n.p.[http://www.moles.org]

"Hotspots: Philippines." *Drillbits and Tailings* 6:9 (November 30, 2001): n.p. [www.moles.org]

Jaimet, Kate. "Placer Dome Blamed in 'World-calibre Disaster'; Canadian Mining Giant says Philippine Government Blocking Cleanup." *Ottawa Citizen*, January 29, 2002, n.p.

Kakuchi, Suvendrini. "Protests Hold Back Philippine Dam Project." Interpress Service, Corporate Watch in Japanese [cwj@corpwatch.org], November 28, 2000. [http://www.jca.ax.apc.org/web-news/corpwatch-jp/132.html]

"The Marcopper Toxic Mine Disaster: Philippines' Biggest Industrial Accident." N.p., 1996. [http://www.tigerherbs.com/eclectica/earthcrash/subject/mining.html]

"Mindex Moves in on the Peoples of Mindoro." *Drillbits and Tailings* 4:18 (November 9, 1999): n.p. [http://www.moles.org/ProjectUnderground/drillbits/4_18/4.html]

"New Digging: Placer Dome, Inc. Returns to Exploit Another Area of Indonesia." *Kerebok* 2:10 (May 2001): n.p. [http://www.jatam.org/xnewsletter/kk10.html]

"The New Gold Rush: The 1995 Philippine Mining Act Lures a New Wave of Profit-hungry Gold Diggers." KASAMA (Solidarity Philippines Australia Network) 10:2 (April–May–June, 1996): n.p. [http://www.cpcabrisbane.org/Kasama/V10N2/Gold Rush.htm]

NikiForuk, Andrew. "Still a Fine Mess; The Controversy over a Philippine Mine Cleanup Rages On." *Canadian Business*, February 22, 2002, n.p.

"Philippine Province Bans Mining for Twenty-five Years." *Drillbits and Tailings* 7:2 (February 28, 2002): n.p. [http://www.moles.org]

"Philippines: Local People against the San Roque Dam." *World Rainforest Movement Bulletin* 42 (January 2001): n.p. [http://www.wrm.org.uy/bulletin/42/Philippines.html]

"San Roque Dam, Agno River, Philippines." International Rivers Network, 2000. [http://www.irn.org/wcd/sanroque.shtml]

Sarfati, Gigi. "The Dam and the Women of Dalupirip." *KASAMA* (Solidarity Philippines Australia Network) 12:2 (April–May–June 1998): n.p. Reprinted from *CHANEG* (September–December 1997), published by the Cordillera Women's Education and Resource Center. [http://www.cpcabrisbane.org/Kasama/V12n2/Dalupirip.htm]

"The Situation of Indigenous Peoples in Mindanao Culled from the Regional Reports of Lumad Organizations and Support Groups in Southern Mindanao, Far Southern Mindanao, Caraga Region and Western Mindanao." N.d. [http://www.mindanow.com/text/articles/people/intro_people.htm]

U.S. Geological Survey. "An Overview of Mining-Related Environmental and Human Health Issues, Marinduque Island, Philippines: Observations from a Joint U.S. Geological Survey–Armed Forces Institute of Pathology Reconnaissance Field Evaluation, 12–19 May 2000. Washington, D.C.: U.S. Government Printing Office.

"The Valley of the Dammed: The Ibaloi People's Struggle Against the San Roque Dam; Ibaloi Cultural Heartland under Threat, Philippines." N.d. [http://www.philsol.nl/B99/San-Roque-Dam-1.htm]

RUSSIA (SIBERIA)

INTRODUCTION

Siberia, part of the largest land mass on Earth, is home to many native peoples who share one major problem: the Russian State's appetite for resources—oil, natural gas, uranium, and others—which lie on or under their lands. From the Yamal Peninsula on the Arctic Circle northeast of the Ural Mountains to Sakhalin Island, north of Japan, indigenous peoples of Siberia have been confronted with large-scale development of oil and gas, uranium, and other resources that has been carried out largely heedless of environmental contamination.

Twenty-six distinct indigenous peoples live in Siberia, ranging in numbers from fewer than 200 (the Oroks) to as many as 34,000 (the Nenets); their total population is more than 160,000. Two other, larger indigenous peoples, the Sakha (formerly called the Yakuts) and the Komi, have their own republics within the Russian federation. Most of the indigenous peoples who live in the tundra rely on reindeer herding, while those who live in the forest and near the sea subsist on hunting and fishing. The reindeer herders are nomadic, following the reindeer in a cyclical pattern of migration, living in movable houses made from reindeer skins (*urangas*). The hunters live in permanent settlements (which traditionally have been made of bark) or in semisubterranean huts, covered with earth and moss. Today, roughly 10 percent of the indigenous peoples continue to migrate (compared to 30 years ago, when 70 percent migrated); most of the others live in settlements, with almost half the population involved in herding, fishing, and hunting ("Peoples of the Frozen North," n.d.).

THE EVENK AND THE KHANTY: OIL AND REINDEER DON'T MIX

The Khanty and affiliated Evenk have inhabited northwestern Siberia's forests and swamps for thousands of years, maintaining their traditional ways of life, which includes hunting and fishing. They also herd reindeer, the basis of their traditional economy, used for food and transportation. Traditional Khanty families live in *chums*, which are similar to North American Indian tipis, as they herd reindeer in winter pastures. The Khanty inhabit land in the West Siberian Taiga, north of the middle Ob on the river Pim. For centuries, reindeer breeders and fishers knew how to sustain themselves within this barren land, how to withstand the chilling cold of wintertime and swarms of mosquitoes in the summer. Now their traditional way of life is in danger due to exploitation of fossil fuels ("Reindeer Herders," 1997).

Oil was found in the area first during the 1960s; by the late 1990s, according to one account, "derricks, roads, pipelines and the workers' estates are eating their way through the land of the Khanty people" ("Reindeer Herders," 1997). The importation of oil drilling is antithetical to the reindeer culture. In the Khantys' homeland, oil spills blacken the wetlands, raised roads trap water causing flooding and ruining the forests, and fires caused by oil workers' carelessness and petroleum-soaked debris send columns of smoke into the air.

Any student of the Plains Indians and the buffalo during the late nineteenth century in North America will recognize parallels in the Vershina Khandy's people's description of changes in the land endured, nearly a century later, by her band of Siberia's Evenk people, who are neighbors of the Khanty. The following sounds very much like the nineteenth-century changes across the U.S. Great Plains. Whether due to the thrust of young industrial capitalism across North America or the final colonizing gasp of the so-called New Soviet Man, the outcome was similar for indigenous peoples' environments:

> The ancestral lands of the Vershina Khandy [Upper Khanda] Band of the Evenk were very strongly affected in the late 1970s and early 1980s by the construction of the Baikal-Amur Railroad. Migrations of wild ungulates from northern territories into the Khanda River Valley for wintering first reduced and then discontinued altogether, as the railroad [was] built without taking into account the ecological peculiarities of the territory. In addition, the construction project attracted numerous people (tens of times more numerous than the local population), who started hunting and fishing. As a result, the Evenk found it much more difficult to gain their daily bread, since all their incomes used to come from hunting and fishing. Then, in the 1980s, along the entire eastern boundary of the Evenk traditional territory, the KI-450 Correctional Labor Colony Administration cut the forest clean, and roads were built in areas previously inaccessible, ones that had served as reserves. As a result, there was the Baikal-Amur Railroad and actively developed areas along it on the north and the easily accessible territory cut and burned by forest fires on the east. (Khandy, 2001)

Following these intrusions, only a 37-kilometer (23-mile) strip of the Khanda River Valley remained within the southern portion of the Upper Khanda Evenk Band's traditional lands, allowing fewer wild animals to pass freely to their winter feeding grounds, where the Evenk traditionally hunted them. When the RUSIA Petroleum Company built a road from the settlement of Magistralnyy to its gas field, this migration route was cut off. Therefore, the band's traditional lands were totally isolated from the migration routes of their food sources (Khandy, 2001).

The RUSIA Petroleum Company allocated some compensation for the harm it had done to the Evenks' traditional economy, but the funds were only sufficient to employ two Evenks as professional hunters. The company's executive dealing with environmental matters, F. T. Selikov, said that the Evenk had been offered relocation at the expense of the company to other population centers of the district, but they had refused to move. The Evenk understand that, by moving, their traditional way of life would be lost (along with ownership and use of the land that once supported it) as the Evenk would be forced to assimilate into the general population.

The response of the indigenous people sounded much like that of some American Indians on the plains of North America a century earlier. The Evenk had, in fact, been reading indigenous American history. "As for the professional hunter salaries, the opinion is unanimous," read one statement composed by a Khanty spokesman. "It is not even an attempt to solve the problem, but glass beads of the 17th- [and] 18th-century merchants and industrialists in a modern interpretation" (Khandy, 2001).

> If the condensate field development [of RUSIA Petroleum] begins reaching industrial proportions and no measures are taken, the Upper Khanda Band will cease to exist. Their southern neighbors, the Evenk band living in Vershina Tutury [Upper Tutura], Kachug District, Irkutsk Oblast, will not be left in peace either, if the field is developed and a gas pipeline is built. Being involved in public monitoring of the development of the Kovyktinskoye gas-condensate field as a member of the Baikal Environmental Wave, an Irkutsk regional non-governmental organization, I am trying to help the Evenk, but I am afraid that this will not be enough and therefore am asking you to step in with whatever assistance is possible. (Khandy, 2001)

The scope of Siberia's environmental problems were outlined by Alexi Yablokov, writing in the *Washington Post*. The southern and central Volga regions, Bashkiria, the central and southern Urals, the Kuzbass, the oil-producing region of western Siberia, the Lake Ladoga basin, and many other regions are officially classified as ecological crisis zones. In the past several years, environmental refugees have appeared in Russia. "People have fled such heavily polluted areas as Prokopyevsk, Nizhny Tagil, Kirishi and Angarsk, moving elsewhere in the nation in search of clear air to breathe and clean water to drink," Yablokov wrote. "Throughout the republic, according to 1989 analyses of fresh-water fish, 69 percent were extremely contaminated by mercury-based pesticides" (Yablokov et al., 1991, C-3).

Oil and gas drilling in northern Siberia has destroyed large expanses of reindeer pasturage on which indigenous peoples built their traditional economies. Millions of barrels of oil are lost each year due to spills and other accidents related to drilling and transportation of petroleum. A team of Russian ecologists commented: "We drill more and more oil, destroy more and more wilderness areas, deprive the local peoples of all possibility of supporting themselves, then turn around and spend the dollars we make selling oil abroad to buy food for those very same people" (Yablokov et al., 1991, C-3). The same group attributed much of Russia's environmental abuse to "the long supremacy of the totalitarian Soviet system with its innate hostility to humans and the world around them, [causing] three generations of citizens [to be] raised with a utilitarian, consumer approach to nature" (Yablokov et al., 1991, C-3).

In western Siberia, according to Survival International, pollution from oil and gas industries has polluted large areas of traditional indigenous lands. Huge flares burn off excess natural gas day and night. Oil also frequently spills into rivers, killing fish and plant life. Forests have been cut down and reindeer pastures have been devastated by industrial vehicles. Many important fish-spawning grounds have been destroyed. The Khanty, for example, watched in horror as their sacred spawning riverbeds were dug up to mine gravel. The Evenks' and Yukagirs' lands also have been contaminated by radiation from failed nuclear tests ("Peoples of the Frozen North," n.d.).

Andrew Wiget and Olga Balalaeva, two researchers who described oil development's impacts on the Eastern Khanty, summed up the situation this way: "Siberia, like the [American] Appalachian coal fields at the beginning of this century, has become a national sacrifice area" ("Putin's Oil," 2000). *Drillbits and Tailings* reported that the amount of oil spilled annually in Russia is equal to 350 accidents the size of the 1989 Exxon Valdez tanker disaster in Alaska, in which 40,000 tons were spilled ("Putin's Oil," 2000).

The Khanty and other indigenous peoples of Siberia are very aware of the fact that their land is being despoiled to enrich other people. "If you take 100 pounds of my gold, then why can't you leave me just two?" Yaloki Nimperov wrote in a letter to Senur Markianovich Khuseinov, director of the Kamynskoye oil field that belongs to the company Surgutneftegaz. Yaloki was angry, so angry that he has threatened some of the oil workers with his gun. He also was angry because local buses that carry oil workers never allowed Khanty to board until the indigenous people threatened to blow up a bridge. After that, the Khanty not only were allowed to ride the buses, but they were charged only half the regular price ("Reindeer Herders," 1997).

Between 1989 and 1997, 14 new production platforms and exploration bases were erected in the Khantys' indigenous lands, one of them only two kilometers from Nimperov's summer camp. Day and night he could hear its noise. Roads to the production platforms were laid with little regard for the needs of the Khanty or the local environment on which they depend. Some roads were built through marshes and lakes for several miles. Oil workers sometimes tried to correct for this by cutting culverts to allow water exchange between parts of lakes, but the culverts were too small to do much good. Water in the culverts often froze during the winter, killing fish for lack of oxygen. When the earth thawed in the spring, the area smelled of rotten fish. Some Khanty were paid compensation while others withdrew in disgust, turning their backs on the oil platforms, moving as far as possible into what remained of the countryside, seeking shrinking pastures on which to feed their reindeer.

A Khanty elder described an earlier, more pristine, time:

> I don't want anything, only my land. Give me my land back where I can graze my reindeer, hunt game and catch fish. Give me my land where my deer are not attacked by stray dogs, where my hunting trails are not trampled down by poachers or fouled up by vehicles, where the rivers and lakes have no oil slicks. I want land where my home, my sanctuary and graveyard can remain inviolable. I want land where I [will] not be robbed of my clothes or boots in broad daylight. Give me my own land, not someone else's—just a tiny patch of my own land. ("Peoples of the Frozen North," n.d.)

In the meantime, several once-vibrant tribes of reindeer herders across Siberia succumbed to alcoholism and suicide as their way of life was obliterated by oil exploitation. *Drillbits and Tailings* reported in its September 30, 2000 edition that Demitri, a 37-year-old reindeer herder, was 1 of only 6 out of 27 pupils in his class at school who had survived. With fewer reindeer to herd, and with hunting grounds limited and rivers polluted, 2 of his classmates, filled with despair, hanged themselves. According to this account, "The others died of alcohol-related incidents from drinking the vodka brought in by the oil workers" ("Putin's Oil," 2000).

"While British motorists complain about the price of petrol, the exploitation of oil is a matter of life or death to many Khanty people. If the people here [Great Britain] knew the true costs to the Khanty people of the petrol they put in their cars, they would put more energy into campaigning for a fairer deal for Siberia's tribes," wrote Stephen Corry, the director of Survival International, a human rights organization based in England ("Putin's Oil," 2000).

In 1994, local Russian administrators distributed parcels of land to the Khanty who were living in traditional fashion. This land was not chosen because the owners' families had owned it or because it was suitable for herding reindeer. Often, the land was miles from traditional homes and ranges and entirely unsuited to traditional Khanty herding and hunting. Some of the allocated land had been destroyed by fire, denuded of moss that once made it useful for feeding animals. As a result, many Khanty moved to land they did not own (but which was still useful for herding and hunting), fearful that it could be turned into oil fields at any time ("Reindeer Herder's Tale," 2000, 8).

THE ENVIRONMENTAL LEGACY OF SOVIET-ERA POLICIES

According to one observer, environmental degradation across the former Soviet Union is a result of decisions "taken decades ago by the leaders of the Communist Party to give priority in the use of raw materials to the military-industrial complex and heavy industry without regard to the effect on human life [that has] turned many of Russia's cities into gas chambers under the open skies" (Yablokov et al., 1991, C-3). Environmental destruction is said by Russian environmentalists to have been an inevitable result of command-economy decisions based on an ideology of gigantism pursued in the absence of any system for environmental protection. Ironically, Marxism (or state socialism, was it was sometimes called) turned out to be as growth oriented and heedless of environmental consequences as its ideological opposite, industrial capitalism.

During the Soviet era, a heavy emphasis was placed on industrialization regardless of environmental consequences. Many very dirty industrial plants were constructed across Siberia. As a result, most of Siberia's large rivers were heavily contaminated. Marine mammals (polar bears, walrus, bearded seals, ringed seals, white whales, and gray whales) have been found to be polluted by industrial waste bearing heavy metals, DDT, and PCBs. Some samples of plankton taken in the Arctic Ocean contain high levels of organochloric pollutants.

During the 1930s and 1940s, indigenous lands were seized by Soviet state-run industries. Reindeer pastures and fishing sites were disrupted, "depriving Siberian indigenous peoples of food and severely disrupting their way of life" ("Peoples of the Frozen North," n.d.). With industrialization came nonindigenous Russians from other parts of the Soviet Union.

Logging, nickel mining, and gold mining became established in Siberian indigenous peoples' lands between 1930 and 1950, during Stalin's regime. During the 1950s, logging operations increased rapidly, encroaching on many indigenous hunting grounds. Large swaths of Siberia, up to 40 percent of some areas, were clear-cut. Because of inefficient planning and transport, large amounts of harvested timber were left to rot (Slavic Research Center, 1999). Exploration for oil and gas and the building of roads and railroads soon followed, accelerating during the 1960s.

Siberian indigenous peoples suddenly found themselves becoming minorities in their own homelands. Until the mid-1980s, the Soviet State pursued aggressive assimilative policies that have been summarized by Survival International, an international group based in England, which advocates for indigenous peoples worldwide.

According to observers in Siberia, "From 1950 to the mid-1980s the authorities in Moscow attempted to suppress all ethnic, linguistic and cultural differences." In schools, indigenous languages ceased to be taught, and children were punished for speaking their own languages. This led to a breakdown between the generations; it is now common for the older people in a community to speak only their own languages, and the youngest to speak only Russian. "In the same period," according to the same account, "many indigenous villages were closed down, and the people were forced to move to larger ones. Villages of different peoples were amalgamated in an attempt to turn the country into a homogeneous Soviet state." Nomads were forced to settle in areas that were impractical for hunting or grazing. Bans on hunting and fishing led to a breakdown of indigenous economies, and forced dependence on the state for subsidies and salaries. The result was destruction of native ways of life: "This loss of the indigenous peoples' ways of life resulted in unemployment, alcoholism and high suicide rates, problems that still plague the peoples of the north" ("Peoples of the Frozen North," n.d.). Following the collapse of the Soviet Union, much of the assimilative infrastructure dissolved. Many indigenous peoples returned to the land, to herd and hunt for sustenance.

Under the Soviet State, indigenous peoples in Siberia had possessed no legal communal title to their lands. In fact, according to a law passed in 1968, immigrants' communal farms and industries were granted land (much of it originally used by indigenous peoples) free of charge. Under the prevailing legal system, land was regarded as being without value until it was utilized for economic improvement as defined by the Soviet State. Indigenous sustenance was given no value under the Soviet system. Compensation for environmental damage also was unknown under this system. Following the collapse of the Soviet Union, some indigenous groups demanded legal guarantees of land ownership and compensation for industrial damage to their homelands.

An analysis by the Slavic Research Center of Hokkaido University, Japan, reported that: "The [Russian] government has tried resettling indigenous peoples as compensation, attempting to provide a 'civilized' mode of living with heated houses and electricity. Such measures have had certain benefits in industrialization, but they have accelerated the decline of indigenous cultures" (Slavic Research Center, 1999).

Disregard for the environment in Russia did not end with the collapse of the Soviet Union. Russian Federation President Vladimir Putin, for example, during 2001, disbanded the State Committee for Environmental Protection (*Goskomekoligii*) soon after assuming office. This agency, which previously had functioned as an independent regulatory body, was merged with the Natural Resources Ministry; environmental protection was subsumed by that agency's prodevelopment mission. According to *Drillbits and Tailings*, the Russian federal government and oil companies also subverted a proposal for a biosphere reserve that would have shielded the land of more than 800 Khanty people from the intrusion of oil and gas development.

As the Russian oil industry continued to expand into the lands of indigenous peoples in Siberia, some of the same consortia were following retreating polar ice caps, provoked by global warming, north into areas of the Arctic that earlier had been inaccessible most of the year because of ice and snow. A report in *Drillbits and Tailings* commented that "further melting of Arctic ice promises to be a tragic humanitarian and environmental disaster, but could result in a huge windfall of natural gas for the Russian petroleum industry" ("Russian Gas," 2000).

The Russian Arctic shelf could hold as much as 2.3 trillion cubic feet (65 trillion cubic meters) of natural gas, more than present known reserves on the Russian mainland. This amount of gas could fulfill natural gas needs for Russia, Europe, and Turkey for as many as 20 years. Already, roughly 25 oil fields have been established on the northern Yamal Peninsula, most extending offshore

Salaspils, Latvia. May 12. The only research nuclear reactor in the Baltic Countries (shown above), belonging to the Latvian Academy of Science in Salaspils, was stopped in June 1998 because a lack of fuel appeared after the disintegration of the Soviet Union. (NewsCom.com/ITAR-TASS)

into the Kara Sea. "The greater potential," commented this *Drillbits and Tailings* report, "lies in the currently inaccessible but rapidly melting Arctic Ocean" ("Russian Gas," 2000).

"It [Arctic sea ice] lost an average of 34,300 square kilometers—an area lager than the Netherlands—each year," said Lisa Mastny, a researcher at the Worldwatch Institute, which has compiled scientific reports on the melting of ice on a worldwide scale. Not only has ice coverage shrunken, but it also has thinned. "Between this period [the 1960s] and the mid-1990s, the average thickness dropped from 3.1 meters to 1.8 meters—a decline of nearly 40 percent in less than 30 years," said Mastny ("Russian Gas," 2000). With pack ice in retreat, an opportunity arises for Gazprom, and other Russian petroleum firms, to drill in the relatively shallow—and increasingly ice-free—waters. Gazprom has already built the German, Polish, and Belarussian sections of a pipeline that stretches from Yamal to Europe. This links into Russia's existing pipeline network.

"There's a dangerous pathology at work here," said Rory Cox, communications director at the U.S.-based Pacific Environment and Resources Center. "By drilling for oil in the Far North and building these pipelines across pristine Siberian landscapes, Russian oil companies are creating a suicidal feedback loop. Global warming provides access to more fuel; burning it brings on more warming. The receding Arctic underscores the need to develop alternatives, not revel in the profits that Gazprom will make" ("Russian Gas," 2000).

THE NENET OF THE YAMAL PENINSULA: A FLOOD OF UNWANTED NATURAL GAS

For decades under Soviet rule, the Nenet (which means "the people" in their language) resisted harsh assimilationist policies of the former Soviet Union, which attempted to organize herders into collectives by redistributing reindeer and pasture lands, with many people coerced into resettlement. Soviet authorities also tried to eradicate indigenous religions by killing shamans and destroying sacred sites.

About 35,000 Nenet live on and to the south and east of the Yamal Peninsula; about 9,000 live on the Peninsula itself. The Khantys live to their south, in the taiga of the middle and lower Ob River. "Yamal" in the Nenets language translates to "end of the earth," or "back of the beyond." Yamal, a large Siberian peninsula in the Arctic Ocean that juts into the Arctic Ocean north of the Urals, is extremely rich in petroleum as well as natural gas. Today, the Nenet face new challenges to the survival of their culture, notably exploitation of their homeland's natural gas reserves, which are among the richest in the world.

According to Andrei V. Golovnev and Gail Osherenko, writing in *Siberian Survival* (1999), "The traditional lands and waters of the Khanty and the forest Nenets have been devastated by oil production and plundered by thousands of newcomers who helped themselves to wild mushrooms and berries, game and fish while the indigenous people were forced to survive in increasingly marginal circumstances" (Golovnev and Osherenko, 1999, 105–6).

Outsiders began to arrive in large numbers during the 1930s, including about 50,000 peasants who had been removed from other parts of Stalin's Soviet Union by force and used as forced labor to develop industrial infrastructure on the Nenets' land. The Nenet, Khanty, and other indigenous peoples rebelled against forced collectivization and other impositions of Soviet authority. Their leaders were arrested, and meetings were broken up by the Red Army, as reindeer herds were confiscated.

Oil and natural gas were discovered in the Nenets' homeland during 1964; substantial production began a year later. Since the 1960s, the Nenets' territory, especially the Yamal Peninsula, has been contaminated in many places by oil, oil products, and other chemical agents, with damage to birds, fish, and mammals, including the Nenets' reindeer herds. Pastures have been despoiled by oil development and transportation, which also has caused destabilization of the permafrost in some areas. Many more immigrants from Russia were drawn to the Nenets' homelands by the exploitation of natural gas reserves during the 1970s.

The Soviet State offered high salaries and bonuses to workers who moved to the Nenets' country to develop oil and gas resources, and to build cities. A million tons of oil was produced across northwestern Siberia during 1965, 28.5 million tons in 1970, 143.2 million in 1975, and 307.9 million in 1980. Gas production, 3.3 million cubic meters in 1965, rose to 9.5 million in 1970, 38 million in 1975, and 160 million in 1980. By the 1980s, thousands of miles of pipelines, roads, and railroads laced the taiga and tundra of northwestern Siberia. One million tons of oil spilled from poorly maintained pipelines in an average year by the 1980s. By 1988, 42,460 square miles of reindeer pastures had become unusable because of oil and gas transport and spills in the Nenet and Khanty *okrugs* (administrative districts).

At least 30 Nenet men, women, and children died within a few years "from causes related to oil exploration, and many reindeer were killed by automobile accidents on heavily traveled roads. In 1990, an oil truck struck and killed an old man's reindeer, provoking families from five clans to raise a tent on a bridge that blocked traffic in both directions. The protest became a popular cause across Russia, one of many ecological confrontations that helped speed the dissolution of the Soviet Union. The response was so great because reindeer were being killed intentionally by outsiders, "decapitated and their carcasses left to rot" (Golovnev and Osherenko, 1999, 106).

The Russian city of Nadym, a company town of 50,000, became known as the gas capital of Russia, as it was built up by GAZPROM, the Russian natural gas monopoly. By the 1990s, Nadym housed thousands of workers who built a road and railway along the eastern side of the Yamal Peninsula, preparing the area for construction of a gas pipeline (Gasperini, 1997). A hot-oil pipeline and tanker terminal were proposed, but thick wintertime ice and the shallowness of Ob Bay limited the size of tankers that could dock there (Golovnev and Osherenko, 1999, 11–13).

By 1990, the Nenet, having been inundated by 500,000 outsiders, comprised only 6 percent of the population in their homeland (Golovnev and Osherenko, 1999, 77). Amidst the development of the gas fields on the Nenets' homelands, an estimated 2,500 indigenous nomadic families live in the area. "We wouldn't have it any other way, we've always lived just with our deer," said Anatoli Vanuito, who expressed concern about ongoing oil and gas development in the Yamal by GAZPROM. "In recent years there have been fewer fish, and the reindeer get sick more often now," he lamented. "If there are no fish and the deer are sick, that's it for us. We have no other way to live" (Gasperini, 1997).

For several years, Nenet herders have said that their reindeer are becoming smaller and more prone to illness. "The tundra is the kind of environment where a single vehicle's track will remain for decades," said Bruce Forbes, an American ecologist who spent three years studying the Yamal ecosystem (Gasperini, 1997).

SIBERIAN INDIGENOUS PEOPLES AND URANIUM POISONING

Siberian indigenous peoples face yet another legacy of the Soviet era: illnesses (often leading to early death) caused by uranium poisoning in an area where thermonuclear devices were tested as weapons, as well as for peaceful uses. Radioactive wastes in large quantities—some of the greatest of any area of the world—have been dumped into the watersheds of the Ob and nearby rivers. Containers, barges, and submarines containing spent nuclear fuel were dumped into the waters off Novaya Zemlya following 132 nuclear tests in the area between 1955 and 1990, 87 of which were set off in the atmosphere. This area became favored by Soviet scientists studying the effects of nuclear radiation on human beings and ecosystems. Radioactivity spread eastward on prevailing winds from nuclear bomb test sites over the Yamal Peninsula (Golovnev and Osherenko, 1999, 100–101).

In this region, according to the Slavic Research Center, nuclear explosions often were used for civilian purposes, such as mining, seismic sounding, and controlling river flows. Some of the worst contamination was reported in Chukotka. Background radiation levels in 1990 were still the same as in the controlled zone around Chernobyl following its nuclear accident (Slavic Research Center, 1999).

Much of the radioactive pollution comes from military bases, sites for testing of nuclear bombs, and wastes from nuclear power plants. Each summer, according to reports by L'auravetl'an, a Russian group that advocates for indigenous rights, "About 5,000 ships in the Arctic leave behind 430,000 cubic meters of oil containing liquid emissions" ("L'auravetl'an Information Bulletin #2," 1997).

Pollution has become so pervasive in the Khantys' lands that most of the reindeer meat exported from western Siberia to Scandinavia has been returned as inedible, due to contamination by heavy metals and radionucleids. "Due to pollution," according to reports of L'auravetl'an, "the sturgeon population in the Ob river has decreased by 10 times [90 percent] within the last 15 years" ("L'auravetl'an Information Bulletin #2," 1997). Parts of the riverbed have been saturated by petroleum pollution. Levels of Cesium-137 and Strontium-90 in the bodies of indigenous reindeer herders have been detected at levels 100 times the average in Russia.

The average life expectancy of indigenous people in Russia was 43 to 45 years by the late 1990s, one of the lowest in the world. None of the indigenous peoples across the Russian republic, from the small nations of the Russian north to the Far East, had an average life expectancy of more than 55 years by 1990 (Yablokov et al., 1991, C-3). Some of the early deaths are said to have been provoked by cancers caused by radiation from nuclear testing carried out in Siberia during the early 1970s. Alcoholism and alcohol-related accidents, as well as upper respiratory infections, also contribute to mortality in these areas (Yablokov et al., 1991, C-3).

THE OIL RUSH ON SAKHALIN ISLAND

Roughly 4,000 Nivkh indigenous peoples on the island of Sakhalin, on the far-eastern edge of the Russian Federation, believe that recent development of rich oil and natural gas deposits may reduce their opportunities to hunt and fish, leading to devastation of their traditional economy. By the year 2005, resource developers on Sakhalin expect that oil and gas development will become the major source of cash income for the island. The Nivkh people on Sakhalin Island have been entirely excluded from negotiations on proposed oil development on their lands and offshore. The indigenous peoples for several thousand years have earned their livelihood from hunting and salmon fisheries that could be devastated by planned oil and gas development. The fishing industry comprises one-third of Sakhalin's economic activity and employs more than 50,000 people ("Putin's Oil," 2000).

During the fall of 2001, an oil-exploration group led by Exxon announced plans to spend as much as U.S. $12 billion on an offshore oil and gas project near Sakhalin Island. Several other multinational oil firms, including B.P. (British Petroleum) and Royal Dutch Shell, also announced plans to seek licenses from the Russian government to explore for oil and gas in the same area. All were said to be "anxious to get a toehold on Sakhalin," principally in so-called Block 5—so anxious that initial exploration was set to begin without required permits (Whalen, 2001, A-10). A *Wall Street Journal* report said that the oil companies were attracted by Sakhalin's proximity to Asian markets, as well as its rich potential—as much oil and gas as Alaska's Prudhoe Bay (Whalen, 2001, A-10).

Federal Russian law requires that "If an agreement is made over economic activities on territories of traditional habitat and activities of the small-numbered ethnic populations, the investor must take all the necessary steps to protect the traditional environment and the way of life of these populations" (Igrain, 1997). As oil and gas development began during the 1990s, the specific nature of these protections was left undefined; profit sharing by the indigenous peoples of the island also was not provided. Additionally, no mechanism was provided to compensate the indigenous peoples of the island for damage to their traditional way of life by pollution from oil and gas development.

REFERENCES

"L'auravetl'an Information Bulletin #2: Report On Siberian Health And Environment." March 13, 1997. [http://www.hartford-hwp.com/archives/56/004.html]

"L'auravetl'an: An Indigenous Information Center by Indigenous Peoples of Russia." United Nations Information Center, Moscow, April 27, 2001. [http://www.indigenous.ru/english/english.htm]

Gasperini, William. "In a Land 'Back of the Beyond' Reindeer Rule the Nomad's Life." *Christian Science Monitor*, April 25, 1997. [http://www.britannica.com/magazine/article?content_id=24234&pager.offset = 60]

Golovnev, Andrei V., and Gail Osherenko. *Siberian Survival: The Nenets and Their Story*. Ithaca, N.Y.: Cornell University Press, 1999.

Igrain, Konstantin. "Sakhalin: Oil and Indigenous Peoples." L'auravetl'an Indigenous Information Center (Moscow) on Native Net, April 24, 1997. [http://www.nativenet.uthscsa.edu/archive/nl/9704/0103.html]

Khandy, Vershina. "[Speaking for the] Upper Khanda Band of the Evenk," April 27, 2001. Cited in "L'auravetl'an: An Indigenous Information Center by Indigenous Peoples of Russia." United Nations Information Center, Moscow, April 27, 2001. [http://www.indigenous.ru/english/english.htm]

"Peoples of the Frozen North." Survival International, n.d. [http://www.survival.org.uk/siberiabg.htm]

"Putin's Oil Politics Threaten Siberia and Sakhalin's Indigenous Peoples." *Drillbits and Tailings* 5:16 (September 30, 2000): n.p. [http://www.moles.org/ProjectUnderground/drillbits/5_16/1.html]

"A Reindeer Herder's Tale." *Survival Newsletter* 42 (2000): 8.

"Reindeer Herders under Siege by Oil Industry In Siberia." Institute for Ecology and Action Anthropology; Report from a Fact-finding Mission to Khanty-Mansi Autonomous Region," March 18, 1997. [http://www.hartford-hwp.com/archives/56/003.html]

Russian Gas Companies Follow Receding Arctic Ice." *Drillbits and Tailings* 5:15 (September 19, 2000): n.p. [http://www.moles.org/ProjectUnderground/drillbits/5_15/1.html]

Slavic Research Center, Hokkaido University, Japan. "Survival's Campaign for the Khanty. Economic Development and the Environment: The Sakhalin Offshore Oil and Gas Fields." 1999. [http://src-h.slav.hokudai.ac.jp/sakhalin/eng/71/kitagawa3.html]

Whalen, Jeanne. "Exxon's Russian Oil Deal Makes Other Firms Feel Lucky." *Wall Street Journal*, December 13, 2001, A-10.

Yablokov, Alexi, Sviatoslav Zabelin, Mikhail Lemeshev, Svetlana Revina, Galina Flerova, and Maria Cherkasova. "Russia: Gasping for Breath, Choking in Waste, Dying Young." *Washington Post*, August 18, 1991, C-3.

THE SOUTH PACIFIC

INTRODUCTION

North Americans of the early twenty-first century usually encounter the islands of the South Pacific in newspaper travel sections, as fantasies of paradise—palms waving on sandy beaches, portending release and relief from the cares of industrial life. While such enclaves do exist, some of the South Pacific's scenery might shock the uninformed. Entire small islands have been shaved down to mud and stubs of dead trees, due to strip mining for various metals and minerals. The Mataiva, Nauru, and Banaba islands, for example, have become open-pit phosphate mines. The Kanak mine of New Caledonia is the fourth-largest producer of nickel in the world. Indigenous people on these islands often have been relocated to other islands that do not contain salable minerals. A number of the indigenous inhabitants of the Solomon Islands, for example, have been removed to make way for gold mining. In some cases, native peoples have stayed at home, to be drafted into the industrial-mining work force.

NEW CALEDONIA'S KANAK NICKEL MINE

The Kanak mine, on New Caledonia island in the South Pacific, has become the fourth-largest producer of nickel in the world, containing nearly a third of the world's nickel deposits. More than 150 million tons of nickel ore have been extracted from the island since 1864, with widespread environmental impacts. Indigenous opposition to nickel mining and smelting has become widespread on the island ("Mining in the South Pacific," n.d.).

The Thio mine on New Caledonia sits atop the world's largest nickel deposit, which has been mined for more than a century. The Thio River and its tributaries have been polluted by massive generation of mining waste, as has the Kouaoua river basin, which flows into red-stained seas and offshore shoals ("Mining in the South Pacific," n.d.).

In addition to the Thio mine, nickel is smelted in New Caledonia at the Domiambo smelter in Noumea, which is owned by Société Le Nickel. Airborne nickel is a class-one carcinogen, considered so toxic that the World Health Organization has ruled that no safe level exists in the air. Kanak has the world's highest level of asthma mortality and the highest levels of lung cancer in the Central and South Pacific. Pollution control is minimal, with local emissions from smelters permitted at levels more than 1,000 times those deemed safe under international standards ("Mining in the South Pacific," n.d.).

Early in 2002, indigenous peoples of New Caledonia attempted to rally against the opening of yet another nickel mine. *Drillbits and Tailings* reported that Canada's INCO Corporation had announced plans to mine the company's Goro property, in the southern reaches of the island, which is said to contain some of the world's richest, as-yet-undeveloped nickel and cobalt deposits. According to *Drillbits and Tailings*, "INCO chief executive officer Scott Hand wants to build a U.S. $1.4-billion commercial nickel-cobalt project on the site" ("Inco Threatens," 2002). An organization of clan chiefs called the Senat Coutumier, considered the voice of New Caledonia's Kanak people, opposes the project. Requested materials arrived in French (New Caledonia is a French colony), so most of the chiefs could not read them. Many indigenous people are concerned that mine wastes may be submerged in the ocean around the island, damaging the world's second-largest coral reef (the only larger one is the Great Barrier Reef of Australia).

THE MATAIVA, NAURU, AND BANABA ISLANDS: SACRIFICED FOR PHOSPHATE MINING

Phosphate has been mined from the Mataiva, Nauru, and Banaba islands in the South Pacific for more than a century, providing feedstock for the agricultural industries in Germany, Britain, Australia, and New Zealand. Nauru first was mined under German control, and later (after 1906) by joint British, Australian, and New Zealand trusteeship, "leaving two-thirds of the island a barren and lifeless wasteland of jagged rock pinnacles" ("Mining in the South Pacific," n.d.). Indigenous islanders had to take legal action to force the trustees' governments to pay compensation for ruining their homelands.

Mining began on Banaba in 1900 after the Pacific Islands Company signed an agreement with two senior landowners. According to a local source, "Like many such agreements, the local community leaders did not and could not sign on behalf of all the islanders and apparently did not understand the implications of the document they were signing" ("Mining in the South Pacific," n.d.). Several indigenous families were resettled on Rabi Island during the 1940s to make way for expansion of phosphate mining on Banaba.

In 1968, the Banabans petitioned the United Nations regarding their grievances. They subsequently sued the British government in 1971 for $120 million. Although the islanders lost in court on most arguments, the United Kingdom, New Zealand, and Australian governments eventually agreed in 1977 to pay $10 million in compensation, principally in the form of unpaid mining royalties. Ninety percent of Banaba Island's 1,500 acres had been strip mined by the mid-1990s ("Mining in the South Pacific," n.d.).

On Mataiva Island, exploratory mining by Australian-owned G.I.E. Raro Moana was halted in 1982 by local peoples' protests. Indigenous Mataiva Islanders said they were never consulted regarding the proposed mine, which they said poisoned their lagoon, preventing them from eating fish, formerly a staple of their diet, for seven years. Mining is overwhelmingly opposed by the local population because of its environmental destructiveness ("Mining in the South Pacific," n.d.).

THE SOLOMON ISLANDS: INDIGENOUS PEOPLES RELOCATED FOR GOLD MINING AND LOGGING

Environmental conflicts over gold mining and industrial-scale logging have brought violence, including two deaths, to the Solomon Islands, which lie to the northeast of Australia and east of Papua New Guinea. A land area of nearly 30,000 square kilometers is spread over 992 islands, with a combined population of about 340,000 people. Roughly two-thirds of the Solomon Islands is covered with tropical rainforest, and many of the people still live in villages whose residents depend on the forests for survival.

Since 1994, Gold Ridge (a subsidiary of Ross Mining N.L., Australia) has been mining gold on the Solomon Islands, relocating indigenous villagers. Local observers have noted that "controversy has surrounded the potential impact of tailings on the local water supply and the supply of electricity, with opposition to the mine generating its own power rather than buying from the Solomon Island Electricity Authority, as originally agreed" ("Mining in the South Pacific," n.d.). Some members of the local community, such as David Thuguvada, a Guadalcanal community leader, have expressed concern regarding the impacts of mining, saying, "The villagers on the plains below Gold Ridge have been excluded from negotiations over the environmental agreement, and we are deeply concerned" ("Mining in the South Pacific," n.d.).

Local indigenous landowners, represented by the Australian legal firm Slater and Gordon (which also represented land owners at the Ok Tedi mine in Papua, New Guinea), filed a constitutional challenge to the mine in the Solomon Islands' High Court. The suit sought to have the compensation agreement set aside on the grounds that it is unreasonable because of its probable environmental impacts.

While gold mining has become a subject of contention on some of the Solomon Islands, logging has sparked controversy on others. Maving Brothers, a Malaysian company, already has logged half the islands' remaining forests, and would like to clear-cut the rest, as indigenous peoples protest.

Community tension over government-supported logging of disputed lands on Pavuvu Island has led to one murder and another death that has been characterized as suspicious. Martin Apa, a well-known local opponent of logging, was killed during early November 1995, in Yandina, Pavuvu's main port. A postmortem found that his neck had been broken and pierced by a sharp object.

The *Solomon Star* reported November 10, 1995 that U.S. $2.2 million was paid in bribes by the logging company Integrated Forestry Industry (a subsidiary of Malaysian company Kumpulan Emas) to ministers and other government employees. Disclosure of these bribes provoked calls for the resignation or discharge of the ministers during a public rally and via a petition signed by churches, unions, and environmental advocates ("Solomon Islands," 1995). Journalist Duran Angiki, who reported the story for the *Solomon Star*, was fired following pressure from the government and logging companies. Even though Angiki was fired, seven Solomon Island government ministers were charged on December 4, 1995 with having received bribes from Maving Brothers.

Despite the indictments, logging in the Solomon Islands continued into the last years of the past millennium at three times the estimated sustainable level, a rate that will eradicate the islands' commercial forests by the year 2010, according to the 1994 annual report of the Central Bank of Solomon Islands ("Solomon Islands," 1995). Greenpeace asserted that "logging practices by the mainly Malaysian and South Korean companies are uncontrolled and destructive, and supply the Japanese and Korean log market. With more than 60 percent of government revenue derived from log export levies, forest depletion means a looming disaster for the economy" ("Solomon Islands," 1995).

During the early 1990s, some Pavuvu Island landowners countered industrial-scale logging with their own program of sustainable development, working with Greenpeace to set up a village-based ecoforestry project. About 90 percent of the land in the Solomon Islands is customarily owned by family groups, providing an economic base for small-scale, sustainable forestry.

Ecotimber is the local, indigenous answer to industrial-scale logging. Since the 1980s, "Australian and Asian logging companies have swept through the Solomon Islands, leaving a trail of disintegrating communities, flattened and degraded forests and silted coral reefs from runoff of exposed fragile soils" (Singh, 2000). Some of the industrial-scale logging continues side-by-side with the indigenous peoples' more cautious approach.

Production and marketing of the Solomon Islands so-called ecotimber began in 1997. According to a report from the Environment News Service, the Solomon Islands' ecotimber "is managed in a way which causes minimal damage to the forests," harvesting timber under local control in a manner and at a speed that will not strip the land. The export of ecotimber arose as a ray of economic and ecological hope "in a country torn by a tribal conflict, with nearly 100 people dead and the economy in tatters" (Singh, 2000). A third of the ecotimber was exported to New Zealand. In 2000, the first shipment of the Solomon Islands' ecotimber reached the port in Sydney, Australia.

"Here is a community that has been trying to provide a future for their forests and children, only to be thwarted by aggressive logging. Martin Apa was a key supporter of eco-forestry and we fear that he may have been targeted for the stand he took. Pavuvu landowners are being intimidated into agreeing to further unsustainable logging," said Grant Rosoman, Greenpeace–New Zealand Forests campaigner ("Solomon Islands," 1995).

REFERENCES

"Inco Threatens Indigenous Kanuks and Environment of New Caledonia." *Drillbits and Tailings* 7:3 (March 29, 2002): n.p. [www.moles.org]

"Mining in the South Pacific: On the Other Side of the Forests." N.d. [http://parallel.acsu.unsw.edu.au/mpi/docs/niart.html]

Singh, Rowena. "Eco-Timber Export Brings Hope to Solomon Islands." Environment News Service, December 8, 2000. [http://ens.lycos.com/ens/dec2000/2000L-12-08-02.html]

"Solomon Islands Murder and Corruption: Logging Takes Its Toll. Martin Apa Murdered; Greenpeace Calls for Investigation." Background Briefing by Greenpeace New Zealand," December 1995. [http://www.greenpeace.org/~comms/forestry/solomo.html]

SRI LANKA

THE WANNIYALA-AETTO: HYDROPOWER AND LOGGING

Introduction

Wanniyala-Aetto ("forest beings") are indigenous to Sri Lanka, an island nation in the Indian Ocean south of India. Known as "Veddahs" in Sri Lanka's main language, Sinhalese, these gentle hunter-gatherers have lived in a sustainable relationship with their tropical forest environment for at least 28,000 years. Following intrusion by loggers and developers of hydroelectric dams, the forest beings' ancient way of life is being threatened. The Wanniyala-Aetto people themselves may soon face extinction.

Dam Inundates Wanniyala-Aetto's Best Lands

In 1955, with funding from the World Bank, the government of Sri Lanka (then known as Ceylon) began construction of the Gal Oya dam, which flooded the Wanniyala-Aetto's best hunting-and-gathering lands, including their best honey-gathering sites and favorite forest caves. Most of the people were resettled in rehabilitation villages in agricultural areas. A significant minority, however, followed their wisdom-keeper and spokesperson, Uru Warige Tissahamy, deeper into the forest. In 1977, the World Bank provided the Sri Lankan government with funding to construct another large hydroelectric power and irrigation project involving the island's largest river system, the Mahaweli Ganga, flooding the homes of some indigenous peoples. At about the same time, large tracts of Sri Lankan rainforest were being logged; 11,000 hectares of the Wanniyala-Aetto's last hunting grounds were clear-cut as thousands of Sinhalese and Tamil settlers moved into their diminished homelands ("Forest Beings," 1999).

On November 10, 1983, the government evicted the Wanniyala-Aetto from the last remaining portions of their forested homeland. The evictions were said to have been forced to set land aside for the Maduru Oya National Park. The evictions had other objectives, however. The land served as a catchment area for three new reservoirs financed with assistance from various foreign-donor agencies, including the United States Agency for International Development. These reservoirs were created to provide irrigation water for so-called green revolution, intensive agroproduction of rice in paddy fields at the edge of the forest. The last 900 Wanniyala-Aetto families were thus forced to leave their homelands ("Forest Beings," 1999). According to a report from the scene by Global Vision, a nongovernmental organization accredited by the United Nations,

> Approximately 5,300 men, women, and children were forced to resettle into three different districts, splitting up their community and destroying the highly integrated social structure on which the Wanniyala-Aetto traditionally depend. These resettlement areas are situated outside the forest, in rice-growing areas totally unfamiliar to the Wanniyala-Aetto and unsuitable for their traditional smidden agriculture. The forest beings are now considered trespassers in their own forest. ("Forest Beings," 1999)

Effects of Relocation

One group of Wanniyala-Aetto, having been moved to the Pollonaruwa District System B resettlement area, became virtual hostages in the civil war between the government and the Tamil Tiger rebels. Wanniyala-Aetto who were removed from the forest and housed in resettlement camps suffered a wide range of problems that are familiar to many residents of American Indian reservations. According to World Vision, "The impact of their eviction has also caused dietary changes, and health problems. The ravages of depression, alcoholism, obesity, diabetes, and fatherless children are now showing up for the first time. A whole generation of Wanniyala-Aetto now lives on public welfare. It is a classic, tragic story ("Forest Beings," 1999).

By the end of the twentieth century, new constitutional measures guaranteeing increased autonomy for Sri Lanka's ethnic minorities were being used to make a case that the forest beings have a right to return to their aboriginal homelands. Despite these promising legal directives, according to an analysis prepared by Global Vision, "The Wanniyala-Aetto continue to suffer human-rights abuses and maltreatment of every kind. Their very ancient culture, spiritual traditions, ethno-botanical medical knowledge, and ecological expertise in the management of tropical forest fauna and flora are on the brink of being lost forever. The Wanniyala-Aetto themselves are now in acute danger of extinction" ("Forest Beings," 1999). Only 2,500 Wanniyala-Aetto survive in a total Sri Lankan population of 20 million.

For more than the 10 years, four Sri Lankan presidents (J. R. Jayawardene, Premadhasa, D. B. Wijetunge, and Chandrika Bandaranaike-Kumaratunga) each promised to let the Wanniyala-Aetto return to Maduru Oya National

Park, and each failed to deliver. In 1998, several Wanniyala-Aetto returned to the park following an official edict allowing them access. Because park guards had not been not informed of the new policy, the returning Wanniyala-Aetto were arrested and shot. One man who was shot by park guards was paralyzed for life. In the meantime, nonnative poachers continued to loot the park of wildlife and timber.

Loss of Traditional Knowledge

Loss of forests in Sri Lanka (as in many other regions) also implies loss of an irretrievable treasure of indigenous knowledge preserved mainly by women. For centuries, women have woven knowledge of the forest into their maintenance of households, and provision of food, fuel, and fodder for their families. Women have passed from generation to generation knowledge of the forests' edible and medicinal species. The spread of the cash economy, usually in male hands, along with increasingly formal systems of land title, is destroying this indigenous knowledge base. In Sri Lanka, as elsewhere, forests are becoming less a source of food than a source of profit via logging.

Because of deforestation, women have been forced to spend more time and energy than ever before gathering firewood and food. According to a report in the *World Rainforest Movement Bulletin*, not only do women have to walk further to find less, but they carry heavy weights for longer distances (as much as 35 kilograms for 10 kilometers), damaging their health. "The need to conserve firewood then affects the family diet," the *Bulletin* wrote, "decreasing variety and nutritional content, with a further deleterious effect on [their] health. This is just one of a range of tasks made more difficult by encroaching deforestation" (Carrere, 2001).

In Sri Lanka, indigenous women have found themselves in direct competition for firewood with tea and tobacco plantations. As their families are drawn into the cash economy, many women also have taken jobs on the plantations, reducing time formerly spent gathering wood and providing for their families. Women have been reacting to increasing deforestation by planting their own gardens, trying to preserve some of the biodiversity that the land, in general, is losing to monocultures and logging.

REFERENCES

Carrere, Ricardo. "Sri Lanka: Deforestation, Women and Forestry." *World Rainforest Movement Bulletin*, September, 2001. [http://www.wrm.org.uy]

"Forest Beings: Indigenous Peoples of Sri Lanka." 1999. [http://www.global-vision.org/srilanka/]

SURINAME

THE SARAMAKA MAROONS: GOLD MINING AND LOGGING

Introduction

The few relatively untouched rainforest areas in Suriname, home of rich biodiversity and thousands of indigenous people and Maroons (descendants of former African slaves), are shrinking in the face of increased mining activity and industrial-scale logging.

The Saramaka indigenous people have lived in this area since the early eighteenth century. Their enslaved ancestors escaped plantations on the coast and established communities in the forests that were isolated from outside contact for many years. During the later eighteenth and nineteenth centuries, their political and cultural autonomy and rights to land and territory were recognized by treaties with the Dutch colonial government ("Suriname: Logging," 2001).

Lands Removed by Nature Reserve

According to their traditions, Saramaka territory is divided among a number of matrilineal clans. Members of given clans possess rights to hunt, fish, farm, and gather forest produce within an area communally recognized by each family group ("Suriname: Logging," 2001). The clans' ownership system is not recognized by Suriname's government, which asserts the state's right to ownership of resources, as well as rights to grant concessions to cut timber and establish mines.

In 1998, the Surinamese government—jointly with U.S.-based Conservation International—established the Central Suriname Nature Reserve, the largest area of protected tropical forest in the world. The establishment of this reserve removed at least a third of the Saramakas' ancestral lands without their consent. At the same time, logging and mining rights to large areas of neighboring rainforests were granted to multinational logging and mining companies. Again, the Saramaka were not consulted. They learned of the concessions only as employees of a Chinese logging company, NV Tacoba Forestry Consultants, arrived in the area and began operations. Soon the area was laced with roads, and water supplies became polluted, killing fish. Game animals were driven away, and subsistence farms were destroyed.

The Canadian companies Golden Star Resources and Cambior have been planning to mine gold in the Gross Rosebel concession, requiring relocation of the Maroons of Nieuw Koffiekamp, who do not want to move. Golden Star has reportedly threatened and harassed community members by using live ammunition to frighten them away from areas in which the company is exploring for gold (Chatterjee, 1998). Golden Star and Cambior are joint partners in the Omai mine in Guyana that dumped between three and four million liters of cyanide and heavy-metal wastes into the Essequibo River when a dam ruptured there August 19, 1995.

Access to Gardens Denied

Since 1994, when construction of its mining camp began, Golden Star has denied the Maroons access to their gardens and traditional hunting and fishing areas. The indigenous peoples have no legal recourse against the harassment because Suriname does not legally recognize any form of traditional native-land tenure. "Our lands are of fundamental importance for our survival as indigenous and tribal peoples. Without the land, forest and rivers there are no trees, birds, animals and fish and we as indigenous and Maroon peoples will not be able to survive," declared a Gran Krutu (Great Gathering) of indigenous and Maroon leaders held during November 1996 ("Gold Fever," 1997).

Suriname has no environmental protection laws. The government maintains no monitoring of pollution. David Fagin, chairman of Golden Star, has been quoted as saying that the company "has looked increasingly at the Guyana Shield because of the increased pressure from environmentalists and the government in the United States" ("Gold Fever," 1997).

As native peoples in Suriname face gold mining, industrial-scale logging also has increased. Chinese logging companies (forced to expand beyond domestic markets by logging bans at home) for the first time have begun operations in Suriname that are affecting the Saramaka people, one of the six Maroon tribes living within Suriname's borders. The Saramaka are one of the largest Maroon tribes, with about 20,000 persons in more than 70 villages along the Suriname River.

The *Philadelphia Inquirer* reported May 20, 2001 that

the company [Jin Lin, a Chinese concessionaire] had plowed large, muddy roads about 45 feet wide into the forest, churned up huge piles of earth, and created fetid pools of green and brown water. Upended and broken trees were everywhere and what were once plots of sweet potatoes, peanuts, ginger, cassava, palm and banana crops—planted in the forest by Maroon villagers—were muddy pits. ("Suriname: Logging," 2001)

According to a report in the *World Rainforest Movement Bulletin*, three complaints were submitted to the government by the Saramaka between October 1999 and October 2000, all of which were ignored. The Indians then took their case to the Inter-American Commission on Human Rights, filing a petition there in October 2000, which was still pending two years later.

REFERENCES

Chatterjee, Pratap. "Gold, Greed and Genocide in the Americas: California to the Amazon." *Abya Yala News: The Journal of the South and Meso-American Rights Center* (1998): n.p. [http://saiic.nativeweb.org/ayn/goldgreed.html]

"Gold Fever Threatens Forests and People in Suriname." *World Rainforest Movement Bulletin* 3 (August 1997): n.p. [http://www.wrm.org.uy/bulletin/3/Suriname.html]

"Suriname: Logging and Tribal Rights." *World Rainforest Movement Bulletin* 53 (December 2001): n.p. [http://www.wrm.org.uy/deforestation/logging.html]

THAILAND

INTRODUCTION

Indigenous peoples in Thailand, like many others around the world, have been finding their traditional lands inundated by increasing numbers of intrusive activities. In the case of Thailand, settlers' invasions, often backed by police and troops, have been accompanied by construction of large hydroelectric projects. Some indigenous peoples have responded by marching en masse to the grounds of Thailand's national assembly, camping outside in large numbers for months at a time, where they have endured police assaults. Additionally, some indigenous peoples, including the Karen, have been exposed to unhealthy levels of lead from a mine and separation plant, dumped into creeks used for drinking water and bathing. As a result of the lead poisoning, most households in a village of 200 people had at least one deformed child. Some children were born without genitalia. One boy showed signs of suffering from Down's syndrome, and another child was born with 12 fingers and 12 toes.

INDIGENOUS PEOPLES IN THE POWERHOUSE

Thai villagers have occupied the powerhouses of several dams in response to inundation by rising reservoir waters, which are destroying some of the best fishing grounds in the country. Thousands of villagers also have marched on the national legislature in Bangkok and camped near its chambers, only to be routed by police with truncheons and tear gas. Critics assert that the dams are not necessary for electricity because Thailand already is oversupplied.

On the night of the Buddhist Lent, July 16, 2000, and continuing into the afternoon of the following day, 223 villagers (including many elderly, women, and a child) were arrested during a demonstration at Government House in Bangkok. The villagers were protesting the Pak Mun dam's adverse effects on their villages. They organized under the aegis of the Assembly of the Poor, the largest and most influential people's organization in Thailand, which advocates for people who have been disenfranchised by so-called development projects.

Reports in the Bangkok news media said that several of the protesters were beaten by police before they were arrested for trespassing on government property. One report said that "although representatives of the Assembly of the Poor have camped in front of Government House time and again to demand their constitutional rights, their demands have fallen on deaf ears" (Korkeatkachorn, 2000).

The protesters replied that their trespass was a minor issue "compared with the government's intrusion through its top-down development projects into the villagers' lives nationwide. Such intrusion of the state has brought about the villagers' poverty, desperation, illness, family disintegration and the selling of their daughters into prostitution" (Korkeatkachorn, 2000).

The demonstration at Government House in Bangkok was the culmination of several acts of civil disobedience staged by indigenous peoples whose lives had been disrupted by dam construction in Thailand. On May 15, 2000, for example, villagers blocked a turbine engine to stop the operation of Pak Mun dam, managed by the Electricity Generating Authority of Thailand. They demanded, according to one account, "that the government open the dam gates for four months to allow fish to swim up-river to spawn, as fish species have been disappearing ever since the completion of the dam in 1994" (Korkeatkachorn, 2000).

After a two-week occupation, on June 2, the government established a committee to study the villagers' complaints, including five members from the government and five from the Assembly of the Poor. After a month of study, the committee recommended opening the gates of the Pak Mun and Rasi Salai dams, halting construction of dam projects, and compensation of villagers affected by the dams. The report was forwarded to the government, which took no action. The result was the July 16 march on Government House, described above.

On July 25, nine days after the beatings and arrests of the villagers, the cabinet agreed to open the gates of the Pak Mun and Rasi Salai dams, but it refused to pay compensation to more than 2,000 families affected by this Sirindhorn dam, which had been completed in 1972. The Assembly of the Poor deemed the response inadequate, after which roughly 300 of its members began a hunger strike.

Dam-site occupations have been utilized several times by Thai native peoples who find their lands threatened. On April 20, 1999, such a tactic was utilized by more than 1,000 villagers from the Assembly of the Mun River Basin and the Assembly of the Poor, who had been affected by the Rasi Salai dam. Five villagers were arrested. "We demand the Government determine how many people are affected by the dam through a participatory process, not a bureaucratic process, and pay compensation for the affected villagers. In the near future, more and more people will join this demonstration," said Boonmee Sopang, a leader of the Assembly of the Mun River Basin ("Affected Villagers," 1999). The demonstrators persisted despite lack of food, heavy rain, and oppressive heat.

The Rasi Salai dam, on which construction began in 1992, is part of the Kong-Chi-Mun Water Diversion Project, which plans to build 13 dams on the Chi and Mun rivers within 42 years, at a total cost of U.S. $5.7 billion. The dams are being built to resolve water shortages in northeast Thailand by diverting water from the Mekong River to the Chi and Mun rivers. Local people assert they were told that a rubber weir would be built. Instead, a concrete dam was built, which raised the water level above the riverbank and inundated more than 100 square kilometers of land. The dam caused the loss of 80 square kilometers of fertile farmland, while it provided irrigation to only one-quarter of that area ("Affected Villagers," 1999).

The same dam also destroyed a freshwater swamp forest along the banks of the Mun River, which had provided fish and herbal medicines to local people. In addition, the dam blocked migration of fish. Local people also noticed an increased salt level in the water, because the dam sits on top of a salt dome that lacks natural drainage and leaches into the reservoir. Most important to local people, the salty water inundated farmland used by more than 3,000 families, who were not compensated.

Following protests, the Electricity Generating Authority of Thailand paid only for the private property of about one-third of the families, ignoring the important role of communal lands in the native villagers' economic system. In an attempt to mitigate flooding impacts on local people, the Electricity Generating Authority built a dike, only to discover that it made the problem worse: "It does not allow the water to flow into the reservoir in monsoon season, and so it floods a vast area of farmland" ("Affected Villagers," 1999). As government inaction continued to anger the villagers, they met and began to discuss ways to destroy the dam.

During April 1999, roughly 3,000 members of the Assembly of the Poor seized land around the Pak Mool dam in Ubon Ratchathani, 600 kilometers from the capital, following protests in Bangkok during 1995, 1996, and 1997. In 1997, around 20,000 people camped outside Government House for 99 days. Following these protests, the government agreed to pay compensation to more than 7,000 families who had been evicted by dam construction. Once the demonstrators went home, however, the government withheld most of the money, leading to more protests.

The Pak Mool dam, which was completed in 1994, "has proved to be a social and environmental disaster." according to a report in the *Nation*, a Bangkok newspaper ("Changnoi," 1999). The dam destroyed a stretch of rapids which contained one of the best fisheries in the country. "It was so beautiful," said one of the leaders of the Pak Mool fishing community taking part in the protest. "There were several sets of rapids, dropping down level by level" ("Changnoi," 1999). Swimming these rapids was said to give local fish a special texture, taste, and value, according to the *Nation*'s report.

The fishing community flourished. Restaurants in the area were nationally famous. Now the dam has disrupted migration patterns. The beautiful rapids have been dynamited into an industrial-ugly channel. The fish have disappeared from a stretch of 30 to 50 kilometers of the Mool river. Some members of the camp are grilling the catch-of-the-day—a couple of 2-inch tiddlers. If you order fish in a nearby restaurant and enquire [sic] about its origin, the answer is likely to be: Laos. ("Changnoi," 1999)

A fish ladder was built into the dam, but few fish were able to find it. The few fish that found the ladder were unable to negotiate its steep steps. Protesters illustrated the ladder's uselessness by using it to dry their washing. "When they first told us about the dam," said a fishing-community leader, "they promised us we would have more fish, more money, better roads and houses, everything would be better" ("Changnoi," 1999).

The *Nation* commented that the Pak Mool dam is a relatively small hydro-electricity project, with output that "would barely power one Bangkok shopping mall" ("Changnoi," 1999). The dam had been running at only around 20 percent of its full capacity. According to the *Nation*, "The rapids, fishery and community have been sacrificed for electricity capacity which is not needed or can be replicated elsewhere" ("Changnoi," 1999). During 1999, the World Bank, which helped finance the Pak Mool project, joined in establishing a World Commission on Dams to investigate projects where the cost-benefit ratio differed widely from the plan. Pak Mool was selected as one of several dams around the world to be reviewed.

THE LAHU AND HMONG: FACING AN INUNDATION OF "LOWLANDERS"

The Lahu and Hmong peoples of Thailand have migrated freely between a number of Southeast Asian countries, including Burma, Laos, Thailand, and Vietnam. Recognizing their disregard of political borders, the name "Hmong" is sometimes translated as "free people" (Magagnini, 2000). Lacking official immigration documents, many Lahu and Hmong have been evicted from Thailand as illegal immigrants by police and troops, who, together with forestry department officials, businessmen, and land-hungry lowlanders, have been driving the indigenous peoples from their highland homelands under cover of national immigration laws.

Julian Gearing, writing in *Asiaweek*, described one such eviction:

It was midday when they burned down Tungpaka village. Thirty armed men walked casually from house to house, torching the tinder-dry buildings. "They didn't say anything. They just set the houses on fire while most of the people were in the fields," said a local member of the Lahu tribe, standing in the weeds that now grow where his family home once stood. (Gearing, 1999)

On March 29, 1999, 13 houses, as well as crops, were destroyed in another raid on the village in Chiang Mai province, in the northern highlands of Thailand, leaving 60 people homeless. In addition to the forcible evictions at Chiang Mai, within a few months, people residing in another village were evicted from their homes in Chiang Rai province. Additionally, community members were forced out of their homes by the army in Prachuab Khiri Khan Province, and farmers' shelters were burned to the ground near the town of Tak. More than 600 hill-tribe people were evicted from their villages during 1999 (Gearing, 1999).

Thai authorities justified the evictions with assertions that the mountain people, many of whom had lived in these areas for centuries, were wrecking the ecological balance of the region. The authorities accused the native people of using slash-and-burn agricultural practices and cutting down trees on public land to plant cash crops such as lychees and cabbages. The authorities also contended that increasing populations of indigenous peoples in the highlands were draining the watershed and depriving lowland farmers of irrigation water.

A Paklang villager gestured at the houses of the forestry rangers next to his lychee plantation. "The authorities gave us these lychee trees a decade ago," he said. "The forest rangers have watched us tend them for years. Now, suddenly, they are planting trees on our land and saying we can't use it any more. Our fathers helped the government fight the communists in these hills. This is how we are now being repaid" (Gearing, 1999).

Plodprasop Suraswasdi, director-general of Thailand's Royal Forestry Department, told *Asiaweek* magazine that no one should underestimate the severity of the risk posed by the tribespeople's activities. "This is a case of do or die for the country," he said (Gearing, 1999). On August 21, 1999, Plodprasop was at the head of a column of about 10,000 forestry rangers, soldiers, and lowlanders who marched onto Paklang village land in Nan Province. Trampling through the lychee and rice plantations of the 3,000 Hmong people, they planted tree saplings—the first of a series of reforestation exercises designed to return the area to its original state. "Five years from now, we will have closed this land," says Plodprasop, who contended that the native peoples' small-scale farming was devastating the land (Gearing, 1999).

The indigenous peoples and their supporters insisted that the evictions had more to do with commercial greed and land hunger than with the reforestation required by the government, which was being enforced at gunpoint. The native peoples asserted that the highlands had been targeted by timber companies, as well as mining and quarrying interests, the tourist industry, and lowlanders seeking land. Said one of the villagers who was forced out of Tungpaka: "The reason the forestry officials and lowlanders burned our village was to clear the area so that a tourist resort can be built at a waterfall nearby" (Gearing, 1999).

Thirteen highland tribes are officially recognized by Thailand, notably the Akha, Karen, Hmong, Lahu, Lisu, and Mien peoples. Most of their members have lived in Thailand for several generations during the past 50 years or have immigrated, principally from Myanmar (Burma) and Laos. Although they are legally recognized as groups, many individual tribespeople have been denied registration for official identification cards or issued with papers that label them as foreigners lacking state-sanctioned citizenship. As more tribespeople have been immigrating from Myanmar (Burma) and Laos, whole groups have been identified as illegal immigrants and denied rights to live on the land, as officials contend that their increasing numbers are overtaxing the environment.

Without proof of citizenship, hill people find themselves economically marginalized. According to *Asiaweek* magazine, "With no national ID card, youngsters from minority groups have poor employment opportunities. Some slip into the illicit economy, selling drugs or, in the case of girls, being forced by circumstances or their parents into prostitution in seedy brothels" (Gearing, 1999).

During May 1999, roughly 3,000 members of the hill tribes rallied in an unprecedented three-week protest in front of the Chiang Mai governor's office, calling for citizenship, land rights, and an end to evictions. The demonstration was broken up after nightfall on May 19 by police, who were joined by about 1,000 flare-throwing forestry rangers (Gearing, 1999).

In the meantime, evictions continued. In Prachuab Khiri Khan Province, at dawn on July 15, 1999, about 350 hill-tribe people were forced out of their homes, then forced to hike 40 kilometers over dense, forested mountains and swollen rivers to a refugee camp near the Myanmar border, where those lacking proper documents were expelled from Thailand. One of the evictees gave birth to a baby boy, who died soon afterward (Gearing, 1999).

THE KAREN: THE TOLL OF LEAD POISONING

Klity Creek in Lower Klity village, Kanchanaburi Province, Thailand, has been polluted for several decades by toxic discharges from a lead-separation plant and a lead mine upstream. "The water has been spoiled for years. We have suffered for so long. Since we are organizing a ceremony to express our gratitude to the Rice Goddess, we thought we might as well conduct a renewal rite for the creek, to drive all the hardship and misery away," explained Somchai Thongphaphumepatavee, 46, a native of Lower Klity, a village of Karen indigenous people (Achakulwisut, 2001).

On March 31, 2001, Thailand's Ministry of Public Health was urged to offer detoxification medication to residents of Lower Klity, following the deaths from lead poisoning of three villagers. Suraphong Kongchantuk of the Karen Studies and Development Center, representing nongovernmental organizations campaigning against lead poisoning in the Mae Klong river basin, delivered a letter with signatures of 62 Mae Klong villagers to Deputy Public Health Minister Suraphong Suebwonglee, urging health authorities to provide them with lead-detoxification medication (Bhatiasevi, 2001).

Atiya Achakulwisut, a reporter for the *Bangkok Post*, described the scene:

In a mournful voice, Somchai described the once-vibrant Klity creek he used to know, a stream that was full of fish, of large trees that towered over its two banks, and how, at night, the croaking of frogs and toads could be heard echoing in the woods. All this was lost once the lead-separation plant, run by Lead Concentrate Thailand Co., went into operation upstream from the village some 20 years ago. Since 1994, hundreds of cattle have died after drinking water from the stream. At least seven residents of the Lower Klity village were believed to have died of causes related to lead poisoning. (Achakulwisut, 2001)

The Thai Department of Pollution Control, which analyzed lead contamination in the village three years after news of the villagers' suffering came to light, confirmed that the level of lead contamination in the Klity creek was "startling and scary," the worst ever documented in Thailand. A study by the department revealed that lead content in the creek was 10 times higher than levels deemed safe. The study also discovered about 15,000 tons of lead-laden sediment piled up as high as 30 centimeters along the course of the 19-kilometer-long creek (Achakulwisut, 2001).

Many of the children in the village showed signs of abnormal physical and mental development, said Suraphong. He added that at least two year-old children have weak hands and legs and still cannot walk, despite having reached walking age. In addition, two girls, one seven years old and the other one month old, were born without ovaries.

Achakulwisut described children in Karen villages suffering from Down syndrome; other children with abnormally high concentrations of lead in their blood had hanged themselves. Many children were very slow learners, and nearly every household in the village of 200 people included at least one deformed child. Some children were born without genitalia. One child was born with 12 fingers and 12 toes. All these children were found to have abnormally high levels of lead in the blood.

In another case, Mamia Thongphaphumcharoen, 32, who had become ill during 1998, was diagnosed and unsuccessfully treated for kidney problems at Thong Phaphum Hospital in Kanchanaburi. She did not recover from her illness until she received medication for lead poisoning from Rajavithi Hospital in Bangkok. The villagers said that many of their relatives had died from a similar condition, which they believe was caused by lead poisoning (Bhatiasevi, 2001).

The lead-separation plant was ordered closed in 1998, after which water quality in the village improved slightly. To restore the creek as a source of livelihood, however, required extensive dredging and several years of natural cleansing. According to Achakulwisut's account, "The creek is deeper in some parts, with less scum-like sediment and no foul smell. Still, the stream is far from being the source of life it used to be" (Achakulwisut, 2001). Even three years after the plant was closed, Thai health authorities told villagers not to drink the water or eat fish, shrimp, or crab from the creek for fear it would increase the lead content in their bodies. Many Karen had no choice but to eat from the creek, a major source of food. "I tried not to eat fish from the creek, but I have to sometimes, otherwise I would have nothing to eat," one man told Achakulwisut (Achakulwisut, 2001).

REFERENCES

Achakulwisut, Atiya. "Deadly River: Despite a Government Promise to Clean up the Contaminated Klity Creek, the Thai-Karen Villagers Who Live Along It Are Still Haunted by Health and Food Problems." *Bangkok Post,* January 30, 2001. [http://www.bangkokpost.net/outlookwecare/300101_Outlook01.html]

"Affected Villagers Occupy the Rasi Salai Dam." Environmental News Network, April 6, 1999. [http://lists.isb.sdnpk.org/pipermail/eco-list-old/1999-June/002061.html]

Bhatiasevi, Aphaluck. "N.G.O.s Urge Authorities to Treat Lead Poisoning Victims." *Bangkok Post,* April 1, 2001. [http://www.ecologyasia.com/NewsArchives/Apr_2001/bangkokpost_040401_News18.htm]

"Changnoi: Not a Picnic for Pak Mool Refugees." *The Nation* (Bangkok), April 21, 1999. [http://www.nextcity.com/ProbeInternational/Mekong/articles/thai9904ii.html]

Gearing, Julian. "The Struggle for the Highlands: Accused of Endangering the Environment, Thailand's Tribespeople Face Eviction and an Uncertain Future." *Asiaweek* 25:43 (October 29, 1999): n.p. [http://www.Asiaweek.com]

Korkeatkachorn, Wipaphan. "'Son of a Commoner' Faces the Assembly of the Poor." *Focus on the Global South.* Bangkok, Thailand: 2000. [http://www.focusweb.org/publications/2000/Son%20of%20a%20commoner%20faces%20the%20Assembly%20of%20the%20Poor.htm]

Magagnini, Stephen. "Orphans of History: A Special Report by the *Sacramento Bee.*" *Sacramento Bee,* December 31, 2000. [http://www.sacbee.com/static/archive/news/projects/hmong/123100/diary.html]

THANKSGIVING CEREMONIAL CYCLES OF NATIVE AMERICANS: ECOLOGICAL PERSPECTIVES

Ceremonies of thanksgiving for the bounty of nature are a common element in many Native American cultures, among them the people who introduced the custom to English colonists who celebrated what later became known as the first Thanksgiving in 1621. A fall thanksgiving holiday, usually accompanied by feasting on traditional Native American foods (turkey, corn, yams, squashes, cranberry sauce, etc.) has been widely practiced since about 1800 by most nonnative people in the United States and Canada. Thanksgiving was declared a national holiday by President Abraham Lincoln in 1863, in the midst of the Civil War. Canada declared an official Thanksgiving holiday in 1879; this day is celebrated six weeks before its counterpart in the United States.

Contrary to popular belief, the Pilgrims did not eat turkey for the first time at the original Anglo-Indian feast. Turkeys were first domesticated by the Aztecs, and they were imported to Spain during the conquest more than a century before English colonists arrived in New England. Aztec turkeys had been imported to England by the time the Pilgrims sailed, and the immigrants had some of the birds on board the Mayflower.

A spirit of thanksgiving to the natural world is common in many Native American cultures. The Haudenosaunee (Iroquois), of what is now upstate New York, maintain a yearlong thanksgiving cycle. Mohawk Nation Council Subchief Tom Porter offered a traditional thanksgiving prayer:

> [Before] our great-great grandfathers were first born and given the breath of life, our Creator at that time said the earth will be your mother. And the Creator said to the deer, and the animals and the birds, the earth will be your mother, too. And I have instructed the earth to give food and nourishment and medicine and quenching of thirst to all life. . . . We, the people, humbly thank you today, Mother Earth.
>
> Our Creator spoke to the rivers and our creator made the rivers not just as water, but he made the rivers a living entity. . . . You must have a reverence and great respect for your mother the earth. . . . You must each day say "thank you"

[for] every gift that contributes to your life. If you follow this pattern, it will [be] like a circle with no end. Your life will be as everlasting as your children will carry on your flesh, your blood, and your heartbeat. (Johansen, 1993, 2)

A tribute to the Creator and a reverence for the natural world is reflected in many Native American greetings that span the North American continent. More than 2,500 miles from the homeland of the Mohawks, the Lummi of the Pacific Northwest Coast often begin a public meeting this way: "To the Creator, Great Spirit, Holy Father: may the words that we share here today give the people and [generations] to come the understanding of the sacredness of all life and creation" (Grinde and Johansen, 1994, 35).

REFERENCES

Grinde, Donald A., Jr., and Bruce E. Johansen. *Ecocide of Native America: Environmental Destruction of Indian Lands and Peoples*. Santa Fe, N.M.: Clear Light, 1994.

Johansen, Bruce E. *Life and Death in Mohawk Country*. Golden, Col.: North American Press/Fulcrum, 1993.

TIBET

INTRODUCTION

Environmental colonization can be on a large scale, as evidenced by Chinese plans to build a railway into Tibet, over some of the world's most rugged terrain. Oppression also may be very personal, as in the case of one Tibetan monk who protested gold mining near his monastery. As a result he was ousted from his home, tortured by Chinese authorities, and his sanctuary was turned into a Chinese center for patriotic education.

CHINA'S TIBETAN RAILWAY

The Dalai Lama, in midsummer 2001, said he believes that China's plans to connect the Tibetan capital Lhasa with three Chinese cities by rail is an attempt to forcibly change Tibet's indigenous demography and environment. Speaking to journalists at a Delhi hotel on July 14, 2001, the fourteenth Dalai Lama, Tenzin Gyatso, the head of state (in exile) as well as the spiritual leader of the Tibetan people, said: "[The] Chinese are setting up the railway tracks, but it is not for economic development. They have plans to transfer 20 million Chinese into Tibet. The purpose of the railways is basically to facilitate the transfer of population." The Dalai Lama also said that the railway will "do serious damage to the environment" ("Dalai Lama," 2001). The Dalai Lama said that the railway could devastate traditional Tibet, because it will strengthen Chinese rule, inundate Tibet with ethnic Chinese settlers, and draw Tibet's natural resources into China.

The Dalai Lama did not deny reports that young Tibetans inside the country might organize armed resistance to the railroad and other manifestations of Chinese domination. He said: "There are individuals who do not believe in my [nonviolent] approach. In fact some of them, who are staying inside Tibet, say 'that family [China] does not understand truth, but only violence'" ("Dalai Lama," 2001).

The Chinese government plans to construct the world's highest railway, at elevations sometimes exceeding 4,500 meters (14,500 feet). The 1,120-kilometer stretch of steel rail will advance the Chinese government's 50-year-old goal of linking Tibet to China's interior, economically, politically, and culturally. The railway was started in June 2001, and at that time was scheduled for completion in 2007, at an estimated cost of U.S. $2.5 billion. Engineering and construction have been challenged by high altitudes, steep grades, frigid temperatures, howling winds, and soils that can expand or contract more than a meter depending on the season ("Tibet Railway," 2001). The region's fragile ecosystem could be damaged if promised protective measures fail. According to a report by the Knight-Ridder News Service, "Some Tibetan groups have complained that the railway will also accelerate China's exploitation of Tibet's natural resources and the destruction of Tibetan culture" ("Tibet Railway," 2001).

The line will run from Tibet's capital, Lhasa, to the Chinese Qinghai provincial city of Golmud, near the Kunlun Pass. The Chinese government's official position is that the rail line "will bring prosperity to impoverished Tibet, where average annual incomes are about half of what they are in the rest of China" ("Tibet Railway," 2001.) High altitudes will require specially adapted engines that will allow the train to function at low levels of oxygen and pressurized cars that will shield passengers from altitude sickness. The trains also will have to be powerful enough to negotiate some of the steepest terrain of any railroad on Earth. Parts of the rail bed also are prone to earthquakes and landslides. The rail line also will traverse 286 bridges and 10 tunnels (one of them about a mile long), according to the project's senior engineer, Zhang Xiuli ("Tibet Railway," 2001).

Before it is completed, the Chinese estimate that its Tibetan railway will tap the time and talents of about 50,000 laborers and engineers, "many working just a few hours a day because of weakness and health problems caused by lack of oxygen" ("Tibet Railway," 2001). Chinese officials expect eight trains a day to ply the line once it opens, with many of the passengers being tourists. The Chinese also have been rather open about the Dalai Lama's objections, telling reporters that the railway will hasten Chinese access to Tibetan timber and minerals, as well as allowing easier transport of troops "to control . . . a still-restive population" ("Tibet Railway," 2001). The railway will replace a rocky, partially paved, two-lane road that is known locally as one of the world's most treacherous highways. The drive from Lhasa to Golmud routinely consumes six bone-rattling days.

Project director Zhao Xiyu said extensive efforts would be made to protect delicate plants as well as wildlife in the area. He said new technology would be used to stabilize frozen soils, and that "some areas of track would be elevated to allow migratory species, including the endangered Tibetan antelope, to pass under the line" ("Tibet Railway," 2001).

Chinese President Jiang Zemin said the government would seek to develop Tibet's economy while at the same time cracking down on separatist activities. "We will continue to pay attention to two major issues in Tibet—economic development and social stability," Jiang was quoted as telling Xinhua, China's state-run news agency. He added that anyone who uses "religion as camouflage" to support independence for Tibet would be punished ("Tibet Railway," 2001). Exiled Tibetans, in response, pointed to a historical irony: the first railroads were introduced into China a century ago by European powers, notably the British, cooperating with the Japanese, to strengthen their imperialist control over China, including foreign military control and transit of Chinese resources out of the country.

A MONK PAYS THE PRICE OF PROTESTING GOLD MINING

Kabukye Rinpoche, head of Nubzur monastery in Kardze, Tibet, was jailed by Chinese authorities during 1999 for expressing his opposition to gold mining near his monastery. The mine, opened in 1992, quickly brought more than 300 Chinese miners to the area. According to a report in *Drillbits and Tailings*, Tibetans believe that the area is a holy mountain and the dwelling place of many protective earth spirits.

Rinpoche, who suffers heart problems, sent letters of protest to local authorities, raising concerns that explosions used to loosen rock on the hillside were causing problems for the nomads herding their livestock. He also asserted that mining was eroding nearby grasslands, which were necessary for local herders.

On July 10, 1996, Rinpoche was arrested, according to the Tibet Information Network (TIN), a support group in London. Immediately after his arrest, Chinese security personnel visited the monastery and broke into Rinpoche's room, where they confiscated posters and a camera. The monk wasn't charged with opposing gold mining, but for hanging posters advocating the independence of Tibet from China ("Tibetan Monk," 1999). The Chinese prohibited displaying photographs of the Dalai Lama and listening to Tibetan-language broadcasts on Voice of America.

Two days after Kabukye Rinpoche's arrest, Chinese security personnel returned to Nubzur monastery and arrested Pantsa, the caretaker of the monastery. Both men then were reportedly interrogated and tortured. A source told TIN that the men were held in a small cell without windows or light, were given very poor food, and were repeatedly beaten and denied sleep at night ("Tibetan Monk," 1999).

The same report added that Rinpoche also was forced to strip naked and stand next to a burning fire, "on one occasion losing consciousness and falling into the fire, causing severe burns to his left arm" ("Tibetan Monk," 1999). In addition, Rinpoche was forced to kneel for hours, damaging his knees so badly that he could not stand up properly. Rinpoche remained in detention until he was formally charged and sentenced on October 27, 1996 to six years' imprisonment, with a fine equal to almost U.S. $1,000, for the crime of "counterrevolutionary splittism [dissent]." A Chinese government so-called work committee has taken up residence in the dissidents' monastery and turned it into a center for "patriotic education" ("Tibetan Monk," 1999).

REFERENCES

"Dalai Lama Slams Lhasa Rail Link; Basharat Peer in New Delhi." July 21, 2001. [http://www.tibet.net/eng/diir/enviro/news]

"Tibetan Monk Tortured for Protesting Chinese Gold Mining." *Drillbits and Tailings* (January 21, 1999): n.p. [http://www.moles.org/ProjectUnderground/drillbits/990121/99012103.html]

"Tibet Railway to Fast-track Political and Economic Goals." Knight-Ridder News Service, July 23, 2001.

TURKEY

THE KURDS: DAM BUILDING

Balfour Beatty, an international engineering, construction, and services group, decided during November 2001, to pull out of the controversial Ilisu dam project in Turkey. The dam, which would have displaced about 25,000 indigenous Kurds, is part of a $1.6 billion hydroelectric and irrigation project 65 kilometers (40 miles) upstream of Turkey's border with Syria and Iraq. Balfour Beatty's chief executive officer, Mike Welton, told the Environment News Service: "The urgent need for increasing generating capacity to meet Turkey's development needs and for social and economic development in the region remains. We have, however, clearly reached a point where no further action nor any further expenditure by Balfour Beatty on this project is likely to resolve the outstanding issues in a reasonable time scale" ("British Engineering," 2001).

Italian builder Impreglio also withdrew from the Swiss-led consortium that holds a contract to build the dam. In Britain, the International Development Committee of the House of Commons said: "The Ilisu Dam was from the outset conceived and planned in contravention of international standards and still does not comply" ("British Engineering," 2001). The World Bank also has refrained from further investments in the project. The Swedish company Skanska also withdrew, following revelations of damage that the dam's construction would cause to indigenous communities and the environment.

The planned dam, if constructed, would disrupt the flow of the Tigris River to Iraq and Syria. The reservoir behind the Ilisu dam, as proposed, would have flooded 52 villages and 15 small towns, and as many as 25,000 people would have been displaced, including Hasankeyf, a Kurdish town of about 5,500 people that dates back at least 10,000 years. The town contains ruins of considerable archeological value, which have been at the center of an international campaign to stop the dam's construction.

The International Rivers Network has been campaigning to stop the Ilisu dam. In a statement released during November 2000, the group said that the Ilisu project violates all seven strategic priorities of the World Commission on

Dams: "gaining public acceptance, comprehensive options assessment, addressing existing dams, sustaining rivers and livelihoods, recognizing entitlements and sharing benefits, ensuring compliance, and sharing rivers for peace, development and security" ("British Engineering," 2001).

REFERENCES

"British Engineering Company Withdraws from Ilisu Dam Project." Environment News Service, November 14, 2001. [http://www.ens-news.com/ens/nov2001/2001L-11-14-02.html]

UNITED STATES OF AMERICA

INTRODUCTION

When most American Indians were assigned reservations in the late nineteenth century, many of their lands were assumed to be relatively worthless for activities deemed valuable to Anglo-American settlers, whose main interest was agriculture. The reservations were assigned at the dawn of fossil-fueled industrialism—ironically, as it turned out, because areas that appeared so barren to the untrained eye held, underground, a wealth of mineral and metal resources. In particular, Indian reservations in the United States possess a substantial proportion of the uranium and coal within U.S. borders. During the ensuing one-and-a-quarter centuries, the circumstances of industrialization and technical change have made many of these treaty-guaranteed lands very valuable.

According to a Federal Trade Commission (FTC) report distributed during October 1975, an estimated 16 percent of the United States' uranium reserves that were recoverable at market prices were on reservation lands; this was about two-thirds of the uranium on land under the legal jurisdiction of the U.S. government. There were, at that time, almost 400 uranium leases on these lands, according to the FTC, and between 1 million and 2 million tons of uranium ore a year, about 20 percent of the national total, was being mined from reservation land.

Americans Indians living within the borders of the United States have, therefore, developed an intimate relationship over the years with exploiters of resources. Environmental provocations afflicting Native American peoples in the United States range from uranium to kitty litter. Even this brief summary of environmental issues facing indigenous peoples who reside within the borders of the United States reveals a range of problems equal to those of any Third World nation—from the toll of uranium mining on the Navajos, to the devastation wrought by dioxin, PCBs, and other pollutants on the agricultural economy of the Akwesasne Mohawk reservation in northernmost New York State. As with the Akwesasne Mohawks, some of the most serious problems span international borders. The Yaquis, whose homelands span the U.S.-Mexican border, have been afflicted with some of the same pesticides as the Mohawks on the U.S-Canadian border.

Some of the environmental problems faced by indigenous peoples in the United States strain one's sense of credulity. Native lands repeatedly have become targets for proposals whose sponsors seem to ignore the fact that these lands have human inhabitants. Witness the Eskimos of Point Hope, Alaska, who have learned that their land had once been proposed as the site of a new harbor to be created with nuclear weapons. The harbor was never created, but the Point Hope Eskimos still found themselves hosting uninvited nuclear waste. Other Alaskan Eskimos have found their reindeer rendered inedible, polluted with heavy metals.

The Western Shoshone of Nevada have come to call themselves "the most bombed nation on Earth," a reference to a neighboring test range for nuclear weapons and (most recently) a proposal to open a national uranium-waste repository at Yucca Mountain. Yet another dump proposal was defeated by a human blockade in Ward Valley, California. The United States' lineup of environmental hot spots even includes one tiny Indian nation, the Goshutes of Utah, who have decided to accept toxic (including radioactive) wastes from the highest bidder, as long as the price is deemed to be high enough.

AKWESASNE: THE LAND OF THE TOXIC TURTLES

Within the living memory of a middle-aged person in the 1990s, Akwesasne (the Mohawk name for the Saint Regis Mohawk reservation that straddles the U.S.-Canadian border in New York State) has become a toxic dumping ground riskier to human health than many urban areas. These environmental circumstances have, in two generations, descended on a people whose whole way of life once was enmeshed with the natural world, a place where the Iroquois' creation story says the world took shape on a gigantic turtle's back. In our time, environmental pathologists have found turtles at Akwesasne that qualify as toxic waste.

The Akwesasne reservation was abruptly introduced to industrialism with the coming of the St. Lawrence Seaway during the mid-1950s. Soon after the Seaway opened, industry began to proliferate around the Mohawks' homeland. A General Motors foundry opened, followed by a aluminum plants and steel mills that provided raw materials and parts for the foundry. Akwesasne has been declared the most polluted Indian reserve in Canada, and the largest nonmilitary contamination site in the United States (Lickers, 1995, 11) By the mid-1980s, the Mount Sinai School of Medicine, having studied pollution at Akwesasne, advised residents to eat no more than a half-pound of locally caught fish per week. Women who were pregnant or nursing, and children under 15 years of age, were told not to eat any local fish at all.

When Ward Stone, a wildlife pathologist for the New York State Department of Environmental Conservation, began examining animals at Akwesasne, he found that the PCBs, insecticides, and other toxins were not being contained in designated dumps. After years of use, the dump sites had leaked, and the toxins had gotten into the food chain of human beings and nearly every animal species in the area. The Mohawks' traditional economy, based on hunting, fishing, and agriculture, had been literally poisoned out of existence.

Stone's environmental tour of Akwesasne began with a visit to one of the General Motors waste lagoons, a place called "unnamed tributary cove" on some maps. Stone gave it the name "contaminant cove" because of its amount of toxic pollution. One day in 1985, at contaminant cove, the environmental crisis at Akwesasne assumed a whole new, foreboding shape. The New York State Department of Conservation caught a female snapping turtle that contained 835 parts per million of PCBs. The turtle carries a special significance among the Iroquois, whose creation story describes how the world took shape on a turtle's back. To this day, many Iroquois call North America "Turtle Island."

While no federal standards exist for PCBs in turtles, the federal standard for edible poultry is three parts per million, or about one-third of one percent of the concentration in that snapping turtle. The federal standard for edible fish is two parts per million. In soil, on a dry-weight basis, 50 parts per million is considered hazardous waste, so the turtle contained roughly 15 times the concentration of PCBs necessary, by federal standards, to qualify its body as toxic waste.

During fall 1987, Stone found another snapping turtle, a male, containing 3,067 parts per million in its body fat—1,000 times the concentration allowed in domestic chicken, and 60 times the minimum standard for hazardous waste. Contamination was lower in female turtles because they shed some of their own contamination by laying eggs, while the males stored more of what they ingested.

In 1985, Stone, working in close cooperation with the Mohawks, found a masked shrew that somehow had managed to survive in spite of a PCB level of 11,522 parts per million in its body, the highest concentration that Stone had ever seen in a living creature, and 250 times the minimum standard to qualify as hazardous waste. Using these samples, and others, Stone and the Mohawks established Akwesasne as one of the worst PCB-polluted sites in North America. In 1986, pregnant women were advised not to eat fish from the St. Lawrence, historically the Mohawks' main source of protein. Until the 1950s, Akwesasne had been home to more than 100 commercial fishermen and about 120 farmers. By 1990, fewer than 10 commercial fishermen and 20 farmers remained.

The Environmental Protection Agency released its Superfund cleanup plan for the General Motors foundry during March 1990. The cleanup was estimated to cost $138 million, making the General Motors dumps near Akwesasne the costliest Superfund cleanup job in the United States, and first on the EPA's most wanted list as the United States' worst toxic dump. By 1991, the cost was scaled down to $78 million, but the General Motors dumps were still ranked as the most expensive toxic cleanup.

"We can't try to meet the challenges with the meager resources we have," said Henry Lickers, a Seneca who is employed by the Mohawk Council at Akwesasne. Lickers has been a mentor to today's younger environmentalists at Akwesasne. He also has been a leader in the fight against fluoride emissions from the Reynolds plant. "The next ten years will be a cleanup time for us, even without the money," said Lickers (Johansen, 1993, 19).

According to Lickers, the destruction of the Akwesasnes' environment was the catalyst that spawned the Mohawks' deadly battle over high-stakes gambling and smuggling. "A desperation sets in when year after year you see the decimation of the philosophical center of your society," Lickers said (Johansen, 1993, 19).

The Mohawks are not alone. Increasingly, restrictive environmental regulations enacted by states and cities are bringing polluters to Native reservations. "Indian tribes across America are grappling with some of the worst of its pollution: uranium tailings, chemical lagoons and illegal dumps. Nowhere has it been more troublesome than at . . . Akwesasne," wrote Rupert Tomsho, a reporter for the *Wall Street Journal* (Tomsho, 1990, 1).

Katsi Cook is a Mohawk midwife, who has studied the degree to which mothers' breast milk has been laced with PCBs at Akwesasne. Cook said that "this means that there may be potential exposure to our future generations. The analysis of Mohawk mother's milk shows that our bodies are, in effect, part of the [General Motors] landfill" (LaDuke, 1994, 45).

—John Kahionhes Fadden

INDIGENOUS CONCERNS REGARDING THE GREAT LAKES

During 2001, a group of indigenous people and non-Native supporters took part in a 2,200-mile journey from the eastern seacoast of North America to the western shore of Lake Superior. The walkers were part of the Migration Journey for the Seventh Generation, following an ancient Anishinabek migration route. Beginning in mid-July 2001, the 2,200-mile Journey for the Next Seven Generations followed the path taken about 1,200 years ago by the Anishinabek (Ojibways) from the gulf of the St. Lawrence River (at Burnt Church, New Brunswick), to Madeline Island in Lake Superior. The Anishinabek (Ojibway, Odawa, and Potawatomi) are Native American peoples of the Great Lakes region.

According to publicity for the march, "The Migration Journey is a call for unity to stand against the pollution of the Great Lakes and the misuse and waste of Great Lakes water. Our mission is to reach out and educate all people about protecting the waters, and restoring the earth's natural balance for seven generations to come" ("2,200 Mile," n.d.). "Now is the time that people wake up, make a stand and speak out for the water to continue the life of the future generations," said Migration Journey walker Corrine Tooshkenig, Walpole Island First Nation, Ontario ("2,200 Mile," n.d.).

According to the same statement, "In the 34 kilometers between Walpole Island and Sarnia, there are more than two-dozen of the world's largest chemical and petrochemical companies, whose discharges into the St. Clair River have made the water undrinkable for residents at Walpole Island" ("2,200 Mile," n.d.). "Scientists tell us that women and children are now considered 'at risk populations' for environmental diseases in the Great Lake region. When the givers of life and our future generations are at risk it is nothing short of environmental genocide. We are all complicit in a cultural suicide," said Migration Journey organizer Kevin Best ("2,200 Mile," n.d.).

The marchers' journey retraced the steps of the Anishinabeks' historical migration route, with seven stopping places that are described in their oral histories. Walpole Island was the third of these seven stopping places. The Migration Journey is regarded by its participants as "a widening of the circle of awareness and a continuation of the work that began during last year's 'Walk to Remember' around Lake Superior. That walk gave a voice to many communities affected by the contamination of their waterways and watersheds, from human and animal waste to P.C.B. and mercury contamination" ("2,200 Mile," n.d.).

Along the way, participants called for cessation of industrial activity that pollutes the lakes' watershed. By mid-September, the blistered and callused marchers pulled into Owen Sound, Ontario. Their numbers ranged from a minimum of five people to a hundred or more on various segments of their trek. Native people along the way, on both sides of the Canadian–United States border, often supplied the trekkers with food and lodging. "We've been astounded by the generosity and support we've received," said Migration Journey coordinator Kevin Best (Avery, 2001, C-1).

In 1970, the Great Lakes were so polluted that fish were dying in large numbers, and some birds that hatched near the lakes were born with crossed beaks. One of the rivers flowing into the lakes, through Cleveland, was so contaminated that it caught fire. In the early 1970s, Lake Erie was declared dead. Such gross pollution made the Great Lakes infamous in the 1970s. A crackdown on chemical dumping into the lakes paid off, however. Bans on the worst compounds and improved industrial practices have witnessed declines in levels of dioxins, DDT, and other toxins. The amount of dioxin entering Lake Superior plummeted 95 percent during the 1990s. The load of phenanthrene, a toxic byproduct of burning coal in Lake Ontario, also dropped more than 90 percent.

By the year 2000, most of the Great Lakes' water was safe for swimming. Bald eagles and peregrine falcons were returning, and fish appeared to be thriving. Officials said, however, that appearances could be deceiving. Decades' worth of contaminants, long buried in the lake bottoms, continued to bioaccumulate in fish. New problems included an expanding number of animal farms (especially large-scale hog farms) that generated more sewage than 100 million people (Munro, 2001). "The future of the [Great Lakes] basin is at risk," Canada's federal environment commissioner, Johanne Gèlinas, said in a report issued during 2001. The Environment Canada report highlights the staggering growth of livestock operations in Ontario and Quebec. Nitrogen, phosphorous, and microbes from the manure can leak into groundwater and nearby lakes and rivers that feed into the Great Lakes. "The problem of how to manage it safely is getting worse," said Gèlinas (Munro, 2001).

By far the highest-volume pollutant entering the lakes was human and animal sewage. Many municipal sewers were not designed to treat the household cleaners and drugs flowing into them, so they run straight through the plants and into the lakes. Gèlinas reported that 40 percent of the municipal effluents of cities examined received only primary treatment. Primary treatment does little more than remove the lumps from the smelly slurry (Munro, 2001).

Sites including parts of Hamilton Harbor, the Detroit and Niagara rivers, and Toronto's waterfront are laden with heavy metals and persistent toxins such as PCBs, furans, and dioxins. Contaminants will cycle through the lakes' food chains for decades to come if they are not removed, Gèlinas said. Ballast water dumped in the lakes by foreign ships traveling on the lakes is another concern. The water, taken on board in distant harbors, could introduce deadly microbes, such as cholera, Gèlinas warned (Munro, 2001).

Scientists also are turning up several worrisome new menaces in the Great Lakes' water. Many of the new threats come not from industry but from homes—"chemicals and pharmaceuticals flushed down toilets, pesticides from backyard gardens and hydrocarbons and other pollutants running off roadways" (Munro, 2001). Some fish caught in the lake are unsafe to eat. More than 360 chemicals have been identified in the water, sediment, and animals of the Great Lakes basin, according to the report. Roughly one-third of these chemicals can have acute or chronic toxic effects. People are exposed to toxins mainly in the fish they eat, and to a lesser extent in drinking water. Environment Canada, which tried to keep tabs on the chemicals going into Great Lakes waters, by the late 1990s was monitoring fewer than half of the 58 most common pollutants in the lakes' watershed.

THE PENOBSCOT: ORGANOCHLORINE CONTAMINATION

Rebecca Sockbeson, a Penobscot, described "the devastating impact of dioxins in my community," to international negotiators of a treaty to eliminate or ban the most widespread persistent organic pollutants (POPs) (Sockbeson, 1999). She said that her nation of nearly 500 people lives on an island in the river, close to seven pulp-and-paper mills.

Dioxins were being created as a byproduct of the chlorine bleaching process in making paper, discharged from all seven of these mills. Dioxins, which are highly potent toxic chemicals that may cause cancer and other health problems, were being poured daily into an adjacent river. Sockbeson said that her people have survived on the fish from this river, but "now we are dying from it." She continued:

Neither dioxin nor cancer is indigenous to the Penobscot people, however they are both pervasive in my tribal community. My people face up to three times the state and national cancer rate, moreover, those that are dying of cancer are dying at younger and younger ages, our reproductive generation. This means that unless you take action to eliminate dioxin and other persistent organic pollutants, there will be no Penobscots living on the island by the end of the next century. (Sockbeson, 1999)

The health and survival of Sockbeson's Penobscot band also is threatened by a choice mothers must make: Should they breast-feed their children (imparting superior nutrition) and thus pass on to them their body burdens of PCBs and dioxins? "With this," Sockbeson concluded, "I humbly, respectfully and desperately urge you to draft a treaty that insures the existence of the Penobscot and other indigenous peoples who are so disproportionately impacted by dioxin." Such a treaty is required, she said, so "that the breast and spoon we feed our babies with is not filled with cancer, diabetes, learning disabilities, and attention deficit [disorder]" (Sockbeson, 1999).

THE YAQUI: PESTICIDE CONTAMINATION

The indigenous Yaqui are a farming people, living and working in the environs of the Yaqui Valley in Sonora, Mexico, spanning the border between the United States and Mexico. After World War II, due to a lack of available water and financing, many of the Yaqui became unable to support their own farms, as they had for centuries. The Yaqui were then forced to lease their lands to outsiders, mainly corporate farmers, who were heavy users of pesticides, herbicides, and fungicides. The use of these chemicals, usually applied by aerial spraying, by tractor, and by hand, brought widespread contamination of land, water, and people.

Concurrently, valley farm operations became mechanized, and irrigation and transport systems were established. The result was a green revolution, with farming becoming big business. Yaqui families from the nearby mountain foothills moved into the valley for employment, while some valley residents moved into the foothills to maintain family-scale farms (Guillette et al., 1998).

Farmers in the valley reported that usually two crops a year were planted, with pesticides applied as many as 45 times per crop. Compounds included multiple organophosphate and organochlorine mixtures, as well as pyrethroids. Thirty-three different compounds were used for the control of cotton pests alone between 1959 and 1990. This list includes DDT, dieldrin, endosulfan, endrin, heptachlor, and parathion-methyl, to name a few examples. As recently as 1986, 163 different pesticide formulations were sold in the southern region of the state of Sonora. Substances banned in the United States, such as lindane and endrin, are readily available to farmers living in the Mexican parts of the valley.

In the valley, pesticide use is widespread and continues throughout the year, with little governmental control. Contamination of the resident human population has been documented; after one month of lactation, women's breast milk concentrations of lindane, heptachlor, benzene hexachloride, aldrin, and endrin were all above limits set by the Food and Agricultural Organization of the United Nations (Guillette et al., 1998). During 1990, high levels of multiple pesticides were found in the cord blood of newborns and in breast milk of valley residents. Children are usually breast-fed, then weaned onto household foods.

Household insect sprays usually were applied each day throughout the year in the lowland homes. In contrast, the foothill residents maintained traditional intercropping for pest control in gardens. They usually controlled insects in their homes by swatting them. Most of these people were exposed to pesticides only when the government sprayed DDT each spring to control malaria (Guillette et al., 1998).

Angel Valencia, a spiritual leader of the Yaqui tribe in the village of Potam in Sonora, Mexico, described the effects of these chemicals among valley residents. Valencia spoke as a representative of the Arizona-based Yoemem Tekia Foundation, an affiliate of the International Indian Treaty Council.

> I have seen with my own eyes the effects of daily contact with these pesticides— it burns their skin, they lose their fingernails, develop rashes and in some cases they have died as a result of exposure to these poisons. . . . The tragedy of this situation makes me both sad and angry—to think of what has been done to the innocent children who are the future of the Yaqui people. They will not be able to grow and develop, as they deserve to. (Valencia, 2000)

During the 1990s, Elizabeth Guillette, an anthropologist and research scientist at the University of Arizona, studied the impacts of pesticide exposure on Yaqui children. Guillette's studies confirmed the observations of Valencia, who said that exposure to pesticides had "a serious impact on the health and physical and mental development of the children of our villages" (Valencia, 2000). Prior to Guillette's research, researchers at the Technological Institute of Sonora in Obregón, Mexico had shown that children in Sonora's Yaqui Valley often were born with detectable concentrations of many pesticides in their blood and were exposed again through consumption of their mothers' breast milk.

"I know of no other study that has looked at neurobehavioral impacts—cognition, memory, motor ability—in children exposed to pesticides," said neurotoxicologist David O. Carpenter of the State University of New York at Albany. "The implications here are quite horrendous," he said, because the magnitude of observed changes "is incredible—and may prove irreversible" (Raloff, 1998). "Although the children exhibited no obvious symptoms of pesticide poisoning, they're nevertheless being exposed at levels sufficient to cause functional defects," observed pediatrician Philip J. Landrigan of Mount Sinai Medical Center in New York (Raloff, 1998).

In Guillette's study, children of the agrarian region were compared to children living in the foothills, where pesticide use is avoided. The study selected two groups of four- and five-year-old Yaqui children who resided in the Yaqui Valley of northwestern Mexico. These children shared similar genetic backgrounds, diets, water-mineral contents, cultural patterns, and social behaviors. The major difference was the level of their exposure to pesticides. Guillette adapted a series of motor and cognitive tests into simple games the children could play, including hopping, ball catching, and picture drawing.

The study was constructed in this manner to minimize variables that can affect the outcome of a pesticide study on children's growth and development. The population had to meet the requirements of similar genetic origins, living conditions, and related cultural and social values and behaviors, all of which are necessary for comparable study and reference groups.

Guillette had assumed that any differences between the two groups would be subtle. Instead, she recalled, "I was shocked. I couldn't believe what was happening" (Luoma, 1999). According to an account by Jon R. Luoma in *Mother Jones*,

> The lowland children had much greater difficulty catching a ball or dropping a raisin into a bottle cap—both tests of hand-eye coordination. They showed less physical stamina, too. But the most striking difference came when they were asked to draw pictures of a person. . . . Most of the pictures from the foothill children looked like recognizable versions of a person. The pictures from most of the lowland children, on the other hand, were merely random lines, the kind of unintelligible scribbles a toddler might compose. . . . It appeared likely they had suffered some kind of brain damage. (Luoma, 1999)

During a follow-up trip in 1998, two years after her initial visit, Guillette found that both groups of children (who at that time were in primary school) had improved in drawing ability. While the lowland children's drawings looked more like people than before, the foothill children were drawing far more detailed images. The lowland youngsters were still evidencing some motor problems, particularly with balance. "Some of these changes might seem minute, but at the very least we're seeing reduced potential," Guillette said. "And I can't help wondering how much these kinds of chemicals are affecting us all" (Luoma, 1999).

No differences were found in physical growth patterns of the two groups of children. Functionally, however, Guillette and colleagues wrote, "the exposed children demonstrated decreases in stamina, gross and fine eye-hand coordination, 30-minute memory, and the ability to draw a person" (Guillette et al., 1998). Guillette gave children red balloons for successful completion of tasks. "Well over half of the lesser-exposed children could remember the color in the object, and all remembered they were getting a balloon. Close to 18 percent of the exposed children could not remember anything," and only half could remember they were getting a balloon. "It was quite a contrast," she said (Mann, 2000, C-9).

Guillette said she noticed that exposed Yaqui children would walk by somebody and strike them without apparent provocation. Otherwise, they tended to sit in groups and do nothing. Foothill children, by contrast, were always busy, engaged in group play. "I'd throw the ball to a group of kids. In the valley, one child would get the ball and just play with it himself," she says. The foothill children played with the ball as a group (Mann, 2000, C-9). Yaqui mothers from the valley also reported more problems getting pregnant and higher rates of miscarriages, stillbirths, neonatal deaths, and premature births.

It was impossible to tell which specific pesticides to which the Yaquis had been exposed, Guillette said. "We know for sure there has been D.D.T. exposure. . . . [The Mexican government] does not know what's being used. The farmer does not give out the information. Pesticides are tied to bank loans, and the banks won't reveal what is being used with certain crops. I just assume everything. The other problem is they get a little of this and a little of that and mix it up. It is very important to remember that the situation is no different agriculturally than what you find in California, the Midwest or the East Coast in the U.S." (Mann, 2000, C-9).

"Many of these contaminants have similar reactions in the body," Guillette said. "Many disrupt the endocrine system, which regulates body functions, and that's the main reason I looked at subtle changes. The shifts may seem slight, but when they occur within a total society, they can have major implications. To me, the approach should not be treatment of the disease or trying to teach compensation for the deficit but to look at the basic problem of contamination" (Mann, 2000, C-9).

According to Guillette, children from the valley appeared less creative in their play. "They roamed the area aimlessly or swam in irrigation canals with minimal group interaction," she said. "Some valley children were observed hitting their siblings when they passed by, and they became easily upset or angry with a minor corrective comment by a parent. These aggressive behaviors were not noted in the foothills. "Some valley mothers stressed their own frustration in trying to teach their child how to draw" (Guillette et al., 1998).

Concluding her study, Guillette raised a question that summarized concerns of parents in the lowlands of the pesticide-ridden valley: "Environmental change has placed the children of the agricultural area of the Yaqui valley at a disadvantage for participating in normal childhood activities. Will they remain at risk for functioning as healthy adults?" (Guillette et al., 1998).

THE POINT HOPE ESKIMOS: AN ATOMIC HARBOR AND A NUCLEAR DUMP

Since the 1950s, the U.S. Atomic Energy Commission (AEC) has proposed to demonstrate the peaceful uses of atomic energy by blowing open a new harbor at Point Hope, Alaska, with a series of underground atomic blasts. The Inuit (Eskimo) people, who live on the far northwest coast of Alaska, have been engaged in ongoing resistance to these plans. In 1962, the government shelved the plan, which was called Operation Chariot. After that, without informing the Inuit, the federal government turned parts of their homeland into a nuclear dump.

Forty-three pounds of radioactive soil was stored near Point Hope, originating from within a mile of ground zero of a nuclear blast in Nevada. The soil contained strontium-85 and cesium-137 (Badger, 1992, B-5). The strontium typically would have lost all its radioactivity years before its deposit at Point Hope, and the cesium would still have had about half its radioactivity after 30 years, according to government officials. The purpose of the experiment, according to the AEC, was to test the toxicity of radiation in an arctic environment. The dump experiment was carried out by the U.S. Geological Survey under license from the AEC.

The government seemed indifferent to the fact that the area was occupied by Native people. Point Hope, the closest settlement to the dump, is an Inuit village of about 700 people, most of whom make a living as whalers. It is one of the oldest continuously occupied towns in North America. The Inuit did not learn of the nuclear dump until Dan O'Neill, a researcher at the University of Alaska, made public documents he had found as he researched a book on the aborted plan to create a harbor on the Alaskan coast with nuclear weapons. O'Neill, using the Freedom of Information Act, learned that the nuclear waste had been stored in the area as part of Project Chariot, which was declassified in 1981. According to O'Neill, the nuclear dump was clearly illegal, and contained "a thousand times . . . the allowable standard for this kind of nuclear burial" (Grinde and Johansen, 1995, 238–39). The nuclear waste that was buried near Point Hope remained unmarked for 30 years, during which time hunters crossed it to pursue game, and caribous migrated through it. Not until September 1992 did the U.S. federal government admit that it had buried 15,000 tons of radioactive soil at Cape Thompson, 25 miles from Point Hope, on the Chukchi Sea in northwestern Alaska.

For many years, the Inuit in the area had suffered cancer rates that far exceeded national averages. The government acknowledged that soil in the area contained "trace amounts" of radiation, but denied that its experimental nuclear dump had caused the Inuits' elevated cancer rate. Until the dump was disclosed during the late 1990s, the Inuit in Port Hope had no clue as to why the incidence of cancer in their village had jumped to 578 per 100,000 within two generations. Some doctors blamed the rise in cancer rates on smoking by the Inuit. In 1997, the chief medical officer of the borough of Barrow, Alaska, published findings linking the increase in cancer incidence to the burial of nuclear waste near Port Hope (Colomeda, 1998).

"I can't tell you how angry I am that they considered our home to be nothing but a big wasteland," said Jeslie Kaleak, mayor of the North Slope Borough, which governs eight Arctic villages, including Point Hope. "They didn't give a damn about the people who live up here." When Senator Frank Murkowski (R-Alaska) visited the village, an elderly woman threw herself at him and shouted, "You have poisoned our land!" (Egan, 1992, A-26).

Energy department spokesman Tom Gerusky acknowledged that the U.S. Geological Survey erred in burying the waste, but he said that a person standing on the mound for a year would be exposed to only a small fraction of radiation received in a single cross-country jet flight (Badger, 1992, B-5). A series of studies in the 1980s by the federal Centers for Disease Control and the Indian Health Service concluded that while radiation from Soviet tests was detected in a number of Alaska villages during the 1960s, the bulk of cancers in the region involved lung and cervical cancer, types that are not generally associated with exposure to radiation.

Radiation afflicting the Eskimos of this area also may stem from dumping by the former Soviet Union. It is unclear how much radioactivity from above-ground nuclear blasts by the Soviet Union may have drifted into the Arctic, but some U.S. officials have said the amount could be considerable. There is also concern that the Bering and Chukchi seas are contaminated by radioactivity: Russia has acknowledged that over the last three decades the Soviets dumped old submarines with damaged nuclear reactors, and more than 10,000 containers of nuclear waste, in waters of the Far North (Egan, 1992, A-26).

The environmental devastation of remote areas in the Arctic is only part of the social disintegration that is afflicting the Inuit of the Far North. In 1960, before widespread energy development on Alaska's North Slope, the suicide rate among Native people there was 13 per 100,000, comparable to averages in the United States as a whole. By 1970, the rate had risen to 25 per 100,000; by 1986, the rate had risen to 67.6 per 100,000. Homicide rates by the mid-1980s were three times the average in the United States as a whole, between 22.9 and 26.6 per 100,000 people, depending on which study was used. Death rates from homicide and suicide reflected rising rates of alcoholism. In the mid-1980s, 79 percent of Native suicide victims had some alcohol in their blood at the time of death. Slightly more than half (54 percent) of the people who committed suicide were legally intoxicated (Grinde and Johansen, 1995, 238–39).

THE GWICH'IN: CARIBOUS AND OIL IN THE ARCTIC NATIONAL WILDLIFE REFUGE

During 2001, the Gwich'in generally opposed President George W. Bush's request to Congress to open oil drilling on the Arctic National Wildlife Refuge 1.5-million-acre coastal plain. The Gwich'in consider the Arctic National Wildlife Refuge to be sacred. In their language, they call it *Vadzaih googii vi dehk'it gwanlii*, which translates to "the sacred place where life begins." Another indigenous group in the same area, the Eskimos, traditionally have lived off seals, polar bears, and bowhead whales. Because oil pollution of their life-giving waters could ruin their livelihood, most Eskimos adamantly oppose offshore oil drilling, which might threaten the marine environment. Oil production in the Arctic National Wildlife Refuge, however, makes good sense to them. This stance brings them into conflict with the Gwich'in.

According to a report in the *Los Angeles Times*,

> The Gwich'in turn up their noses at whale meat but live in perfected rhythm with the mysterious, wandering herd of about 140,000 caribou that migrates

between Gwich'in lands in Canada, through the snow-and-emerald peaks of the Brooks Range and, in the animals' birthing cycle, down to the tundra of the Arctic coastal plain. To drill oil wells into the calving grounds of the caribou herd, most tribe members believe, is to pierce the soft place that a Gwich'in thinks of as himself. . . . The Gwich'in's fight against oil development in the Arctic refuge is really about far more than caribou, the village's livelihood. In their minds, it is about their right to live off a healthy land—and their right to decide what constitutes a healthy land. (Murphy, 2001, 1)

Residents of Kaktovik, a mostly Eskimo community of about 260 people on a coastal island at the northern edge of the proposed drilling area, are generally in favor of the push to open the Arctic National Wildlife Refuge. They deride environmentalists who describe the coastal tundra as "pristine" and laugh at antidrilling ads, which show spruce-forested hillsides deep within the Brooks Range where no drilling is being considered (Lane, "Refuge in a Storm," 2001, A-3). Leaders in Kaktovik see drilling as a source of income, because regional Eskimo corporations can tax the facilities of oil companies and their contractors.

Parts of the area are hardly pristine. Several Cold War–era Distant Early Warning (DEW) radar stations deposited toxic chemicals that still blight the local environment. One spit of land was known locally as Drum Island, because until 1998 it contained 7,000 to 8,000 abandoned fuel barrels.

Evon Peter, a 25-year-old Gwich'in leader who lives in Arctic Village, in the Arctic National Wildlife Refuge on the foothills of the majestic Brooks Range, is convinced that oil production will doom the caribou, which "pass like a moving carpet across Timberline Mountain every spring on their way to lay down calves on the Arctic coastal plain" (Murphy, 2001, 1). "The whole history of Alaska is of white people coming in for natural resources, oppressing native people and becoming rich," Peter said. "We are in a dynamic relationship with all other things: animals, land, spirits, us. If you have these things in balance, you're being human" (Murphy, 2001, 1).

The coastal area east of the Canning River is a vast birthing ground for caribous and other species. It is a largely unspoiled plain that, if left alone, according to a report in the *Seattle Times,* "would be one of the ideal places in the polar north where the impact of global climate change could be studied without an overlay of industrial activity" (Lane, "Refuge in a Storm," 2001, A-3).

Greg Gilbert, one of about 250 people living in Arctic Village, worries that oil exploration will draw more human activity into the coastal plain and harm the caribou, his family's main food source. Gilbert is a member of the 7,500-member Gwich'in Indian Nation living in the boreal forests of northwest Canada and north-central Alaska, and he hunts caribou to feed his family of seven. Until about 50 years ago, the Gwich'in trekked to follow the porcupine herd. Now they live in settlements of one-story clapboard houses heated by wood stoves, often without running water. "Caribou are sensitive," Gilbert said. "The oil companies want to drill in the refuge, but the elders say the caribou will disappear. They will change their route. "I think drilling will eventually happen. The oil companies have the power to do it. But I worry about the future, my kids, their kids" (Banerjee, 2000, C-1).

The Gwich'in of Arctic Village, however, vehemently oppose drilling because of its possible effects on the coastal tundra. They also hold the area sacred because it sustains the caribou, which provide much of the village's country food. "A lot of life begins up there," said Gideon James, the secretary of the local tribal council. "It's not a wise thing to go disturbing an animal in its calving area." "We believe the Arctic refuge is too wild to waste," said J. Scott Feierabend, an Anchorage-based vice-president of the National Wildlife Federation. Any effort to drill there, he said, "would forever alter the wilderness character of the place" (Lane, "Oil in the Wilderness," 2001, A-4).

Earl Lane, reporting for *Newsday*, described the home of the caribou:

> By early June, the pregnant caribou will have completed their 700-mile migration to the calving ground after scrambling along high ridges and through the river valleys that penetrate the Brooks Range. Shortly thereafter, most of the rest of the herd of 130,000 will join the cows and calves. The tundra, soon to be covered with the blooms of dozens of species of sedges, mosses, wildflowers, shrubs and other plants, provides rich forage for the caribou. (Lane, "Oil in the Wilderness," A-4)

The Arctic National Wildlife Refuge, which is not presently accessible by road, comprises nearly 20 million acres. Controversy over oil exploration involves a 1.5-million-acre coastal plain at the edge of the Beaufort Sea. Proponents of exploration maintain that at least 16 billion barrels of oil await exploitation beneath the Refuge and adjacent state waters, based on a 1998 study by the U.S. Geological Survey. The 16-billion-barrel figure is said by the U.S. Geological Survey report to have low odds (one chance in 20) of recovering oil given expected costs of production and market prices. The most likely amount of recoverable oil, according to the report, is closer to 5.7 billion barrels (Lane, "Refuge in a Storm," 2001, A-3). Looking only at the onshore federal lands in the Refuge, the U.S. Geological Survey estimated a 19-in-20 chance of finding at least 2 billion barrels of economically recoverable oil (assuming a market price of $24 a barrel). Drilling critics say that is less than six months of U.S. supply, at present rates of consumption (Lane, "Oil in the Wilderness," 2001, A-4).

The oil industry maintains that, compared to development of the Prudhoe Bay site 30 years ago, it now employs new technology to minimize environmental impacts of oil exploration. Proponents of drilling assert that the "footprint" of exploration in the Arctic National Wildlife Refuge probably will be no more than 2,000 acres. Industry sources also contend that advances such as "ice roads," which are laid down over the frozen tundra in winter and melt away during the warmer months, also minimize impacts on the local flora and fauna. Environmentalists criticize the use of "ice roads" because they require a great deal of fresh water (as much as 1.5 million gallons per mile) to build and maintain.

Opponents of drilling assert that the oil companies have not become as ecofriendly as they profess to be. Oil spills are still a problem, they say, pointing to an incident on April 15, 2001, during which 92,400 gallons of an oil-saltwater mixture leaked from a 10-inch pipeline in the Kuparuk field west of Prudhoe Bay—the fourth significant spill in a year. Another spill, in February 2001, spread about 11,550 gallons of crude oil and methanol from an oil-processing facility in Prudhoe Bay (Lane, "Refuge in a Storm," 2001, A-3). The oil companies point out that they maintain emergency-response teams to handle oil spills, while opponents of expanded oil exploration believe that the existence of these teams shows the perilous nature of oil drilling and transport.

Another worry is sabotage of pipelines. During early October 2001, a single gunshot into the existing Alaska pipeline caused 285,000 gallons of oil to spill onto the tundra during three days. The size of the pipeline network makes ruptures (accidental or not) a constant threat. The Prudhoe Bay oil field, discovered in 1968, by the year 2000 included a sprawling web of 500 miles of roads and 1,100 miles of pipelines, rigs, well-heads, processing plants, living quarters, and other facilities, along with roughly 4,000 oil wells in various stages of exploration or production. The North Slope complex emits about 56,400 tons of smog-producing nitrogen oxide each year, more than twice as much as the city of Washington, D.C. (Lane, "Refuge in a Storm," 2001, A-3).

Environmentalists and indigenous opponents of drilling also "suspect that permanent roads and other facilities, such as ports and airstrips, would inevitably follow if several substantial oil deposits were found beneath the refuge" (Lane, "Refuge in a Storm," 2001, A-3). According to the U.S. Geological Survey, much of the area's oil is located in several different sites scattered under the northwest section of the coastal plain, rather than in a single giant field (as at Prudhoe Bay), thus requiring more roads and pipelines.

Some oil workers on the North Slope wrote an open letter expressing concern about staff cutbacks and inadequate oversight of some old technology, "such as the surface-shutoff valves that are designed to halt well-head ruptures before they allow thousands of barrels of crude oil to spill onto the tundra" (Lane, "Oil in the Wilderness," 2001, A-4). Recent inspections by the Alaskan government found failed components in about 10 percent of the safety valves examined in British Petroleum Exploration's western Prudhoe Bay fields, twice the usual rate.

Ken Whitten, a recently retired specialist on caribou with the Alaska State Department of Fish and Game, said studies show that oil development already has displaced some calving females from their traditional grounds in the Prudhoe Bay area. Caribou cows that spent more time in the oil fields gained less weight and had fewer calves than cows that seldom encountered development, he said (Lane, "Oil in the Wilderness," 2001, A-4). Supporters of drilling say disruption of caribou migrations can be minimized by doing exploratory drilling only during the winter when the caribou are absent. Once year-round oil production begins, the caribou can be respected with restrictions on ground and air transport while the animals are calving. Plans also call for construction of pipelines five feet off the ground so that large animals may pass under them.

Canada will fight the plan by U.S. President George W. Bush to drill for oil in the Arctic National Wildlife Refuge, according to Environment Minister David Anderson. "I'm utterly opposed [to the plan]," Anderson said (Jaimet, 2001, A-1). The Bush administration's proposal to open the Refuge to oil drilling also has met with little success in the U.S. Senate. Following a defeat in October 2001 of an attempt to open the Refuge, the Senate voted on December 3, 2001 against a proposed amendment to the Railroad Retirement Bill, which would have opened the Arctic National Wildlife Refuge to energy exploration.

In an effort to push the measure through the Senate, Senate Minority Leader Trent Lott, a Missouri Republican, and Alaska Republican Senator Frank Murkowski attempted to attach the controversial energy legislation, allowing for oil and natural gas drilling in the Refuge, to an unrelated railroad bill ("Senate Defeats," 2001). Despite a weekend of lobbying by the energy industry and supporters of the energy bill, and after hours of floor debate, the Senate overwhelmingly rejected Murkowski's effort, 94–1 ("Senate Defeats," 2001).

Opposition has held on subsequent votes as well. After more lobbying by the Bush administration, Senate Republicans again voted on April 18, 2002. The vote was 54–46—14 votes short of the 60 required to break a Democratic filibuster of an amendment, offered by Alaska's senators, to open the Arctic National Wildlife Refuge to oil drilling.

SEWARD PENINSULA OF ALASKA: DON'T EAT THE REINDEER

The Seward Peninsula of Alaska has been extensively mined during the last 100 years for cadmium and various lead-bearing ores, which are easily absorbed by plants that provide food eaten by ungulates, concentrating in their liver, kidney, and muscle tissues. In some cases, health officials have warned local Native peoples to avoid eating reindeer, once a dietary staple. In addition, contamination from weapons testing, accidental pollution, or illegal dumping may have found its way into the lichens of northwestern Alaska, thereby accumulating in reindeer and caribou tissue (Heavy Metal, 2000).

The people on the Seward Peninsula live a subsistence lifestyle in which a high percentage of their diet comes from local plants and animals. The incidence of cancer and other diseases appears to be rising among the indigenous people in this region, who subsist on contaminated reindeer and other "country food." The people in local villages are particularly concerned that contamination from air pollution, mining operations, and dumpsites are concentrating in the tissues of subsistence animals, posing a health risk (Heavy Metal, 2000). The University of Alaska's Reindeer Research Program has detected high levels of cadmium and lead in several species. If similar concentrations were to be found in meat, consumption of 40 to 60 grams of meat per week would exceed the recommended intake rate (*Heavy Metal*, 2000).

BOMBS AWAY IN THE ALEUTIAN ISLANDS

Six decades ago, a military base was established on Adak Island in the Aleutians (1,200 miles west of Anchorage, Alaska) to fend off possible Japanese attacks during World War II. The U.S. army established a military presence on the island to counter Japanese forces, which briefly occupied Attu and Kiska islands in the Aleutians. In 2001, the U.S. navy joined with the Environmental Protection Agency (EPA) to remove unexploded bombs on Adak Island, near an indigenous Aleut community. The removal of unexploded ordnance will allow the Aleuts to develop industry on the site.

The abandoned army base was designated a Superfund site during 1994, and it was closed in 1997. The cleanup is a combined effort by the navy, the EPA, the U.S. Fish and Wildlife Service, the Aleutian/Pribilof Island Association, the Aleut Native Corporation, and the Adak community. The cleanup was a part of a 47,000-acre land exchange from the federal government to the Aleut Corporation, which will allow the local Aleut community to develop a fish-processing industry, fueling facility, and a regional hub for air cargo traffic.

"Addressing unexploded ordnance contamination has stymied cleanups at other sites in the U.S.," said Aleut Corporation Commissioner Michele Brown. "We needed to put Adak back to good use, so endless delay was not an option. The project team devoted long hours and ingenuity to identify and remediate risks, resulting in a responsible, effective, environmentally protective decision that allows the Aleut Corporation to move a step closer to creating good jobs for [indigenous] Alaskans" ("Navy, Environmental," 2001).

THE HOPI AND NAVAJO: TURNING BLACK MESA TO COAL SLURRY

More than half of the electricity consumed in the United States is generated by burning coal. A sizable percentage of that coal comes from strip mines on American Indian reservations. One large example is the electricity-generating complex on the Navajo Nation at the "four corners"— the conjunction of Arizona, New Mexico, Colorado, and Utah. "Since 1974," according to one observer, "the Mojave Generating Station and the Navajo Generating Station . . . have been polluting the world's air. The Mojave Generating Station alone uses 18,240 tons of coal per day at full load. Combined, the two plants require 12 million tons of coal a year and are the largest polluters in the country. Astronauts saw the pollution cloud from these coal fired plants from the Moon" (Giuliano, 2001). The six coal-fired power plants that comprise Four Corners, completed during the 1970s, emit pollution that on some days fills the local atmosphere with higher levels of sulfur and nitrogen oxides than the air of New York City or Los Angeles (Schneider, n.d.).

A significant proportion of the coal that powers these plants is strip mined from Black Mesa, Arizona, home of the Hopi Indian Reservation (which is also home to several thousand Navajo). Black Mesa is also the site of a growing Peabody coal mine, which conveys water resources out of this arid area as coal slurry. On Black Mesa, the Peabody Western Coal Company uses more than 3 million gallons a day (1.4 billion gallons a year) of once-pristine, potable groundwater that leaves the aquifer as coal slurry.

The aquifers once provided 60,000 people on Black Mesa with water, but by the year 2000 they were being depleted much faster than nature could replenish them. Computer models run by hydrologist Ron Morgan indicate that by 2050, not even allowing for Peabody's coal-slurry needs, "virtually every spring on the Hopi homelands will be bone dry" (Vandevelder, 2000, 14). By 2050, Morgan estimates that the aquifer will be drawn down at ten times the rate that it is being recharged. "All the computer models tell us that these depletions are right around the corner," Morgan said (Vandevelder, 2000, 14).

The slurry pipeline, which is owned by Enron Corporation, transports its coal-water mix 273 miles to the Mojave Generating Station in Laughlin, Nevada, where it is converted into electricity for use by consumers in Nevada and California, as well as parts of Utah and central Arizona. In the meantime, many of the Dineh (traditional Navajo) living at Black Mesa have no electricity.

When Peabody was confronted during the late 1990s with complaints that it wasn't paying enough for its coal, the Hopi Nation government renegotiated to increase the company's payments by 10 percent, or about $1 million more per year. The deal was sweetened by a $1 million bonus paid at the signing of the agreement. By 1998, coal revenues accounted for nearly 40 percent of the Navajo Nation's governmental budget and 80 percent of the Hopi Nation's budget ("Figure This," 1998). One critic asked, "Will this increase help the 50 percent unemployed Navajo out on the Rez, and will it raise the yearly average income of $750? Since the wells under the Rez are almost dry, from whence will come the water needed to move this coal to Albuquerque and Nevada?" ("Figure This," 1998).

Many Navajo and Hopi have moved away from the path of the growing Peabody mine, accepting the company's offer of new homes with plumbing and solar power. Others, however, have decided to remain in their traditional homes. Maxine Kescoli is one of a number of elders who have refused to move to make way for expansion of the coal mine. Peabody has mining rights to the land under her traditional Navajo hogan, which is one mile from the Kayenta mine. "My umbilical cord is buried here," Kescoli said ("Figure This," 1998).

To illustrate why she prefers to remain in her traditional home, Kescoli called on *hozho,* the Navajo foundation of belief, which means "to walk in beauty." This belief runs counter to exploitation of coal and other resources. This clash is evident not only at Black Mesa, but in many areas of the Navajo Nation where coal is mined. Emma Yazzie, another elder, herded sheep and goats in the shadow of the Four Corners' strip-mine draglines until she died, even as pollution fouled her home. She walked to the bottom of the strip mines near her home and impeded the draglines—a one-woman, antimining protest movement (Johansen and Maestas, 1979, 143–46).

The clash of cultures is very sharp:

> During the past 30 years, 10,000 Navajo sheepherders have been removed from land near Black Mesa, some at gunpoint. Four thousand graves and sacred places have been desecrated. Charges are that the Peabody Coal Company, U.S. agents, and Hopi police have impounded livestock illegally. . . . All of the wells may be dry within four years. Peabody Coal plants trees as "restoration." All of the trees are dead. This is not "walking in beauty." ("Figure This," 1998)

Before the mid-1970s, Black Mesa was home to several thousand Navajo and Hopi sheepherders, weavers, silversmiths, and farmers, whose families had lived there for several hundred years. Over the years, many of these people were forced off Black Mesa by the encroaching strip mine. During the late 1980s, a United Nations report described the case of the forced relocation as "one of the most flagrant violations of indigenous peoples' human rights in this hemisphere" ("A Brief History," n.d.).

Coal-fired power plants emit more toxic pollution than any other form of electricity production. For every megawatt hour of electricity produced, coal generates 2,071 pounds of carbon dioxide, 13.8 pounds of sulfur oxides, 4.8 pounds of nitrogen oxides, and 3.2 pounds of particulate matter. By comparison, natural gas emits 1,205 pounds of carbon dioxide per megawatt hour, 0.008 pounds of sulfur oxides, 4.3 pounds of nitrogen oxides, and negligible particulate matter (Giuliano, 2001).

The scope of Peabody's operation at Black Mesa can be sketched using statistics: During 1998, for example, Peabody Western Coal Company (a subsidiary of the Peabody Group) removed just under 11.8 million tons of coal from its Arizona mines. Since mining began on Black Mesa three decades ago, close to 40 billion gallons of groundwater have been pumped, all to feed Peabody's coal-slurry pipeline. With the pipeline conveying as much as 43,000 tons of slurried coal per day, the company pumps as much as 120,000 gallons of water per hour ("Clean Coal's," n.d.). The coal slurry competes with local residents for the only source of water available to the Hopi and Western Navajo. By the late 1990s, Peabody's strip mine at Black Mesa had expanded to 100 square miles, "the largest privately-owned coal mine in the world" ("A Brief History," n.d.). Into this pit have fallen burial and sacred sites, religious structures, and Anasazi ruins.

The U.S. federal government has relocated 9,000 Navajo to "new lands" at Sanders, Arizona, to land contaminated by the largest radioactive-waste spill in North America, in and near the Rio Puerco (see "The Navajo: The High Price of Uranium," below). According to one source, "Some people living there have died from cancer or are dying from it now. The birth defect rate is

Black Mesa residents protest underground pumping by Peabody Western Coal Company after the annual horse ride by Navajo Nation Council Delegates, Monday, July 15, 2002, in front of the Navajo Nation Council Chambers in Window Rock, Arizona. (AP Photo/*Gallup Independent*/Douglas Tesner)

outstanding. . . . The suicide rate is outstanding as well" ("A Brief History," n.d.). A radiation meter outside a school on the relocation lands registers 700 rads. The relocation agents have resisted complaints that the "new lands" are contaminated, saying that people will be unaffected if they do not drink the water (Schneider, n.d.). There is, however, no other water source in the area. People live in trailers, their traditional economy and way of life destroyed.

Other families of Navajo refugees from Black Mesa have drifted from place to place for many years. Some live in shacks, others live in vehicles, while the lucky ones squeeze in with other family members ("A Brief History," n.d.).

Others [have] found themselves having to pay for water, heat, food, electricity, [and] taxes, things they never had to deal with before. Many of the elders speak little or no English—people who had no experience with a cash economy have been moved to border towns. These Navajos were warehoused in substandard housing. . . . While the relocation law required the federal government to provide community facilities and services and to minimize the adverse social, cultural, and economic effects of relocation, that promise remains unfulfilled almost two decades later. Many find it impossible to get jobs, and they are forced into homelessness. The genocide is complete. ("A Brief History," n.d.).

Iyawata Britten Schneider wrote in *Country Road News* that "in Arizona, the Department of Interior and the Bureau of Indian Affairs created a land dispute between the Hopi and Dineh people by acquiring land and paying the Dineh large sums and the Hopi next to nothing. This land dispute was then used to support the passage of the Relocation Act of 1974 that forced the move of 10,000 Dineh to 'new lands.' This relocation brought with it increased health risks, high suicide rates, and severe depression among the Dineh. The movement of the Dineh allowed for the leasing of more land for the coal mines" (Schneider, n.d.).

Coal mining is antithetical to traditional herding and agriculture in the region. According to Schneider:

> Because the southwest is arid and has high winds, coal dust can be seen on the sheep herds, on the land, in the water supply and the air. Residents of the coal mining areas suffer from chronic lung disease and high cancer rates. Livestock drinking water from nearby ponds die within a few hours and traditional crops fail. In its reparation [sic] of land, Peabody does not separate the topsoil from the bottom, leaving the soil with high saline content. Vegetation has not survived in this soil. (Schneider, n.d.)

The Black Mesa coal-slurry pipeline also is prone to occasional spills, such as one that occurred over a two-and-a-half-year period between 1999 and 2001. Black Mesa Pipeline agreed to pay penalties of $128,000 for discharging almost 485,000 gallons of coal slurry in northern Arizona. The Arizona Department of Environmental Quality discovered the violations during a series of inspections of Black Mesa's facilities.

"Had the pipeline been properly maintained, these spills would not have occurred," said Alexis Strauss, water division director for the U.S. Environmental Protection Agency's Pacific Southwest Office. "Desert ecosystems are quite fragile and filling arroyos with crushed coal is unnecessary and unacceptable" ("Black Mesa Spill," 2001). Corrosion of the pipeline sometimes results in ruptures; the resulting spread of slurry can harm local wildlife.

Increasing numbers of Hopis and Navajos have been protesting the fact that Peabody's use of local water is destroying their way of life. Former Hopi Chairman Ferrell Secakuku joined Navajos from Big Mountain during 2001 to protest the use of aquifer water for the Black Mesa coal slurry and to urge creation of new, sustainable forms of energy. "We found out the water table is being depleted," Secakuku said, denying reports by the federal government and Peabody Western Coal that have claimed otherwise (Norrell, 2001). Secakuku spoke to a crowd of American Indians and other environmentalists, including longtime opponents of Navajo relocation at Big Mountain. Secakuku and other protesters entered the offices of Black Mesa Pipeline to urge officials to halt the slurry pipeline, which annually depletes 1.3 billion gallons from the Navajo aquifer. He said using water to transport coal threatens to leave the Hopi village of Moenkopi without water by the year 2011. "Every time you breathe, Peabody is pumping 50 gallons," Secakuku told the crowd (Norrell, 2001). During the protest, Roberta Blackgoat held a sign proclaiming, "The Creator Is the Only One Who Is Going to Relocate Me" (Norrell, 2001).

Secakuku added that there had been no response from the Department of the Interior to a proposal that would provide Lake Powell water to the Hopi for drinking purposes. (Lake Powell has its own environmental problems; see chapter below, "The Navajo Medicine Men's Association, Lake Powell, and Carbon Monoxide Poisoning.") He said that if no alternative water source is provided by the year 2004, he and Bucky Preston would march to the Hopi Tribal Council and demand termination of contracts with Peabody, ending coal mining on Black Mesa. "By the year 2004, we will stop the mining," he said. "Soon the aquifer will crumble and no longer take the recharge" (Norrell, 2001).

In defense of the mine, Peabody Coal's John Wasik, executive for southwestern operations, said: "Continued operation of the Black Mesa Mine is in the public interest and will provide long-term economic benefits to the Hopi Tribe and the Navajo Nation. It also provides an essential and secure energy supply for more than 1 million Southwest families who rely on [it for] electricity." Peabody further maintains that the mine "will inject about $1.5 billion in direct economic benefits into reservation economies in royalties, taxes, wages, and vendor contracts over the proposed extension [of mine operations]" for 15 years, from 2005 to 2020 (Wanamaker, 2002, C-1). The mine employed about 250 people in 2002; 96 percent of the mine's work force and 82 percent of its supervisory staff were Native American, according to the Peabody company (Wanamaker, 2002, C-1).

THE QUECHAN: GOLD MINING

During 2000, Bruce Babbitt, President Bill Clinton's Secretary of the Interior, denied a permit requested by Glamis Gold for a 1,600-acre open-pit, cyanide-leaching gold mine on public land in eastern Imperial County, adjacent to the Quechans' Fort Yuma reservation, near San Diego, California. A year later, however, President George W. Bush indicated that his appointees might approve the mine.

The U.S. Bureau of Land Management (BLM), which controls the land in question, at first tentatively rejected the Canadian mining company's bid (Perry, 2000). This case was regarded as a major test of a 1996 order by President Bill Clinton, requiring federal agencies to show greater sensitivity to the "sacred geography" of Native Americans. Opponents of the mine also believe that it could threaten endangered species, including the desert tortoise, which is considered sacred by the Quechan and other Native Americans in the area (Fernandez, 1998, 6). Before Clinton issued that order, federal policy was governed by the Mining Act of 1872, which leaned heavily toward granting permission for mining on public property whenever possible, regardless of how Native Americans regarded the land.

According to an account in the *Los Angeles Times,* the Quechan Indian Nation "asserts that the rocky, wind-swept area bounded by Picacho Peak, Pilots Knob and Muggins Peak contains the spirits of the Creator and two other mythological figures who passed through the region a thousand years ago. The area is mentioned prominently in Quechan songs" (Perry, 2000). In its tentative decision, the BLM concluded that the proposed gold mine "would result in significant adverse effects to prehistoric cultural resources, Native American traditional cultural uses and values, and visual resources" (Perry, 2000).

Glamis Imperial Corporation, which wants to open the gold mine, may challenge its rejection by the BLM in federal court. Glamis Imperial has been seeking to strip mine roughly 130,000 tons of ore a day. The gold would be extracted from ore by bathing mounds of rock with sodium cyanide. The company at one point volunteered to move one of the ore heaps away from an area called the Trail of Dreams, "where Quechans believe that they can communicate with spirits and receive visions" (Perry, 2000). The proposed mine would operate about six miles from the Quechan reservation's population center, which includes a casino and extensive agricultural areas.

The proposed mine site, 45 miles northeast of El Centro, California, is located in a region that has experienced gold mining for more than 100 years. Three of the state's most productive gold mines of the twentieth century (Picacho, Mesquite, and American Girl) were opened on adjacent federal land.

By 2002, President Bush's Interior Secretary, Gale Norton, was considering whether to issue a new permit for the proposed gold mine. The Quechans protested the move, saying that the open-pit mine would "severely impact some of their sacred cultural and historical sites" (May, 2001, B-1). The company first applied for a permit in 1994, but dropped plans after Babbitt's ruling. Glamis Imperial spokesman Dave Hyatt, vice-president for investor relations, said the company would reapply for a permit, but that it might not mine the site (if the permit was approved) until gold prices rose from then-present levels (about $280 per troy ounce at the time). Hyatt scoffed at assertions by the Quechans that the area is sacred. "Half of southwestern California is sacred to them," he said. "The railroads and pipelines already cut through their spiritual trails" (May, 2001, B-2).

In the meantime, Norton announced that her department would rewrite a legal opinion issued by Babbitt that denied permission to mine, citing preference for the archaeological, religious, and cultural resources sacred to the Quechan tribe and other peoples of the Colorado River. According to Roger Flynn, a lawyer for the Western Mining Action Project, rewriting of this legal opinion does not approve the mine, but opens the issue to reconsideration. "Today's [October 25, 2001] announcement is an affront to all American Indians. It indicates the direction this department may be headed in implementing executive orders and handling trust assets so important to Indian peoples. They appear destined to break another promise—their promise to protect this sacred area from certain destruction. This is an outrage," stated Mike Jackson Sr., president of the Quechan Tribal Council ("Hotspots: California," 2001).

BLOCKING A NUCLEAR DUMP IN CALIFORNIA'S WARD VALLEY

A village of tents and tipis grew in the desert 22 miles west of Needles, California during the last half of February 1998, as several hundred Native American and non-Indian environmentalists put their bodies on the line in an attempt to stop construction of a dump for low-level nuclear waste on land they regard as sacred.

By the third week of February, roughly 250 people were camped at "ground zero" of Ward Valley, 80 acres of federally owned land designated for the waste site. The encampment prevented soil testing necessary for planning of the dump, which had been proposed to receive waste from hospitals, nuclear plants, and various industries that cannot dispose of radioactive materials in other landfills.

By February 19, occupants of the encampment had defied two sets of federal orders to leave the site, and had blocked roads leading to it. A 15-day order that had been issued in late January expired on February 14. A five-day eviction order issued that day expired on February 19, with protesters still confronting U.S. Bureau of Land Management (BLM) vehicles that had surrounded their camp. Protesters said that the BLM and other federal agents had been rumbling around the camp in large land rovers and flying over it in small aircraft and helicopters at odd hours of the night in an attempt to intimidate the campers and deprive them of sleep. Religious ceremonies continued in the camp amid the glare of headlamps and the drone of aircraft overhead. Some of the protesters had chained themselves together. After complaints (and arrival of news media reporters), the vehicles were withdrawn from the perimeter of the camp.

Steve Lopez, a Fort Mojave Native spokesman, said that the Ward Valley is central to the creation stories of many Native American peoples in the area, and is also the habitat of endangered tortoise species. "Taking away the land is taking away part of ourselves. They used to use bullets to kill our people off. Now it's radioactive waste," Lopez said (Johansen, "Ward Valley," 1998, 7). Ward Valley also is sacred to many Native peoples in the area because of its proximity to Spirit Mountain, the birthplace of their ancestors.

The protesters included a number of elders affiliated with five tribes of the Colorado river basin (Fort Mojave, Chemehuevi, Cocopah, Quechan, and Colorado River Indians), who refused to move, along with the rest of the protesters. Instead, people in the encampment sent out appeals for supplies and more visitors, asking, according to a Colorado American Indian Movement posting on the Internet, "for the physical presence of anyone willing to travel to Ward Valley and participate in the protection of this sacred land and the people defending it" (Johansen, "Ward Valley," 1998, 7). Donations of food, water, blankets, batteries, and rain gear were requested. Wally Antone of the Colorado River Native Nations Alliance said, "Our ceremonies will continue here and our elders will not move. You will have to drag us out, and I say those words with honor" (Johansen, "Ward Valley," 1998, 7).

On February 18, as the BLM's second deadline was set to expire, leaders of the camp invited five BLM officials to their central fire for a religious ceremony, after which the officials were told that the protesters would not move. Andy Mader of the Arizona American Indian Movement (AIM) said via Internet that all the cellular phones in the camp were malfunctioning. Some suspected that the government had shut them down. Meanwhile, supporters of the protest outside the camp had received phone calls indicating that several other AIM chapters were sending people to Ward Valley.

Tom Goldtooth of the Indigenous Environmental Network reported from the camp via Internet that the protest had become very vigorous, not only because the land is regarded as sacred by Native peoples, but because many southern California and Arizona urban areas rely on the Colorado River for water that could be polluted by the proposed waste dump. The same water is used to irrigate crops in both the United States and Mexico.

A position paper circulated by Save Ward Valley, an environmental coalition, stated that the proposed dump lies above a major aquifer, 18 miles from the Colorado River. Furthermore, the report asserted that all six of the presently active nuclear waste dumps in the United States are leaking. U.S. Ecology, the contractor selected for the Ward Valley site, presently operates four of those six dumps, the position paper said. The coalition also asserted that not all the waste buried at such sites is low-level waste. As much as 90 percent of the radioactivity proposed for burial would come from nuclear power plants, including cesium, strontium, and plutonium, the statement said.

Speaking on behalf of the coalition, Goldtooth, who coordinated protests by non-Native supporters, evaluated the protest, which eventually caused plans for the dump to be shelved:

> Incorporating the importance of your traditional ways and the use of the sacred Fire as the foundation for guidance and resistance as we fight for environmental justice and Native rights has proven successful at Ward Valley. . . . I applaud the many non-natives from the peace movement to the anti-nuclear movement and the global human family that raised the consciousness of the world that the sacred tortoise and the ecology of a desert environment must not be sacrificed anymore by the whims of the nuclear waste industry. I witnessed the coming together of the non-Native supporters and the Colorado River tribal communities and Tribal Nations in an historical moment where everyone agreed to fight together with one mind and one spirit to defend the sacredness of the Mother Earth and to defend the sovereignty of the Fort Mojave Tribal Nation. (Goldtooth, 2001)

THE SEMINOLE OF FLORIDA: A BUILDING CODE AS ASSIMILATIVE TOOL

During the early nineteenth century, an army under the command of Andrew Jackson chased the Seminoles' ancestors from present-day Georgia into Florida, when the area was still claimed by Spain. The Seminoles hid in the Everglades for nearly a half-century, resisting repeated attempts at subjugation by the U.S. army as Jackson's presidency came and went, and as the other four of the "Five Civilized Tribes" (Cherokees, Creeks, Chickasaws, and Choctaws) were removed to Indian Territory, now Oklahoma.

Until the mid-twentieth century, some of the Seminoles lived nearly isolated in the Everglades. When Seminole land claims were settled, the traditional Seminoles refused to take part, insisting that land cannot be bought and sold under Native law. They insisted on their right to occupy the land that had belonged to their ancestors under natural law, not U.S. civil law. The traditional Seminoles, wrote Catherine Caufield in the magazine *South Florida*, "[are] the true descendants of the 'unconquered Seminoles' of Florida schoolbook cliche" (Johansen, 1996, 44).

Today's Reality and Schoolbook Cliches

The distance between schoolbook cliche and present-day reality can be measured in time (one-and-a-half centuries) and in space (several million acres). In 1996, the assimilative weapon of choice against one small traditional Seminole community at Immokalee, Florida, became a county electrical and plumbing code. Collier County, which includes Naples, Florida, attempted to disperse one of the last surviving traditional Seminole settlements—already down to their last five acres of land—because their chickee dwellings (which are designed with four cypress poles and a thatched roof of palmetto fronds) did not conform to late-twentieth-century Anglo-American regulations for plumbing and electrical wiring. At least two children were removed from traditional Seminole villages by county officials on grounds that their parents' homes were said to be substandard.

Mark Madrid, information director of the American Indian Movement for South Florida, said that Collier County had proposed to open a toxic-waste dump adjacent to the Immokalee settlement (the word "Immokalee" means "place where we camp" in the Seminole language). On February 7, 1996, Collier County narrowed the number of possible sites for its landfill to five, all of them in the Immokalee area adjacent to or near the Seminoles' settlement. The final site was expected to cover 1,200 acres and to have a useful life of about 50 years.

Madrid, who is Creek (and thus related to the Seminoles by language and family lineage), said that the county's ambition to install a waste dump adjacent to the present Immokalee Seminole camp is "an open secret." "If you understand Florida politics," he said, "you'll understand how this is being done"

(Johansen, 1996, 44). Madrid said that the Seminoles at Immokalee had decided to move rather than endanger other traditional Seminole camps (roughly a dozen), most located in remote areas of Collier County.

Reacting to Collier County's assertion of authority, traditional Seminoles marched from their homes in Immokalee to the county commissioners' offices in Naples, 39 miles away, on May 12, 13, and 14, 1996. The traditional Seminoles' struggle also spread onto the Internet, from which they received support from Mohawks at Kahnawake, who likened their struggle to the confrontation at Oka, Quebec, in 1990 (see Canada, "The Kanesatake Mohawks: Debating Niobium Mining," for a description of the Oka confrontation).

On June 27, 1996, the Seminoles and their supporters again visited the Collier County commission's chambers. This time, the county postponed any decision regarding application of the building code to the Seminoles (or siting of the waste dump) until at least August 22. The county postponed its hearing because of two developments. First, the Seminoles had filed two lawsuits and an injunction based on federal laws protecting Native American religious freedom. Second, the Pacific Land Company had turned over ownership of the land under the Seminoles' village to a coalition of original owners and tenants, including the Seminoles. For the first time, the traditional Seminoles now had an ownership stake in the land under U.S. law.

The "Taming" of the Everglades

For nearly a hundred years after the Seminole wars ended in the mid-nineteenth century, the surviving Seminoles were left more or less alone, because immigrants regarded the Everglades as hostile and nearly impenetrable. In the mid-twentieth century, however, roads and canals with picturesque names like Alligator Alley and the Tamiami (Tampa to Miami) Trail began to pierce the wild Everglades. Ironically, many of these early routes could not have been built without Seminole labor. The Seminoles could withstand the tropical heat and had a natural immunity to mosquito-borne encephalitis.

As the tendrils of development closed in around them, many of the Seminoles settled a land claim stemming from 1842 by agreeing to move to designated reservations. One such group (among five Seminole reservations) became known as the Seminole Tribe of Florida, Inc., a body known to patrons of Everglades jungle tours, alligator wrestling, and gambling. This federally recognized Seminole tribe opened one of the first Indian bingo halls in the late 1970s, and has since figured in several major lawsuits defining the legal status of Indian gaming.

Unrecognized by the government, small bands of traditional Seminoles retreated into the forest. The traditional Seminoles pointed to the 1842 agreement, which was signed by President James Polk, as evidence that, as traditional Seminole leader Bobby Billie said, "We have a right to live on this land, same as everyone else" (Johansen, 1996, 45). In the mid-1950s, however, as the federal government pressed the Seminoles to accept reservation land in exchange for extinguishment of claims to five million acres of southwest Florida, the traditional Seminoles refused to participate, because, as Billie said, "we don't sell the land; we don't buy the land, and we don't say a person can own the land, because it doesn't belong to man. It belongs to the Creator" (Johansen, 1996, 45).

For many years, the traditional Seminoles had subsisted largely from hunting and trading, as they remained nearly unknown to other Floridians, wishing to be left alone. As long as their land was held to be without value as defined by mainstream capitalism, the traditional Seminoles were allowed to live outside the dominant culture. As the twentieth century passed its midpoint, however, the tendrils of asphalt, the wakes of power boats, and the attention of county government breached their cherished solitude.

When the federal government prepared to pay off Seminole land claims with $16 million in the late 1970s and early 1980s, the traditional Seminoles engaged legal help from the Indian Law Resource Center (ILRC) to ensure the government would not get their land. The request was lodged by Guy Osceola, a descendant of the famous Seminole chief of the same name who led a resistance movement during conflicts with the U.S. army in the early nineteenth century. The ILRC persuaded Congress to overrule the state of Florida and treat the traditional Seminoles as a group separate from the federally recognized tribe, which collected the money and surrendered its land claim.

The traditional Seminoles also detested attempts to memorialize their ancestors in Anglo-American fashion. In 1995, a number of them protested the unveiling of a life-sized bronze statue of Seminole medicine man Sam Jones, who also fought in the Seminole wars, at Florida's Tree Tops Park. Bobby Billie told the *Fort Lauderdale Sun-Sentinel* that a statue of a holy man was sacrilegious. The Seminole Tribe of Florida, Inc. had paid for $15,000 of the statue's $60,000 cost.

Bingo Comes to the Seminoles

The villages of the traditional Seminole contrast sharply with the garish development that characterizes the Seminole Tribe of Florida, Inc.'s reservation near Hollywood, Florida, which straddles State Route 441 amidst discount-smoke shops, a tourist attraction called the Magical Indian Village, and the facade of the original Seminole bingo hall. Tribal chairman James Billie operates his own tourist attraction, Camp Billie Safari, where, according to an article in the magazine *South Florida*, "tourists looking for an Everglades experience can spend the night in a chickee and ride in a swamp buggy" (Johansen, 1996, 46). An alligator-wrestling arena with a wet bar recently has been added to the camp.

It was Billie who came up with the idea of bingo in the 1970s. He calls it "the best thing that ever happened to the Seminoles," and "sweet revenge" on non-Indians (Johansen, 1996, 46). The money is all the sweeter, says Billie, because the federal and state governments cannot tax it. With Billie in the lead, the

Seminole Tribe of Florida, Inc. also has negotiated a royalty on the sales of sports clothing marketed by the University of Florida. Instead of rejecting the Seminole mascot as a stereotype, Billie decided to get in on the fiscal action.

After bingo was introduced in 1979, the Seminole Tribe of Florida, Inc.'s corporate income increased from $1 million to $40 million a year. The state of Florida refused to negotiate the necessary compacts required by the Federal Indian Gaming Regulatory Act to permit casino gambling by the Seminoles. With only bingo, enrolled Seminoles each received monthly per capita checks of about $1,000. Most recently, the Seminole Tribe of Florida, Inc. went to court against the state of Florida on grounds that the state was not negotiating a gambling compact "in good faith." The U.S. Supreme Court ruled 5 to 4 that a state cannot be sued by an Indian tribe under the Federal Indian Gaming Regulatory Act.

Traditional Seminoles Resist Assimilation

As one group of Seminoles joined the cash economy, the traditionals continued to live in small villages of traditional chickee huts. Because they had refused to deal with the federal government, the traditional Seminoles had no treaty, no reservation status, and no protection under a body of federal law that defines Native American communities' semisovereign status. Because the Seminoles had held to their traditional law, the U.S. legal system had defined them out of existence. The present-day traditional Seminoles' villagers live on land owned by the Pacific Land Company, which was leased to them at $1 a year. Many of the Seminoles took work in the vast agricultural fields that surround their settlements.

Late in December 1995, Collier County assistant attorney Ramiro Manalich announced that his office would prosecute the Pacific Land Company for building-code violations unless the traditional Seminoles could demonstrate that they were exempt from the county building code by federal law or treaty. The pretext for the prosecution was a fire in one of the Billies' chickees, which the county attributed to unsafe electrical wiring. Because of that fire, the county said it was imposing its building code on the traditional Seminoles for their own health and safety.

The following spring, Bobby Billie reacted to the county's threat by becoming one of the main organizers of a 750-mile environmental protest walk from the Everglades to Florida's state capital, Tallahassee—the Walk for the Earth. Along the way, the walkers visited with Mexican farmworkers who were being exposed to pesticides in central Florida, poor people who lived near large phosphate mines, and others who were fighting development of a major landfill near Newberry. They made a point of visiting burning sugar-cane fields and pulp mills, as well as portions of the Ocala National Forest that serve as a bombing range for the U.S. navy. About 30 people walked all or most of the 750 miles as others came and went. The statewide walk was organized by the Florida Coalition for Peace and Justice, based in Gainsville. Walkers included local residents, a number of traditional Seminoles, and other people from such countries as Venezuela, Mexico, and Belgium. Walkers averaged 18 miles on foot a day, with an occasional day off to rest.

Along the way, Billie and his cowalkers held press conferences and issued statements telling all who would listen how deeply they were distressed by the environmental destruction they witnessed:

> You cannot overpower the Creator's law. You are part of the creation. I'm telling you to stop destroying his creation. You are part of the creation, and you are destroying yourselves—which means your [children, and their] kids, and their grand kids beyond the future. You are not thinking about them. . . . I know you white people who look like a human being. I am asking you to act like human beings and do the right thing for generations yet to come. . . . [Y]ou call yourselves human beings, but I don't think so. (Johansen, 1996, 47)

Julie Hauserman, columnist for the *Tallahassee Democrat*, wrote that "no cheering crowds greeted Bobby Billie as he walked up Monroe Street toward the Old Capitol" (Johansen, 1996, 47). Billie and 200 others who took part in the Walk for the Earth seemed hardly noticed by the thousands of people who milled along the street in mid-April. Legislators in the Florida statehouse, where Billie arrived with a petition for the governor, seemed even less concerned. The petition called an end to pollution, unrestrained development, and corporate control of resources, especially in Florida wilderness areas. It pointed out that the greatest impact of pollution fell on the poor, many of whom, like the traditional Seminoles, were members of minority groups. The 200 traditional Seminoles and their supporters were lost in the din of 700 fully leathered motorcyclists, who were lobbying the legislature for a repeal of the state's mandatory helmet law. The president of the National Football League was also in town the same day asking for tax favors.

Bobby Billie arrived home after the Walk for the Earth to his family's hamlet of chickee huts near Immokalee, in the Everglades roughly 30 miles northeast of Naples, to catch up on the "battle of the building code." Contrary to the impressions of some people who had never lived in them, the Seminoles' traditional houses are quite sturdy; many of them are said by their occupants to have withstood hurricanes. As they defend their right to live in chickees, the traditional Seminoles have adapted other aspects of early-twenty-first-century U.S. society, economy, and culture. Many of the chickees now have electricity; some have computers, telephones, and fax machines, which were enlisted in the effort to maintain the traditional Seminoles' small island of personal and communal sovereignty. The Seminoles were adamant that change should be accommodated on their own terms. English is spoken side-by-side with Seminole in the settlements. Some Seminoles drive trucks and consume soft drinks, but they also maintain schools in their villages in a conscious effort to maintain their language and culture. While the best-known traditional village is Bobby Billie's settlement of about 30 people on five acres near Immokalee, other traditional settlements nestle in the Everglades near Naples and Tampa, as well as along the Tamiami Trail across the Everglades.

"They are trying to take the last things that are part of our lifestyle," said Danny Billie, referring to the struggle to retain traditional chickee dwellings. "What's happening to us is a continuation of what happened five hundred years ago when Europeans arrived," Billie told the *Fort Lauderdale Sun-Sentinel*. "One more time, the traditional Seminole camps are being forced to move for the expanding capitalist mentality that that has nothing to do with the finer things in life," said Madrid, who said that Collier County is one of the fastest-developing areas of south Florida. "These are the last holders of the independent camps. They are the inheritors of a long line of medicine that goes back to the Pleistocene age" (Johansen, 1996, 47).

THE COEUR D'ALENE: MINING WASTE IN IDAHO

In October 2001, the U.S. Environmental Protection Agency proposed a $459 million cleanup of accumulated toxic mining waste in Idaho's Coeur d'Alene Basin. This 1,500-square-mile area was devastated in a flood during 1999, when 1 million pounds of lead flowed into Lake Coeur d'Alene. On another occasion, according to a report in *Drillbits and Tailings*, a million pounds of zinc flowed into the same lake.

The Coeur d'Alene tribe has been directly affected by this environmental disaster. The cleanup is expected to take 20 to 30 years at a cost of up to $4 billion ("Hotspots: Idaho," 2001). Since 1983, a cleanup has been ongoing near Kellogg, Idaho, which was declared a Superfund site by the EPA. By 2001, more than $200 million had been spent at this site, most of it to replace contaminated soil in local 'residents' yards and to cap mining tailings ("Hotspots: Idaho," 2001, 3).

THE NAVAJO: THE HIGH PRICE OF URANIUM

During the 1940s and 1950s, Navajo uranium miners hauled radioactive uranium ore out of the earth as if it was coal. Some ate their lunches in the mines and slaked their thirst with radioactive water. Their families' homes sometimes were built of radioactive earth, and their neighbors' sheep may have watered in small ponds that formed at the mouths of abandoned uranium mines, called "dog holes" because of their small size. On dry, windy days, the gritty dust from uranium waste-tailing piles covered everything in sight.

The Navajo language has no word for "radioactivity." No one told the miners that within two or three decades many of them would die of radiation-induced cancers. In their rush to profit from uranium mining (and the lives of the miners), very few mining companies provided ventilation in the early years. Some miners worked as many as 20 hours a day, entering the mines just after the blasting of sandstone had filled the mines with silica dust. Many mine owners didn't even provide toilet paper. When miners relieved themselves, they wiped with fistfuls of radioactive "yellowcake"—uranium ore.

About half the recoverable uranium within the United States lies within New Mexico—and about half of that is beneath the Navajo Nation. Because a large number of former Navajo uranium miners have died of lung cancer, many Navajos have come to oppose any further uranium mining. They have joined forces with non-Native environmentalists who oppose the mining of uranium and its use in power plants and weapons.

Uranium has been mined on Navajo land since the late 1940s; the Indians dug the ore that started the United States' stockpile of nuclear weapons. For 30 years after the first atomic explosions in New Mexico, uranium was mined much like any other mineral. More than 99 percent of the rock the mines produced was waste, cast aside as tailings near mine sites after the uranium had been extracted. One of the mesalike waste piles grew to be a mile long and 70 feet high. On windy days, dust from the tailings blew into local communities, filling the air and settling into water supplies. At that time, the beginning of the 1950s, the Atomic Energy Commission assured worried local residents that the dust was harmless.

During February 1978, however, the Department of Energy released a Nuclear Waste Management Task Force Report that disclosed that people living near the tailings ran twice the risk of lung cancer as the general population. The *Navajo Times* carried assertions that one in six Navajo uranium miners had died, or would die prematurely, of lung cancer. For some, the news came too late. As Navajo miners continued to die, children who played in water that had flowed over or through abandoned mines and tailing piles came home with burning sores.

The First Navajo Uranium Mines

The Kerr-McGee Company, the first corporation to mine uranium on Navajo Nation lands (beginning in 1948), found the reservation location extremely lucrative. There were no taxes at the time; there were no health, safety or pollution regulations; and there were few other jobs for the many Navajos who had recently arrived home from service in World War II. Labor was cheap, and uranium, in demand for stockpiles of nuclear armaments, was profitable.

The first uranium miners in the area, almost all of them Navajos, remember being sent into shallow tunnels within minutes after blasting. They loaded the radioactive ore into wheelbarrows and emerged from the mines spitting black mucus from the dust, coughing so hard that many of them experienced headaches, according to Tom Barry, energy writer for the *Navajo Times* (June 15, 22, and 29, 1978), who interviewed the miners. Such mining practices exposed the Navajos who worked for Kerr-McGee to between 100 and 1,000 times the level of radon gas later considered safe. Officials for the Public Health

Service estimated these levels of exposure after the fact; no one was monitoring the Navajo miners' health in the late 1940s.

When mining was initiated, no one considered environmentally appropriate ways to deal with its tailings piles. Even if the tailings were to be buried—a staggering task—radioactive pollution could leak into the surrounding water table. A 1976 Environmental Protection Agency report found radioactive contamination of drinking water on the Navajo Reservation in the Grants (New Mexico) area near a uranium-mining and milling facility (Eichstaedt, 1995, 208).

Among Indians and non-Indians alike who oppose uranium mining in the area, there runs a deep concern for the long-term poisoning of the land, air, and water by low-level radiation. These concerns provoked demands from both Indian and non-Indian groups for a moratorium on all uranium mining, exploration, and milling until the issues of untreated radioactive tailings and other waste-disposal problems are faced and solved. Doris Bunting of Citizens Against Nuclear Threats—a predominantly non-Native group, which joined efforts with the Coalition for Navajo Liberation and the National Indian Youth Council to oppose uranium mining—supplied data indicating that radium-bearing sediments had spread into the Colorado river basin, from which water is drawn for much of the Southwest.

The specter of death that haunted former Navajo uranium miners came at what company public-relations specialists might have deemed an inappropriate time. By late 1978, more than 700,000 acres of Indian land were under lease for uranium exploration and development in an area centering on Shiprock and Crownpoint, both on the Navajo Nation. Atlantic Richfield, Continental Oil, Exxon, Humble Oil, Homestake, Kerr-McGee, Mobil Oil, Pioneer Nuclear, and United Nuclear were among the companies exploring, planning to mine, or already extracting ore. During the 1980s the mining frenzy subsided somewhat as recession and a slowing of the nuclear arms race reduced demand. Some ore was still being mined, but most of it lay in the ground, waiting for the next upward spike in the market.

Even as the boom ended, many of the miners who had been condemned to slow death by lung cancer did not know that the yellow ore they had mined was killing them. Peter Eichstaedt's painstakingly detailed study (*If You Poison Us: Uranium and American Indians*) described how the miners learned what had been done to them, and how they went about winning at least a possibility of compensation from the U.S. government.

Using a combination of interviews and scientific data, Eichstaedt demonstrated that the U.S. government (particularly the Atomic Energy Commission) knew that uranium mining was poisoning the Navajos almost from the beginning, a key point in the debate over compensation, which has thus far resulted in a small number of miners collecting $100,000 each. The government and the mining companies said they kept medical knowledge from the miners out of concern for national security and profits.

The Effects of Pervasive Radioactivity

Uranium-mine dust produced silicosis in the miners' lungs, in addition to lung cancer and other problems associated with exposure to radioactivity. As early as 1950, government workers were monitoring radiation levels in the mines that were as much as 750 times over the limits deemed acceptable at that time. By the 1960s, nearly 200 of the miners already had died of uranium-related causes. That number had doubled by 1990.

Radioactivity contaminated drinking water in parts of the Navajo reservation, producing birth defects and Down's syndrome, both previously all but unknown among the Navajo. Of all infant deaths in areas served by the Navajo Indian Health Service (NIHS) area for the years 1990 through 1992, 35.0 percent were caused by congenital anomalies. Mortality attributed to malignant neoplasms (age-adjusted rate) was 78.5 percent for the NIHS unit in the years 1990 through 1992 (Colomeda, 1998).

The most wrenching parts of Eichstaedt's book are the personal stories of miners who watched members of their families die because of radiation poisoning that permeated their entire lives. Some miners were put to work packing thousand-pound barrels of "yellowcake." They carried the radioactive dust home on their clothes. Some of the miners ingested so much of the dust that it was "making the workers radioactive from the inside out" (Johansen, citing Eichstaedt, 1995, 11). Downwind of uranium-processing mills, the dust from yellowcake sometimes was so thick that it stained the landscape a half-mile away. Dry winds blew dust from tailings piles through the streets of many Navajo communities. "We used to play in it," said Terry Yazzie, referring to an enormous tailings pile behind his house. "We would dig holes and bury ourselves in it" (Johansen, citing Eichstaedt, 1997, 11).

The neighbors of this particular tailings pile were not told it was dangerous until 1990, 22 years after the mill that produced the tailings pile closed, and 12 years after Congress authorized the cleanup of uranium-mill tailings in Navajo country. Abandoned mines also were used as shelter by animals, who inhaled radon and drank contaminated water. Local people milked the animals and ate their contaminated meat.

The Largest Nuclear Accident in the United States

Thanks to its location between the United States' media capital, New York City, and its political capital, Washington, D.C., as well as the coincident opening of the movie "*The China Syndrome*," Three Mile Island was America's best-publicized nuclear accident. It was not the largest such accident. The biggest expulsion of radioactive material in the United States occurred July 16, 1978, at 5 A.M. on the Navajo Nation. On that morning, more than 1,100 tons of uranium-mining waste tailings gushed through a packed-mud dam near Church Rock, New Mexico. With the tailings, 100 million gallons of radioactive water gushed through the dam before the crack was repaired.

While no one in New York or Washington, D.C. had much to worry about, the Navajo and non-Indian residents of the Rio Puerco River area did. The area is high desert, and the Rio Puerco is a major source of water. By 8 A.M., radioactivity was monitored in Gallup, New Mexico, nearly 50 miles away. The contaminated river, the Rio Puerco, showed 6,000 times the allowable standard of radioactivity below the broken dam shortly after the breach was repaired, according to the Nuclear Regulatory Commission. The few newspaper stories about the spill outside of the immediate area noted that the area was "sparsely populated" and that the spill "poses no immediate health hazard" (Grinde and Johansen, 1995, 211).

The *Los Angeles Times* sent a reporter, Sandra Blakeslee, to the area a month after the spill occurred. By that time, United Nuclear Corporation, which owned the dam, had cleaned up only 50 of the 1,100 tons of spilled waste. Workers were using pails and shovels because heavy machinery could not negotiate the steep terrain around the Rio Puerco. And where were cleanup crews going to put 1,100 tons of radioactive mud, when the next substantial rain would carry it back into the river?

Along the river, officials issued press releases telling people not to drink the water. They had a few problems; many of the Navajo residents could not read English and had no electricity to power television sets and radios. Another consumer of the water—cattle—did not read them either. In the meantime, the local newspaper in Grants, New Mexico, which displayed an atomic symbol on its masthead at the time, blamed the publicity that followed the spill on "Jane Fonda and the anti-nuclear weirdos [who] have scared the hell out of people" (Johansen, 1997, 11).

John Bartlitt, of New Mexico Citizens for Clean Air and Water, expressed perplexity over the lack of attention paid to the accident. About 80 percent of the radioactivity in uranium ore remains in the tailings, he said. "The radioactivity which remains in a pile of tailings after 600 years is greater than that remaining in [nuclear] power-plant water after 600 years," Bartlitt said (Johansen, 1997, 11).

"Grants Enchants"

Since 1950, when a Navajo sheepherder named Paddy Martinez found a strange-looking yellow rock in nearby Haystack Butte, Grants, New Mexico had boomed with uranium mining. Grants styled itself as "the Uranium Capital of the World," as new pickup trucks appeared on the streets and mobile-home parks grew around town, filling with non-Indian workers. For several years, before the boom abruptly ended in the early 1980s, many workers in the uranium industry earned $60,000 or more a year.

After the Rio Puerco spill and the collapse of demand for uranium in the early 1980s, Grants dropped the nickname "Uranium Capital of the World," and began promoting itself as a haven for retirees under a new slogan—"Grants Enchants." A report from the New Mexico Environmental Improvement Division said that while the 1978 spill had been "potentially hazardous . . . its short-term and long-term impacts on people and the environment were quite limited" (Johansen, 1997, 11). While it issued these soothing words, the same report also recommended that ranchers in the area avoid watering their live-stock in the Rio Puerco.

The Environmental Improvement Division's report noted that since the river water was not being used for human consumption, "the extent to which radioactive and chemical constituents of these waters are incorporated in live-stock tissue and passed on to humans is unknown and requires critical evalua-tion" (Johansen, 1997, 12). The report also said that the accident's effect on groundwater should be studied more intensely. Several Navajos said that calves and lambs were being born without limbs, or with other severe birth defects. Other livestock developed sores, became ill, and died after drinking from the river. Tom Charley, a Navajo, told a public meeting at the Lupton Chapter House that "the old ladies are always to be seen running up and down both sides of the [Rio Puerco] wash, trying to keep the sheep out of it" (Johansen, 1997, 12). The Centers for Disease Control examined a dozen dead animals and called for a more complete study in 1983, then dropped the subject.

More problems began to appear. A waste pile at the United Nuclear mill, which had produced the wastes that gushed down the Rio Puerco in 1978, was detected leaking radioactive thorium into local groundwater. On May 23, 1983, the state of New Mexico issued a cease-and-desist order to United Nuclear to halt the radioactive leakage. The company refused to act, stating that its leak did not violate state regulations. Allendale and Appalachian, two insurance companies that were liable for about $35 million in payments to United Nuclear because of losses related to the accident, sued the company on the belief that it knew that the dam was defective before the spill. The dam was only two years old at the time of the accident.

Along the Rio Puerco, several ranchers reacted to state assurances that the spill left no long-term effects by selling their land, for millions of dollars, to the federal government. The ranchers sold out under the 1974 Relocation Act, meant to move Navajos from the former "joint-use area" claimed by the Hopis. The land was purportedly acquired to relocate Navajos who had lost their homes in the land dispute with the Hopis. The Navajos asked the federal Envi-ronmental Protection Agency, the Bureau of Indian Affairs, and the Reloca-tion Commission for assurance that the land was safe. All three declined to provide the requested written assurance to the Navajos. The Navajos raised several questions, including the extent of contamination in underground aquifers, the extent of remaining radioactivity in surface waters and soils, the effects of wind-blown dust from the contaminated area, and the long-term effects of the contamination on livestock and people in the area.

As a result of mining for uranium and other materials, the U.S. Geological Survey predicted that the water table at Crownpoint would drop 1,000 feet, and that it would return to present levels 30 to 50 years after the mining ceased. Much of the water that remained could be polluted by uranium residue, the report indicated. The Indians owned the surface rights; the mineral rights in the area were owned by private companies such as the Santa Fe Railroad. "If the water supply is depleted, then this [Crownpoint] will become a ghost town," said Joe Gmusea, a Navajo attorney. "The only people left will be the ones who come to work in the mines" (Johansen, 1997, 11).

John Redhouse, associate director of the Albuquerque-based National Indian Youth Council, said that the uranium boom was "an issue of spiritual and physical genocide." "We are not isolated in our struggle against uranium development," Redhouse said. "Many Indian people are now supporting the struggles of the Australian aborigines and the Black indigenous peoples of Namibia [in South Africa] against similar uranium developments. We have recognized that we are facing the same international beast" (Johansen, 1997, 11).

The Struggle for Compensation

By 1978, the Navajos were beginning to trace the roots of a lung-cancer epidemic that had perplexed many of them, since the disease was very rare among Navajos before World War II. Many of the miners who started America's nuclear stockpile had died of lung cancer. Harris Charley, who worked in the mines for 15 years, told a U.S. Senate hearing in 1979, "We were treated like dogs. There was no ventilation in the mines." Pearl Nakai, daughter of a deceased miner, told the same hearing that "no one ever told us about the dangers of uranium" (Grinde and Johansen, 1995, 214). The Senate hearings were convened by Senator Pete Domenici, R-New Mexico, who was seeking compensation for disabled uranium miners and for the families of the deceased. "The miners who extracted uranium from the Colorado Plateau are paying the price today for the inadequate health and safety standards that were then in force," Domenici told the hearings, held at a Holiday Inn near the uranium boomtown of Grants, New Mexico (Grinde and Johansen, 1995, 214).

Senate hearings during 1979 initiated proposals to compensate the miners for what investigators called deliberate negligence. By 1930, radioactivity in uranium mines had been associated with lung cancer in European tests. Scientific evidence linking radon gas to radioactive illness existed after 1949, but measures to ventilate the Navajo mines never were implemented, as the government pressured Kerr-McGee and other producers to increase the amount of uranium they were mining. The Public Health Service recommended ventilation in 1952, but the Atomic Energy Commission said it bore no responsibility for the mines, despite the fact that it bought more than 3 million pounds of uranium from them in 1954 alone. The Public Health Service monitored the health of more than 4,000 miners between 1954 and 1960 without telling them of the threat to their health.

Dr. Joseph Wagoner, special assistant for occupational carcinogens at the Occupational Safety and Health Administration, a federal agency, said that of 3,500 persons who mined uranium in New Mexico, about 200 had died of cancer by the late 1970s. In an average population of 3,500 persons, 40 such deaths could be expected. The 160 extra deaths were not the measure of ignorance, he said. Published data regarding the dangers of radon was widely available to scientists in the 1950s, according to Wagoner, but health and safety precautions in the mines were not cost-effective for the companies. "Thirty years from now we'll have the hidden legacy of the whole thing," Wagoner told Molly Ivins of the *New York Times* (Johansen, 1997, 12).

For a dozen years, bills that would have compensated the Navajo miners were introduced, discussed, and left to languish in congressional committees. By 1990, the cancer-related death toll among former miners had risen to 450, and was still rising. By the early 1990s, about 1,100 Navajo miners or members of their families had applied for compensation related to uranium exposure. The bureaucracy had approved 328 cases, denied 121, and withheld action on 663, an approval rate that Representative George Miller, chairman of the House Natural Resources Committee, characterized as "significantly lower than in other cases of radiation compensation" (Johansen, 1997, 12). Representative Miller said that awards of compensation were being delayed by "a burdensome application system developed by the Department of Justice" (Johansen, 1997, 12).

Miller's committee was investigating not only the Navajo death toll from radiation poisoning, but many other reports that indigenous peoples were willfully and recklessly exposed to radiation during the Cold War. The geographic range of purported radiation poisonings spans half the globe—from the Navajos in the southwestern United States, to Alaskan Natives whose lives were endangered when atomic waste products from Nevada were secretly buried near their villages (see this entry United States of America: "The Point Hope Eskimos: An Atomic Harbor and a Nuclear Dump"), to residents of the Marshall Islands in the South Pacific, an area in which the United States tested atomic and hydrogen bombs in the atmosphere between 1946 and 1958 (see "Marshall Islands," this volume). As investigations developed, it appeared that the treatment of Navajos was not the exception, but just one example of a deadly pattern of reckless disregard for indigenous life, human and otherwise, in colonized places.

Delays in Compensation

By the beginning of 2001, former Navajo uranium miners and millers, as well as their families, were holding meetings and protest marches in a renewed attempt to collect compensation for cancers associated with radiation exposure under the Radiation Exposure Compensation Act. Meetings and protest marches were held in Grants, New Mexico and Cortez, Colorado, as well as in or near several towns on the Navajo Nation.

National legislation compensating victims of uranium mining and fallout had been enacted in 1990, guaranteeing up to $100,000 to miners and "downwinders" (those who suffered radiation exposure downwind of nuclear weapons tests). The program was not funded until 1992, and even then only at very low levels. It ran out of money during 1993, when the government began issuing IOUs.

During July 2000, President Bill Clinton signed an amendment to the 1990 law, extending its provisions to millers and transporters of radioactive uranium ore, but once again without adequate funding to implement the law's provisions. In the meantime, increasing numbers of former uranium miners were suffering illnesses and deaths due to their exposure. The amendment expanded the list of eligible afflictions and slashed the amount of time a miner had to have spent working with uranium to be eligible for the compensation program. The measure also opened the program to those who worked in open-pit uranium mines and uranium-milling plants, as well as underground mines. Many miners had complained that they were excluded by the overly stringent rules of the original law.

Melton Martinez, who has been instrumental in the struggle to gain compensation for miners and millers, said that the government knew uranium had caused many cancers and organ failures. Even wives of workers who washed the yellow-coated clothes of their miner husbands were experiencing cancers, Martinez said (Purdom, 2001). Elisabeth Lopez-Rael, the mayor of Milan, New Mexico, told a gathering there that she knew a man in Grants who was given an IOU by the government. "He told me he probably wouldn't live long enough to spend it and he has lost a lot of friends already," Lopez-Rael said (Purdom, 2001).

Several former miners raised the possibility that aquifers in the area remain contaminated by radioactivity from the former uranium mines. One man asked: "Where does the contamination go? Into the water" (Purdom, 2001). Because of radioactivity's persistent nature, Martinez said, "We are doing this not only for ourselves and our neighbors, we are doing this for our children too" (Purdom, 2001). "They did experiment on us like guinea pigs. It makes me angry," one former miner said, as he sat on the steps outside the U.S. House of Representatives. "I would have lived longer, but they gave me a shorter life on this Earth" ("Uranium Miners," 2000).

Gilbert Badoni of Shiprock, New Mexico said he and his siblings "played in uranium-mine tailings and drank radioactive water" during the decades his father worked in uranium mines in Colorado in the 1950s and 1960s. Badoni said his father would come home covered in yellow uranium dust, which covered everything in their small home when their mother brushed it off the clothes ("Uranium Miners," 2000). Badoni blames his lung problems and his siblings' cancers on that exposure. "The U.S. government has abused innocent women and children. They have abused my family," Badoni said, choking back tears. "They have abused my Navajo people. That's not right" ("Uranium Miners," 2000).

THE NAVAJO MEDICINE MEN'S ASSOCIATION, LAKE POWELL, AND CARBON MONOXIDE POISONING

Carbon monoxide poisoning from houseboats has killed at least a dozen people at Lake Powell, the United States' second-largest artificial lake (the lake behind the Glen Canyon Dam is the largest), on the border of Utah and Arizona, in an area where summertime atmospheric inversions can make the air dirtier (and deadlier) than that of most urban areas. The Dine (Navajo) Medicine Men's Association, which calls the lake a desecration of nature, believes it should be drained. The Medicine Men's Association also objects to the fact that, according to a development plan to expand recreational facilities at Lake Powell, archeological and ceremonial sites would be harmed by planned construction that would expand recreational activities in the area.

A number of Native and non-Native environmentalists oppose expansion of boat moorage in the area, which could increase air pollution. A coalition of eight Native American and environmental organizations have called on the National Park Service to suspend contractual and other work on development of the proposed Antelope Point Marina near Page, Arizona. According to published plans, as many as 225 new hotel rooms and 300 boat slips are to be constructed at Antelope Point, along with a gas station and fueling docks, a 150-space recreational-vehicle campground, a sewage plant, up to 100 units of commercial housing, food service, and other commercial operations. The National Park Service has proposed a joint development venture with the government of the Navajo Nation.

"We are working to save our sacred sites and protect our cultural heritage," said Thomas Morris Jr. of Window Rock, president of the Dine Medicine Men's Association. "The Navajo Nation should be supporting traditional ways and setting good examples for children, but here they want to sell liquor and promote jet skis" (Orr, 2000).

Since 1994, people have been dying of carbon monoxide poisoning at Lake Powell. The deep, cold lake has become a summer home for increasing numbers of portable houseboats, which emit carbon monoxide and other pollutants in their exhausts. Lake Powell itself offers the perfect environment for asphyxiation. It is situated in a valley that is prone to atmospheric inversions, especially during the summer, when warm air holds pollutants near the lake's cool surface. Some people have increased their chances of carbon monoxide poisoning by "teak surfing," riding the backs of motorboats near their exhausts. Many houseboats also are designed with swimming areas underneath, near their exhausts. Autopsies have revealed as much as 62 percent carbon monoxide in victims' blood, compared to a usual background range of 2 percent. Roughly 40 percent is considered a fatal dose (West, 2001, A-1).

"Building more tourist resorts won't really help people build a sustainable economic future for themselves," said Anna Frazier, executive director of Dine CARE, a Dine (Navajo) environmental group. "We need to focus on providing basic necessities, like developing safe drinking water supplies and renewable energy sources for our rural communities, and educating our people on ways to conserve energy and water" (Orr, 2000).

Critics of the proposed marina also have expressed concern about declining water quality that could result from plans to expand Antelope Point, suggesting an increase in the numbers of personal watercraft on Lake Powell. A press release from the critics' coalition noted that the boats' inefficient two-cycle engines "emit large amounts of unburned oil and gasoline into water bodies through their exhaust systems, potentially endangering drinking water quality for more than 20 million Colorado River water users downstream" (Orr, 2000).

In addition to Lake Powell's other problems, in early 2002 Living Rivers and eight other river-protection groups disclosed that the lake was badly silted. In a letter to the National Park Service, the groups call for federal action to address "the growing problem of river mud, which is interfering with boating activities in the upper reaches of [Lake Powell]" ("Lake Powell," 2002). "This is the beginning of the end for Lake Powell," said John Weisheit, Living Rivers conservation director and a professional river guide with 17 years' experience. "People talk about Lake Powell filling with silt sometime in the future, but the future is now" ("Lake Powell," 2002).

According to a report by the Environment News Service, "Similar impacts are being felt today in the Grand Canyon far downstream, where in summer 2001 the Pearce's Ferry boat take out was closed due to thick layers of oozing sediment clogging the upper reach of Lake Mead Reservoir" ("Lake Powell," 2002).

THE GROS VENTRE AND ASSINIBOINE: GOLD MINING AND CYANIDE POISONING IN MONTANA

The Little Rocky Mountains of Montana, which long have been regarded as sacred by the Assiniboine and Gros Ventre, are now laced with the effluent of open-pit gold mines that have produced toxic acid drainage from cyanide-heap gold mining. Andrew Schneider of the *Seattle Post-Intelligencer* described Gus Helgeson, the president of Island Mountain Protectors, a Native American environmental and cultural organization, standing atop Spirit Mountain as he scanned "the gashes, pits and piles of rock that once was his tribe's most sacred land. . . . The strong man weeps" (Schneider, 2001).

During 1855, the Assiniboine and Gros Ventre were moved to the Fort Belknap Reservation, which was named for a U.S. Secretary of War. The Assiniboine and Gros Ventre gave up 40,000 acres of land in exchange for a government promise to feed, clothe, and care for them. At the time, federal Indian agents said nothing about the gold that was buried in Spirit Mountain, but they made it clear the tribes could either agree to their terms or starve. Spirit Mountain is part of the Little Rockies, an island of mountains in the nearly flat prairie. To Native Americans, the mountains were valued for their deer, bighorn sheep, herbs, natural medicines, and pure water. Gold mining has destroyed all of that.

In 1884, Pike Landusky and Pete Zortman discovered gold on the reservation. Facing starvation, in 1895, the Assiniboine and Gros Ventre signed an agreement negotiated by General George Bird Grinnell, selling portions of their gold-laced land to the federal government for $9 an acre, paid in livestock and other goods, as miners besieged towns that had been named after the gold's discoverers. Under the terms of the General Mining Law of 1872, the government sold the land to individuals and private companies for $10 an acre.

"The first time the mining company let me up here, let me see what they had done to our land, the pain surrounded me. It was like watching our ancestors die, raped of their honor," said Helgeson. "They destroyed this place, took their gold off in armored trucks and left us a wounded mountain spewing poison on the people the mountain was stolen from," he said (Schneider, 2001).

Over several decades, according to Schneider's account, "scores of shafts were driven into the Little Rockies, and an estimated $1 billion in gold and silver was taken out of the ground—more than $300 million by the last owner of the mine, Pegasus Gold Corp. of Canada" (Schneider, 2001). Underground mining continued until the 1950s, after which open-pit strip mining was initiated. In 1979, with gold prices rising rapidly, Pegasus Gold and a subsidiary, Zortman Mining, built mines that extracted gold from heaps of low-grade ore with cyanide solutions.

Pegasus Gold, which owned several mines in the Little Rocky Mountains of Montana, went bankrupt when gold prices fell sharply after 1980, "leaving the state of Montana with a $100 million cleanup liability and the tribes with the prospect of perpetually polluted water" (Huff, 2000). Cyanide-assisted gold mining continued until 1990, during which time the mine was expanded nine times without any substantial environmental review, despite cyanide spills into the water table used by the Indians (Abel, 1997).

By 1990, the Assiniboine and Gros Ventre began to challenge the environmental side-effects of cyanide-heap gold mining, forming a Native environmental-advocacy group, Red Thunder, which joined with non-Indian environmental groups to resist federal permits for the Zortman-Landusky's mine's next requested expansion. The group's appeal was denied. By December 1992, Pegasus applied for another expansion of the mine, as Red Thunder joined with another Native environmental group, Island Mountain Protectors. Both prepared plans to challenge the expansion under the federal Clean Water Act, maintaining that the cyanide-leach method used in the mine was poisoning the reservation's water supply.

During July 1993, an intense thunderstorm brought a flood of acidified mine wastewater into the town of Zortman, after which the Bureau of Land Management (BLM) required the mine's owners to develop a new reclamation plan. At about the same time, an Environmental Protection Agency study found that the mine had been "leaking acids, cyanide, arsenic and lead from each of its seven drainages" (Abel, 1997). The state of Montana soon joined the EPA in a suit based on the Clean Water Act, which was settled out of court in July 1996, with Pegasus and Zortman Mining pledging to pay $4.7 million in fines to the tribes, the federal government, and the state. The mine's owners also pledged to follow a detailed pollution-control plan in the future.

Shortly thereafter, a request to triple the mine's size (from 400 to 1,192 acres) was approved by the Montana Department of Environmental Quality and the BLM (Abel, 1997). In January 1997, the Fort Belknap Community Council and the National Wildlife Information Center sued the Montana Department of Environmental Quality, alleging that the agency's decision to allow an expanded mine violated state law.

During September 1997, federal and state environmental agencies fined Pegasus and Zortman Mining $25,300 for violating the clean-water settlement by polluting a stream in the Little Rockies the previous summer. John Pearson, director of investor relations for Pegasus, asserted that discharges were the result of "acts of God" during "extraordinarily heavy rains" (Abel, 1997). By late 1997, with gold prices (and the company's share prices) declining rapidly, Pegasus warned that its mine would close by January 1, 1998 if the expansion plan was not accepted. The state of Montana and local Native environmental activists wondered whether Pegasus would survive long enough as a corporate entity to complete promised reclamation of existing mines. In January 1998, Pegasus filed for Chapter 11 bankruptcy protection.

In the meantime, Pegasus left behind open pits that were described by Andrew Schneider in the *Seattle Post-Intelligencer*:

> Pegasus dug pits the size of football fields and lined them with plastic or clay. Crushed ore was dumped in mounds as high as 15 feet and soaked with a mist of cyanide. It was the largest cyanide heap–leach operation in the world. . . . The heavily contaminated water trickled and flowed through fissures in the mountain, into the surface streams and underground aquifers that supply drinking water for 1,000 people who live in and around Lodge Pole and Hays, reservation towns north of the mountains. (Schneider, 2001)

Streams flowing off the mountain smell of rotten eggs (the chemical signature of sulfide), cloudy and lifeless. "This is death," said John Allen, a tribal spiritual leader, as he filled his hands with putrid muck. "The mines take millions in gold from our land and leave us poisoned water. The miners and the government experts have argued for years about whether the water is bad. All they have to do is look, but they choose not to see" (Schneider, 2001). Allen, who was 46 years of age in 2001, has thyroid problems, as do three of his siblings. His father has lymphatic cancer. Doctors who specialize in environmental medicine have told the Assiniboine and Gros Ventre that the source of their diseases is contaminated water. Environmental advocates among the two tribes also report a high rate of stillbirths.

THE NORTHERN CHEYENNE: METHANE GAS EXTRACTION

The Northern Cheyenne of eastern Montana have been raising questions about plans for methane gas development on their land, where three decades of strip mining already has contributed to numerous environmental problems. The Northern Cheyenne's reservation rests between Rosebud Creek and the

Tongue River Valley, on top of the Fort Union Coal fields, one of the richest coal deposits in North America.

A new plan to initiate coal-bed methane development, which extracts methane from coal beds, could create as much as eight million gallons of waste-water per day. This wastewater is so high in sodium that it precludes agricultural use and poses a danger to fish and wildlife. Because coal-bed methane development is a new technology, there are no Native American, state, or federal laws regulating the development or protecting the environment from its possible adverse effects.

Gail Small, executive director of Native Action, a grassroots organization on the Northern Cheyenne reservation, asserted that "essentially the federal government is opening up the entire Tongue River Valley for very vested interests. These types of tactics that shut out the Cheyenne from regional energy decisions only serve to make the Cheyenne feel more alienated and vulnerable to actions that our trustee [the Bureau of Indian Affairs] has undertaken. This puts the Northern Cheyenne in a very defensive position, and always sets us up to be the spoiler in Montana's economy. A lot of this could be avoided if the federal government and state government would treat us as a government, and consult us as they do in a government-to-government relationship itself" (LaDuke, 2001).

Coal-bed methane development requires "dewatering" (removal of water) from the fragile aquifer and coal-bed system. The discharged water is not considered pristine: its temperature (70 degrees Fahrenheit) is considerably higher than the temperature of the nearby watercourses, including the Tongue River, into which the water would be discharged (LaDuke, 2001).

The prospective draw-down of the aquifer is a major concern to many Northern Cheyenne. If developed, the 109,000-acre methane coalfield will be "dewatered" at the rate of 150 gallons per minute. On the Northern Cheyenne reservation, new methane development may create as much as 8 million gallons of wastewater per *day*. In comparison, total *annual* reservation water use at present is at 21 million gallons (LaDuke, 2001).

The BLM's monitoring wells in the Powder River Basin have indicated that the water level in the aquifer already has dropped more than 200 feet, largely due to increasing human usage, raising questions regarding how long the aquifer will remain useful, even without coal-bed methane development.

THE POLITICAL ECONOMY OF KITTY-LITTER STRIP MINING

Representatives of the Reno-Sparks Indian Colony paid a surprise visit to the 2001 annual meeting of the Oil-Dri Corporation shareholders' meeting in Chicago, attempting to stop a proposed kitty-litter strip mine at the edge of their 550-person community in Nevada. The residents of the community, in Hungry Valley, want to prevent Oil-Dri from opening two strip mines and a processing plant within a mile of their community. They fear heavy traffic near schools, pollution of groundwater, and dust-laced air pollution from the mines and processing plant. They also present evidence, supported by an array of photographs, that reclamation will never heal the earth in the dry environment.

Ben Felix, a Hungry Valley citizen working with the colony, said during the visit to Chicago: "Six of us came from Reno, Nev., to attend the shareholders' meeting of the Oil-Dri Corp. They are proposing to put a strip mine and manufacturing plant in an already existing residential area. This directly affects the Reno-Sparks Indian Colony. They are trying their utmost best to get this processing plant that will run 24 hours a day, seven days a week for 365 days out of the year" (Pierpoint, 2001, A-1).

"They are claiming that they will be able to reclaim the existing land back to its pristine nature," said Felix, "but it really doesn't address the issues. Arsenic, mercury, selenium and some other minerals will be leached into our ground water. The air, dust and pollution that will be caused in the valley will mean breathing very, very fine particulates, finer than chalk dust off a chalkboard, and this is the air our children are going to have to breathe for the next hundred years" (Pierpoint, 2001, A-1).

The Oil-Dri Corporation asserts that the Reno-Sparks mine is the only commercially viable mine for kitty litter in the United States. "If this [the proposed mine] is exhausted, does that mean the end of kitty litter as we know it?" asked Felix. "Are our children going to have to suffer severe asthma and other health problems just to satisfy the pocketbook of one corporation?" (Pierpoint, 2001, A-1).

Oil-Dri has received reclamation and air permits for the projects during the two-and-a-half years of the Indians' resistance. Oil-Dri plans to open the mines in the fall of 2002. The company plans to spend $10 million on the mines and processing plant, which it says will offer as many as 100 jobs. Washoe County planning commissioners voted December 19, 2001 to deny a permit allowing Oil-Dri to open a kitty-litter processing plant in Hungry Valley, central Nevada.

THE WESTERN SHOSHONE: "THE MOST BOMBED NATION ON EARTH"

Since 1985, many Western Shoshone people have joined with environmentalists to resist nuclear-weapons testing in Nevada, trespassing on the world's only remaining nuclear-weapons proving ground, northwest of Las Vegas. The test site occupies land that President Harry Truman confiscated from the Western Shoshone for "national security" purposes during 1951, forcibly relocating 100 Native American families. The seizure of Western Shoshone land for nuclear-weapons testing has never been fully tested in court.

Traditional Shoshone chief Corbin Harney "came out from behind the bush," as he says, in 1985, initiating protests of the U.S. government's nuclear weapons testing ("Corbin Harney," n.d.). He also has taken hundreds of people (Indian and non-Indian) into sweat lodges to pray against testing. Harney and other Western Shoshone leaders repeatedly have crossed a gated cattle guard onto the 1,350-square-mile test site. On occasion, they have been joined by large groups of people marching in support of a ban on nuclear testing. The marches, which have included several thousand people, sometimes are held on Mother's Day (in honor of Mother Earth). At some of these marches hundreds of people have been arrested.

Harney cited evidence that the underground tests leak radiation aboveground. "Downwinders," people who live downwind from nuclear-weapons test sites, especially in southern Utah, have, according to Harney, "observed extremely high numbers of cancers, leukemia, and other physical deformities in their population" ("Corbin Harney," n.d.). Harney compared these reports to similar findings among citizens of Kazakhstan (in the former Soviet Union) who lived downwind from a nearby test site. They have reported similarly high levels of cancer, leukemia, illness, and birth deformities. Protests later closed the test site at Kazakhstan. During 1993, when Harney visited Kazakhstan, he learned that local water had been irreversibly contaminated with dangerous levels of radiation. "You can't drink a glass of water there anymore," he said. "All they had to drink was Vodka, cartons of juice, and bottled water imported from Europe" ("Corbin Harney," n.d.).

About 700 nuclear tests for Great Britain and the United States have since been conducted at the Nevada Test Site since 1951, on lands belonging to the Western Shoshone according to the 1863 Treaty of Ruby Valley. Until 1963, the tests were conducted aboveground. Atmospheric tests were banned by international treaties during the 1960s; in the 1970s tests also were limited to a 150-kiloton explosive yield.

The Environmental Protection Agency has documented cumulative deposits of plutonium in soil samples more than 100 miles north of the test area. "We are the most-bombed nation in the world," said William Rosse Sr., a Western Shoshone elder. "We've had more than our share of radiation, and now they want to dump more nuclear waste on our land at Yucca Mountain" ("Havasupai Fight," 1992). "We have always been a free tribe, never have we been conquered and never have we sold land to the U.S.," said John Wells of the Western Shoshone National Council. "My people are still suffering from the 828 underground nuclear weapons tests and 105 above-ground nuclear weapons tests at the Nevada Test Site" (Hodge, 2001).

Large-scale Civil Disobedience at the Nuclear Test Site

During 1992, on Easter, several hundred people were arrested for crossing the Nevada Test Site's cattle guard. About 1,000 drum-pounding people protested that year, 761 of whom were arrested. They were packed into the Nye County Jail at Beatty, charged with misdemeanor trespassing, and released the same day (Rogers, 1992, n.p.). Keith Rogers, a reporter for the *Las Vegas Review-Journal*, described the scene at the 1992 Easter Sunday protest:

Dressed in yellow coveralls with a plastic nose that served as a makeshift rabbit suit, Hugh Romney, the 1960's anti-war activist better known as Wavy Gravy, stood before the line of police at the cattle guard. "There's nobody left to blow up but ourselves," he said about nuclear testing moments before his arrest. "It's an insult to our planet." Then in 1960's fashion, the throng chanted, "The whole world is watching," while some threw flowers in the air and [others] put some flowers on a barbed-wire fence. (Rogers, 1992, n.p.)

During the early morning of April 3, 1997, a cold, snowy day, two vans loaded with antinuclear activists tried to blockade Highway 95, five miles east of the Nevada Test Site. They had planned to lock a junked car into freshly poured concrete, along with "custom devices that would link seven humans, six men and seven women, aged 20 to 55, to the car, to concrete-filled barrels, and to each other" (Lee, 1997).

The protesters assembled their blockade under the eyes of a small audience that included several newspaper reporters, a Japanese public-television crew, and official representatives of *hibakasha* (Japanese survivors of atomic bombings at Hiroshima and Nagasaki), who draped the participants with paper-crane necklaces and antinuclear protest buttons. Several peace activists and a representative of Nelson Mandela also looked on.

Blockade plans were disrupted when the protesters were routed by police. Some protesters laid across the highway and locked themselves together to obstruct police buses. Five hours passed before the state highway patrol and commercial security guards cleared the four-lane-wide human blockade.

During those five hours, according to one participant,

we were photographed, fed, watered, massaged, glared at, and in the case of the north-side blockaders and the Greenpeace spokesman, threatened by an irate trucker with a gun he claimed to have in his cab. Luckily one of our peacekeepers and some of the officers . . . heard the threat and surrounded the trucker and calmed him down. . . . Traffic was backed up several miles on each side of the highway, mainly trucks (including one or two hauling low-level radioactive waste to be buried in shallow trenches at the test site). (Lee, 1997)

The protesters were arrested and taken to the Clark County Jail in Las Vegas, charged with trespassing and obstruction of an officer, held overnight, and then released without bail.

Once again, during May 2000, roughly 700 people gathered at the Nevada Test Site "to celebrate Mother's Day and [to] demand an end to the radioactive poisoning of Mother Earth" ("Mother's Day," 2000). Following a rally at the test site's gates, featuring music and speakers from around the world, 198 people crossed into the restricted area and submitted to arrest. As the arrests were taking place, Ian Zabarte, representing the Western Shoshone National Council, told test-site officials that they were trespassing on Shoshone lands in violation of international law, and also in violation of a treaty signed with the

U.S. government in 1863. The rally drew a variety of Shoshone and non-Native peoples intent on stopping nuclear testing.

> Participants attended workshops, discussion groups and nonviolence trainings. The new youth program was thoroughly enjoyed, with activities for families, small children and youth. Mother's Day began with at dawn [with] sweat lodges for women, a Eucharist Service offered by 35 members of the Episcopal Peace Fellowship, and a Grandmothers and Crones Ceremony. Following a brunch served by the men in camp, a march was led by Corbin Harney, Western Shoshone Spiritual Leader, members of the Western Shoshone National Council, and other Native American community leaders. Hundreds of grandmothers, children, and families and supporters of all ages followed the eagle staffs . . . to the test-site gates. (Lee, 1997)

About 175 activists also participated in a Western Shoshone "occupation" of the test site, erecting a tipi on restricted ground and joining in a sunrise prayer ceremony led by Harney. Another tipi was erected more than five miles inside the test-site perimeter, high on a ridge top, where another sunrise ceremony was celebrated "by tired but inspired activists" (Lee, 1997). A third tipi was built well inside the front entrance of the test site, and it was visible to workers at the test site as they arrived at dawn.

Nuclear Storage at the "Serpent Swimming West"

In addition to their status as "the most bombed nation on Earth," the Western Shoshone now find their ancestral lands proposed as the home of the United States' principal nuclear-waste storage facility at Yucca Mountain, which their language calls "The Serpent Swimming West." The Western Shoshone don't like the prospect, and neither do most of the rest of the people of Nevada. The U.S. Department of Energy has proposed to haul by road and rail 77,000 tons of spent nuclear fuel rods and other high-level radioactive waste, transported from 103 storage sites in 39 states to Yucca Mountain, the only site being considered for its permanent storage ("Energy Chief," 2002). The Shoshone's name for their homeland is Newe Segobia (which means "earth mother"). Newe Segobia, which was guaranteed to the Shoshone by the 1863 Treaty of Ruby Valley, includes Yucca Mountain.

By August 2001, an estimated 43,000 metric tons of spent nuclear fuel was being stored in water pools and concrete casks at more than 70 nuclear-plant sites around the United States, awaiting burial at a geologic repository at Yucca Mountain in Nevada, about 90 miles northwest of Las Vegas. Yucca Mountain is also the proposed long-term home for 10,000 metric tons of high-level waste from U.S. military weapons programs, including the navy's nuclear reactors, which has been stored in government installations mainly in Idaho, Washington State, and South Carolina.

The U.S. Congress designated Yucca Mountain as the principal site for a high-level waste facility in 1987. The Yucca Mountain repository may be licensed to hold nuclear waste for 10,000 years. Congress also directed the U.S. Department of Energy (DOE) to determine whether the waste could be safely placed there, after which the DOE deployed teams of scientists to evaluate the site's geology, hydrology, and geochemistry "in what is probably the most comprehensive and systematic assessment ever conducted of a piece of land anywhere on the planet" (Kolar, 2001, A-13).

By mid-2001, the DOE had spent 14 years and $7 billion assessing Yucca Mountain's suitability as a nuclear-waste storage facility (Wastell, 2001, 27). The DOE found that Yucca Mountain meets its definition of a safe place to store nuclear waste. According to the DOE, Yucca Mountain is arid and geologically stable. The DOE's reports say that the chambers holding nuclear waste canisters would be at a safe distance from the underground water table. The site is, according to its advocates, "the perfect place for a nuclear burial ground" (Kolar, 2001, A-13).

Very few people in Nevada agree with the DOE's assessment. On September 5, 2001, hundreds of angry people showed up at DOE-sponsored public hearings in Las Vegas, Carson City, Elko, and Reno to express their objections to proposals for storing nuclear waste in Yucca Mountain. The protesters included a broad array of Nevada residents, from Western Shoshone spiritual leaders to Governor Kenny Guinn and the entire Nevada congressional delegation. According to several polls, roughly 90 percent of Nevadans are opposed to the project.

Protesters paraded to microphones (with some participating by satellite hookup from Washington, D.C.) until after midnight, overwhelming the few people who spoke in favor of establishing a nuclear-waste repository in the mountain, which the Western Shoshone and Paiute hold to be sacred. In Las Vegas about 450 people filled the Department of Energy's National Nuclear Security Administration's meeting room, overflowing the room's capacity and spilling into the hallway. Hundreds of other people in a nearby room watched the proceedings on television. The large number of participants attended the hearing despite its remote location, tight security, late hours, and a limit on comments of five minutes per person. A total of 132 people signed up to speak at the hearing, which began shortly after 6 P.M. Fewer than three dozen people had spoken by 11 P.M.; the last person was heard at 2:10 A.M. (Rogers, 2001).

"This fight transcends party affiliations, transcends socio-economic class, race or gender, and galvanizes all Nevadans from every corner of the state in opposition," Governor Guinn said, to a standing ovation. "We in Nevada will not stand for it" (Lewis, 2001). Many Nevada residents objected to the projected cost of the project (as much as $60 billion), as they argued that the safest and cheapest way to handle the spent nuclear fuel would be to put it in dry-cask storage on the sites where it is generated. Dry-cask storage should last for 100 years; during that time, say the Yucca Mountain storage site's opponents, scientists could explore new nuclear-waste recycling technologies, which might reduce the volume of waste and the period of radioactivity.

One of the tunnels inside Yucca Mountain, Mercury, Nevada, 1988. (U.S. Dept. of Energy)

The crowd heckled and hissed at Gary Sandquist, a professor of mechanical engineering from the University of Utah, who advocated the proposed nuclear-waste storage site. "We must store the nuclear fuel somewhere," said Sandquist, who maintained that Yucca Mountain is the best place (Lewis, 2001). Las Vegas Mayor Oscar Goodman said he will personally arrest any driver of a truck bearing nuclear waste in that city. "Well, if they can't tell us that we're safe, how dare they even consider bringing this crap here?" Goodman asked (Lewis, 2001). Opponents of the Yucca Mountain storage site include many Las Vegas casino and hotel owners, who complain that visitors will not vacation in a city if they believe they may be poisoned by nuclear waste. "People come to Las Vegas to gamble with their money, not their lives," said one gaming-industry source (Wastell, 2001, 27). "If they need support on nuclear power, they won't get it from Nevada," said Senator John Ensign (Wastell, 2001, 27). Nevada has no nuclear power plants of its own.

Opponents assert that the Yucca Mountain site is not suitable for nuclear storage because it contains numerous earthquake-prone geologic faults. Opponents also believe that rapid water-migration pathways (some of them geothermal in origin), which lace Yucca Mountain, will allow water to infiltrate waste containers, corrode them, and possibly carry radioactivity into the area's water table. The DOE wants to store plutonium at Yucca Mountain until methods are found to make it useful in light-water nuclear reactors, perhaps in a half-century. According to this reasoning, plutonium at Yucca Mountain will be a resource in reserve, not stored waste. Harney fears that the plutonium could

View of the ESF South Portal at Yucca Mountain, Mercury, Nevada, 1988. (U.S. Dept. of Energy)

reach critical mass underground, setting off a nuclear reaction that would make local water unfit to drink. Plutonium has a radioactive half-life of 24,000 years.

"Yucca Mountain is not a safe place to put any kind of nuclear waste," Harney said. "It's not a mountain to begin with, like they've been telling us, it's [a] rolling hill. That's a moving mountain. There are seven volcanic buttes there. . . . Underneath it is hot water that's causing a lot of friction in that tunnel, and today they're telling you it's not dangerous. But how come, if it's not dangerous, many, many of my people have died from cancer caused by radiation?" (Lewis, 2001).

Harney said that at least 621 earthquakes have been recorded in the area (at magnitudes of 2.5 or higher on the Richter scale) during the last 20 years. A major earthquake at Yucca Mountain could cause groundwater to surge into the storage area, forcing plutonium into the atmosphere and contaminating the water supply ("Yucca Mountain," n.d.). One earthquake at Little Skull, measuring 5.6 on the Richter scale, was eight miles from the proposed disposal site.

In addition to the geologic perils of Yucca Mountain, project opponents assert that transport of nuclear waste over public highways and rails from across the United States creates potential for a "mobile Chernobyl" while the waste is on the road ("Yucca Mountain," n.d.). Environmentalists who oppose efforts to transport spent fuel to Nevada for storage at Yucca Mountain contend that, if one of the canisters holding the highly radioactive material were to rupture, people living along transport routes would be at risk. Once the Yucca Mountain site is operating, federal officials project that the Yucca Mountain site will receive six to seven shipments of highly radioactive waste daily. Waste will traverse 46 of the 50 states on its way to the site.

The Politics of Nuclear-Waste Disposal

At the end of November 2001, a congressional report on Yucca Mountain's nuclear-waste potential was leaked to the press. The General Accounting Office report was very critical of the DOE's use of misleading and incomplete information on the site. The recommendation of the report was for an indefinite delay of the DOE's site report. Such a delay (which never occurred) probably would have effectively scuttled the plan for a nuclear-waste dump at the site.

United States Representative Shelley Berkley of Nevada, a longtime opponent of the plans, said that "this report has potential to derail the Yucca Mountain project altogether. It details the shocking bias and mismanagement that Nevadans have been alleging for years. This is the smoking gun we've been looking for" ("Leaked Report," 2001). Berkley and Senator Harry Reid, both Nevada Democrats, commissioned the report from the General Accounting Office (Congress's investigative office) after they received what they call "an anonymous whistleblower" ("Leaked Report," 2001). Reid also has been a fierce opponent of the Yucca Mountain project.

The draft report said that the DOE "is unlikely to achieve its goal of opening a repository at Yucca Mountain by 2010 and has no reliable estimate of when, and at what cost, such a repository could be opened" ("Leaked Report," 2001). According to the Environment News Service, "The report characterizes work to determine whether Yucca Mountain can safely contain spent nuclear fuel from the nation's 103 nuclear-power plants as 'a failed scientific process' that has resulted in continual changes to the site suitability criteria" ("Leaked Report," 2001). Senator Reid said that the report indicates that science has taken a back seat to politics at Yucca Mountain from the start. "This report could very well signal the beginning of the end of the Yucca Mountain project," he said ("Leaked Report," 2001).

Speaking for the Bush administration, which later approved the project, Energy Secretary Spencer Abraham called the preliminary General Accounting Office report "fatally flawed" ("Leaked Report," 2001). He said that the report's premature release destroyed its credibility. At any rate, said Abraham, Nevada's congressional delegation had ordered creation of a report that supported a "predetermined conclusion" ("Leaked Report," 2001). Reid fired back: "The D.O.E. has wasted $8 billion of taxpayers' money on this project, and still isn't using sound science as a basis for their recommendations. Apparently, the D.O.E. is actually suppressing science at the expense of the health and safety of Nevadans and all Americans" ("Leaked Report," 2001).

The state of Nevada filed a lawsuit on December 17, 2001 to halt the Yucca Mountain project, "alleging that Energy Department ground rules for judging whether the site is suitable for nuclear waste storage are contrary to what Congress intended" ("Energy Chief," 2002). Governor Guinn said the state has assembled a legal team including nuclear scientists, physicists, and environmental experts, all with law degrees. "For the first time in 18 years," Guinn said, "we now have the wherewithal to enter into the judicial system with very competent attorneys and scientific people on their staff, and we're going to do everything we can to prohibit it from coming here no matter what decision he [Abraham] makes" ("Energy Chief," 2002). The lawsuit, filed with the Federal District Court in Washington, D.C., asks that Abraham be prevented from making recommendations on Yucca Mountain until the ground rules are reviewed by the courts.

Abraham was met by protesters at the Las Vegas federal building when he toured the Yucca Mountain site on January 7, 2002, preparing a recommendation on the site for President Bush. Governor Guinn and Abraham met for an hour during the visit, after which the governor said he had a chance to "reaffirm our adamant commitment against this project" ("Energy Chief," 2002). A day after Abraham visited the site, the DOE recommended approval of Yucca Mountain as the national nuclear-waste facility. Abraham said that within 30 days he intended to recommend to President Bush that the Yucca Mountain site is "scientifically sound and suitable to hold radioactive waste" ("Nevada Outraged," 2002). Furthermore, Abraham said that the development of the Yucca Mountain storage facility "will help ensure America's national security and secure disposal of nuclear waste, provide for a cleaner environment, and support energy security. We should consolidate the nuclear wastes to enhance protection against terrorist attacks by moving them to one underground location that is far from population centers" ("Nevada Outraged," 2002).

Governor Guinn replied tersely:

> I told him [Abraham] that I am damn disappointed in this decision and to expect my veto. I explained to him we will fight it in the Congress, in the Oval Office, in every regulatory body we can. We'll take all of our arguments to the courts. This fight is far from over. I also told him that on behalf of all Nevadans, I am outraged that he is allowing politics to override sound science. I told the secretary that I think this decision stinks, the whole process stinks, and we'll see him in court. ("Nevada Outraged," 2002)

As the Bush administration approved nuclear-waste storage at Yucca Mountain, the Nuclear Waste Technical Review Board (NWTRB) issued a report concluding that scientific uncertainties make it impossible to guarantee that the dump will remain safe for the thousands of years necessary to protect the environment. While the board found that "no individual technical or scientific factor . . . would automatically eliminate Yucca Mountain from consideration as the site of a permanent repository" for the nation's nuclear waste, a variety of problems exist with the studies that aim to ensure the safety of the site (Lazaroff, 2002). The NWTRB study questioned the adequacy of computer models used to project how the site's natural features, including geological and hydrologic formations, will protect the stored wastes. The report also raised concerns about how or whether the casks—designed to contain wastes for the 10,000 years required by lawmakers—will hold up to the potential tests of time, natural disasters, and manmade disasters (Lazaroff, 2002).

"Gaps in data and basic understanding cause important uncertainties in the concepts and assumptions on which the D.O.E.'s performance estimates are now based," the report concluded. "Because of these uncertainties, the Board has limited confidence in current performance estimates generated by the D.O.E.'s performance assessment model" (Lazaroff, 2002). The DOE has spent $4 billion studying the site during the last 24 years. Spent nuclear fuel and high-level radioactive waste is now scattered across 131 sites in 39 states, according to the DOE.

On February 5, 2002, President Bush approved Abraham's recommendation of Yucca Mountain as a storage site for 77,000 tons of high-level nuclear waste, generated to that date by power reactors and nuclear-weapon production across the United States. In a letter to congressional leaders announcing his decision, Bush said that proceeding with the repository program "is necessary to protect public safety, health, and the nation's security because successful completion of this project would isolate in a geologic repository at a remote location highly radioactive materials now scattered throughout the nation" ("Bush Green-lights," 2002). Following Bush's recommendation, the site's licensing proceedings will begin before the Nuclear Regulatory Commission.

The president's decision provoked renewed outrage among Nevada elected officials and environmentalists. It also delighted the nuclear industry. Joe Colvin, president and chief executive officer of the Nuclear Energy Institute, said that "after almost two decades of exhaustive scientific evaluation showing that the site is suitable to isolate and safely dispose of used nuclear fuel, the federal government is acting responsibly and taking steps to fulfill its obligation to the American people" ("Bush Green-lights," 2002).

Nevada Senator Harry Reid, a Democrat, countered: "President Bush has betrayed our trust and endangered the American public." Senator Reid charged that Bush had lied to him and to the people of Nevada because "just last week in a meeting with Senator [John] Ensign, Governor Guinn and me at the White House [he] again vowed to wait until he received and reviewed all of the scientific evidence on Yucca Mountain. Today President Bush has broken his promise" ("Bush Green-lights," 2002). Reid said that transport of the waste will require the use of 20,000 railcars, traveling through 43 states. "The President," he said, "has created 100,000 targets of opportunity for terrorists who have proven their capability of hitting targets far less vulnerable than a truck on the open highway" ("Bush Green-lights," 2002).

Within minutes of Bush's approval, Nevada Governor Guinn announced that he will exercise his Notice of Disapproval to the U.S. Congress, known as the Governor's Veto. Congress would then have 90 legislative days in which it could override Guinn's veto on a simple majority vote. "We will exhaust every option and press our legal case to the limit," Guinn asserted. "The Nevada Legislature, cities, counties and now the private sector have raised $5.4 million toward our fight" ("Bush Green-lights," 2002). On April 8, Governor Guinn vetoed the Bush administration's recommendation to build a permanent repository for radioactive wastes at Yucca Mountain. "Let me make one thing crystal clear—Yucca Mountain is not inevitable, and Yucca Mountain is no bargaining chip," Guinn said at an address at the University of Nevada. "And, so long as I am governor, it will never become one" ("Nevada Governor," 2002). By law, Congress had 90 legislative days from the date of the veto to override Guinn's veto on a simple majority vote. On May 8, 2002, the House of Representatives approved the Yucca Mountain project by a vote of 306–117.

Illustrating the fervor with which a number of Nevadans oppose the Yucca Mountain facility, Kalynda Tilges, nuclear issues coordinator for the Las Vegas–based group Citizen Alert, said that if Yucca Mountain shipments eventually go ahead, "I will be standing in front of the first truck of the first gate they send it from, and I will not be alone. And if I'm not dead, when I get out of jail, I'll go stand in front of the next one. They will bring nuclear waste to Yucca Mountain over my dead body" ("Bush Green-lights," 2002).

Despite President Bush's belief that the Yucca Mountain site's construction rests on sound science, the Nuclear Regulatory Commission's Advisory Committee on Nuclear Waste issued a report during September 2001, which asserted that recommendations favoring the site's safety rely "on modeling assumptions that mask a realistic assessment of risk" and that "computations and analyses are assumption-based, not evidence-supported" (Ewing and Macfarlane, 2002, 659). An analysis of the situation in *Science,* the most prominent general scientific journal in the United States, concluded that "in our view, the disposal of high-level nuclear waste at Yucca Mountain is based on an unsound engineering strategy and poor use of present understanding of the properties of spent nuclear fuel. . . . To move ahead without first addressing the outstanding scientific issues will only continue to marginalize the role of science and detract from the credibility of the Department of Energy effort. As Thomas Jefferson cautioned George Washington, 'Delay is preferable to error'" (Ewing and Macfarlane, 2002, 660).

On July 9, 2002, the U.S. Senate voted to move ahead with the repository at Yucca Mountain by a vote of 60–39. President Bush signed the measure on July 23.

During the ongoing debate vis-à-vis Yucca Mountain, British researchers developed a computer model indicating that a volcanic eruption might cause greater damage than previously thought to a nuclear waste storage facility there.

The study said that Yucca Mountain is located within a long-lived volcanic field. Prior risk assessments have suggested that the probability of volcanic activity occurring during the 10,000-year compliance period of the repository is between 1 in 1,000 and 1 in 10,000. Past eruptions from volcanoes within 12 miles of the proposed repository have produced small volumes of molten rock. While the amount of magma has been relatively small, the high percentage of volatile gases in it indicates that the magma has been very explosive ("Yucca Mountain Volcanic," 2002).

Models developed by Andrew Woods of the BP Institute, University of Cambridge, United Kingdom, and colleagues envision that volcanic magma rising from below Yucca Mountain would form a narrow body of molten rock called a *dike*. According to this study, such a dike could cut through the repository, causing some storage areas to expand rapidly. Based on their models, Woods and his team found that magma in the drifts "could reach speeds up to 600 miles per hour, filling parts of the repository with magma within a matter of hours after the initial eruption" ("Yucca Mountain Volcanic," 2002).

According to an Environment News Service description of the research, "Flowing magma might displace canisters holding radioactive waste, Woods said. And intense heat associated with the magma would be expected to cause extensive damage to the containers. The results suggest that a greater number of canisters could be affected than previously estimated. The researchers suggest that the pressure associated with the magma could be enough to open new and existing fractures at Yucca Mountain, providing a conduit for radioactive material to reach the surface" ("Yucca Mountain Volcanic," 2002).

THE ISLETA PUEBLO TASTES ALBUQUERQUE'S EFFLUENT

By the late 1980s, people of the Isleta Pueblo, who live along the Rio Grande River, six miles downstream from the Albuquerque metropolitan area, began to experience problems that were outside their historical experience. Their corn and bean crops became stunted. Concurrently, five grandmothers, all from the same neighborhood (and all about the same age) were diagnosed with stomach cancers that killed all of them within a few months.

Political authorities on the reservation pressured the state's Department of Environmental Quality for a toxicological evaluation. The state then found that Isleta Pueblo's water supply was being contaminated by a number of sources upstream, including a slaughterhouse that was dumping ground-up animal carcasses into the river. Leakage of petroleum waste products also was detected from a wrecking yard, and the city of Albuquerque was found to be dumping raw sewage into the Rio Grande.

Soil tests at Isleta indicated dangerous levels of benzene (a lethal solvent) and nitrates. They soon discovered the cause: as the people of Isleta Pueblo had blessed the ground for spring planting in their annual Winter Dance, Albuquerque's main sewer line had ruptured. According to a report by Paul Vandevelder in *Native Americas*, "Millions of gallons of raw effluent poured into the river and flowed downstream to Isleta. City officials called to warn the tribe but said there was nothing they could do; it was too late . . . the Rio Grande was percolating with putrid green foam" (Vandevelder, 2001, 43).

The city of Albuquerque later told the U.S. Environmental Protection Agency that correcting the problems that were ruining Isleta's water would cost $300 million and entail $15 million a year annually in additional operating costs. Isleta's right to enforce water-quality standards through the Environmental Protection Agency was upheld by federal courts, a major legal victory for Native American peoples faced with the toll of off-reservation pollution. The case was presented in the context of religious freedom, with sacred ceremonies requiring clean water. In December 1997, the U.S. Supreme Court reviewed the Tenth Circuit's decision in favor of the Isleta Pueblo, requiring cleanup of the water by the city of Albuquerque.

THE LAGUNA PUEBLO AND ANACONDA'S JACKPILE URANIUM MINE

Until 1982, the Laguna Pueblo (50 miles east of the Navajo Nation, in New Mexico) was the site of the world's largest uranium strip mine. The 7,000-acre Anaconda Jackpile mine opened in 1952 and ceased operation when the ore had been sold on the market. By 1979, roughly 650 people, many of them Laguna residents, were working in the mine, aided by Native American hiring preferences. Until it closed, the mine made the Laguna Pueblo people affluent by most reservations' standards. The price of this affluence was the productivity of the agricultural valley that had sustained the Laguna Pueblo for centuries before the advent of the mine.

Before 1952, the Rio Paguate had provided water for a valley full of farms and ranches. Winona LaDuke, writing in *Native Americas*, described the post-mine landscape: "Rio Paguate now runs through the remnants of the strip mine, emerging on the other side a fluorescent green in color" (LaDuke, 1992, 58). Since 1973, the U.S. Environmental Protection Agency has known that the mine was leaching radiation into the Laguna Pueblo's water supply. By 1975, the EPA had issued reports describing large-scale groundwater contamination. The Laguna tribal center, the community center, and many reservation houses also were found to be dangerously radioactive. The problem was compounded when Anaconda used low-grade uranium ore to surface many roads on the reservation (LaDuke, 1992, 58).

COPPER MINING AT PICURIS PUEBLO

A proposed copper mine on Copper Hill in New Mexico was abandoned after a long struggle with the traditional people of the Picuris Pueblo, in northern New Mexico. The project would have resulted in a pit 300 meters deep and 90 meters wide on a 2,100-meter-high peak, which would have been mined for as much as 20 million tons of copper per year ("Picuris Pueblo," 1998).

The Taos–Rio Arriba Mining Reform Alliance (TRAMRA) that helped fight against the mine claimed some measure of the success in stopping the mine. "The message I got was that they recognized that there were some places that were appropriate for mining and some that are not. And they agreed this was not," said TRAMRA president Robert Templeton ("Picuris Pueblo," 1998). Picuris Pueblo land manager and former governor of Picuris, Gerald Nailor, said, "We were very serious about this. We were ready to go to war. It's the concern of people living close to the land. . . . Justice is what I am talking about" ("Picuris Pueblo," 1998).

THE ZUNI: SACRED WATERS AND COAL STRIP MINING

A majority of people at the Zuni Pueblo believe that development of a strip mine in western New Mexico will harm a lake whose waters they hold sacred. In the face of the Zunis' opposition, however, the Salt River Project Agricultural Improvement and Power District has asserted that it will move ahead with plans to produce coal on the site by the end of 2005. Bob Barnard, mine project manager for the Arizona-based Salt River Project, said that the utility must find a new source of coal for its Coronado Generating Station in St. John's, Arizona, before supplies from the McKinley mine near Gallup run out by 2005 or 2006 ("Zuni: Utility," 2001).

The utility, which serves about 190,000 customers (mostly in Phoenix, Arizona), plans to dig about 80 million tons of coal from the 18,000-acre Fence Lake mine during its expected useful life of 50 years. The Zunis oppose state and federal permits for the project. Hydrologists employed by the Zunis and the Bureau of Indian Affairs believe that pumping wells required for the mine will harm the flow of brine at the Zuni Salt Lake, which is on pueblo land 12 miles from the proposed mine's site.

The Salt River Project cannot begin operation without a life-of-mine permit from the Interior Department. The Zunis challenged the state of New Mexico's initial approval of a permit for the mine in 1996. The Zunis' lawyers have asserted that the project's supposed benefits cannot justify "the environmental injuries, cultural dislocations and desecration of Native American sites" ("Zuni: Utility," 2001). The Salt River Project says the proposed mine near Fence Lake "will generate hundreds of jobs and millions of dollars in tax revenue for the state [of New Mexico]" (Rayburn, 2001).

Barnard estimated that about 200 jobs would be created during the construction phase and about 100 during the 25-year projected life of the mine. About $60 million to $70 million in royalty payments from sales of the mine's 81.3 million tons of coal would go to the New Mexico State Education Trust Fund over 25 years. In addition, Barnard said, New Mexico will benefit from about $60 million in various forms of taxes from the mining operation (Rayburn, 2001). He also said that, without approval of the new mine, the Salt Lake utility would be forced to buy coal to generate electricity from Wyoming's Powder River Basin, 1,300 miles away, at a cost, including rail transport, of up to $24 a ton, compared to $19 a ton from the proposed mine (Rayburn, 2001).

Residents in Quemado, population about 450, were concerned about the mine's water usage. "Water is a real precious thing in the Quemado area. We don't know how much they're going to take, and it worries us," said Jerry Armstrong, owner of J&Y Auto Service in Quemado (Rayburn, 2001).

To the Zunis and other Native peoples in the area, the lake is Salt Mother, "a deity responsible for the steady flow of brine from an ancient volcanic cinder cone in the lake. For centuries, the Zunis, Navajos, Hopis, Acomas and other tribes have gathered salt for religious purposes" (Neary, 2001). The Zunis believe that pumping of the lake's water to suppress dust at the lake could cause it to dry up. Even without a mine siphoning its water, the lake averages only four feet in depth. The company proposes to pump an estimated 5,424 acre-feet of water, mostly for dust suppression. An acre-foot is about 325,000 gallons, which would cover an acre to a depth of a foot (Neary, 2001). The Zunis believe that the aquifer from which the utility plans to take water is connected to the lake, and would drain it.

Barnard said that his company is committed to monitoring groundwater in the area to ensure that pumping will have no effect on the lake. "There have been studies by, at last count, I think eight different sets of hydrologists, and all of them have pretty much concluded that the only way to protect the lake was to put in a series of monitoring wells to ensure that if there were any changes in the aquifer, the lake would be protected," Barnard said. "And we would quit using the aquifer before it would have any chance to affect the lake" (Neary, 2001).

George Kanesta, writing in a Zuni newspaper, *The Shiwi Messenger,* described the interrelation of the aquifers and the Salt Lake:

> There are two main aquifers in the area. One is the Dakota Aquifer and the other is the Atarque Aquifer. Both may be connected to the Salt Lake. If the two main aquifers have connections, the water will come from the Salt Lake instead of the aquifers. S.R.P. wants to drill wells to get water for their own use. They say this will have little or no damage to the Zuni Salt Lake. Yeah Right! I may not know for sure where this water will come from, but I do believe there is a strong possibility that it will come from our Salt Lake. (Kanesta, 2001)

"Zuni people will not sacrifice our Salt Woman for the almighty dollar or to provide electrical resources for Arizona or California, because she can never be replaced," Zuni tribal councilor Arden Kucate said during a town hall meeting in Grants, New Mexico (Weaver, 2001).

During July 2001, the New Mexico state mining division granted the Salt River Project a five-year extension of its state permit. The mining division also required the project's managers to complete a pump test before initiating mining to prove that the mine will not harm the lake. In an appeal filed August 10, however, Salt River asked the state mining division to withdraw its requirement for the pump test. Brian Segee, a lawyer with the Southwest Center for Biological Diversity, said the utility "has had years and years to prove their claim that water pumping associated with the Fence Lake Mine won't harm Zuni Salt Lake, and they've failed to do so" ("Zunis Object," 2001).

The state mining division's permit may be useless to the project, however. On October 24, 2001, the Department of the Interior decided not to sign a mining permit for the Salt River Project. Southwestern Sierra Club activist Andy Bessler supported the Interior Department decision, saying the mine would dry up Salt Lake. "You can't mitigate these effects," he said. "This is really not the place for this kind of mine" (May, 2001, B-2). On April 27, 2002, Zuni tribal leaders closed a highway that enters their reservation from Arizona and imposed a curfew, asking residents to turn off their lights as part of a protest of mining development planned for Salt Lake.

On August 4, 2003, the Salt River Project suddenly pulled out of its 20-year-long campaign to utilize the lake's watershed for an 18,000-acre coal strip mine and 44-mile rail line because it had found a more economically viable location in the Powder River Basin of Wyoming.

THE OKLAHOMA CHEROKEE: RESISTING THE DUMPING OF TOXIC ASH

Since 1986, more than 3,000 tons of toxic incinerator ash has been seeking a home, with no country willing to accept it. In 2001, the ash problem again stirred controversy: it seemed destined for the homeland of the Oklahoma Cherokee, with Waste Management, Incorporated trying to convince the tribe to take the ash as part of a money-making enterprise.

The Oklahoma Cherokee Nation operates a toxic landfill, which was built as part an economic development venture. The landfill also has a capacity to manage local municipal waste from local towns. Some tribal members of the Cherokee Nation expressed concern that the landfill was developed "with no input from either the tribal council nor the tribal citizens themselves" ("Toxic Ash," 2001).

The dissidents worried that storage of imported toxic ash eventually could contaminate the Cherokees' environment and affect the health of local residents. A task force was formed, consisting of concerned Cherokee Nation citizens, community residents, and tribal representatives from the United Keetoowah Band of Cherokee Indians. The task force opposes a prospective deal between the Cherokee Nation of Oklahoma and contractors for the "Haiti ash barge," which has floated around the globe for nearly 17 years in search of a burial place.

"This trash has sailed the whole of Earth and has been rejected by even the most impoverished and corrupt of nations. It has been tied to drug dealers and corrupt military leaders in Haiti and organized crime in the U.S. How is this going to look if the enrolled membership of the Cherokee Nation allows it to be buried here?" asked JoKay Dowell, who is of Quapaw, Peoria, and Cherokee descent. Dowell is a longtime community organizer living near Tahlequah, where the federally recognized Cherokee Nation of Oklahoma is headquartered ("Toxic Ash," 2001).

The wayward ash barge had a 15-year history by the time the Oklahoma Cherokee were asked to take custody of its contents. During 1986, the municipal government of Philadelphia contracted to remove about 14,000 tons of incinerator ash, which was loaded onto a ship called the *Khian Sea* and directed to the Bahamas, only to be turned away by the Bahamian government. The barge also approached (and was turned away from) Puerto Rico, Bermuda, the Dominican Republic, Honduras, and the Netherlands before it was mislabeled as topsoil fertilizer and allowed to dock in Haiti in December of 1987 ("Toxic Ash," 2001). Once in Haiti, almost 4,000 tons of the ash was unloaded before the Haitian government stepped in. The ship fled in the middle of the night with 10,000 tons of the Philadelphia ash, which was eventually illegally dumped into the Atlantic and Indian Oceans ("Toxic Ash," 2001).

In Haiti, testing of the 4,000 tons of remaining ash revealed that it contained lead, cadmium, and other heavy metals, as well as dioxins. Haiti's courts eventually ordered the ash to be dug up and returned to the United States, where Philadelphia refused to accept responsibility for it. After that, a plan to dispose of the wayward ash in a Broward County, Florida landfill met with community opposition and was canceled. By the fall of 2001, the ash barge was sitting in a Florida canal awaiting a final resting place. Greenpeace then learned that the wayward waste might be destined for the homeland of the Oklahoma Cherokees.

"I asked him how they [the Cherokee Nation of Oklahoma] could uphold themselves as stewards of the environment and even think of bringing in this waste," Dowell declared ("Toxic Ash," 2001). She then notified residents living near the landfill, who organized a committee they named Don't Waste Indian Lands to alert the Cherokees regarding the proposed deal to bury the waste in the tribal landfill. According to Jamie Clinton, Cherokee Nation landfill manager, representatives of the Oklahoma Cherokees visited the ash barge in Florida, where it had been docked for almost two years. They spent $5,000 testing the waste ash, and found, according to Clinton, that it "exceeded standards for safety," compared to other municipal waste ("Toxic Ash," 2001).

"Never mind that," said Dowell. "What is considered 'safe' is controversial in and of itself. It wasn't safe enough for Waste Management to keep it even though they have landfills throughout the South. It's not safe enough for other states like Louisiana, Georgia or Florida to take it. I wonder is it safe enough for Chief Smith to bury next door to his home in Sapulpa?" ("Toxic Ash," 2001). Dowell expressed concern that the Cherokee Nation of Oklahoma could become a major center for waste dumping. "To the south of us is the contaminated site of the former Kerr-McGee uranium processing facility. To the west is Fansteel, a thorium-contaminated site. The Cherokee Nation owns a landfill to the east that has caused pollution of nearby creek water, according to area residents. All they think about is money, not the people," said Dowell. "Indian land is not a trash dump for the rest of society. Our tribal leaders cannot build tribal economies by turning Indian lands into society's dumping grounds. It's Philadelphia's trash. Let Philadelphia take responsibility for it" ("Toxic Ash," 2001).

The Cherokees' protests ultimately kept the ash from their homelands. It finally was buried, during June 2002, in a landfill near Greencastle, Pennsylvania.

THE CHEYENNE RIVER SIOUX: THE HOMESTAKE MINE'S TOXIC LEGACY

The Cheyenne River Sioux, with legal support from the U.S. federal government, have successfully sued the Homestake Mining Company for polluting waterways in South Dakota with lead-laced mining waste. Homestake was sued during 1997 for dumping 100 million tons of mine tailings from the Lead (South Dakota) mine—which had been contaminated with cyanide and arsenic—into Whitewood Creek, the Belle Fourche River, the Cheyenne River, and the Missouri River during the previous century. The company asserted that the dumping stopped in 1977.

According to a report in *Drillbits and Tailings*, the Lead mine is located northeast of the Cheyenne River Sioux reservation. It was the first major, large-scale gold mine in the United States, containing deposits discovered by George Armstrong Custer's incursion into the Black Hills in 1874. Over a century, this mine probably produced the single largest quantity of gold of any mine in the Western Hemisphere ("Cheyenne River," 1999).

Homestake was ordered to pay a total of $5 million, most of which was directed toward restoring the Whitewood, Belle Fourche, and Cheyenne rivers. One-third of the $4 million was reserved for restoration projects on the Cheyenne River Siouxs' reservation; the rest was be placed in the South Dakota Game, Fish and Parks Fund to be used jointly for restoration by the United States and South Dakota ("Cheyenne River," 1999). "This is the first time Homestake has ever had to pay anything for any of the damage they've caused," said Steven Emery, Cheyenne River Sioux tribal attorney ("Cheyenne River," 1999).

THE OGLALA LAKOTA: URANIUM TAILINGS POLLUTION

Two hundred tons of uranium-mill tailings from Edgemont, South Dakota washed into the Cheyenne River and the Angostora reservoir on June 11, 1962. Radioactive residues from the Edgemont mine thus made their way into the water table at Pine Ridge, the home of the Oglala Lakota (Sioux), downstream from the 7.5-million-ton tailings pile at the abandoned mine.

Water samples taken from the Cheyenne River and a subsurface well on the Redshirt Table (which supplies water to Pine Ridge) "revealed a gross alpha radioactivity of 19 and 15 picocuries per liter respectively" (LaDuke, 1992, 59). Under U.S. federal standards, radiation above five picocuries is regarded as dangerous to human life. In 1980, the Indian Health Service (IHS) tested the water at Slim Buttes, on the Pine Ridge reservation, and found that it was radioactive at 3 times the federal standard, and 10 times the radiation of a nearby uncontaminated well that was being used as a control.

By the late 1970s, health problems related to radiation exposure were becoming commonplace at Pine Ridge. One study released in 1979 described miscarriages by 14 Pine Ridge women, 38 percent of an HIS sample taken at the local hospital. Most of the miscarriages occurred before the fifth month of pregnancy, and many of them were accompanied by profuse bleeding. Winona LaDuke described the effects of radiation on children at Pine Ridge:

> Of the children who were born [at Pine Ridge], some 60 to 70 percent suffered from breathing complications as a result of undeveloped lungs and/or jaundice. Some were born with such birth defects as cleft palate and club foot. Subsequent information secured under Freedom of Information Act requests from the Indian Health Service verified the data. Between 1971 and 1979, 314 babies were born with birth defects in a total Indian population of under 20,000. (LaDuke, 1992, 59)

UTAH'S GOSHUTE: WELCOMING URANIUM FUEL STORAGE, FOR A PRICE

The government of the tiny Goshute band of Indians in Utah, near Salt Lake City, is a rare example of a Native American tribe that has willingly submitted itself to energy colonization for a price. In this instance, the issue is nuclear-waste storage, and the price may be as much as $300 million. The Goshutes' leadership wants to host one of the country's largest nuclear-waste dumps, but federal officials are concerned that the group cannot even keep its own drinking water clean (Miniclier, 2001).

Only 25 people live on the seven-by-eight-mile, 17,700-acre reservation in the desert, located about 70 miles southwest of Salt Lake City; only 15 of 121 enrolled tribal members live on the reservation. They live in a cluster of trailers and other homes on the western slope of the Stansbury Mountains. Pioneers named the place Skull Valley after finding some skulls near a spring (Miniclier, 2001).

The Goshutes' council wants to lease about one square mile of the reservation for a nuclear-waste dump. Ultimately, the plan must be approved by the Nuclear Regulatory Commission. Several Utah groups have combined to oppose the Goshutes' invitation to become a major waste dump. These include Ohngo Guadedah Devia (a grassroots group of residents from the Skull Valley Goshute reservation) and the Environmental Justice Foundation. Both have joined with the state of Utah and a number of other environmental groups to wage a legal battle against the Bureau of Indian Affairs, seeking to declare a lease signed between Private Fuel Storage (a consortium of eight nuclear power utilities) and the Goshute Tribal Council "null and void," on grounds that the provisions of the lease were never put to a referendum vote, nor disclosed to tribal members.

The Utah legislature, Utah Governor Mike Leavitt, and the Southern Utah Wilderness Alliance all have objected strenuously to the deal. Advocates of the arrangement assert that the dump will close after permanent waste storage on a national scale becomes available at Yucca Mountain, Nevada. (See, above, "The Western Shoshone: 'The Most Bombed Nation on Earth.'") Private Fuel Storage (PFS) signed the deal in December 1996 with the three-member Skull Valley Tribal Council and local landowners, offering $300 million in exchange for their support of the dump. The deal committed the Goshutes to host a nuclear-waste facility for 25 years with an option to extend the agreement for another quarter-century.

The Goshutes' land, which is already surrounded by several industries that emit toxic effluents, has been chosen by Private Fuel Storage as the site for a temporary, aboveground private storage area. "I will deploy every tool I can to fight the storage of high-level nuclear waste in our state," Governor Leavitt said. "We don't produce this waste; we shouldn't store it. We are engaging in serious legal warfare to keep this lethally hot waste out of Utah" ("Governor Leavitt," 2001).

The Goshute reservation sits in the midst of activities that nearly no one else would want to host. Nearby are the Utah Test and Training Range, where the U.S. Air Force tests F-16 fighters and cruise missiles; the Dugway Proving Grounds, a test center for chemical and biological weapons; the Deseret Chemical Depot, with its stockpile of nerve and blistering agents; and the Tooele Chemical Demilitarization Facility, where military chemicals are destroyed (Miniclier, 2001).

According to the PFS contract with the Goshutes, an 840-acre site on the reservation will be designated to receive as much as 40,000 metric tons of spent uranium for temporary storage (20 or 40 years). Pending federal approval of the plan, PFS will remove spent uranium fuel rods from nuclear power plants coast-to-coast in the United States, place them on specially designed railcars, and ship them to Skull Valley. Under the agreement, the Skull Valley reservation would house the first private, high-level radioactive storage site in the history of the United States. The government of Tooele County, which surrounds the Goshutes' reservation, has come out in support of the plan in exchange for promises of up to $200 million in "mitigation fees."

Leon Bear, chairman of the Goshutes' tribal council (which signed the agreement with PFS), has ignored three deadlines set by the Denver regional office of the Environmental Protection Agency to clean up the reservation's dangerously polluted drinking water, according to the *Denver Post*. Bear's unwillingness to clean up the reservation's water supply has raised concerns about the tribe's ability to safely store large amounts of nuclear waste. The EPA's tests indicate that the Goshutes' water system, which uses untreated surface water, contains coliform and E. coli bacteria, "whose presence indicate the water may be contaminated with human or animal waste" and "is a threat to human health" (Miniclier, 2001). Most of the reservation's residents drink bottled water supplied by the EPA. Dianne Nielson, director of Utah's Department of Environmental Quality, asked, "If the tribe can't provide safe drinking water, how is it going to handle high-level nuclear waste?" Bear replied that "tribal members have been drinking the surface water for years [and that] the [Goshute tribe] will have no management responsibilities for the nuclear storage facility, but will merely lease the site" (Miniclier, 2001).

Bear dismissed such concerns, saying: "Little minds have little thoughts. . . . I'm always fighting with the EPA about water" (Miniclier, 2001). "I'm more afraid of the nerve gas at Tooele Army Depot or the chemical/biological experiments at Dugway Proving Ground than I am of the storage facility," said Bear, who envisioned the waste dump as economic salvation for his shrinking tribe. When he signed a lease with PFS, Bear reasoned that the area was surrounded with toxic industries, so nuclear waste storage "would provide a lucrative, safe, nonpolluting source of income" (Miniclier, 2001).

As a limited liability company, PFS wrote a contract excusing itself from any responsibility for damage caused by accidents or other unforeseen circumstances. Given the fact that Bear also has excused the Goshutes from any similar responsibilities, state officials have wondered who or what would be responsible should the stored nuclear waste cause pollution or any other type of accident. Environmentalists point out that the area is prone to earthquakes, and also that roughly 4,000 F-16 fighter planes a year fly over the area as they go to and from nearby test-bombing ranges.

Plans to install a nuclear-waste dump on the Goshutes' reservation have provoked dissension within the tribe. Bear's opponents insist that he was ousted in a recall vote called by antinuclear activists on August 25, 2001, but Bear asserted he was still leader of the 112-member Goshute Tribe. The local Bureau of Indian Affairs (BIA) superintendent, Allen Anspach, agreed with Bear. The leadership question is key to whether the Goshutes will honor the dump agreement (which the BIA has conditionally approved) (Taliman, 2002, A-1).

The Goshutes' insistence that their site was safe failed to convince regulators, however. Citing the risks of military aircraft operations over Skull Valley, the Atomic Safety and Licensing Board, an independent judicial arm of the Nuclear Regulatory Commission, issued a ruling during March 2003 that blocked a license for the Private Fuel Storage consortium's plans to build its nuclear fuel storage facility on Goshute tribal lands.

THE BLACKFEET: FOOT-HIGH MUSHROOMS AND TOXIC MOLD

Blackfeet reservation families in northern Montana have contracted a number of ailments, some of them serious, from a combination of substandard construction practices and toxic chemicals used in reservation housing. Some of the houses, on rotting wooden foundations, have become home to black mold that covers their walls, and also foot-tall mushrooms in some of their basements.

Jamie LaPier, who lived in one of the houses 30 miles east of Browning, Montana, said that the mushrooms growing out of the basement carpet were easy to remove, but toxic mold on the walls of bedrooms was impossible to scrub away. Finally, LaPier sealed the rooms off with duct tape and blocked the door with a bookcase to keep children out, as plumbing leaks fed the mushrooms and mold (Seldon, 2002, B-1).

Roughly 150 reservation houses constructed during the 1970s and 1980s contained wooden foundations treated with chromated copper arsenate (CCA), a toxic wood preservative. The houses were built with funds from the federal Department of Housing and Urban Development (HUD) and the

Blackfeet Housing Authority. Twenty years later, a HUD spokesperson said that the structural problems were a tribal, not a federal, responsibility (Seldon, 2002, B-1). Most of the wooden foundations were guaranteed for 50 years, but they lasted barely a decade before floors buckled and sank. Walls bowed, as windows and doors popped out of their frames, exposing residents to the extreme cold of Montana winters. Most of the houses are only minimally insulated, making some health problems worse (Seldon, 2002, B-1).

In the meantime, many of the homes' residents complained of chronic headaches, nosebleeds, asthma (as well as other breathing difficulties), general fatigue, dizziness, kidney problems, and several forms of cancer.

THE MAKAH: TESTING WHALING RIGHTS

During the late 1990s, the Makah, who live on Washington State's far northwest coast, came to fisticuffs with environmentalists over their right to harvest as many as five gray whales a year. Permission to hunt and kill the whales was obtained from the International Whaling Commission pursuant to the Makahs' right, under the Medicine Creek Treaty of 1854, under which the Makah and other signatories relinquished claims to 200,000 square miles of their territory in exchange for a small reservation and the right to fish "at all usual and accustomed grounds and stations" (Bristol, 1998, 3).

According to an account by Tim Bristol in the newsmagazine *Native Americas*, "The Makah said the hunt is an effort to restore a broken link with their past—[until this event], no living Makah had participated in a whale hunt—but the practices of their ancestors still permeate Makah life" (Bristol, 1998, 3).

Animal-rights activists, led by Sea Shepherd, often have resorted to physical violence to stop fishing and whaling that they believe harms the species. Sea Shepherd's representatives asserted that the Makah were harvesting whales for commercial purposes, not for cultural preservation. According to *Native Americas*, "Whale-protection advocates have yelled insults at Makah whalers, [and] blared rock music from speakers on their boat. The Makahs replied on one occasion with a hail of rocks and chunks of concrete. Makah tribal police arrested four of the animal-rights protestors who set foot on a dock within the reservation. The Makah generally regard the protestors as 'eco-colonists'" (Bristol, 1998, 3).

On May 17, 2002, a federal judge rejected a request for an injunction against whaling by the Makah tribe in Washington State, three years to the day after the Makah returned to whaling for the first time in 70 years. United States District Court Judge Franklin Burgess declined to order a halt to the tribe's whaling, pending the outcome of a lawsuit filed by the Fund for Animals, the Humane Society of the United States, and other groups and individuals. "While the court is sensitive to plaintiffs' concern, these concerns are outweighed by the Makah Tribe's rights under the Treaty of Neah Bay," Burgess wrote in his decision ("Makah Whaling," 2002).

After the gray whale was removed from the federal Endangered Species List in 1994, the tribe petitioned for permission to resume the hunt. Under require-

In the waters off of Neah Bay, Washington, on December 16, 1988, Makah whaler Donny Swan harpoons a float in preparation for the Makah Indians' first whale hunt in some 70 years. So much time had passed since the last Makah hunt in 1920, there is no one still alive who knows how to whale. And hunting the whale, especially in small boats and in winter seas, is dangerous. (AP Photo/Jacob Henifin)

ments set by the National Marine Fisheries Service (NMFS), the tribe is allowed to hunt legally in traditional dugout canoes (although boats with motors are most often used now). They are required to kill the whales with guns, a method the federal agency calls more "humane" than spear hunting ("Makah Whaling," 2002). Michael Markarian, executive vice-president of the Fund for Animals, replied to the ruling by saying, in part, that "whaling may have been a tradition in the past, but there is nothing traditional about cruelly shooting these majestic creatures with high-powered rifles" ("Makah Whaling," 2002).

Late in December 2002, a federal appeals court ruled that the Makahs' hunting of gray whales off the coast of Washington violates the Marine Mammal Protection Act. The ruling reversed a previous trial-court decision, finding that in approving the hunt, the government had failed to comply with its own regulations governing activities that may harm the environment.

WASHINGTON STATE FISHING RIGHTS: THE USUAL AND ACCUSTOMED PLACES

When the Romans colonized Gaul about the time of the birth of Christ, they found a fish-eating people. Soon demand rose in Roman markets for the pink-fleshed fish they named *salmo,* "the climber." The waters were overfished;

stocks were depleted, and during the ensuing centuries, salmon became extinct in many European rivers.

The taste for salmon traveled with the European colonists to the New World. By the turn of the century, descendants of those fishermen who feasted on salmon in Europe found the same fish in many streams of the Pacific Northwest. When the whites first set eyes on the area during the last century, it was said that one could walk across the backs of salmon from one stream bank to another. Fifty thousand Indians took 18 million pounds of salmon a year from the Columbia River watershed before European-Americans began to flood into the area about 1850.

To the Indian nations of the Pacific Northwest, the salmon was as central to economic life as the buffalo was to the Native peoples of the Great Plains; 80 to 90 percent of the Puyallup diet, for example, was derived from fish. The salmon was more than food; it was the center of a way of life. A cultural festival accompanied the first salmon caught in the yearly run. The fish was barbecued over an open fire and bits of its flesh parceled out to all. The bones were saved intact, to be carried by a torch-bearing, singing, dancing, and chanting procession back to the river, where they were placed into the water, the head pointed upstream, symbolic of the spawning fish, so the run would return in later years.

The Reserved Right to Fish

Washington was declared a territory of the United States on March 2, 1853, with no consent from the Indians who occupied most of the land. Isaac Stevens was appointed governor and superintendent of Indian affairs for the territory. Stevens had been born in Andover, Massachusetts, and graduated from West Point military academy in 1839. He then served in the U.S. Army as an engineering officer and, during the war with Mexico (in 1846–1847), he served on the staff of General Winfield Scott. Stevens was named Indian agent for Washington Territory. His main charge at the time was to survey the area as a prospective route for a new transcontinental railroad. Stevens served as Washington's territorial delegate to Congress in 1857, then returned to active duty with the Union army during the Civil War. During that war he was promoted to major general before he was killed at the Battle of Chantilly (Virginia) during 1862.

As territorial governor, Stevens possessed ambitions to build the economic base of the territory; this demanded attraction of a proposed transcontinental railroad, which, in turn, required peaceful cession of land occupied by Native peoples. Stevens worked with remarkable speed; in 1854 and 1855 alone, he negotiated five treaties with 17,000 Indians east and west of the Cascades, by which Native peoples ceded much of the land that became Washington, Oregon, and Idaho. Stevens was known as a relentless bargainer. On one point, however—the right to fish—Native Americans everywhere stood fast because they relied on fishing for a livelihood.

For Puget Sound, the Strait of Juan de Fuca, and the Pacific Coast, between 1854 and 1857, Stevens negotiated several treaties (Medicine Creek, Point Elliott, Point No Point, Neah Bay, and Quinault River). The importance of fishing in the area is attested by the fact that Native American negotiators retained the right to fish, usually from "usual and accustomed places" in most of these treaties. Each treaty contained minor variations on the following clause from the Treaty of Point Elliott:

> ARTICLE 5. The right of taking fish at usual and accustomed grounds and stations is further secured to said Indians in common with all citizens of the Territory, and of erecting temporary houses for the purpose of curing, together with the privilege of hunting and gathering roots and berries on open and unclaimed lands; provided, however, that they shall not take shell fish from any beds staked or cultivated by citizens. (American Friends Service Committee, 1970, 28)

By signing the treaties, the Indians ceded to the United States 2,240,000 acres of land. During negotiations, Stevens himself had pointed out that "it was also thought necessary to allow them to fish at all accustomed places, since this would not in any manner interfere with the rights of citizens and was necessary for the Indians to obtain a subsistence" (Johansen and Maestas, 1979, 184; American Friends Service Committee, 1970, 23).

At the Medicine Creek treaty, the first of Stevens's negotiations, he sat with 600 to 700 members of the Nisqually, Puyallup, and Squaxin Island peoples. George Gibbs, a member of Stevens's party, recounted that the treaty drew hundreds of Indians who arrived in convoys of cedar canoes, wearing "all kinds of fantastic clothing" (Wilkinson, 2000, 11). The negotiations took place on marshy ground during three days of rainy weather, as the wind whispered through the tall fir trees surrounding the site. Negotiations took place using the truncated 500-word vocabulary of the Chinook jargon that was usually used for trade.

By the 1850s, Anglo-American immigration was surging across the continent; Seattle and Omaha, for example, were started as urban areas at about the same time, shortly after 1850. Under these conditions, treaty making had lost much of the state-to-state equity it had possessed before 1800 in the eastern reaches of the continent. Isaac Stevens, who had graduated first in his class at West Point, often arrived at treaty negotiations with the instruments of "negotiation" at which he planned to arrive already written.

Chief Leschi Hung in Seattle

Stevens's plans to pacify the Pacific Northwest were not accomplished without bloodshed. Some Native peoples refused to sign, notably an alliance led by the Nisqually Leschi, who waged a brief war of resistance. Leschi eventually paid for this resistance with his life in the aborning town of Seattle, which was named for the Duwamish/Suquamish chief Sea'th'l, who had removed his people from its site with little protest, aside from a now-famous farewell speech.

The Yakamas, claiming that Stevens had bought off their leaders, recruited allies from several other tribes and, as allies of Leschi's band, waged a guerilla war with the U.S. army that spread westward, over the Cascade mountains, to the Pacific Coast. The army brought the rebellion under control after three years of occasionally bloody fighting. During the rebellion, Dr. Marcus Whitman (who was said to have come to the Walla Walla area to civilize the Indians with a Bible in one hand and a whip in the other) was murdered, along with his wife and 12 other whites. A freelance posse formed immediately and hung five captured Cayuse in retribution (Emmons, 1965).

Leschi, the Nisquallies' most prominent leader at the time, refused to sign the Medicine Creek treaty because it allowed his people only two square miles of land. Leschi's mark on the treaty probably was forged. The treaty allowed fishing, but the land Stevens proposed to allot to them was on a thickly wooded bluff west of the Nisqually River, with no ready access to the Nisquallies' usual fishing grounds. Stevens's arrogance, and Leschi's insistence on adequate living room with river access, led to a brief war after which the Indian leader was hanged.

Leschi was characterized at this time as a strong leader "with a commanding presence. He carried himself well—accounts typically emphasize his penetrating gaze" (Wilkinson, 2000, 14). Leschi's people made common cause with disgruntled Native peoples east of the Cascades. Leschi and his allies attacked the newly established settlement of Seattle on February 29, 1856 with about 1,000 warriors. The warriors might have overrun the settlement, except that a U.S. Navy ship that happened to be stationed in the harbor opened fire on them. After his allies were defeated, Leschi surrendered to the U.S. Army. Initially, Leschi was offered amnesty, but Governor Stevens took up his case as a personal matter, determined to bring Leschi to trial for deaths that had occurred during the raid on Seattle.

The war ended when a nephew of Leschi, Sluggia, betrayed him. Stevens is said to have bribed Sluggia with 50 blankets to betray Leschi. Leschi was then arrested and taken into custody by the whites. Wa He Lut, also a Nisqually, quickly avenged this betrayal by killing Sluggia. Three days after Leschi's arrest, he was tried for the murder of Colonel A. Benton Moses, a U.S. army soldier who had been killed during the uprising. Leschi had not personally killed Moses, and his death was clearly an act of war, not a murder suited for trial in a civilian court. For these reasons, Stevens's first attempt to convict Leschi resulted in a deadlocked jury.

A second jury was selected, comprised totally of non-Indians, which took less than a day to convict Leschi. Leschi protested that if he was convicted, then the U.S. army soldiers who had killed Indians during the uprising also should be charged with murder. On February 19, 1858, roughly 300 people gathered at Fort Steilacoom around a gallows. Leschi rode to the gallows on horseback as Nisqually drums sounded in the distance. Charles Grainger, the hangman, said later, "I felt like I was hanging an innocent man, and I believe it yet" (Wilkinson, 2000, 18).

Stevens Declares Martial Law

While settlement of the coastal Northwest began in tandem with California, accelerating after 1850, most of the people who traversed the continent along the Oregon Trail were not looking for gold or other quick riches. The majority sought to set up farms (some planned utopian communes). Some of the white settlers aided Indians who were being pursued by Stevens's militia. When his white opponents sought the protection of local courts, Stevens called up a militia of a thousand men, declared martial law, closed the courts, and arrested the chief justice of the territory—a rare example of government by fiat in the face of settlers' resistance to ill-treatment of Indians.

At about the same time that Chief Leschi was hanged, Stevens's militia also arrested several settlers suspected of aiding renegade Indians. Lion A. Smith, Charles Wren, Henry Smith, John McLeod, Henry Murray, and another man asserted that they were taken from their land claims in Pierce County without due process of law, and without any complaint or affidavit being lodged against them. The men were escorted against their wills to Fort Steilacoom (near Tacoma), where they were held, at Stevens's request, on charges of treason. Following complaints by attorneys for the men, Stevens issued a martial law declaration suspending civil liberties in Pierce County (including Tacoma), accusing the arrested settlers of giving aid and comfort to the enemy. A few days later, Stevens ordered the men back to Olympia, out of Pierce County, because a judge there had issued a writ of *habeas corpus* on their behalf.

The Modern Fishing-Rights Struggle Begins

The state of Washington arrested the first Native American for fishing it deemed illegal (in violation of its fish-and-game regulations) during 1913. Addressing the Washington State Legislature during 1915, Charles Buchanan, Indian agent at the Tulalip Reservation on Puget Sound, observed that the salmon runs had

> naturally lessened with the advent of the white man; more recently, the use of large capital, mechanized assistance, numerous great traps, canneries, etc., and other activities allied to the fishery industry have greatly lessened and depleted the Indians' natural sources of food supply. In addition . . . the stringent application to Indians of state fish-and-game laws has made it . . . increasingly precarious for him to procure his natural food. . . . One by one [the Indians'] richer and remote fishing locations have been stripped from him while the law held him helpless and resourceless. (Brown, 1982, 139–40)

Buchanan's point of view represented one pole of contemporary non-Indian opinion *vis-à-vis* Native Americans and their fishing rights. The other extreme was expressed in 1916 by the Washington State Supreme Court:

> At no time did our ancestors in getting title to this continent ever regard the Aborigines as other than mere occupants, and incompetent occupants of the soil. . . . Neither Rome nor sagacious Britain ever dealt more liberally with their subject races than we with these savage tribes whom it was generally tempting and always easy to destroy and whom we have often permitted to squander vast areas of fertile land before our eyes. (Brack, 1977, 6)

European-American immigration to the Northwest was surging early in the twentieth century. So was their industries' impact on the fisheries on which the Indians had depended for thousands of years. By 1914, 16 million salmon were caught annually in waters adjoining or enclosed by Washington State. By the 1920s, mainly because of overfishing and pollution of spawning streams by industry and timber harvesting, annual catches had declined to an average of 6 million. By 1929, the state had gone into the business of leasing Native American fishing rights to private companies. During that year, the state, for a payment of $36,000, denied the Quinaults' fishing rights and leased their "usual and accustomed places" to Bakers Bay, a local corporation.

In the late 1930s, following construction of several large hydroelectric dams on the Columbia River and its tributaries, the annual salmon catch declined to as low as 3 million, about one-sixth of what Native American peoples alone had been harvesting a century earlier. By the 1970s, with more aggressive conservation measures in place (including construction of fish ladders at most major dams), the annual catch rose to between 4 and 6 million. (The fish ladders are necessary because the dams obstruct salmon migrations to their spawning grounds. Nearly all salmon, by unerring instinct, make their way from the ocean back to the stream in which they were born.)

During 1962, the Washington State Supreme Court ruled against a Swinomish fisherman who had asserted his right to fish under the Treaty of Point Elliott (1855). The court ruled that the state had a right to subject Indians to "reasonable and necessary regulations" for conservation of the resource (Brown, 1976, 5). "Conservation" thus became a code word for upholding non-Indian fishing interests, as a state fisheries police was assembled to enforce the ruling. Indian fishermen remarked that they had not built the dams, logged the streams, or taken boats with SONAR to sea, all factors in the decline of salmon runs to the point where they required state-imposed "conservation."

In 1963, state fishing officials acknowledged that Indians were catching only 6 percent of the salmon returning to Washington State waters. Non-Indian sports fishermen were taking 12 percent and commercial interests were taking 82 percent at that time (Brown, 1982, 155). The state maintained, however, that the Indians were responsible for the declining salmon runs. Its rationale was that the Indians, fishing from shore, were taking fish that had returned to spawn. Conveniently ignored by the state were the many fish that were taken by whites in boats offshore, which would have returned to the same streams to spawn if they had had a chance.

By 1965, the U.S. Supreme Court had ruled that Indians had a right to fish at their "usual and accustomed places," but that the state had the right, through its courts, to regulate Indian fishing. That ruling, and a few federal court rulings after it, had little practical effect as long as the state, whose fishery managers were adamantly opposed to any Indian fishing at all, held the power to define and enforce what was meant by "conservation."

"Fish Wars"

In January 1857, with Chief Leschi's rebellion at its height, President Franklin Pierce invoked Article Six of the Medicine Creek Treaty (allowing the Nisquallies a replacement reservation) and allotted them 4,700 acres, four times Stevens's original two square miles. This reservation also offered river frontage, occupying both banks of the Nisqually River. With access to the river, the Nisqually Frank family never needed government rations. Puget Sound was only eight miles downriver from the Franks' family home, so the elder Billy Frank sometimes took a canoe to the Sound's mud flats for clams, geoducks, and oysters. "When the tide goes out," he would say, "our table is set" (Wilkinson, 2000, 22).

By the early 1960s, state fisheries police were conducting wholesale arrests of Indians, confiscating their boats and nets. Indians took their treaty-rights case to state courts and found judges solidly in support of non-Indian commercial interests. Denied justice in the state courts, the tribes pursued their claim at the federal level. During the 1960s and early 1970s, they also militantly protected their rights in the face of raids by state fisheries authorities. A nucleus of fishing-rights activists from Franks Landing, living only a few miles from the site at which the Medicine Creek Treaty had been signed, continued to fish on the basis of the treaty, which gave them the right to fish as long as the rivers ran.

Billy Frank Jr. recalled that it was "nearly a daily event to get hassled by those guys. It was a good day if you didn't get arrested. After a while I didn't even bother to tell people at the Landing because they already knew: 'If we don't come back home, call C. J. Johnson.' He was the bail bondsman" (Wilkinson, 2000, 33–34). Frank, who was first arrested for taking fish in 1945, had been going to jail for nearly three decades by the time Judge Boldt ruled that what he had been doing was perfectly legal.

Indian fishing that the state deemed illegal soon spread to the Puyallup River, near Tacoma. On one occasion during 1970, more than 300 state police outfitted as soldiers invaded a Puyallup camp on the river and arrested all 62 people they found there. The Indians' ability to catch fish, and to sell them, became more of a challenge as the state's repression intensified. Shortly after nightfall, residents of Franks Landing sometimes loaded a pickup truck with freshly caught salmon bound for San Francisco's Union Square, with "muffled laughter and *sotto voce* whispers of 'Go get 'em,' and 'You're doing the Great Spirit's work!'" (Wilkinson, 2000, 42).

Fishing-rights activists took to their boats in several locations along Puget Sound, from Franks Landing, near Olympia, to Tulalip, near Everett, north of Seattle. State fisheries police continued to descend on the Indian fishermen. At times, vigilante sports fishermen joined state fisheries police in harassing the Indians, stealing their boats, slashing their nets, and sometimes shooting at them. The elders and women stood with the younger men. Many non-Indians lent support to the protests, including many from a Latino service center in Seattle, El Centro de la Raza, who pulled nets alongside Marlon Brando, Jane Fonda, and Dick Gregory.

Billy Frank Jr. went to jail roughly 90 times as he defended the Nisquallies' treaty fishing rights:

> I went to jail when I was fourteen years old. That was the first time I ever went to jail for treaty rights. The State of Washington said I couldn't fish on the Nisqually River. . . . Ninety times I went back to jail. . . . The State of Washington said: "You can't go on that river and you can't go fishing anymore." That's what they told us Indians: "If you go on that river, you're going to jail." We went back fishing and we went to jail over and over again until 1974." (Grinde and Johansen, 1995, 152)

Frank later became a widely recognized leader in efforts to implement the court-ordered solutions to conflict in Washington State fisheries. At one point, he was a key organizer of a boycott against Seattle's First National Bank and other businesses that had opposed the Indians' right to fish. By the 1990s, Frank was chairman of the Northwest Indian Fish Commission and a leading regional environmental activist. He was awarded the Albert Schweitzer Prize for humanitarianism by John Hopkins University in 1992.

After many years in state custody, a cedar dugout canoe in which Frank had been arrested many times was released by state fisheries authorities. The canoe was installed in what Frank called a "spot of honor," along the riverbank where his quest to fish in accordance with the treaties had begun (Grinde and Johansen, 1995, 166–67).

The "Boldt Decision"

The fish-ins continued until February 12, 1974, when U.S. District Court Judge George H. Boldt ruled that Indians were entitled to an opportunity to catch as many as half the fish returning to off-reservation sites, the "usual and accustomed places" mentioned in the treaties. Judge Boldt's ruling affected anadromous fish (notably salmon), which hatch in fresh water and journey to the ocean as young fish to spend most of their lives, returning to fresh water to spawn.

Lawyers for Indian fishing people did not think much of their prospects with Judge Boldt when he was first appointed to handle the fishing-rights case. He was an Eisenhower appointee, a man known for upholding law and order and proper courtroom decorum. Boldt had no prior background or judicial record in Native American law or treaties.

After intense study of the treaties and their historical milieu, Judge Boldt interpreted the Medicine Creek Treaty of 1854, focusing on nineteenth-century dictionaries' definitions of "in common with." Boldt said the words meant "to be shared equally." Judge Boldt put three years into the case; he used 200 pages to interpret one sentence of the treaty in an opinion that some legal scholars say is the most carefully researched and thoroughly analyzed opinion ever handed down in an Indian fishing-rights case.

Fish caught at reservation sites, on which Indians have an exclusive right to fish, were not to be counted in this apportionment. The state's right to regulate off-reservation fishing by Indians at their usual and accustomed places was limited by Judge Boldt to that which is reasonable and necessary for conservation, meaning the perpetuation of a run or species of fish. During the next three years the Ninth Circuit Court of Appeals upheld Boldt's ruling, and the U.S. Supreme Court twice let it stand by refusing to hear appeals.

In his initial findings of fact, Judge Boldt wrote that the Washington departments of fish and game had systematically discriminated against Indian fishermen by allowing non-Indian sportsmen and commercial interests to catch an overwhelming majority of the available fish. The judge flatly rejected state assertions that Indians were responsible for declining salmon runs, as the state had maintained. "Notwithstanding three years of exhaustive trial preparation, neither Game nor Fisheries has discovered and produced any credible evidence showing any instance, remote or recent, when a definitely identified member of any plaintiff tribe exercised his off-reservation treaty rights by any means detrimental to any species of [salmon]," wrote Boldt. "Indeed, the near-total absence of substantive evidence to support these apparent falsehoods was a considerable surprise to this court" (Brown, 1982, 157).

The Washington State Supreme Court barred the state fish and game departments from enforcing any aspect of Boldt's ruling. With the state refusing to enforce Boldt's ruling in the face of angry white fishing interests, Boldt took control of the Indian fishery himself, and administered the ruling through the Federal District Court in Tacoma. In the meantime, Boldt's ruling in *United States v. Washington* was elaborated in *Washington v. Washington State Commercial Passenger Fishing Association* (1979). The court's ruling in this case held that the figure of 50 percent should be a "maximum and not a minimum allocation." The central premise of its decision, the U.S. Supreme Court ruled, was that Native people should be entitled to enough fish to provide them with a livelihood.

The Supreme Court also held that special benefits accorded Indians under treaties do not violate the equal protection clause of the Fourteenth Amendment. Furthermore, the Court ruled that the treaties should be understood in the sense in which they would naturally be understood by Indians. The words "in common with" and "right of taking fish" in the original treaties indicated to the court's majority that the Indians who signed them were entitled to a share of the harvest, over and above an equal opportunity to fish.

Lost in the fishing-rights fray were a number of small, landless western Washington tribes that were not recognized by the federal government and therefore not entitled to participate in the federally mandated solution. A few such tribes, such as the Upper Skagit and Sauk-Suiattle, were recognized after the Boldt decision. A number of others remained in legal limbo with no fishing rights under federal law.

While the commercial interests raged against Boldt's rulings, the Indians were catching nothing close to the 50 percent allowed by the Boldt ruling; in 1974 they caught between 7 and 8 percent, in 1975 they caught between 11 and 12 percent, and in 1977 they caught 17 percent, depending on who did the counting, the Indians or the state.

Non-Indian Backlash and the Rise of Senator Slade Gorton

Judge Boldt's ruling had a profound effect, not only on who would be allowed to catch salmon in Puget Sound but also on general relations between non-Indians and Indians. Non-Indians, who had long come to regard the salmon harvest as virtually their own, were suddenly faced with large potential reductions in their harvests. Hostility became so serious that Indians armed their fishing camps after enduring attacks. Many non-Indians displayed their reaction to the decision with bumper stickers that said, "Can Judge Boldt, not salmon."

State officials and the fishermen whose interests they represented were furious at Boldt. Rumors circulated about the sanity of the 75-year-old judge. It was said that he had taken bribes of free fish and had an Indian mistress, neither of which were true. Judge Boldt was hung in effigy by angry non-Indian fishermen, who on other occasions formed "convoys" with their boats and rammed coast guard vessels that had been dispatched to enforce the court's orders. At least one coastguardsman was shot.

A backlash against Indian fishing rights evolved the nucleus for a nationwide non-Indian campaign to abrogate the treaties. Washington State's attorney general (later U.S. senator) Slade Gorton called Indians "supercitizens" with "special rights," and proposed that constitutional equilibrium be reestablished—instead of the state openly violating the treaties (Boldt had outlawed that), it would purchase the Indians' fishing rights.

Gorton had been Washington State's attorney general while the fishing-rights cases were adjudicated by Judge Boldt. Before he was defeated in the 2000 election, Gorton's agenda included attaching "riders" to various Interior Department funding bills. One such rider required Native American governments to waive their sovereign immunity in exchange for $767 million in federal funds designated for tribal operations, a figure representing half the operating budgets of 550 Native American governments nationwide. Senator Gorton complained that Indians were using their treaty rights as a shield against civil lawsuits and as shelters for tax-free profits from casinos, sales of cigarettes, and other goods. The tribes replied that they have the same protection against civil lawsuits as state, federal, and county governments.

Another Gorton "rider" would have denied federal funds to Indian tribes which earned more than a designated amount of money from gambling or other forms of economic development. This measure was developed from an assumption that reservation residents were becoming suddenly rich, despite the fact that unemployment for Native Americans during the 1990s was still about 15 percent, three times the national average for non-Indians.

Another Gorton initiative, to tax profits from Indian casinos, was narrowly defeated in the Senate during 1998 after a General Accounting Office (GAO) study disclosed that only a few reservations had casino windfalls worth taxing. Ten casinos owned by Indian tribes generated almost half the $7 billion in annual revenues, with a net income of $1.6 billion a year, according to the GAO. Nearly all of them are located close to major urban areas that supply a large number of non-Indian gamblers. Most reservations in rural areas have small gaming operations, or none at all.

Senator Gorton—who is a member of the New England fishing family of the same name, whose Gorton's frozen products are sold in grocery stores nationwide—chaired the Senate's interior appropriations subcommittee, which supervised the budget of the Bureau of Indian Affairs. He used a special privilege of his office, the "chairman's mark," to insert his riders after public debate on the spending bill. Gorton said that he had found nothing in any Indian treaty saying that Indians must be continually supported by federal taxpayers. When Gorton called treaty-based sovereign immunity an anachronism, he may not have realized that the same language was used by President Andrew Jackson in 1830 to support the Removal Act, which legally authorized the Trail of Tears.

Even some Senate Republicans argued that Gorton's riders, if enacted, would not survive constitutional tests in court. Senator John McCain, chairman of the Senate Indian Affairs Committee, opposed the riders on grounds that they would deny the validity of treaties, viewing them as part of a rising anti-Indian movement in Congress based on assumed Indian affluence from gambling operations. McCain said that the treaties are clear in what they say. They were written, McCain said, in exchange for a significant portion of America.

Gorton's riders echoed his tactics at home, especially with regard to the Lummi. Gorton has supported non-Indian homeowners near Bellingham, on northern Puget Sound, who live on the Lummi reservation and use roughly 70 percent of its water. When the Lummi announced plans to cut water supplies to non-Indians, Gorton retaliated by introducing a special measure in the U.S. Senate, which failed to pass, designed to cut Lummi federal appropriations by half.

Senator Gorton's initiatives thus accomplished little for the antitreaty forces. Gorton's antitreaty crusade in the Senate came to a halt when he narrowly lost a bid for reelection in the 2000 election, during which treaty rights played a major role.

Fishing Rights East of the Cascades

Fishing-rights controversies did not end in Washington State with the 1974 Boldt decision, which restored recognition of treaty rights only in the treaty areas west of the Cascade mountain range. East of the Cascades, during the 1980s, the fishing-rights battle continued in a form that reminded many people of the "fish-ins" of the 1960s. Many Native people along the Columbia River and its tributaries also fished for a livelihood long before European-Americans migrated to their land, but their right to do so had not been judicially recognized. For years, Wanapam David Sohappy (1925–1991), his wife Myra, and their sons erected a riverbank shelter and fished in the traditional manner.

The Sohappys' name came from the Wanapam word *souiehappie*, meaning "shoving something under a ledge." David Sohappy's ancestors had traded fish with members of the Lewis and Clark expedition. The Wanapam had never signed a treaty, wishing only to be left in peace to live as they had for hundreds, perhaps thousands, of years. By the early 1940s, Sohappy's family was pushed from its ancestral homeland at Priest Rapids and White Bluffs, which became part of the Hanford Nuclear Reservation, in the middle of a desert that Lewis and Clark characterized as the most barren piece of land that they saw between St. Louis and the Pacific Ocean. Still, David Sohappy fished, even as his father, Jim Sohappy, warned him that if he continued to live in the old ways, "the white man is going to put you in jail someday" (Keefe, 1991, 4).

Industrial development and urbanization during the 1950s devastated the Celilo Falls, which had been one of the richest Indian fishing grounds in North America. Most of the people who had fished there gave up their traditional livelihoods, moving to the nearby Yakima reservation or into urban areas. David Sohappy and his wife Myra moved to a sliver of federal land called Cook's Landing, just above the first of several dams along the Columbia and its tributaries. They built a small longhouse with a dirt floor. Sohappy built fishing traps from driftwood. As the "fish-ins" of the 1960s attracted nationwide publicity west of the Cascades, Sohappy fished in silence, until state game and fishing officials raided his camp, beat family members, and, in 1968, detained Sohappy in jail on charges of illegal fishing. He then brought legal action; the case, *Sohappy v. Smith*, produced a landmark federal ruling that was supposed to prevent the states of Washington and Oregon from interfering with Indian fishing, except for conservation purposes.

The state ignored the ruling, continuing to harass Sohappy and his family. Usually under cover of darkness, state agents sunk the Sohappys' boats and slashed their nets. Raids sometimes interrupted family meals. During 1981 and 1982, the states of Washington and Oregon successfully (but quietly) lobbied into law a federal provision that made the interstate sale of fish taken in violation of state law a felony—an act aimed squarely at the Sohappy family.

Eight months before the law was signed by President Reagan, the state enlisted federal undercover agents in a fish-buying sting that the press called "Salmonscam," intended to entrap Sohappy. He was later convicted in Los Angeles (the trial had been moved from the local jurisdiction because of racial prejudice against Indians) of taking 317 fish, and sentenced to five years in prison. One of Sohappy's sons also was sentenced to five years in prison for having sold 28 fish. During the trial, testimony about the Sohappys' religion and their own practice of conservation were not allowed.

Sohappy became a symbol of Native American rights across the United States. Myra Sohappy sought support from the United Nations Commission on Human Rights to have her husband tried by a jury of his peers in the Yakama Nation's court. The new trial was arranged with the help of Senator Daniel Inouye, chairman of the Senate Select Committee on Indian Affairs. The Yakama court found that the federal prosecution had interfered with Sohappy's practice of his Seven Drum religion. "When I first got convicted in Los Angeles," Sohappy recalled later, "while going through Lompoc [federal penitentiary], they asked me: 'What are you in here for?' I told them 'fishing.' 'Fishing?' they asked. They couldn't believe that anyone could be put in federal prison for fishing" (Sohappy, 1991, 7).

Released after 20 months in prison, Sohappy had aged rapidly. Confinement and the prison diet had sapped his strength. Sohappy suffered several strokes during the months in prison, when he was even denied the use of an eagle prayer feather for comfort (it was rejected as "contraband" by prison officials). Back at Cook's Landing, Sohappy found that vindictive federal officials had tacked an eviction notice to his small house. Sohappy took the eviction notice to court and beat the government's rap for what turned out to be his last time. He died in a nursing home in Hood River, Oregon, on May 6, 1991.

A few days later Sohappy was buried, as his Wanapam relatives gathered in an old graveyard. They sang old songs as they lowered his body into the earth, having wrapped it in a Pendleton blanket. He was placed so that the early morning sun would warm his head, facing west toward Mount Adams. Tom Keefe Jr., an attorney who had been instrumental in securing Sohappy's release from prison, stood by the grave and remembered:

> And while the sun chased a crescent moon across the Yakima Valley, I thanked David Sohappy for the time we had spent together, and I wondered how the salmon he had fought to protect would fare in his absence. Now he is gone, and the natural runs of Chinook that fed his family since time immemorial are headed for the Endangered Species Act list. "Be glad for my dad," David Sohappy, Jr. told the mourners. "He is free now, he doesn't need any tears." (Keefe, 1991, 6)

Conclusion: A Land-Based Fishery?

Some of the fishing-rights debate has centered on the utility and fairness of a fishery in which large numbers of salmon have been taken by non-Indian commercial fishing people far at sea, long before they return to their spawning grounds on the streams and rivers, and long before Native fishermen can catch them at their treaty-designated "usual and accustomed places."

A study by Russel L. Barsh, published in 1977, concluded that land-based fishing by net and trap is more cost efficient, more energy efficient, and more protective of sustained fishing runs than marine-based fishery. Despite its rationality and utility, land-based fishing has never caught on among non-Indians. The problem, wrote Barsh, is that the salmon and other fish are regarded as a "common good," free to whoever gets them first. Fishing boats may be wasteful of energy and other resources, but the boats reach the fish first (Barsh, 1977).

Salmon are blissfully ignorant of human national boundaries. They do not carry passports, and they cannot be branded, like cattle, as they roam the world's largest "open range." Marine-based fishing encourages the capture of salmon before they are fully mature. The most efficient fishery, argued Barsh, would restrict the harvest to the streams and rivers where salmon return to spawn. This is the traditional Native American method of fishing, using traps and spears. Barsh maintained that those who own or lease streamside fishing grounds would have an incentive to maintain the runs, which is lacking in a marine-based fishery where large boats and fancy gear are used to chase immature fish across the open ocean. At nearly every turn, argued Barsh, the state of Washington legislated inefficiency into salmon fishing. In 1934, the state even made non-Indian fish traps illegal. Most Indian fishing people lack the capital necessary to buy and maintain oceangoing boats.

Requiring everyone to fish from land in the traditional Native American manner would certainly save energy, and would allow more fish to return to their streams of origin. However, a land-based fishery as a practical fact may not appear anytime soon. Non-Indian interests will not easily relinquish their remaining advantages in this contested fishery.

THE YAKAMA: HANFORD'S RADIOACTIVE LEGACY

The Hanford Nuclear Reservation, in eastern Washington State, released more than 440 billion gallons of irradiated water 30 miles upstream from the Yakama reservation between 1945 and 1989. Due to this decades-long bath of radiation dumped directly into the river system, oysters caught at the mouth of the Columbia River were so toxic by the early 1960s that when one Hanford employee ate them and returned to work the following day, he set off the plant's radiation alarm (Weaver, 1996, 49). An investigation revealed that the day before he had eaten a can of oyster stew contaminated with radioactive zinc. The oysters had been harvested in Willapa Bay, along the Pacific Coast in Washington State, 25 miles north of Astoria (Schneider, 1990, A-9).

The 560-square-mile Hanford Nuclear Reservation was established during 1943 on lands traditionally used for hunting, fishing, and gathering by the Yakama and Umatilla. The same area is adjacent to the homeland of the Nez Percé. The Hanford facility produced the plutonium used in the first atomic bombs dropped on Hiroshima and Nagasaki. Thus, releases of radioactive materials have been contaminating these peoples, as well as the Coeur d'Alene, Spokane, Colville, Kootenai, and Warm Springs Tribes for many years. In 1986, after disclosure that radiation was secretly released into the air and water from 1944 until January 1971 (when the last of the eight reactors was closed), Yakama leaders were among the first to call for a thorough study of the danger.

For nearly 30 years, ending in 1971, eight of the nine nuclear reactors at the Hanford complex were cooled by water from the Columbia River. Millions of gallons of water pumped directly through the reactor cores picked up large amounts of nuclear material, making the Columbia downstream the most radioactive river in the world, according to state and Federal authorities (Schneider, 1990, A-9). In July 1990, a federal panel said some infants and children in the 1940s absorbed enough radioactive iodine in their thyroid glands to destroy the gland and cause an array of thyroid-related diseases.

Although Hanford has stopped producing plutonium, recently released documents indicate that radioactive materials leaking from storage tanks there have continued to contaminate groundwater in the area. The Indigenous Environmental Network reported that "the Hanford site also has eight nuclear reactors that have . . . contaminated the Columbia River on which many of the Tribes depend for basic sustenance" ("Uranium/Nuclear," 2001). Unaware of the contamination, indigenous people collected berries near the Columbia, hunted for eels in its tributaries, and took salmon from its waters.

New York Times reporter Keith Schneider described the Columbia below the Hanford Reservation:

> Dammed in the 1950's below Hanford and developed by industrial companies, the river's water is green and gray now. Salmon runs are much smaller than they were before World War II, and Johnny Jackson, a 59-year-old fisherman . . . said some fish he catches were marred by deep, infected welts and growths. . . . Documents declassified beginning in 1986 said that "radiation spread to the river's bacteria, algae, mussels, fish, birds and the water used for both irrigation and drinking." (Schneider, 1990, A-9)

Radiation in the river has caused concern at Hanford for decades. In 1954, the situation was reviewed in secret meetings at the Washington, D.C. headquarters of the Atomic Energy Commission, which operated the plant. Lewis L. Strauss, the commission's chairman, flew to Hanford in the summer of 1954 and was told that "levels of radioactivity in some fish in the Columbia River, particularly whitefish, were so high that officials were considering closing sport fishing downstream" (Schneider, 1990, A-9). Ducks, geese, and crops irrigated with Columbia River water also were said to pose a potential health threat.

The public never was alerted. In a memorandum prepared on August 19, 1954, Dr. Parker urged the Government to keep the problem secret because the radiation levels were still within safety guidelines, and closing sport fishing would compel the plant to discuss the issue publicly and compromise secrecy. "The public relations impact would be severe," he wrote (Schneider, 1990, A-9).

THE WISCONSIN CHIPPEWA: SULFIDE MINING VERSUS TREATY RIGHTS

Native Americans in northern Wisconsin have united with neighboring non-Indians to fight an Exxon bid to develop what could have been one of the world's largest sulfide mines near an area in which Native peoples exercise their treaty rights to hunt, fish, and gather wild rice. The political movement that has grown out of mining resistance in this area, which calls itself the Watershed Alliance to End Environmental Racism (WATER), went head-to-head with Exxon lobbyists in the Wisconsin statehouse over a proposed mining moratorium. Opposition to mining intensified as Wisconsin residents discovered what the mining wastes could do to the earth from which they wrest their lives. Metallic sulfides, combined with air and water, can create sulfuric acid, as well as residues of several toxic metals, including mercury, lead, and arsenic.

Exxon proposed to dump its mine wastes into a 350-acre waste pond that would reach, in some locations, to within 15 feet of the water table, raising fears that sulfuric acid and heavy metals would leach into local water supplies. These waters are used not only for bathing, drinking, and cooking; they also sustain the local Chippewa with fish, as well as wild rice, a cash crop as well as a sacred part of Sokaogon Chippewa religious life. Native Americans in the area realized that if the mine is constructed, their drinking water and their harvest could become highly toxic.

The conflict over sulfide mining in Wisconsin began in 1975, when Exxon Minerals Company located one of the largest zinc–copper sulfide deposits in North America adjacent to the Mole Lake Sokaogon Chippewa reservation near Crandon, Wisconsin. Company plans anticipated extraction of roughly 50 million tons of sulfide ore over the life of the mine, about 30 years. This volume of waste could fill Egypt's Great Pyramid about 10 times.

The environmental alliance against sulfide mining in the area began in earnest during the 1980s when Kennecott Copper, a subsidiary of Rio Tinto Zinc, developed plans for an open-pit copper sulfide mine near Ladysmith, Wisconsin. In 1993, a small mine was started on this site, while local opponents feared that a larger mine operated by Exxon would follow. The proposed mine site lies on land guaranteed to the Sokaogon Chippewa by treaty in 1855.

Once word of the mine's possible effects spread, sportsmen's groups, many of which had opposed local Native Americans' fishing rights a few years earlier, allied with Native Americans because their access to healthy wildlife was threatened. The Chippewas also have been joined by several non-Indian environmental groups, including the Green Party.

Exxon has maintained that new technologies will mitigate the spread of sulfide byproducts from the mine wastes, but few people in the area bought the company's argument. Mine opponents worked with Wisconsin State Representative Spencer Black, Democrat of Madison, to fashion a bill that would prohibit the opening of a sulfide mine in Wisconsin until a similar mine in another state had been closed for 10 years with no contamination of surrounding groundwater. The bill passed the state Senate 29–3 during March 1997; late in 1997 the bill was reported out of the assembly's environment committee by a 6–4 vote. "There are enough votes to pass it; the question is rather the bill's opponents will be able to attach weakening amendments during the assembly floor debate," said Al Gedicks, a professor of sociology at the University of Wisconsin–Lacrosse (Johansen, 1998, 6).

Faced with unyielding local opposition, BHP Billiton pulled out of Nicolet Minerals in Crandon early in 2003, abandoning plans to build an underground zinc and copper mine near the headwaters of the Wolf River in northern Wisconsin. Dale Alberts, the president of subsidiary Nicolet Minerals Co., told reporters, "A number of companies have indicated an interest in this deposit. I would expect it to be sold within the next few months" ("Hotspots: Wisconsin," 2002).

NATIVE PEOPLES LINE UP AGAINST YELLOWSTONE NATIONAL PARK'S "BUFFALO CULL"

More than 1,200 buffalo were killed between 1997 and 2001 by state and federal officials, one-third of the last free-roaming herd in America. The rationale for the killings is that the buffalo may infect cattle that graze adjacent to Yellowstone with brucellosis, despite the fact that no buffalo has ever transmitted this disease to a cow in the wild.

According to Winona LaDuke, a White Earth Anissinabe (and vice-presidential candidate for the Green Party), the Yellowstone herd are descendants of the last buffalo, which survived great massacres of millions of buffalo a century ago. "Every time a buffalo is killed here, that brings back sharp pains of what happened in the late 1800's to our people," said James Garrett of Cheyenne River. "I descend from the people that were killed at Wounded Knee, and my family still feels that pain. We were killed for no reason, the same as when a buffalo is killed here" (LaDuke, n.d.).

LaDuke described the annual buffalo cull:

> Each year, as the snow starts to fall at Yellowstone National Park, the last free buffalo in America, the descendants of the Great Buffalo Nation, face death. Driven by their survival instincts, Yellowstone buffalo attempt to leave the park during harsh winters in search of food. As they cross an invisible border into the state of Montana, they are captured and killed. (LaDuke, n.d.)

The U.S. federal government has all but excluded Native American participation in this issue, despite the fact that federal law requires Native participation "in a decision-making process of such cultural and spiritual significance as the fate of Yellowstone's buffalo" (LaDuke, n.d.). Eighty different Native American nations have formally opposed the killings and requested a role in the decision making. National organizations such as the National Congress of American Indians, representing more than 300 Native American nations, and the Intertribal Bison Cooperative, with a membership of 50-plus nations, also have requested a voice. By law, all potential decisions require agreement of federal agencies and Native American governments.

To give an appearance of participation, according to LaDuke, the government has held three rather hastily called "tribal consultation" meetings. Native American representatives received less than two weeks' notice of these events; one meeting was scheduled at the same time as the National Congress of American Indians annual gathering, the largest gathering of tribal leaders in the country.

"Despite the insulting process," wrote LaDuke, "Tribes forwarded a specific set of requests. To date, federal agencies have given no response as to any of these requests, including, at the centerpiece, a request for actual tribal participation in the E.I.S. [Environmental Impact Statement] team" (LaDuke, n.d.).

LaDuke wrote that the Native community is frustrated and angry with a tribal consultation process that has "marginalized indigenous peoples' knowledge of buffalo" (LaDuke, n.d.). LaDuke continued: "The Nez Perce, Blackfeet, Crow and others whose treaties actually encompass part of Yellowstone National Park have no decision-making voice. There is no seat at the table for

A bull bison from Yellowstone National Park crashes through chest-deep snow in West Yellowstone, Montana, January 6, 1998, while being hazed back into a "Bison Safe Zone" by volunteers for a group called Buffalo Nations. The bull was foraging for food in a subdivision north of West Yellowstone. Buffalo Nations volunteers patrol outside Yellowstone to haze bison away from Montana Department of Livestock shooters. Last year nearly 1,100 Yellowstone bison were shot or captured outside the park. (AP Photo/*Bozeman Daily Chronicle*, Doug Loneman)

the Ho Chunk (Winnebago), Lakota, Kiowa, Gros Ventre, Cheyenne, Shoshone Bannock, and others whose spiritual practices, cultural practices and languages are entirely intertwined with buffalo" (LaDuke, n.d.).

The rationale for Native American involvement is ecological as well as political. The basic effort is to renew the ecology and cultures of the Great Plains, the largest biological region in North America, and, in that process, to restore Native peoples' collective humanity, according to LaDuke. The Native organizations suggest acquisition of additional lands for the buffalo, even if this means purchase of grazing allotments at fair-market value, cattle vaccination programs in grazing areas adjacent to Yellowstone, and the removal of buffalo families to tribes and federal grasslands, encouraging the restoration of widespread buffalo herds.

On March 4, 2002, four women were arrested after locking themselves together in the office of Montana Department of Livestock Executive Director Marc Bridges to protest the killings of wild buffalo. Abbi Dunlap, Emily Kodama, Julia Piaskowski, and Jennifer Schneider called for a moratorium on the culling of the animals. "The D.O.L. has consistently shown that they will not obey the rules they helped to write and they refuse to be honest with the public about their actions. I am fed up," wrote Kodama in a statement. "These are the last wild buffalo in America and if the government won't do its duty to protect the buffalo, then I believe we have a moral obligation to act on their behalf" ("Bison Activists," 2002). During the preceding winter, 26 bison were killed, 19 of them bulls, which the U.S. Department of Agriculture says pose a low risk of disease transmission. A group of nine yearlings and pregnant females was captured and five were killed ("Bison Activists," 2002).

REFERENCES

"2,200 Mile Walk to Protect the Great Lakes Coming to Detroit and St. Clair Rivers Area This Week." Schedule program, August 21–25, n.d. Indigenous Environmental Network, [http://www.ienearth.org/subject.html#publications] [http://migrationjourney.cjb.net]

Abel, Heather. "The Rise and Fall of a Gold Mining Company." *High Country News* 29:24 (December 22, 1997): n.p. [http://www.hcn.org/servlets/hcn.Article?article_id=3860]

American Friends Service Committee. *Uncommon Controversy: Fishing Rights of the Muckleshoot, Puyallup, and Nisqually Indians.* Seattle: University of Washington Press, 1970.

Anquoe, Bunty. "House Begins Investigating Possible Radiation Exposure." *Indian Country Today*, June 9, 1993, A-8.

Avery, Roberta. "Migration Journey Feels Pain of Mother Earth; Saving Planet for the Seventh Generation and Beyond. " *Indian Country Today*, September 19, 2001, C-1.

Badger, T. A. "Villagers Learning a Frightening Secret: U.S. Reveals that it has Buried Radioactive Soil Near Alaska Town 30 Years Ago. Residents Fear that Atomic Testing may have Damaged the Food Chain." *Los Angeles Times*, December 20, 1992, B-5.

Baker, James N. "Keeping a Deadly Secret: The Feds Knew the Mines Were Radioactive." *Newsweek*, June 18, 1990.

Banerjee, Neela. "Can Black Gold Ever Flow Green?" *New York Times*, November 12, 2000, C-1.

Barreiro, Jose, ed. "Readings On James Bay." *Akwe:kon Journal* 8:4 (winter 1991): n.p.

Barry, Tom, and Beth Wood. "Uranium on the Checkerboard: Crisis at Checkpoint." *American Indian Journal*, June 1978, n.p.

Barsh, Russel L. *The Washington Fishing Rights Controversy: An Economic Critique*. Seattle: University of Washington School of Business Administration, 1977.

"Bison Activists Arrested." Environment News Service, March 5, 2002. [http://ens-news.com/ens/mar2002/2002L-03-05-09.html]

"Black Mesa Spill Nets $128,000 Fine." Arizona Water Resource, News Briefs, June 2001. [http://ag.arizona.edu/AZWATER/awr/mayjune01/news.html]

Brack, Fred. "Fishing Rights: Who is Entitled to Northwest Salmon?" *Northwest Magazine, Seattle Post-Intelligencer*, January 16, 1977, 6.

"A Brief History of Relocation on Black Mesa." N.p., n.d. [http://www.blackmesais.org/cultural_sen_history.html]

Bristol, Tim. "Eco-Colonialism" at Makah." *Native Americas* 15:4 (winter 1998): 3.

Brown, Bruce. "A Long Look at the Boldt Decision." *Argus* (Seattle), December 3, 1976, 4–5.

———. *Mountain in the Clouds: A Search for the Wild Salmon*. New York: Simon & Schuster, 1982.

"Bush Green-lights Yucca Mountain Nuclear Waste Dump." Environment News Service, February 15, 2002. [http://ens-news.com/ens/feb2002/2002L-02-15-01.html]

"Cheyenne River Sioux Tribe Wins Suit Against Homestake." *Drillbits and Tailings* 4:11 (July 23, 1999): n.p. [http://www.moles.org/ProjectUnderground/drillbits/4_11/2.html]

Churchill, Ward, and Winona LaDuke. "Native America: The Political Economy of Radioactive Colonialism." *The Insurgent Sociologist* 13:3 (spring 1986): 51–84.

"Clean Coal's Dirty Facts." N.p., n.d. [http://www.blackmesais.org/clean_coals_dirty_facts.html]

Cohen, Faye G. *Treaties on Trial: The Continuing Controversy over Northwest Fishing Rights*. Seattle: University of Washington Press, 1986.

Colomeda, Lori A. (Salish Kootenai College, Pablo Montana). "Indigenous Health." Speech delivered in Brisbane [Australia], September 9, 1998. [http://www.ldb.org/vl/ai/lori_b98.htm]

"Corbin Harney, *Shoshone Medicine Man*." Blue Dolphin Publishing, 1996. [http://liteweb.org/wildfire/corbin.html]

Egan, Timothy. "Eskimos Learn They've Been Living Amid Secret Pits of Radioactive Soil." *New York Times*, December 6, 1992, A-26.

Eichstaedt, Peter. *If You Poison Us: Uranium and American Indians*. Santa Fe, N.M.: Red Crane Books, 1995.

Emmons, Della Gould. *Leschi of the Nisquallies*. Minneapolis: T. S. Denison, 1965.

"Energy Chief Over Yucca Mountain." Environment News Service, January 7, 2002. [http://ens-news.com/ens/jan2002/2002L-01-07-02.html]

Ewing, Rodney C., and Allison Macfarlane. "Nuclear Waste: Yucca Mountain." *Science* 296 (April 26, 2002): 659–60.

Fernandez, Lydia. "Gold Mine Threatens Quechan Sites." *Native Americas* 15:2 (summer 1998): 6.

"Figure This One!" *Four Corners Clamor* 1:6 (summer–fall 1998): n.p. [http://www.ausbcomp.com/redman/clamor6.htm#newdeal]

Forberg, S., O. Tjelvar, and M. Olsson. "Radiocesium in Muscle Tissue of Reindeer and Pike from Northern Sweden Before and After the Chernobyl Accident: A Retrospective Study on Tissue Samples from the Swedish Environmental Specimen Bank." *Science of the Total Environment* 115 (1992): 179–89.

Foushee, Lea. "Acid Rain Research Paper." Unpublished paper presented at Indigenous Women's Network; Testimony before the International Council of Indigenous Women, Samiland, Norway, August 1990.

Gedicks, Al. *The New Resource Wars: Native and Environmental Struggles Against Multinational Corporations*. Boston: South End Press, 1993.

Giuliano, Jackie Alan. "Killing Tomorrow for a Few Megawatts Today." Environment News Service, March 2001. [http://ens.lycos.com/ens/mar2001/2001L-03-30g.html]

Goldtooth, Tom. "Indigenous Environmental Network Statement to Nora Helton, Chairwoman, Fort Mojave Indian Tribe, Other Lower Colorado River Indian Tribal Leaders, Tribal Community Members, Elders and Non-Native Groups and Individuals, in Reference to the Ward Valley." Unpublished paper presented Victory Gathering to Celebrate the Defeat of a Proposed Nuclear Waste Dump, Ward Valley, California, February 16, 2001. [http://www.ienearth.org/ward_valley2.html]

"Governor Leavitt Opens Office High-level Nuclear Waste Opposition." News release issued by Office of the Governor, State of Utah, Salt Lake City, December 7, 2000. [http://www.governor.state.ut.us/html/nukewaste.html]

Grinde, Donald A., Jr., and Bruce E. Johansen. *Ecocide of Native America: Environmental Destruction of Indian Lands and Peoples*. Santa Fe, N.M.: Clear Light, 1995.

Guillette, Elizabeth A., Maria Mercedes Meza, Maria Guadalupe Aquilar, Alma Delia Soto, and Idalia Enedina Garcia. "An Anthropological Approach to the Evaluation of Preschool Children Exposed to Pesticides in Mexico." *Environmental Health Perspectives* 106:6 (June 1998): n.p. [http://www.anarac.com/elizabeth_guillette.htm]

Havasupai Fight to Save Grand Canyon From Uranium Mining. July 21, 1992. [http://www.ratical.org/ratville/native/havasuEFN.txt]

"Heavy Metal Levels in Reindeer, Caribou, and Plants of the Seward Peninsula." Current Research Programs, Reindeer Research Program: University of Alaska at Fairbanks, April 2000. [http://reindeer.salrm.alaska.edu/research.htm]

Hodge, Damon. "Dumpy Meeting: Final Vegas Hearing on Yucca Mountain Plagued by the D.O.E." *Las Vegas Weekly*, September 11, 2001.

"Hotspots: California." *Drillbits and Tailings* 6:9 (November 30, 2001): n.p. [www.moles.org]

"Hotspots: Idaho." *Drillbits and Tailings* 6:7 (October 31, 2001): 3.

"Hotspots: United States." *Drillbits and Tailings* 6:10 (December 30, 2001): n.p. [www.moles.org]

"Hotspots: Wisconsin." *Drillbits and Tailings* 7:8 (October 2, 2002): n.p. [http://www.moles.org]

Huff, Andrew. "Gold Mining Threatens Communities." *The Progressive*, July 11, 2000. [http://www.progressive.org/mpdvah00.htm]

Jaimet, Kate. "Canada to Fight Bush over Arctic Oil Drilling: Environment Minister 'Utterly Opposed' to Exploitation of Wildlife Refuge." *Ottawa Citizen*, January 4, 2001, A-1.

Johansen, Bruce. "A Sensible, Impossible Fishing Solution." *Argus* (Seattle), April 7, 1978, 3, 12.

————. "Ward Valley: A 'Win' for Native Elders." *Native Americas* 15:2(Summer, 1998):7–8.

Johansen, Bruce, and Roberto Maestas. *Wasi'chu: The Continuing Indian Wars*. New York: Monthly Review Press, 1979.

Johansen, Bruce E. *Life and Death in Mohawk Country*. Golden, Col.: North American Press/Fulcrum, 1993.

————. "Victims of Progress: Navajo Miners." [Review of Eichstaedt, *If You Poison Us*.] *Native Americas* 12:3 (fall 1995): 60–61.

————. "The Right to One's Own Home: The Seminole Chickee Sustains Despite County Codes." *Native Americas* 13:3 (fall 1996): 44–47.

————. "The High Cost of Uranium in Navajoland." *Akwesasne Notes*, n.s. 2:2 (spring 1997): 10–12. [http://ratical.com/radiation/UraniumInNavLand.html]

————. "Exxon vs. Chippewas in Wisconsin." *Native Americas* 15:1 (spring 1998): 6.

Johansen, Bruce E., and Donald A. Grinde Jr. *The Encyclopedia of Native American Biography*. New York: Henry Holt, 1997.

Kanesta, George. "No Mine, No Way—Stay Away SRP!" *The Shiwi Messenger: A Public Forum By, Of and For the Zuni Community*, August 3, 2001. [http://www.zunispirits.com/messenger/080301editorial.html]

Keefe, Tom, Jr. "A Tribute to David Sohappy." *Native Nations* (June–July 1991): 4–6.

Kettl, Paul A. "Suicide and Homicide: The Other Costs of Development," *Akwe:kon Journal* 8:4 (winter 1991): n.p.

Kolar, Michael J. "Waste That Won't Go Away; It's Time for Washington to Store Nuclear Leftovers at Yucca Mountain." *Pittsburgh Post-Gazette*, August 15, 2001, A-13.

LaDuke, Winona. "Environmental Work: An Indigenous Perspective." *Akwe:kon Journal* 8:4 (winter 1991): n.p.

————. "Indigenous Environmental Perspectives: A North American Primer." *Native Americas* 9:2 (summer 1992): 52–71.

————. "Breastmilk, P.C.B.s, and Motherhood: An Interview With Katsi Cook, Mohawk." *Cultural Survival Quarterly* 17:4 (winter 1994): 43–45.

————. "Commentary: Blast from the Past: Bush's Energy Policy." *The Circle: Native American News and Arts* 22:9 (September 2001): n.p. [http://www.thecirclenews.org/092001/news2.html]

————. "The Future of Yellowstone Buffalo." *Honor the Earth*. N.p., n.d. [http://honorearth.org/buffalo/future.html]

"Lake Powell Is Silting In." Environment News Service, January 15, 2002. [http://ens-news.com/ens/jan2002/2002L-01–15–09.html]

Lane, Earl. "Oil In The Wilderness: The Battle over a Land of Beauty; Bush's Proposal to Explore Alaskan Refuge for Oil Fuels Economic, Environmental Clash." *Newsday*, May 6, 2001, A-4.

————. "Refuge in a Storm: Nature Lovers and Oil Drillers Clash, with Little Room for Compromise." *Seattle Times*, May 20, 2001, A-3.

Lazaroff, Cat. "Independent Review Questions Approval of Yucca Mountain." Environment News Service, January 25, 2002. [http://ens-news.com/ens/jan2002/2002L-01-28-07.html]

"Leaked Report Blasts Yucca Mountain Nuclear Waste Site." Environment News Service, November 30, 2001. [http://ens-news.com/ens/nov2001/2001L-11-30-01.html]

Lee, Susan. "Blockade of Nevada Test Site Highway." *Synthesis/Regeneration* 14 (fall 1997): n.p. [http://www.greens.org/s-r/14/14-24.html]

Lewis, Sunny. "Angry Nevadans Pack Yucca Mountain Hearing." Environment News Service, September 6, 2001. [http://ens.lycos.com/ens/sep2001/2001L-09-06-01. html]

Lickers, Henry. "Guest Essay: The Message Returns." *Akwesasne Notes*, n.s. 1:1 (April–May–June 1995): 10–11.

Luoma, Jon R. "System Failure: The Chemical Revolution has Ushered in a World of Changes. Many of Them, It's Becoming Clear, Are in Our Bodies." *Mother Jones*, July–August, 1999. [http://www.motherjones.com/mother_jones/JA99/endocrine/ html]

"Makah Whaling Can Proceed, Judge Rules." Environment News Service, May 21, 2002. [http://ens-news.com/ens/may2002/2002L-05-21-09.html#anchor4]

Mann, Judy. "A Cautionary Tale about Pesticides." *Washington Post*, June 2, 2000, C-9.

Manning, Mary. "Shoshone Leader Describes Deadly Nuclear Legacy." *Las Vegas Sun*, May 29, 1996. [http://www.lasvegassun.com/sunbin/stories/archives/1996/mar/29/ 504524152.html]

———. "Shoshone Move Anti-nuke Office to Nevada." *Las Vegas Sun*, January 29, 1997. [http://www.lasvegassun.com/sunbin/stories/archives/1997/jan/29/505534873. html]

May, James. "Quechans Oppose Proposed Gold Mine: Site Located on 'Spirit Trail' Where Youths Learn Tribal Conditions." *Indian Country Today*, November 7, 2001, B-1, B-2.

Miniclier, Kit. "Nuclear-storage Plan Targeted; E.P.A. Cites Indians' Inability to Keep Water Supply Clean." *Denver Post*, June 17, 2001. [http://www.denverpost. com/Stories/0,1002,53%257E47549,00.html]

"Mother's Day Gathering at the Nevada Test Site, May 12–15, 2000. 700 Protest Nuclear Testing on Native Land—198 Arrested." Report on Healing Global Wounds "Honoring the Mother" Gathering at Nevada Test Site. [http://www. groundworkmag.org/nuke/nuke-nts-news.html]

Munro, Margaret. "Are The Great Lakes Safe or Sorry? The Great Lakes Appear to Have Made A Triumphant Recovery; But Looks Can Be Deceiving." *National Post* (Canada), October 6, 2001. [http://www.nationalpost.com/features/1001/ 100601lake.html]

Murphy, Kim. "Two Views of the Dynamics of Oil." *Los Angeles Times*, April 16, 2001, 1.

"Navy, Environmental Protection Agency to Clean Unexploded Ordinance off Adak Island." Environment News Service, December 14, 2001. [http://ens-news.com/ ens/dec2001/2001L-12-14–09.html]

Neary, Ben. "Mine Could Affect Zuni Salt Lake." *Boulder (Colorado) Daily Camera* in *Santa Fe New Mexican*, February 5, 2001. [http://www.thedailycamera.com/news/ statewest/05azuni.html]

"Nevada Governor Vetoes Yucca Mountain." Environment News Service, April 8, 2002. [http://ens-news.com/ens/apr2002/2002L-04-08-09.html]

"Nevada Outraged: Yucca Mountain O.K.'d for Nuclear Waste." Environment News Service, January 10, 2002. [http://ens-news.com/ens/jan2002/2002L-01-10-04.html]

Norrell, Brenda. "Hopi and Navajo Protest Pipeline." *Indian Country Today* in *Asheville (North Carolina) Global Report* 129 (July 5–11, 2001): n.p. [http://www.agrnews.org/ issues/129/nationalnews.html]

Orr, David. "Native Americans, Environmentalists Question Proposed Lake Powell Marina; Groups Call for E.I.S. on Project." *Wild Wilderness* (November 30, 2000); n.p. [http://www.wildwilderness.org/aasg/powell.htm]

Parman, Donald L. *Indians and the American West in the Twentieth Century.* Blooming-ton: Indiana University Press, 1994.

Perry, Tony. "U.S. Backs Tribe, Rejects Gold Mine Proposal." *Los Angeles Times,* November 17, 2000. [http://www.ienearth.org/mining_campaign_1c.html#nov 17-2000]

"Picuris Pueblo Wrestles with Summo on Copper Hill and Wins." *Drillbits and Tailings,* August 7, 1998. [http://www.moles.org/ProjectUnderground/drillbits/981207/98120703. html]

Pierpoint, Mary. "Reno-Sparks Fights Kitty Litter Mine; High Traffic, Dust, Potentially Toxic Minerals All Concerns." *Indian Country Today,* December 12, 2001, A-1.

Purdom, Tim. "Radiation Victims Rally; Marchers Complain about IOUs, Delays in Payments.'" *Gallup Independent,* March 13, 2001. [http://www.cia-g.com/~gall pind/3-13-01.html#anchor1]

Raloff, J. "Picturing Pesticides' Impacts on Kids." *Science News* 153:23 (June 6, 1998): 358.

Rayburn, Rosalie. "Coal Mine Plan Goes to Public for Comment." *Albuquerque Jour-nal,* June 22, 2001. [http://www.albuquerquejournal.com/news/366027news06-22-01.htm

Rogers, Keith. "Nevada Anti-Nuclear Protest." *Las Vegas Review-Journal,* April 20, 1992, n.p.

———. "Political Leaders, Residents Speak Out Against Yucca Mountain Project." *Las Vegas Review-Journal,* September 6, 2001. [http://www.commondreams.org/head lines01/0906-01.htm]

Schneider, Andrew. "'A Wounded Mountain Spewing Poison:' The Mining of the West: Profit and Pollution on Public Lands: A *Post-Intelligencer* Special Report." *Seattle Post-Intelligencer,* June 12, 2001. [http://seattlep-i.nwsource.com/specials/ mining/27076_lodgepole12.shtml

Schneider, Iyawata Britten. "Environmental and Human Rights Devastation in the Southwest." *Country Road News,* n.d. [http://www.countryroadchronicles.com/ Articles/CountryRoadNews/devastation.html]

Schneider, Keith. "Washington Nuclear Plant Poses Risk for Indians." *New York Times,* September 3, 1990, A-9.

Seldon, Ron. "Families Sicken and Die in Mold-Plagued HUD Housing; Residents Fight Back Against Substandard Construction." *Indian Country Today,* April 3, 2002, B-1, B-2.

"Senate Defeats Energy Rider which Would Open Arctic National Wildlife Refuge." Environment News Service, December 3, 2001. [http://ens-news.com/ens/ dec2001/2001L-12-03-09.html]

Skogland, T. "Radiocesium Concentrations in Wild Reindeer at Dovrefjell, Norway." *Rangifer* 7 (1987): 42–45.

Sockbeson, Rebecca. "Statement by Rebecca Sockbeson for IRATE (Indigenous Resis-tance Against Tribal Extinction), and I.E.N. (Indigenous Environmental Net-work)." International POPs Elimination Network (I.P.E.N.), P.C.B. Working Group, September 8, 1999. [http://www.ipen.org/irate.html]

Sohappy, David. "This has been a Long Road . . ." *Native Nations* (June–July 1991): 7–8.

Stevens, Hazard. *The Life of Isaac I. Stevens.* 2 vols. Boston: Houghton-Mifflin, 1900.

Taliman, Valerie. "N.R.C. Reviews Goshute Nuke Plan." *Indian Country Today,* April 10, 2002, A-1, A-2.

Tomsho, Rupert. "Reservations Bear the Brunt of New Pollution." *Wall Street Journal*, November 29, 1990, 1.

"Toxic Ash 'Burnt Offerings' Going to American Indian Lands; Environmental Injustice in Oklahoma Cherokee Nation." Indigenous Environmental Network, September 25, 2001. [http://www.ienearth.org/alerts.html#cherokee]

"Tribe, Environmentalists Fear Opening of Uranium Mine near Grand Canyon." Associated Press, August 1, 2001. [http://www.ienearth.org/mining_campaign_1a.html#0801-tribe]

Udall, Stewart. *The Myths of August*. New York: Pantheon, 1994.

United States v. Washington. 520 F. 2d 676 (9th Cir. 1975); cert. denied 423 U.S. 1086 (1976)

"Uranium Miners, Families Bring Tales of Pain to Washington." *Arizona Republic*, April 15, 2000. [http://www.ienearth.org/mining_campaign_1c.html#nov17-2000]

"Uranium/Nuclear Issues and Native Communities." Indigenous Environmental Network, 2001. [http://www.ienearth.org/nuciss.html]

Valencia, Angel. "Statement of Angel Valencia, Yoemem Tekia Foundation, Tucson, Arizona POPs Negotiations, March 23, 2000." *Native News*, March 27, 2000. [http://www.mail-archive.com/natnews@onelist.com/msg00572.html]

Vandevelder, Paul. "Between a Rock and a Dry Place." *Native Americas* 17:2 (summer 2000): 10–15.

———. "A Native Sense of Earth." *Native Americas* 18:1(spring 2001): 42–49.

Wald, Matthew L. "Handicapping Reactors by the Numbers." *New York Times*, June 19, 2001, C-1.

Wanamaker, Tom. "Hopi Tribe Opposes Lease Extension on Water-thirsty Black Mesa Mine." *Indian Country Today*, February 13, 2002, C-1.

Washington v. Washington State Commercial Passenger Fishing Association. 443 U.S. 658, 686 (1979).

Wastell, David. "Fear and Loathing in Las Vegas over Underground Nuclear Dump." *London Sunday Telegraph*, May 20, 2001, 27.

Weaver, Jace, ed. *Defending Mother Earth: Native American Perspectives on Environmental Justice*. New York: Maryknoll, 1996.

Weaver, Kasey. "Sacred Sites Threatened." *We Have Many Voices* 1:13 (July 21, 2001): n.p. [http://www.turtletrack.org/ManyVoices/Issue_13/SacredSites_721.htm]

West, Maureen. "Fatal Ski-boat Fumes Sound Alarm." *Arizona Republic*, July 5, 2001, A-1.

Wilkinson, Charles. *Messages from Frank's Landing: A Story of Salmon, the Treaties, and the Indian Way*. Seattle: University of Washington Press, 2000.

"Yucca Mountain: No Place for Nuclear Waste." *Honor the Earth*. N.p., n.d. [http://honorearth.org/nuclear/no_place.html]

"Yucca Mountain Volcanic Hazard Greater Than Thought." Environment News Service, July 31, 2002. [http://ens-news.com/ens/jul2002/2002-07-31-09.asp]

"Zuni: Utility Plan Would Desecrate Sacred Lake." *Salt Lake City Tribune*, June 9, 2001. [http://members.tripod.com/TopCat4/news/4jun01.txt]

"Zunis Object to Coal Mine Ruling." *Gallup Independent*, August 27, 2001. [http://www.cia-g.com/~gallpind/8-27-01.html]

VENEZUELA

INTRODUCTION

Venezuela's government has been an international laggard in recognizing indigenous land tenure, as well as in consulting native peoples' rights when infrastructure development and resource extraction intrude upon their lands and ways of life. The Pemon have expressed their frustration at this lack of consultation by threatening to topple pylons of a new high-tension electricity line linking Venezuela and Brazil. Elsewhere in Venezuela, the Waraos have been casting a skeptical eye at oil exploration, which has sullied the rainforests that they call home.

THE PEMON: LAND RIGHTS AND POWER TRANSMISSION

The Canaima National Park and the nine-million-acre Imataca Forest reserve, which are among Venezuela's most popular ecotourist attractions, have become home to high-tension electricity pylons carrying power to Brazil. The heavy transmission lines crackle as they traverse the tropical forest that comprises the habitat of 20,000 indigenous Pemon, Akawaio, Arawako, Karina, and Warao peoples. The 400-mile, 200-megawatt, $400 million power line began transmission during the summer of 2001, 3 years behind schedule and $85 billion over budget, having been plagued by work stoppages and sabotage by native peoples (Webb-Vidal, 2002, n.p.).

Venezuela's indigenous communities were generally ignored as the line was being built; the Venezuelan government and Edelca, the state-owned power company directing the project, have been sharply criticized for disregarding indigenous rights and blighting the environment. Construction of the power lines has caused deforestation, erosion, and loss of forests, including many animal and plant habitats.

"Angered at having been ignored during the project's execution," reported Andy Webb-Vidal in the *London Financial Times*, "some of the indigenous communities are threatening to disrupt the supply to Brazil by pulling down pylons, more than 30 of which were knocked down during construction" (Webb-Vidal, 2002, n.p.). "The pylons could be knocked down at any time," said Silviano Castro, a leader among the Pemons. "All of us who are affected can see that the power line we never asked for, or were asked about, is at fault" (Webb-Vidal, 2002, n.p.).

With tensions rising, parts of the Canaima National Park have been militarized. Troops have been stationed in the park to protect the pylons, arriving in trucks belonging to the Zurich-based A.B.B. Group, the main construction contractor for the power lines. Local communities allege ill-treatment by the army. "Militarization of the area is what we least like," Castro said (Webb-Vidal, 2002, n.p.).

"The Venezuelan government has betrayed the indigenous people," said Stephen Corry, director-general of Survival International, an English group that advocates for the rights of indigenous peoples worldwide. "Not only has it failed to consult and listen to the Pemon over the building of the electricity line, but it has violated their most fundamental constitutional right by not recognizing collective ownership titles to their land" (Webb-Vidal, 2002, n.p.). Edelca contends that although mistakes were made, steps also were taken to minimize deforestation and limit the visual impact of the unsightly pylons.

Pemon leaders expect that roads and other infrastructure necessary to service the power lines will attract various large-scale economic activities, including legal and illegal mining, logging, and tourism. The Pemon assert that "the economic dynamics that will surely flourish around the power lines will not only disfigure the landscape but will also attack the ecological and cultural stability of the zone" (Carrere, 2001). The Coalition Against the Electric Transmission Line has pointed out that "developmentalism is a political and economic model that intervenes violently, with large-scale technology on the ways of living, fragmenting, out-casting and disjointing individuals and their habitat, under the alibi of giving the population a better quality of life in the future" (Carrere, 2001). Many holes and ditches cut to install the pylons for the transmission line have been ravaged by erosion. Roads built to aid construction of the power lines have brought motor vehicles to the area.

The Pemon themselves have fallen into factions between those who are willing to negotiate and others who want to retain their sacred sites, lands, and resources. One observer described the situation:

> For the former, blackmail and psychological pressure: the basic rights set out in the new constitution are conditioned to the laying of the power lines. For the latter, intimidation: firing practice near the communities, distribution of pamphlets stating that the Indians would be bombed and that they would be excluded from state benefits. (Carrere, 2001)

Environmentalists in Venezuela are concerned by the possible effects of the new power lines on Sierra de Lema, "which is still largely unexplored and known for the uniqueness of its plant life" (Hamilton, 1997, 9). Fears have been expressed that power line construction will seriously affect animal life in the area.

The Venezuelan government has not recognized any Indian land rights, and it never consulted people before making development plans, said Jose Poyo, an indigenous leader in the area. Indigenous groups in Venezuela "also fear that the line will lead to processing and manufacturing facilities in the Amazon as it brings electricity to new gold mining and logging concessions in the Sierra Imataca Forest Reserve" (Knight, 1998).

Indigenous peoples in Venezuela cite as precedent the introduction of a paved highway between Manaus and Caracas. The highway, which first connected the Amazon Valley and the Caribbean coast during the early 1970s, cut through indigenous land, after which roughly half of the Waimiri-Atroari Indians died of introduced diseases (Knight, 1998).

THE WARAO: RESISTING OIL DEVELOPMENT

The Warao, an indigenous people living on the Orinoco Delta of northeastern Venezuela who are world-renowned for their basket weaving as well as their palm-frond and balsa-wood crafts, have expressed alarm at the impact of proposed intensive oil drilling on their habitat and way of life. The 22,000 Warao (meaning "canoe people" in their language) convened in general congress to draft a common position on transnational corporations' plans to drill for oil on their fragile delta homeland, one of the world's largest wetlands. The Orinoco is the world's eighth-largest river, in terms of volume. Its delta includes an intricate system of tributaries and wetlands that stretches over 40,000 square kilometers, one of the largest such areas in the world.

Oil drilling in the area has been promoted by a state consortium, Petroleos de Venezuela (PDVSA), with participation by several foreign companies whose plans will impact Venezuela's 28 indigenous communities. "We have already begun to see the damages in the delta," said Dalia Yanez, the coordinator of the Network of Warao Women (Gutierrez, 1997). During 1995, British Petroleum obtained a concession to a 48,000-hectare tract in Pedernales, an abandoned oil field in the delta, where production could reach 200,000 barrels a day.

The Warao have expressed alarm at the "dizzyingly swift deterioration" of their living conditions, as oil exploitation has advanced, according to Jesus Jimenez, one of their leaders (Gutierrez, 1997). Yanez added that "the delta is our social as well as physical space, where a river and a plant—palms—give us life" (Gutierrez, 1997). According to an account by Interpress Service, Yanez said problems of health, education, land ownership, and social development among her people have been getting worse, and that oil has already leaked into the water during oil prospecting in Pedernales, injuring the local flora and fauna. Furthermore, since seismic activity began in the oil fields two years ago, the Isla de Plata (Silver Island), previously home to 200 people, has sunk (Gutierrez, 1997).

While oil development could bring cash income to the Waraos, "What good is the money if it brings us death as a people?" asked Yanez (Gutierrez, 1997). The Warao possess ancestral rights to the delta, which (unlike for most other indigenous peoples in Venezuela) are recognized, at least in theory, by Venezuela's constitution, local legislation, and common law, as well as by international conventions signed by Venezuela.

The Waraos' opposition to oil drilling on their lands has been shaped by historical circumstances. For example, when the Manamo, a major tributary of the Orinoco, was suddenly closed off by oil exploration, about 6,000 Warao were killed by malaria-carrying mosquitoes that bred in its newly stagnant waters, said Yanez. Oilwatch Venezuela said that oil exploration in the swamps has included explosions that have caused the leaks and environmental damage described by Yanez. The president of the state consortium, Luis Giusti, denied assertions of oil-related environmental damage. He told the Interpress Service that strict environmental norms were followed in Pedernales, and that British Petroleum—as did the previous operator Lagoven—used a costly system of wooden platforms and boardwalks to avoid damage to the ecosystem (Gutierrez, 1997).

The Warao have split over whether to accept oil-company proposals for joint ownership, or to reject them outright. "Some of our leaders are led astray as they are acculturated" by mainstream society, complained Yanez (Gutierrez, 1997).

REFERENCES

Carrere, Ricardo. "Venezuela: A Power Line That Kills." *World Rainforest Movement Bulletin* (August 2001): n.p. [http://www.wrm.org.uy]

Gutierrez, Estrella. "Venezuela—Indigenous: Waraos Analyze Impact of Oil Drilling." Interpress Service, March 20, 1997. [http://www.oneworld.org/ips2/mar/venezuela.html]

Hamilton, Dominic. "Pemon Indians Paralyze Powerlines." *Native Americas* 14:3 (fall 1997): 9.

Knight, Danielle. "Amazon Indigenous Groups Oppose Infrastructure Projects." Interpress Service, January 18, 1998. [http://www.amazonwatch.org/newsroom/mediaclips/jan1898ips.html]

Webb-Vidal, Andy. "Venezuela's Deal to Help Ease Power Shortage in Brazil Sparks Controversy: Indigenous Groups Say Their Wishes Have Been Ignored in the Rush to Build the New Power Line." *London Financial Times*, August 14, 2002, n.p. (in LEXIS)

YEMEN

THE JAHM PIERCE PIPELINES

On November 21 and again on December 2, 1998, Jahm tribesmen blew up a pipeline in northern Yemen to protest of the lack of basic infrastructure in their communities. The pipeline, which ran from Marib province to the Red Sea port of Ras Isa, was operated jointly by the Hunt Oil Corporation and Exxon. According to a report in *Drillbits and Tailings*, about 35 tribesmen were arrested after the November incident. Tribal sources said they were trying to force the government to build power plants and a water distribution system in their region ("Yemeni Tribes," 1998). After punching six holes in the pipeline, sheiks from seven clans of the Jahm tribe met with Yemeni president Ali Abdullah Saleh to present their grievances.

Nomadic tribes in Yemen also have kidnapped visiting foreigners, including some oil-company executives, to bring attention to their demands. During 1997, the Murad tribe kidnapped an American engineer employed by Halliburton (an oil-services construction company) at the Saad al-Kamel oil field, demanding compensation for land lost to oil exploration. During 1998, the Bani Dhabyan tribe kidnapped a British family to highlight its lack of electricity, roads, running water, schools, and medical facilities. "They said that, having tried for years to win basic concessions from the government for their community without success, hostage taking was for them a last desperate tactic," said Rageh Omaar, a journalist with the British Broadcasting Corporation, who was arrested by the Yemeni government for interviewing the tribe but then released ("Yemeni Tribes," 1998).

REFERENCES

"Yemeni Tribes Blow Up Oil Pipeline." *Drillbits and Tailings* (December 7, 1998): n.p. [http://www.moles.org/ProjectUnderground/drillbits/981207/98120703.html]

ZAMBIA

BLAMING "THE POOR" FOR DEFORESTATION

Introduction

Deforestation is considered a major environmental issue in Zambia, where woodland conversion to agriculture and wood harvesting for charcoal production are major causes of forest loss. Because charcoal is used mainly by poor, rural people, "the simplistic conclusion is therefore that 'poverty' or 'the poor' are to be blamed for deforestation" (Carrere, 2001).

Colonialism and Traditional Crop Production

During the first half of the twentieth century, traditional crop production in Zambia was dominated by shifting cultivation (the "citemene" system), in which the tropical soil was effectively husbanded for hundreds of years by tribes of African indigenous peoples. For many years, the farmers of Zambia selectively logged trees, burned the branches, and used ash as a fertilizer for the soil. Due to the nature of the soil, this method worked well and land could be used productively for about five consecutive years before being left fallow (Carrere, 2001).

This indigenous agricultural system was ignored by colonial intruders, who never realized its ecological nature and believed it to be backward and destructive. Indigenous farmers were pushed into settled, European-style agriculture. With the "green revolution" and increasing European and urban influences during the 1970s, chemical fertilizers were promoted and hybrid maize was introduced, making farmers dependent on subsidized fertilizers. Overuse of fertilizers raised the carrying capacity of the land, but it also resulted in soil erosion, acidification, and loss of fertility (Carrere, 2001).

Charcoal as a Fuel Source

Privatization of electricity generation (imposed on many countries by the International Monetary Fund and the World Bank) increased electricity prices and affected the electrification policy. Many local people could not pay for electricity and were driven to use charcoal as a fuel source. Charcoal was introduced as a cooking-energy source in the city of Lusaka, creating a new incentive among rural communities in central Zambia to clear woodlands to supply fuel wood to the urban market. Income from charcoal production was used to buy household products in the cash economy. In some cases, income was invested in agricultural production after the removal of subsidies. Under the traditional agricultural system, trees were cut and burned; but with the commodification of charcoal, cut trees were converted to charcoal for sale and the land was also cultivated to produce both food and cash crops (Chidumayo, 2001).

Eventually, both agriculture and charcoal production destroyed large areas of Zambia's forests. Uncontrolled or poorly controlled commercial exploitation of timber came to be a major cause of deforestation in many parts of Zambia. Few of the profits reaped from this activity (which was supported by the government) benefited the local communities, especially those with no timber to exploit. Most of the money realized from timber sales went abroad or to Lusaka, the capital ("I.D.R.C.," n.d.).

"As a result," according to the *World Rainforest Movement Bulletin*,

government policies (and not "the poor") are at the root of deforestation in Zambia. It was government policies that made people switch from sustainable swifting cultivation to unsustainable "green revolution" crop production. High electricity tariffs have pushed people to use charcoal instead of electricity. Government promotion of certain cash crops—such as sunflower, soybeans and cotton—have incentivated [*sic*] further forest destruction. The government thus needs to be made responsible, not only for the past and current destructive process but, more importantly, for taking the necessary steps to address the problem. (Carrere, 2001)

REFERENCES

Carrere, Ricardo. "Zambia: Causes of Deforestation Linked to Government Policies." *World Rainforest Movement Bulletin* (September 2001): n.p. [http://www.wrm.org.uy]

Chidumayo, E. N. "Land-cover Transformation in Central Zambia: Role of Agriculture, Biomass, Energy and Rural Livelihoods." Unpublished paper presented at the Biological Sciences Department, University of Zambia, 2001. [http://coe.asafas. kyotou.ac.jp/news/PastExper2001Jan/PastExper2001_Paper_Chidumayo.htm]

"I.D.R.C., Knowledge and Development." n.d. [http://idrinfo.idrc.ca/Archive/ ReportsINTRA/pdfs/v19n1e/108939.pdf]

ZIMBABWE AND BOTSWANA: AN ALLIANCE FOR WILDLIFE

Elephants are thriving in parts of southern Africa, notably in Zimbabwe, where a novel program called "Campfire" has been initiated to ease conflicts between indigenous peoples and their large neighbors. According to Campfire's count, in 1900, Zimbabwe's elephant population was less than 4,000; by the 1990s, it had risen to more than 64,000, twice the number that the land can support, "without resulting in severe environmental degradation and the loss of other species" ("Sharing the Land," 1995). Human deaths from encounters with elephants are as common in Zimbabwe as traffic-accident deaths in the United States.

Under the Communal Areas Management Program for Indigenous Resources (Project Campfire), district councils apply for authority to manage wildlife outside national parks. Rural councils that join the program sell a hunting quota to a safari operator who in turn arranges for wealthy hunters, mostly from the United States and Europe, to shoot big game. The hunting quota is set annually and must be approved by Zimbabwe's Department of National Parks to ensure that the harvest is sustainable. More than 90 percent of Project Campfire's profits ($1.6 million in 1995) come from sport hunting, and two-thirds come from hunting elephants (Pye-Smith, n.d.). According to an Internet Web page maintained by Project Campfire, "Hunters are considered the 'ultimate ecotourists' in Zimbabwe, as they have a much lower impact on the environment than other tourists. In addition, their presence in remote areas acts as an anti-poaching deterrent, and hunters pay much higher fees than other tourists" ("Sustainable," 1995).

"Every year between 1992 and 1997," said Chief Lux Masule, one eye warily tracking three elephants ambling past him, "someone was killed by an elephant in the Chobe Enclave." One year it was a vet; another year it was a policeman. Frequently it has been peasant farmers working their fields. Such deaths are to be expected, as the Enclave's 7,000 inhabitants live on a strip of land that lies between the Chobe River—a magnet for game—and a forest that supports a huge concentration of elephants (Pye-Smith, n.d.).

For centuries, the villagers harvested wild animals for meat and hide, but during colonial times the relationship changed, with the British government effectively assuming ownership of game. After independence, the Botswanan government adhered to the policies established by its predecessors: the state, not the people, determined the fate of wildlife (Pye-Smith, n.d.).

Project Campfire was established to produce cash income from the hunting of elephants and other large game for local communities. "In 1989," explained Masule, "we asked the government to give us back the right to manage the wildlife ourselves." Four years later the Chobe Enclave Conservation Trust was established. The program's principal aim is to generate profits for local communities through the sale of hunting quotas (Pye-Smith, n.d.). Male elephants sell for about 4,000 British pounds each. Fifteen percent of the revenue from these sales is retained by the trust; the rest is divided equally between local villages. Since this program was initiated, the villagers have built grinding mills, classrooms, a campsite, a gas station, and other facilities with their share of the revenues.

Project Campfire was first proposed by conservationists who believed that wildlife was unlikely to survive outside protected areas without an incentive to tolerate and protect it. "In the old days," said Champion Chinoyi of the Zimbabwe Trust, "poachers were considered as heroes, and the villagers would give them shelter from the authorities," because wild animals raided crops, occasionally killed people and brought no benefits to the impoverished farming communities (Pye-Smith, n.d.). With profits from the safaris, some Zimbabwe indigenous villages have built new schools, grinding mills, and workshops that manufacture paper from elephant dung. Revenue from Project Campfire pays the salaries of several people in many villages, from game wardens to teachers. In some cases, Project Campfire money has been used to build solar-powered fences around agricultural land and villages to keep out foraging animals.

Roughly 65 percent of Project Campfire's revenues come from the hunting of elephants; their trophy fee is far higher than for other species, because most sportsmen who come to hunt in Zimbabwe desire an elephant trophy. In some districts, such as Bulilima Mangwe and Muzarabani, elephant harvesting is the only source of income. With a trophy fee of up to U.S. $12,000 or more, together with a daily hunting fee of $1,000, one elephant can produce $33,000 in cash income for a community during the course of an average 21-day hunt. By comparison, according to Project Campfire's estimate, a rural family of eight subsists on about $150 per year in Zimbabwe ("Sharing the Land," 1995).

In 1980, before Project Campfire, the Binga district only had 13 primary schools and no secondary schools. By 1995 the district was home to about 56 primary schools and 9 secondary schools ("Sustainable," 1995). According to Project Campfire, "Five years ago, [the] Kanyurira ward was a community of no-hopers. Since the introduction of Campfire, they have built a school, are building a clinic, and have purchased a tractor to plough the fields. Their managerial skills have improved dramatically" ("Sustainable," 1995).

Some Zimbabweans would like to be able to sell ivory from elephants they kill and the horns of rhinos killed by safari hunters under Project Campfire's aegis. The sale of wild-animal projects is now banned under international law, however. "For example," according to a Web page maintained by Project Campfire, "in 1991, Hwange district residents could have quadrupled their income had they been legally allowed to sell the ivory and hides from three elephants that were hunted after persistently raiding crops. The following year, the [ban on ivory sales] cost Campfire communities an estimated U.S. $93,000 in lost revenues from elephant products. Rural communities have U.S. $1.6 million in stockpiled unsold ivory in Zimbabwe" ("Sharing the Land," 1995).

In 1999, Japan became the first country in 10 years to legally import African ivory. The United Nations allowed a relaxation of its international ivory-trade ban on condition that tusks from Namibia, Zimbabwe, and Botswana were harvested only from elephants that had died of natural causes or were killed to control overpopulation. Japan was allowed to import 50 tons of ivory, with proceeds directed toward elephant conservation ("Japan Imports," 1999). Ivory is an important part of Japanese life and art, and is used to sculpt figures and make seals for use on official documents. Until the world trade in ivory was banned during the 1980s, Japan was the largest national purchaser.

Poaching also has declined because of game scouts' vigilance. Local councils cull animal herds as a collective enterprise and sell meat at cost to villagers. Villagers, for their part, dig holes to provide elephants with water during the dry season. Some also arrange food deliveries for them in hungry times. Before Project Campfire, many elephants died of thirst during droughts. The number of elephants in Zimbabwe has risen from 46,000 in 1980 to 64,000 in 1995 (Pye-Smith, n.d.).

Despite Project Campfire's successes, elephants still stir occasional enmity by raiding fields and stripping villagers' corn and other crops. Animal-rights groups in the United States, led by the Humane Society of the United States, have asserted that the program is "biologically unsustainable" (Pye-Smith, n.d.). Despite the criticism, elephant numbers continue to rise in Zimbabwe. The criticism is especially troublesome to Project Campfire's proponents, however, because the program has required "seed money" from the United States and European nations to get started. Animal-rights lobbyists in the United States, for example, have put considerable pressure on the U.S. Agency for International Development (USAID), which has helped to fund Project Campfire.

On a Web site maintained by the U.S. National Wildlife Federation, Dan Hodges of Martell, Nebraska wrote: "Campfire is a small 'band-aid' trying to cover a much larger gash. Tourist and photo safaris seem to be a more viable, longer-lasting solution. Preserving valuable wildlife resources to be enjoyed by a more diverse group makes more conservation sense than depleting them" ("Is This the Way?" 1995). Gail McMahon of Birmingham, Alabama, wrote to the same Web page: "Campfire may be working—but for whom? Certainly not for the unlucky individual animals selected for the kill. Hunting is morally and ethically repugnant, regardless of the end it is trying to achieve. This program is merely another means of poaching" ("Is This the Way?" 1995). Robert E. Holtz of Saint Paul, Minnesota, wrote that an issue that is never addressed in

Project Campfire is "what I see as the key issue—human population growth. Unless that growth is curbed, people will continue to squeeze animals into smaller and smaller areas. Therefore, programs such as Campfire, good as they are and well-intended as they may be, are doomed in the long run" ("Is This the Way?" 1995). Mary E. Picardi of Belle Haven, Virginia agreed: "One native mentioned he had 11 children. With this kind of population explosion, no amount of increased agriculture or killing of more animals will feed the expanding populace. Along with schools and hospitals and other good things, a program to teach family planning seems appropriate" ("Is This the Way?" 1995).

Zimbabweans have reminded animal-rights activists that elephants can be cruel to humans:

A month after I saw Chief Masule he wrote to remind me of the annual elephant kills in Chobe between 1992 to 1997. His letter read: "The record for 1998 has been completed. On Monday 15 June, my own uncle of my age was killed by elephants while looking for his stray cattle about 1.5 kilometers from home. The deceased was severely mutilated, leaving him partially identifiable." (Pye-Smith, n.d.)

REFERENCES

"Is This the Way to Save Africa's Wildlife? National Wildlife Federation: Reader Follow-up." July/August, 1995. [http://www.nwf.org/internationalwildlife/ltrja95.html]

"Japan Imports African Ivory." British Broadcasting Corporation News, July 16, 1999. [http://news.bbc.co.uk/hi/english/world/africa/newsid_396000/396123.stm

Maveneke, T., J. Hutton, and E. Kawadza. "Campfire in Zimbabwe: Integrating African Wildlife into the Local Economy Through Sustainable Use." Africa Resource Trust, n.d. [http://www.art.org.uk/articles/art_integratingwildlife.html]

Pye-Smith, Charlie. "Living with Elephants." One World, United Kingdom, n.d. [http://www.oneworld.org/patp/pap_7_4/pye.htm]

"Sharing the Land: People and Elephants in Rural Zimbabwe." Project Campfire, Zimbabwe, 1995. [http://www.campfire-zimbabwe.org/facts_07.html]

"Sustainable Rural Development: Driven by Campfire." Project Campfire, 1995. [http://www.campfire-zimbabwe.org/facts_02.html]

SELECTED BIBLIOGRAPHY

"Aboriginal Landowners Fight Off Barrick Homestake and Gold Mining." *Drillbits and Tailings* 7:1 (January 31, 2002). [englishdrillbits@topica.com]

Abrash, Abigail. "The Amungme, Kamoro, and Freeport: How Indigenous Papuans Have Resisted the World's Largest Gold and Copper Mine." *Cultural Survival Quarterly* 25, no. 1 (spring 2001): 38–43.

———. "The Victims in Indonesia's Pursuit of Progress." *New York Times*, March 6, 2001, A–21.

Achakulwisut, Atiya. "Deadly River: Despite a Government Promise to Clean Up the Contaminated Klity Creek, the Thai-Karen Villagers Who Live Along it Are Still Haunted by Health and Food Problems." *Bangkok Post*, January 30, 2001. [http://www.bangkokpost.net/outlookwecare/300101_Outlook01.html]

Achieng, Judith. "Kenya: Civic Group to Sue over Titanium Project." Interpress Service, October 12, 2000. [http://www.afrika.no/index/update/archives/2000October 13.shtml]

"Aerial Herbicide War on Drugs Poisons Land, Water." Environment News Service, January 15, 2001. [http://ens-news.com/ens/jan2002/2002L-01-15-04.html]

"Affected Villagers Occupy the Rasi Salai Dam." Environmental News Network, April 6, 1999. [http://lists.isb.sdnpk.org/pipermail/eco-list-old/1999-June/002061.html]

Akwesasne Notes. *Basic Call to Consciousness*. Akwesasne, Mohawk Nation, via Rooseveltown. New York: Akwesasne Notes, 1978.

Ali, Saleem, and Larissa Behrendt, eds. "Mining Indigenous Lands: Can Impacts and Benefits be Reconciled?" *Cultural Survival Quarterly* (spring 2001). [http://www.cs.org/publications/CSQ/index.htm]

All Burma Students' Democratic Front (ABSDF). "Wet Shu Concentration Labour Camp." April 20, 1995, New Delhi, India. [http://www.rainforestrelief.org/reports/teak_tort.html]

Allen, William. "Preserving, Exploring 'Central America's Amazon': St. Louis Zoo Biologist's Mission Is to Save the World's Endangered Wildlife." *St. Louis Post-Dispatch*, January 28, 2001, A–1.

Althaus, Dudley. "The Fated Forest: Nature's Way; Upsetting the Balance; Deforestation Threatens Thousands of Plant and Animal Species as Well as the Earth's Atmosphere; Studies Show Rain Forests Are Declining at Alarming Rates." *Houston Chronicle*, September 30, 2001, 6.

"Amazonian Deforestation Is Accelerating, U.S. Study Finds." Environment News Service, January 15, 2002. [http://ens-news.com/ens/jan2002/2002L-01-15-09.html]

"Amazon Watch Launches Mega-Project Report: New Pipelines Threaten Intact Amazon Rainforest in Brazil." Amazon Watch, August 10, 2001. [http://www.amazonwatch.org/newsroom/newsreleases01/aug1001_br.html

American Friends Service Committee. *Uncommon Controversy: Fishing Rights of the Muckleshoot, Puyallup, and Nisqually Indians.* Seattle: University of Washington Press, 1970.

Anane, Mike. "Another Cyanide Spillage in Ghana." F.I.A.N. Coordination, Ghana, October 24, 2001.

———. "Ghana: Cyanide Spill Worst Disaster Ever in West African Nation," Environment News Service, October 24, 2001.

Anquoe, Bunty. "House Begins Investigating Possible Radiation Exposure." *Indian Country Today,* June 9, 1993, A-8.

Antonowicz, Anton. "Baking Alaska: As World Leaders Bicker, Global Warming Is Killing a Way of Life." *London Mirror,* November 28, 2000, 8–9.

Aquila, Richard. *The Iroquois Restoration: Iroquois Diplomacy on the Colonial Frontier, 1701–1754.* Detroit: Wayne State University Press, 1983.

"Aracruz: Indigenous Peoples Refuse to Plant Eucalyptus." *World Rainforest Movement Bulletin* 32 (March 2000): n.p. [http://www.wrm.org.uy/bulletin/32/plantations.html]

"Argentina: The Struggle of the Kolla People." *World Rainforest Movement Bulletin* 5 (October, 1997): n.p. [http://www.wrm.org.uy/bulletin/5/Argentina.html]

Arias, Natalia. "Environmentalists Affirm Position in Defense of Life as Construction for O.C.P. Pipeline is Given Green Light." Amazon Watch and Accion Ecologica, n.d. [http://www.amazonwatch.org/newsroom/newsreleases01/jun0601_ec.html]

Arnold, Philip P. *Eating Landscape: Aztec and European Occupation of Tlalocan.* Boulder: University Press of Colorado, 2001.

Ashton-Jones, Nick. "Causes of Terrorism? Shell Oil in Nigeria, 1993 to 2001. October, 2001. [http://www.shell-terror.net/]

Astor, Michael. "Indian Land Conflict Hits Brazil." Associated Press, June 1, 2002.

Atarah, Linius. "Environment-Ghana: Gold Mine Spills Cyanide in River—Again." Interpress Service, November 1, 2001.

"L'Auravetl'an: An Indigenous Information Center by Indigenous Peoples of Russia. United Nations Information Center, Moscow, April 27, 2001. [http://www.indigenous.ru/english/english.htm]

"L'Auravetl'an Information Bulletin #2: Report On Siberian Health And Environment." March 13, 1997. [http://www.hartford-hwp.com/archives/56/004.html]

"Australia: Aborigines Fight Mine." Survival International Update, July 2001. [http://www.survival.org.uk/about.htm]

Authier, Philip. "Cree Pact Clears Final Hurdle." *Montreal Gazette,* February 5, 2002. [http://www.canada.com/montreal/montrealgazette/story.asp?id={43F9B38E-7577-4579-AEED-90A8F9638936}]

Avery, Roberta. "Migration Journey Feels Pain of Mother Earth; Saving Planet for the Seventh Generation and Beyond." *Indian Country Today,* September 19, 2001, C-1.

"B.B.C. Exposes Scandal of Brazilian Government Neglect: Last Amazonian Hunter-Gatherers Face Extinction." Survival International, August 29, 2002. [http://www.survival-international.org]

Badger, T. A. "Villagers Learning a Frightening Secret: U.S. Reveals that It Buried Radioactive Soil Near Alaska Town 30 Years Ago. Residents Fear that Atomic Testing May Have Damaged the Food Chain." *Los Angeles Times,* December 20, 1992, B-5.

Bailey, Rasul. "India's Banking Sector Feels Enron's Aftershocks." *Wall Street Journal*, December 6, 2001, A-8.

Baker, James N. "Keeping a Deadly Secret: The Feds Knew the Mines Were Radioactive." *Newsweek*, June 18, 1990.

Bales, Kevin. *Disposable People: New Slavery in the Global Economy*. Berkeley: University of California Press, 1999.

Ballantine, Betty, and Ian Ballantine. *The Native Americans: An Illustrated History*. Atlanta: Turner Publishing, 1994.

"Bank Funding for Oil Project and Government Repression Both on the Rise in Chad." *Drillbits and Tailings* 6:5 (June 30, 2001): n.p. [http://www.moles.org/Project Underground/drillbits/6_05/1.html]

Barreda, Andres. "Militarization and Oil in Chiapas," *La Jornada*, August 17, 1999, N.p.

Barreiro, Jose, ed. "Readings on James Bay." *Akwe:kon Journal* 8:4 (winter 1991): 3–48.

Barrionuevo, Alexi, and Thaddeus Herrick. "Wages of Terror: For Oil Companies, Defense Abroad Is the Order of the Day." *Wall Street Journal*, February 7, 2002, A-1, A-12.

Barry, Tom, and Beth Wood. "Uranium on the Checkerboard: Crisis at Checkpoint." *American Indian Journal* (June 1978): n.p.

Barsh, Russel L. *The Washington Fishing Rights Controversy: An Economic Critique*. Seattle: University of Washington School of Business Administration, 1977.

Barton, Robin. "Tourism Threatens Health and Welfare of India's Island Tribes." *London Independent*, April 22, 2001, Sunday.

Bengwayan, Michael. "Luzon Dam Plan Riddled with Flaws, Fault Lines." Environment News Service, October 5, 1999. [http://ens.lycos.com/ens/oct99/1999L-10-05-01.html]

Bhatiasevi, Aphaluck. "N.G.O.s Urge Authorities to Treat Lead Poisoning Victims." *Bangkok Post*, April 1, 2001. [http://www.ecologyasia.com/NewsArchives/Apr_2001/bangkokpost_040401_News18.htm]

"B.H.P. Billiton Runs from Responsibilities in Papua New Guinea." *Drillbits and Tailings* 6:10 (December 30, 2001): n.p. [www.moles.org]

Bigert, Claus. "A People Called Empty." In Rainer Wittenborn and Claus Bigert, eds., "Amazon of the North: James Bay Revisited." Program for presentation, Santa Fe Center for Contemporary Arts, August 4 through September 5, 1995.

"Bison Activists Arrested." Environment News Service, March 5, 2002. [http://ens-news.com/ens/mar2002/2002L-03-05-09.html]

"Black Mesa Spill Nets $128,000 Fine." *Arizona Water Resource: News Briefs* (June 2001): n.p. [http://ag.arizona.edu/AZWATER/awr/mayjune01/news.html]

"Bolivia: Indigenous Peoples' Forests Menaced by Oil Exploration." *World Rainforest Movement Bulletin* 35 (June, 2000): n.p. [http://www.wrm.org.uy/bulletin/35/Bolivia.html]

"Bolivian Rainforests Allocated without Indigenous Consent." Worldwide Forest/Biodiversity Campaign News, August 1, 1999. [http://www.wildideas.net/forest/alerts/1999-08-09-bolivia.html]

The Bonn Declaration. Third International Forum of Indigenous Peoples and Local Communities on Climate Change, July 14–15, 2001, Bonn, Germany. [http://www.ienearth.org/climate_1-p2.html#bonn2001]

Borst, Barbara. "Poaching Threatens Bear Population." Interpress Service, May 7, 1996. [http://www.nativenet.uthscsa.edu/archive/nl/9605/0084.html]

"Botswana: Government Plans to Destroy Bushman Tribes." Survival International, January 30, 2002. [mr@survival-international.org]

"Botswana Leaves Bushmen in Desert without Water." Survival International, February 22, 2002. [http://www.survival.org.uk/about.htm]

"Botswana Persecutes Bushmen." Survival International, August 15, 2001. [http://www.survival.org.uk/about.htm]

"Botswana Tortures Bushmen, then Prosecutes Them." Survival International, February 14, 2002. [http://www.survival-international.org]

Brack, Fred. "Fishing Rights: Who Is Entitled to Northwest Salmon?" *Seattle Post-Intelligencer Northwest Magazine*, January 16, 1977, 6.

Brandon, William. *American Heritage Book of Indians*. New York: Dell, 1961.

"Brazil Cracks Down on the Illegal Mahogany Trade." Environment News Service, December 12, 2001. [http://ens-news.com/ens/dec2001/2001L-12-12-03.html]

"Brazil Indians Free Reporters Taken Hostage." Reuters News Service, October 5, 2001. (In LEXIS)

"Brazil Indians Take Reporters Hostage to Get Land." Reuters News Service, October 5, 2001. (In LEXIS)

"Brazil: Mahogany Loggers Destroying the Amazon Forest." *World Rainforest Movement Bulletin* 53 (December 2001): n.p. [http://www.wrm.org.uy/deforestation/logging.html]

"Brazil: Panara Win Damages for 1970s Genocide." *Native Americas* 17:4 (winter 2000):9.

"Brazil's Amazon Rainforest Shrinking Fast." Environment News Service, May 15, 2001. [http://ens.lycos.com/ens/may2001/2001L-05-15-03.html]

"Brazil's Indigenous People Resist Large River Modifications." Environment News Service, May 30, 2001. [http://ens.lycos.com/ens/may2001/2001L-05–30–01.html]

"Brazil: The Struggle of the Pataxó Indigenous Peoples in Bahia." *World Rainforest Movement Bulletin* 31 (December 2000): n.p. [http://www.wrm.org.uy/bulletin/41/Brazil.html]

"Brazil: Yanomami to Lose Another 20 Per Cent?" Press release, Survival International (London), January 28, 2002.

"A Brief History of Relocation on Black Mesa." N.d. [http://www.blackmesais.org/cultural_sen_history.html]

Bristol, Tim. "Eco-Colonialism at Makah." *Native Americas* 15:4 (winter 1998): 3.

"British Engineering Company Withdraws from Ilisu Dam Project." Environment News Service, November 14, 2001. [http://www.ens-news.com/ens/nov2001/2001L-11-14-02.html]

Brodzinsky, Sibylla. "Colombian Judge Orders Halt to Drug Spraying." *St. Petersburg (Florida) Times*, July 28, 2001, 2-A.

Brown, Bruce. "A Long Look at the Boldt Decision." *Seattle Argus*, December 3, 1976, 4–5.

———. *Mountain in the Clouds: A Search for the Wild Salmon*. New York: Simon & Schuster, 1982.

Brown, DeNeen L. "Culture Corrosion in Canada's North; Forced Into the Modern World, Indigenous Inuit Struggle to Cope." *Washington Post*, July 16, 2001, A-1.

Brown, Paul. "Gift of Life." *The Guardian* (London), February 14, 2001. [http://www.urg.org.au/jabiluka/index.htm]

Bryant, D., D. Nielsen, and L. Tangley. *The Last Frontier Forests: Ecosystem on the Edge*. Washington, D.C.: World Resources Institute, 1997.

Bryce, Robert. "Spinning Gold." *Mother Jones*, September/October, 1996. [http://www.etan.org/news/kissinger/spinning.htm]

"Bulletin: Botswana—Bushmen Tortured for Hunting." Survival International, May 2001. [http://www.survival.org.uk/about.htm]

"Bush Gives Oxy a Hand in Escalating Colombia's War." *Drillbits and Tailings* 7:2 (February 28, 2002): n.p. [http://www.moles.org]

"Bush Green-lights Yucca Mountain Nuclear Waste Dump." Environment News Service, February 15, 2002. [http://ens-news.com/ens/feb2002/2002L-02-15-01.html]

"Bushmen Campaign Spreads." Survival International News Release, May 10, 2002. [http://www.survival-international.org/ad.htm]

Cadbury, Deborah. *Altering Eden: The Feminization of Nature*. New York: St. Martin's Press, 1997.

Calamai, Peter. "Chemical Fallout Hurts Inuit Babies." *Toronto Star*, March 22, 2000. [http://irptc.unep.ch/pops/newlayout/press_items.htm]

Caluza, Desiree. "Ibaloi Leader of Opposition to San Roque Dam Dies." *Philippine Daily Inquirer* (Manila), October 5, 2001. [http://www.inq7.net/reg/2001/oct/05/reg_8–1.htm]

"Cambodia: Timber Concessions vs. Community Forests." *World Rainforest Movement Bulletin* 53 (December 2001): n.p. [http://www.wrm.org.uy/deforestation/logging.html]

"Cambodia: Villagers Defend Their Resin Trees." *World Rainforest Movement Bulletin* 54 (January 2002): n.p. [http://www.wrm.org.uy]

"Cameroon: Social and Environmental Impacts of Industrial Forestry Exploitation." *World Rainforest Movement Bulletin* 53 (December 2001): n.p. [http://www.wrm.org.uy/deforestation/logging.html]

"Cameroon: Unsustainable Forestry for European Benefit." *World Rainforest Movement Bulletin* 52 (November 2001): n.p. [http://www.wrm.org.uy]

"Camisea Gas Field and Pipeline Project." Oxfam America, n.d. [http://www.oxfamamerica.org/advocacy/camisea.html]

"Camisea Natural Gas Project, Peru." Amazonwatch, October 8, 2001. [http://www.amazonwatch.org/campaigns.html]

"Canadian Indians Paddle to New York City to Protest Quebec Power Plant. *Syracuse Post-Standard*, April 5, 1990, A-2.

Capps, Walter Holden, ed. *Seeing With A Native Eye*. New York: Harper & Row, 1976.

Carr, Francis. "Indorayon's Last Gasp?" *Down to Earth*, January 2001. [http://www.gn.apc.org/dte/CInd.htm]

Carrere, Ricardo. "Argentina: Forest Conserved by the Wichí Destroyed by Agricultural Companies." *World Rainforest Movement Bulletin* 49 (August 2001): n.p. [http://www.wrm.org.uy]

———. "Ecuador: Action to Stop the Oil Pipeline Continues." *World Rainforest Movement Bulletin* 50 (September 2001): n.p. [http://www.wrm.org.uy]

———. "Eritrea: Sustainable Forest Use Threatened by Government Policies." *World Rainforest Movement Bulletin* 50 (September 2001): n.p. [http://www.wrm.org.uy]

———. "Indonesia: Mamberamo Dam Threatens Nomadic Tribes." *World Rainforest Movement Bulletin* 49 (August 2001): n.p. [http://www.wrm.org.uy]

———. "Nicaragua: Indigenous People Win Major Legal Battle." *World Rainforest Movement Bulletin* 50 (September 2001): n.p. [http://www.wrm.org.uy]

———. "Philippines: Planting Trees and Terror." *World Rainforest Movement Bulletin* 50 (September 2001): n.p. [http://www.wrm.org.uy]

————. "Sri Lanka: Deforestation, Women and Forestry." *World Rainforest Movement Bulletin* 50 (September 2001): n.p. [http://www.wrm.org.uy]

————. "Venezuela: A Power Line That Kills." *World Rainforest Movement Bulletin* (August 2001): n.p. [http://www.wrm.org.uy]

————. "Zambia: Causes of Deforestation Linked to Government Policies." *World Rainforest Movement Bulletin* 50 (September 2001): n.p. [http://www.wrm.org.uy]

Cattaneo, Claudia. "B.C. Natives Blockade Petro-Can Pipeline Site." *National Post* (Canada), August 14, 2001. [http://www.nationalpost.com/]

"Central Africa: European Union's Major Responsibility over Deforestation." *World Rainforest Movement Bulletin* 53 (December 2001): n.p. [http://www.wrm.org.uy/deforestation/logging.html]

"Changnoi: Not a Picnic for Pak Mool Refugees." *The Nation* (Bangkok), April 21, 1999. [http://www.nextcity.com/ProbeInternational/Mekong/articles/thai9904ii.html]

Chatterjee, Pratap. "Gold, Greed and Genocide in the Americas: California to the Amazon." *Abya Yala News: The Journal of the South and Meso-American Rights Center* (1998): n.p. [http://saiic.nativeweb.org/ayn/goldgreed.html]

————. "Enron In India: The Dabhol Disaster." CorpWatch: Holding Corporations Accountable, July 20, 2000. [http://www.corpwatch.org/issues/politics/featured/2000/enronindia.html

Chesos, Richard. "Carve-Up of Forests Can Go Ahead, Court Rules." *The Nation* (Nairobi, Kenya), October 5, 2001. AllAfrica.com. [http://allafrica.com/stories/200110040519.html]

"Cheyenne River Sioux Tribe Wins Suit against Homestake." *Drillbits and Tailings* 4:11 (July 23, 1999): n.p. [http://www.moles.org/ProjectUnderground/drillbits/4_11/2.html]

Chidumayo, E. N. "Land-cover Transformation in Central Zambia: Role of Agriculture, Biomass, Energy and Rural Livelihoods." Biological Sciences Department, University of Zambia, 2001. [http://coe.asafas.kyotou.ac.jp/news/PastExper2001Jan/PastExper2001_Paper_Chidumayo.htm]

"Chile: Mapuche Indigenous Peoples' March to the Capital City." *World Rainforest Movement Bulletin* 23 (May 1999): n.p. [http://www.wrm.org.uy/bulletin/23/Chile.html]

"Chile: The Struggle of the Pehuenche against the Ralco Dam." *World Rainforest Movement Bulletin* 42 (January 2001): n.p. [http://www.wrm.org.uy/bulletin/42/Chile.html]

Churchill, Ward, and Winona LaDuke. "Native America: The Political Economy of Radioactive Colonialism." *The Insurgent Sociologist* 13:3 (spring 1986): 51–84.

"Clean Coal's Dirty Facts." N.d. [http://www.blackmesais.org/clean_coals_dirty_facts.html]

"Clinton/Gore Administration Backs Occidental Petroleum (Part 2)." *Drillbits and Tailings* 5:8 (May 31, 2000): n.p. [http://groups.yahoo.com/group/graffis-l/message/11105]

"Colombia Faces 'Apocalypse.'" *Native Americas* 18:1 (spring 2001): 4–5.

"Colombia: Flap over Fumigation." *World Press Review*, October 2001, 25.

"Colombian Government Neglect Leaves the Nukak Facing Extinction." *Native Americas* 14:3 (fall 1997): 8.

"Colombian Indigenous People Killed Opposing Dam." Environment News Service, February 10, 2001. [http://ens.lycos.com/ens/feb99/1999L-02-10-01.html]

Colomeda, Lori A. "Indigenous Health." Speech delivered at Salish Kootenai College, Pablo Montana, Brisbane, Australia, September 9, 1998. [http://www.ldb.org/vl/ai/lori_b98.html]

"Communique to the National and International Community in Response to the Discovery of Petroleum in Our Ancestral Territory Kera Chikara." *Drillbits and Tailings* 8:3 (April 11, 2003): n.p. [project_underground@moles.org]

"Community Leader Protesting Dam in Honduras Murdered." Global Response's Quick Response Network. *Cultural Survival News Notes*, July 19, 2001. [http://www.cs.org/main.htm]

Cone, Marla. "Human Immune Systems May be Pollution Victims." *Los Angeles Times*, May 13, 1996, A-1.

"Congo Republic: Increased Logging Activities." *World Rainforest Movement Bulletin* (April 2002): n.p. [http://www.wrm.org.uy]

Connelly, Stephen, and Nikky Wilson. "Stress for Semi-nomadic Farmers: A Key to Resilience." *LEISA Magazine*, April 2001, n.p. [http://www.ileia.org/2/17-1/10-11.PDF]

Connor, Steve. "Death Sentence for the Amazon: Scientists Say $40 Billion Project Is Set to Destroy 95 per cent of Rainforest by 2020." Amazon Watch, 2001. [http://www.amazonwatch.org/newsroom/mediaclips01/010119iuk.html]

Connor, Steve, and Tim Ecott. "Islanders Threatened by Coral Destruction." *London Independent*, October 26, 2000, 19.

"Consortium Officials Violently Remove Journalists and Activists from Company Headquarters." *Drillbits and Tailings* 6:7 (August 31, 2001): n.p. [http://www.moles.org/ProjectUnderground/drillbits/6_07/1.html]

Coon-Come, Matthew. *Remarks of the National Chief Matthew Coon-Come.* Peoples Summit of the Americas Environment Forum, April 18, 2001. [http://www.afn.ca/Press%20Realeses%20%20speeches/april_18.htm]

"Coordinator of Movement against Xingu Dams is Murdered." Amazonwatch, August 25, 2001. [http://www.amazonwatch.org/newsroom/mediaclips01/braz/010825feab.html]

Corbin, Harney. Shoshone Medicine Man. N.p.: Blue Dolphin Publishing, 1996, n.d. [http://liteweb.org/wildfire/corbin.html]

Cornplanter, Edward, and Arthur C. Parker. "The Code of Handsome Lake." In Arthur C. Parker, *Parker on the Iroquois*, with an introduction by William N. Fenton. Syracuse, N.Y.: Syracuse University Press, 1968.

"Costa Rica: Indigenous Territory Threatened by Hydroelectric Dam." *World Rainforest Movement Bulletin* 46 (May 2001): n.p. [http://www.wrm.org.uy/bulletin/46/Costa Rica.html]

"Costa Rica: Opposition to Hydroelectric Dam." *World Rainforest Movement Bulletin* 52 (November 2001): n.p. [http://www.wrm.org.uy]

"Cotingo Dam in Brazil Is Halted, Sparing the Macuxi and the Ingarico." *Native Americas* 12:1–2 (summer 1998): 15.

"Cree Claim Victory in Hydro Hearings." Canadian Broadcasting Corporation, December 1, 2000. [http://www.northstar.sierraclub.org/HYDRO_Cree_Claim%20_%20Victory.htm]

"Cree Indians Win Battle in Their Struggle to Stop Canadian Hydropower Plant." *Omaha World-Herald*, September 12, 1991, 4.

———. *Ecological Imperialism: The Biological Expansion of Europe, 900–1900.* Cambridge, U.K.: Cambridge University Press, 1986.

Cronon, William. *Changes in the Land: Indians, Colonists, and the Ecology of New England*. New York: Hill & Wang, 1983.

Crosby, Alfred W. *The Columbian Exchange: Biological and Cultural Consequences of 1492*. Westport, Conn.: Greenwood Press, 1972.

"Cross Lake Takes Hydro Battle to U.S." *First Perspectives: News of Indigenous Peoples of Canada* (May 29, 2001): n.p. [http://collection.nlc-bnc.ca/100/201/300/first_perspective/2001/05–29/story7.html]

Cruz, Cheryl G. "Runaround Delays Ambuklao, Binga Dams Damages Claims." *Baguio Sun-Star* (Luzon, Philippines), August 10, 2001. [http://www.sunstar.com.ph/baguio/08-10-2001/topstories2.html]

"Dalai Lama Slams Lhasa Rail Link; Basharat Peer in New Delhi." July 21, 2001. [http://www.tibet.net/eng/diir/enviro/news]

"Dam Affected People Occupy Tractebel Headquarters in Rio." Environment News Service, October 31, 2001. [http://ens-news.com/ens/oct2001/2001L-10-30-03.html]

"Dam Protesters Occupy Brazil's Ministry of Energy." Environment News Service, March 14, 2001. [http://ens.lycos.com/ens/mar2001/2001L-03–14–02.html]

"The Damming of the Mamberamo River in West Papua Threatens Villages." *Cultural Survival News Notes*, August 17, 2001 [http://www.cs.org/main.htm]

Daters, Michaela Reuss Daters. "Taking a Job at Jabiluka." *Sydney City Hub*, December 31, 1997. [http://www.big.com.au/film/jab31–12–97.html

Dayaneni, Gopal. "Field Notes and Reflections from a Project Underground Staffer's Trip to the Niger Delta: March 22 to April 10, 2002." *Drillbits and Tailings* 7:4 (April–May, 2002): n.p. [http://www.moles.org]

"De Beers and Anglo Linked to Forced Removal of Last Kalahari Bushmen." *Drillbits and Tailings* (October 7, 1997): n.p. [http://www.moles.org/ProjectUnderground/drillbits/971007/97100704.html

Declaration of the First International Forum of Indigenous Peoples on Climate Change. Lyon, France, September 4–6, 2000. [http://www.ienearth.org/climate_1-p2.html]

Deloria, Vine, Jr. *God is Red*. Golden, Col.: North American Press, 1992.

"Demand Justice for Mapuche Youth Brutally Repressed for Protesting Against REPSOL." *Drillbits and Tailings* 6:9 (November 30, 2001): n.p. [www.moles.org]

Dennis, Matthew. *Cultivating a Landscape of Peace: Iroquois-European Encounters in Seventeenth Century America*. Ithaca, New York: Cornell University Press, 1993.

Dewailly, Eric, P. Ayotte, C. Blanchet, J. Grodin, S. Bruneau, B. Holub, and G. Carrier. "Weighing Contaminant Risks and Nutrient Benefits of Country Food in Nunavik." *Arctic Medical Research* 55, supp. 1 (1996): 13–19.

Dewailly, E., P. Ayotte, S. Bruneau, S. Gingras, M. Belles-Isles, and R. Roy. "Susceptibility to Infections and Immune Status in Inuit Infants Exposed to Organochlorines." *Environment Health Perspectives* 108 (2000): 205–211.

Dewailly, E., S. Bruneau, C. Laliberte, M. Belles-Iles, J.-P. Weber, and R. Roy. "Breast Milk Contamination by PCB and PCDD/Fs in Arctic Quebec. Preliminary Results on the Immune Status of Inuit Infants." *Organohalogen Compounds* 13 (1993): 403–6.

Dewailly, E., S. Dodin, R. Verreault, P. Ayotte, L. Sauve, and J. Morin. "High Organochlorine Body Burden in Breast Cancer Women with Oestrogen Receptors." *Organohalogen Compounds* 13 (1993): 385–88.

Dewailly, E., J. J. Ryan, C. Laliberte, S. Bruneau, J.-P. Weber, S. Gringras, and G. Carrier. "Exposure of Remote Maritime Populations to Coplanar PCBs." *Environmental Health Perspectives* 102, supp. 1 (1994): 205–9.

Dobbyn, Paula. "Contaminated Game has Natives Worried." *Anchorage Daily News*, May 2, 2000. [http://www.adn.com] [http://www.ienearth.org/food_toxic.html]

"Dogrib First Nation Fights Back Diavik's Land Grab." *Drillbits and Tailings* 4:18 (November 9, 1999): n.p. [http://www.moles.org/ProjectUnderground/drillbits/4_18/3.html]

Diaz-Romo, Patricia, and Samuel Salinas-Alvarez. "Migrant Workers and Pesticides. A Poisoned Culture: the Case of the Indigenous Huicholes Farm Workers." *Abya Yala News: The Journal of the South and Meso-American Rights Center* (n.d.): n.p. [http://saiic.nativeweb.org/ayn/huichol.html]

"A Dictatorship-type Action Gives Aracruz a Spurious Victory." *World Rainforest Movement Bulletin* 11 (April 1998): n.p. [http://www.wrm.org.uy/bulletin/11/Brazil.html]

"Dig In! Waste Flow, Blood Flow!" *Kerebok* 2:10 (May 2001): n.p. [http://www.jatam.org/xnewsletter/kk10.html]

"A Disaster Looms for Communities in Mariinduque, Philippines." *Drillbits and Tailings* 7:4 (April–May 2002): n.p.[http://www.moles.org]

Dobbyn, Paula. "Contaminated Game has Natives Worried." *Anchorage Daily News*, May 2, 2000. [http://www.adn.com] [http://www.ienearth.org/food_toxic.html]

"Dogrib First Nation Fights Back Diavik's Land Grab." *Drillbits and Tailings* 4:18 (November 9, 1999): n.p. [http://www.moles.org/ProjectUnderground/drillbits/4_18/3.html]

Donnelly, R. "The Curse of Xawara." *Socialist Standard* (Great Britain), June 1998. [http://www.worldsocialism.org/xawara.htm]

Dozier, Edward P. *The Pueblo Indians of North America*. New York: Holt, Rinehart & Winston, 1970.

Drillbits and Tailings 6:7 (August 31, 2001): n.p. [http://groups.yahoo.com/group/protecting_knowledge/message/1770] [http://www.moles.org/ProjectUnderground/motherlode/chevron/chevinfo.html.]

Duffy, Andrew. "Global Warming Causing Arctic Town to Sink, Says Inuit Leader—'Warning Signal.'" *Montreal Gazette*, April 18, 2000. [http://www.climateark.org/articles/2000/2nd/glwatosi.htm]

Dumont, C. *Proceedings of 1995 Canadian Mercury Network Workshop. Mercury and Health: The James Bay Cree*. Cree Board of Health and Social Services, Montreal, 1995. [http://www.cciw.ca/eman-temp/reports/publications/mercury95/part4.html]

Dumont, Charles, Manon Girard, François Bellavance, and Francine Noël. "Mercury Levels in the Cree Population of James Bay, Quebec, from 1988 to 1993/94." *Canadian Medical Association Journal* 158 (June 2, 1998):1439–45. [http://www.cma.ca/cmaj/vol-158/issue-11/1439.htm]

Dutter, Barbie. "Islanders Plan Their Flight from Rising Sea." *London Daily Telegraph*, July 16, 2001, 8.

"Earth Day with Ka Hsaw Wa: Everybody Belongs." Environment News Service, April 7, 2000. [http://ens.lycos.com/ens/apr2000/2000L-04-07-01.html]

"Earthpulse: Cultural Extinctions Loom." *National Geographic* (September 2001): n.p.

"Ecuador Asks Colombia to Halt Aerial Coca Fumigation." *Cultural Survival News Notes*, July 26, 2001. [http://www.cs.org/main.htm]

"Ecuador: Human Shield in Defense of the Mindo Forest against Oil Pipeline." *World Rainforest Movement Bulletin* 54 (January 2002): n.p. [http://www.wrm.org.uy]

"Ecuadorian President Approves Pipeline Backed by Occidental." *Drillbits and Tailings* 6:5 (June 30, 2001): n.p. [http://www.moles.org/ProjectUnderground/drillbits/6_05/3.html]

"Ecuador: O.C.P. Protesters in Amazon Attacked by Military." *Drillbits and Tailings* 7:2 (February 28, 2002): n.p. [http://www.moles.org]

"Ecuador's Highest Court Upholds $1.1 Billion Dollar Oil Pipeline." *Cultural Survival News Notes*, August 30, 2001. [http://www.cs.org/main.htm]

Egan, D'arcy. "Dam the Ecology if Quebec Goes Full-speed Ahead." *Cleveland Plain Dealer*, July 29, 2001, D-15.

Egan, Timothy. "Eskimos Learn They've Been Living amid Secret Pits of Radioactive Soil." *New York Times*, December 6, 1992, A-26.

Eichstaedt, Peter. *If You Poison Us: Uranium and American Indians*. Santa Fe, N.M.: Red Crane Books, 1995.

Ekine, Sokari. *Blood & Oil: Testimonies of Violence from Women of the Niger Delta*. Port Harcourt, Nigeria: Niger Delta Women for Justice, 2002.

Elton, Catherine. "Pristine Amazon Jungle Threatened by Big Oil Firm This Month; Peru Granted Mobil a Deadline Extension on a Decision." *Christian Science Monitor*, September 30, 1999.

Emmons, Della Gould. *Leschi of the Nisquallies*. Minneapolis: T. S. Denison, 1965.

"Energy Chief Over Yucca Mountain." Environment News Service, January 7, 2002. [http://ens-news.com/ens/jan2002/2002L-01-07-02.html]

"Europe Rejects Brazilian Mahogany Imports." Environment News Service, March 29, 2002. [http://ens-news.com/ens/mar2002/2002L-03-29-01.html]

Ewen, Alex. "Indians in Brazil: Is Genocide Inevitable?" *Native Americas* 13:4 (winter 1996): 12–23.

Ewen, Alexander. *Voices of Indigenous Peoples: Native People Address the United Nations*. Santa Fe, N.M.: Clear Light, 1994.

Ewing, Rodney C., and Allison Macfarlane. "Nuclear Waste: Yucca Mountain." *Science* 296 (April 26, 2002): 659–60.

"ExxonMobil Sued For Atrocities in Indonesia." *Cultural Survival News Notes*, June 21, 2001. [http://www.cs.org/main.htm]

———. "Gold Mine Threatens Quechan Sites." *Native Americas* 15:2 (summer 1998): 6.

Faisal, Agus. "Indonesia: Togean People Defend Their Forests, Lands, and Ocean." *World Rainforest Movement Bulletin* 52 (November 2001): n.p. [http://www.wrm.org.uy]

Fellman, Gordon. *Rambo and the Dalai Lama: The Compulsion to Win and its Threat to Human Survival*. Albany: State University of New York Press, 1998.

Fernandez. Lydia. "Dispute over Indigenous Territory in Colombia Leads to Assassinations and Death Threats." *Native Americas* 16:1 (spring 1999): n.p. [http://www.yvwiiusdinvnohii.net/News99/0399/NAJ990311Colombia.htm]

"Figure This One!" *Four Corners Clamor* 1:6 (summer–fall 1998): n.p. [http://www.ausbcomp.com/redman/clamor6.htm#newdeal]

Fixico, Donald L. *The Invasion of Indian Country in the Twentieth Century: American Capitalism and Tribal Natural Resources*. Niwot, Col.: University Press of Colorado, 1998.

Forberg, S., O. Tjelvar, and M. Olsson. "Radiocesium in Muscle Tissue of Reindeer and Pike from Northern Sweden before and after the Chernobyl Accident: A Retrospective Study on Tissue Samples from the Swedish Environmental Specimen Bank." *Science of the Total Environment* 115 (1992): 179–89.

"Forest Beings: Indigenous Peoples of Sri Lanka." 1999. [http://www.global-vision. org/srilanka/]

Forrester, Vincent. "Uranium Mining and Aboriginal People." The Sustainable Energy and Anti-Uranium Service, Inc., Australia, 1997. [http://www.sea-us.org.au/black uranium.html]

Foushee, Lea. "Acid Rain Research Paper." Indigenous Women's Network, Testimony before the International Council of Indigenous Women, Samiland, Norway, August, 1990.

Free Trade Unions of Burma. "Forced Labour at Indo-Burma Border." BurmaNet, March 1997. [http://www.rainforestrelief.org/reports/teak_tort.html]

"Freeport-McMoran Admits Funding Millions to Indonesian Military." *Drillbits and Tailings* 8:3 (April 11, 2003): n.p. [project_underground@moles.org]

Fujii, Miki. "Life Among Borneo's Ethnic Tribes." *Daily Yomiuri* (Tokyo), April 7, 2001, 10.

Funari, Ricardo, and Jennifer Hanna. "Recovering Lands; Recovering Futures." *Native Americas* 18:1 (spring 2001): 38–41.

Gargan, Edward A. "Lust for Teak Takes Grim Toll; Illegal Logging Decimating Indonesia's Majestic Forests." *Newsday*, June 25, 2001, A-7.

Gasperini, William. "In a Land 'Back of the Beyond' Reindeer Rule the Nomad's Life." *Christian Science Monitor*, April 25, 1997. [http://www.britannica.com/magazine/ article?content_id=24234&pager.offset=60]

Gearing, Julian. "The Struggle for the Highlands: Accused of Endangering the Environment, Thailand's Tribespeople Face Eviction and an Uncertain Future." *Asiaweek* 25:43 (October 29, 1999): n.p. [Asiaweek.com]

Gedicks, Al. *The New Resource Wars: Native and Environmental Struggles against Multinational Corporations*. Boston: South End Press, 1993.

————. *Resource Rebels: Native Challenges to Mining and Oil Corporations*. Cambridge, Mass.: South End Press, 2001.

Gentile, Gary. "Total Denial Continues." EarthRights International, May 2000.

Ghanaian Communities Hit Hard by Two Cyanide Spills." *Drillbits and Tailings* 6:9 (November 30, 2001): n.p. [www.moles.org]

"Ghana: I.M.F., Mining and Logging." *World Rainforest Movement Bulletin* 54 (January 2002): n.p. [http://www.wrm.org.uy]

"Ghana: N.G.O.s Criticise Gov't Handling of Cyanide Spillage." *Ghanaian Chronicle*, November 21, 2001.

Gilday, Cindy Kenny. "A Village of Widows." *Arctic Circle*. N.d. [http://arcticcircle. uconn.edu/SEEJ/Mining/gilday.html]

Giuliano, Jackie Alan. "Killing Tomorrow for a Few Megawatts Today." Environment News Service, March 2001. [http://ens.lycos.com/ens/mar2001/2001L-03-30g.html]

"Globalization Threat to Cultural, Linguistic and Biological Diversity—UNEP." *Businessworld* (Philippines), February 15, 2001, n.p. (in LEXIS)

Global Response: Environmental Education and Action Network. October, 2001. [http://www.globalresponse.org/gra/current.html]

Goldberg, Jeffrey. "The Great Terror: in Northern Iraq, There is New Evidence of Saddam Hussein's Genocidal War on the Kurds, and of its Possible Ties to Al Quaeda." *The New Yorker* (March 25, 2002): 52–73.

"Gold Fever Threatens Forests and People in Suriname." *World Rainforest Movement Bulletin* 3 (August 1997): n.p. [http://www.wrm.org.uy/bulletin/3/Suriname.html]

"Gold Mining in Honduras Project." Center for Economic and Social Rights, May 11, 2001. [http://www.cesr.org/PROGRAMS/honduras.htm]

Goldtooth, Tom. "Indigenous Environmental Network Statement to Nora Helton, Chairwoman, Fort Mojave Indian Tribe, Other Lower Colorado River Indian Tribal Leaders, Tribal Community Members, Elders and Non-Native Groups and Individuals, in Reference to the Ward Valley Victory Gathering to Celebrate the Defeat of a Proposed Nuclear Waste Dump." News release, Ward Valley, California, February 16, 2001. [http://www.ienearth.org/ward_valley2.html]

Golovnev, Andrei V., and Gail Osherenko. *Siberian Survival: The Nenets and Their Story*. Ithaca, N.Y.: Cornell University Press, 1999.

"Governor Leavitt Opens Office High-level Nuclear Waste Opposition." News Release: Office of the Governor, State of Utah, Salt Lake City, December 7, 2000. [http://www.governor.state.ut.us/html/nukewaste.html]

"Great Whale Still Dead." January 22, 2001. [Portfolio@newswire.ca]

Green, Sara Jean. "Fighting a GIANT." *Windspeaker*, n.d.. [http://www.ammsa.com/classroom/CLASS3Lubicon.html]

"Green Smokescreen for Eviction of Forest People." *Survival Newsletter* 42 (2000): 1.

Grinde, Donald A., Jr., and Bruce E. Johansen. *Ecocide of Native America: Environmental Destruction of Indian Lands and Peoples*. Santa Fe, N.M.: Clear Light, 1995.

———. *Exemplar of Liberty: Native America and the Evolution of Democracy*. Los Angeles: American Indian Studies Center, 1991.

Grossman, Zoltan. Personal communication with author. Wisconsin Campaign to Ban Cyanide in Mining and the Midwest Treaty Network, November 6, 2001.

"Guarani-Kaiowa of Brazil Win Land Rights." Survival International Press Release, November 13, 2002. [http://www.survival-international.org/enews.htm]

"Guatemala: A Dam and the Massacre of 400 People." *World Rainforest Movement Bulletin* 42 (January 2001): n.p. [http://www.wrm.org.uy/bulletin/42/Guatemala.html]

"Guatemala Massacre Related to World Bank Project Under Scrutiny." *Native Americas* 13:2 (summer 1996): 7.

"Guatemala: Security for Shrimps; Insecurity for the Local Population." *World Rainforest Movement Bulletin* 48 (July 2001): n.p. [http://www.wrm.org.uy/bulletin/48/Guatemala.html]

Guerette, Deb. "No Clear-cut Answer: Timber Rights Allocation on Lubicon Land a Worrisome Development." *Grande Prairie Daily Herald Tribune*, March 5, 2001. [http://www.tao.ca/~fol/Pa/negp/ht010305.htm]

Gutierrez, Estrella. "Venezuela—Indigenous: Waraos Analyze Impact of Oil Drilling." Interpress Service, March 20, 1997. [http://www.oneworld.org/ips2/mar/venezuela.html]

"Guyana: Transnational Mining Companies' Impacts on People and the Environment." *World Rainforest Movement Bulletin* 43 (February 2001): n.p. [http://www.wrm.org.uy/bulletin/43/Guyana.html]

Hahn, Tamar. "Inuit Activist Puts Face to Global Warming." Earth Times News Service, November 16, 2000. [http://www.earthtimes.org/nov/profileinuitactivistputsnov16_00.htm]

Hamilton, Dominic. "Pemon Indians Paralyze Powerlines." *Native Americas* 14:3 (fall 1997): 9.

Hanna, Jennifer. "The Guarani are Committing Suicide: Mato Grosso de Sul, Brazil." *Native Americas* 14:1 (spring 1997): 32–41.

Harding, Jim. "Uranium Mining: Northern Development or the New Colonialism?" Regina Group for a Non-Nuclear Society, 1981. [http://www.accesscomm.ca/users/kmactaggart/rgnns4.htm]

Hardtke, Marcus. "Cambodia: The Forestry Sector Reform and the Myth of a Sustainable Logging Industry." Global Witness, Phnom Penh, 2000. [http://www.oneworld.org/globalwitness/reports/credibility/credibility.htm]

"Harper's Index." *Harper's Magazine*, October 2001, 13.

"Havasupai Fight To Save Grand Canyon From Uranium Mining." July 21, 1992. [http://www.ratical.org/ratville/native/havasuEFN.txt]

Heavy Metal Levels in Reindeer, Caribou, and Plants of the Seward Peninsula. Current Research Programs, Reindeer Research Program: University of Alaska at Fairbanks, April, 2000. [http://reindeer.salrm.alaska.edu/research.htm]

Herbert, H. Josef. "Inuit Say They Are Witness to Global Warming in the Arctic." *Milwaukee Journal-Sentinel*, November 19, 2000. [http://www.jsonline.com/alive/news/nov00/warming20111900.asp?format=print]

Hernandez, Silvio. "Ngobe-Bugle Want Land and Independence." Interpress Service, March 28, 1995. [http://www.stile.lboro.ac.uk/~gyedb/STILE/Email0002069/m21.html]

Hertz, Noreena. *The Silent Takeover: Global Capitalism and the Death of Democracy.* New York: The Free Press, 2001.

Hill, Miriam. "Iqaluit's Waste Woes Won't Go Away; City Sets Up Bins Where Residents Can Dump Plastics, Metal." *Nunatsiaq News*, July 27, 2001, 5.

Hodge, Damon. "Dumpy Meeting: Final Vegas Hearing on Yucca Mountain Plagued by the D.O.E.." *Las Vegas Weekly*, September 11, 2001. [http://www.lasvegasweekly.com/2001_2/09_13/news_upfront7.html]

Hodgson, Martin. "Colombia Halts Drug Poison Spraying." *The Guardian* (London), July 31, 2001, 13.

Hoffman, J. J. Wells and J. Titus. *Projecting Future Sea-level Rise.* Washington, D.C.: Environmental Protection Agency, 1983.

Horta, Korinna. *Questions Concerning The World Bank and Chad/Cameroon Oil and Pipeline Project. Makings of a New Ogoniland? Corporate Welfare Disguised as Aid to the Poor?* New York: Environmental Defense Fund, 1997. [http://www.edf.org/pubs/Reports/c_chadcam.html]

"Hotspots: Australia." *Drillbits and Tailings* 6:7 (October 31, 2001): 3. [englishdrillbits@topica.com]

"Hotspots: Australia." *Drillbits and Tailings* 7:1 (January 31, 2002): n.p. [englishdrillbits@topica.com]

"Hotspots: Bougainville." *Drillbits and Tailings* 6:9 (November 30, 2001): n.p. [www.moles.org]

"Hotspots: Burma." *Drillbits and Tailings* 8:3 (April 11, 2003): n.p.[project_underground@moles.org]

"Hotspots: California." *Drillbits and Tailings* 6:9 (November 30, 2001): n.p. [www.moles.org]

"Hotspots: Colombia." *Drillbits and Tailings* 6:7 (August 31, 2001): 3. [www.moles.org]

"Hotspots: Ecuador." *Drillbits and Tailings* 6:7 (August 31, 2001): 3. [www.moles.org]

"Hotspots: Ghana." *Drillbits and Tailings* 6:7 (October 31, 2001): 3. [www.moles.org]

"Hotspots: Idaho." *Drillbits and Tailings* 6:7 (October 31, 2001): 3. [www.moles.org]

"Hotspots: Nigeria." *Drillbits and Tailings* 6:9 (November 30, 2001): n.p. [www.moles.org]

"Hotspots: Papua New Guinea." *Drillbits and Tailings* 6:9 (November 30, 2001): n.p. [www.moles.org]

"Hotspots: Wisconsin." *Drillbits and Tailings* 6:9 (November 30, 2001): n.p. [www.moles.org]

"Hotspots: Wisconsin." *Drillbits and Tailings* 7:8 (October 2, 2002): n.p. [http://www.moles.org]

Huey, Lois M., and Bonnie Pulis. *Molly Brant: A Legacy of Her Own.* Youngstown, N.Y.: Old Fort Niagara Association, 1997.

Hughes, J. Donald. *American Indian Ecology.* El Paso, Tex.: Texas Western Press, 1983.

Hulen, David. "Hunt Is on for Pollutant Traces in Bering Sea; Alaska Villagers, Scientists Wonder if Toxic Substances are Endangering Animals and People Who Eat Them." *Los Angeles Times*, August 15, 1994, A-5.

"Human Rights Group Says Myanmar Still Using Forced Labor." Associated Press, March 8, 2001.

Hunt, George T. *The Wars of the Iroquois: A Study in Intertribal Trade Relations.* Madison: University of Wisconsin Press, 1940.

Hutton, D., and L. Connors, L. A *History of the Australian Environmental Movement.* Cambridge, U.K.: Cambridge University Press, 1999.

"Hydroelectric Production Is Anything but Cheap." *Virtual Circle: First Nations' Chronicles*, March 21, 2001. [http://www.vcircle.com/journal/archive/010.shtml]

"I.D.R.C., Knowledge and Development." N.d. [http://idrinfo.idrc.ca/Archive/ReportsINTRA/pdfs/v19n1e/108939.pdf]

Igrain, Konstantin. "Sakhalin: Oil and Indigenous Peoples." L'auravetl'an Indigenous Information Center (Moscow). Native Net, April 24, 1997. [http://www.nativenet.uthscsa.edu/archive/nl/9704/0103.html]

Images Asia. "Forced Labor and Human Rights Abuses." March 1996. [http://www.rainforestrelief.org/reports/teak_tort.html]

"Inco Threatens Indigenous Kanuks and Environment of New Caledonia." *Drillbits and Tailings* 7:3 (March 29, 2002): n.p. [www.moles.org]

"India: Mining Ancestral Lands for Corporate Profits." *World Rainforest Movement Bulletin* (November 2001): n.p. [http://www.wrm.org.uy]

"Indian Police Kill Three Protesting Mining by Canadian Alcan Inc." *Drillbits and Tailings* 6:5 (January 31, 2001): n.p. [http://www.moles.org/ProjectUnderground/drillbits/6_01/4.html]

"Indians and Police Clash in Panama over Mining." *Native Americas* 12:3 (fall 1995): 7–8.

"Indians Take Dam Protest to Chilean President." Reuters News Service, March 8, 2002.

"India: Road Brings Death to Isolated Tribe." Survival International, February, 2002. [http://www.survival-international.org/jarawauab0202.htm]

"India's Supreme Court Closes Isolated Jarawa Tribe's 'Road of Death.'" Survival International, May 27, 2002. [http://www.survival-international.org/enews.htm]

"Indigenous Peoples Fight for Territorial Rights in Guyana." *World Rainforest Movement Bulletin* 17 (November 1998): n.p, [http://www.wrm.org.uy/bulletin/17/Guyana.html]

"Indigenous People Suffering Gross Rights Violation!" *Free Press Journal*, August 9, 2000. [http://www.indiaworld.co.in/news/features/feature495.html]

"Indigenous Peruvians Mobilize while ExxonMobil Further Explores Rainforest." *Drill-bits and Tailings* 5:12 (July 20, 2000): n.p. [http://www.moles.org/ProjectUnderground/drillbits/5_12/2.html]

"Indigenous Protest in Honduras." *Cultural Survival International News Notes*, July 19, 2001. [http://www.cs.org/main.htm]

"Indigenous Small-scale Mining under Threat." *Down to Earth* 48 (February 2001): n.p. [http://www.gn.apc.org/dte/48ssm.htm]

"Indonesia: Low Expectations on Log Export Ban." *World Rainforest Movement Bulletin* 53 (December 2001): n.p. [http://www.wrm.org.uy/deforestation/logging.html]

"Indonesian Man Shot at Australian Gold Mine." Environment News Service, January 22, 2002. [http://ens-news.com/ens/jan2002/2002L-01-23-01.html]

"Indonesian President Calls for Logging Halt." Environment News Service, May 27, 2002. [http://ens-news.com/ens/may2002/2002-05-27-19.asp#anchor2]

"Innu Begin Occupation to Halt Construction At Voisey's Bay." *Drillbits and Tailings* (August 21, 1997): n.p. [http://www.moles.org/ProjectUnderground/drillbits/970821/97082104.html]

"The Innu Nation: Background Information." August 2000. [http://www.ienearth.org/military_impacts.html

"Innu Nation Launches Court Challenge to Military Plans for Supersonic Test Flights over Innu Lands." Indigenous Environmental Network, August 8, 2000. [http://www.ienearth.org/military_impacts.html

"Innu Respond to N.F. Hydro/Alcoa Announcement." *Cultural Survival News Notes*, July 26, 2001. [http://www.cs.org/main.htm]

Interpress Service. *Story Earth: Native Voices on the Environment.* San Francisco: Mercury House, 1993.

Iqrar Haroon, Agha. "Save Kalash Valley Before It Is Too Late." Ecotourism Society Pakistan, n.d. [http://www.ecoclub.com/news/19.html]

Isbister, John. *Capitalism and Justice: Envisioning Social and Economic Fairness.* Bloomfield, Conn.: Kumarian Press, 2001.

"Is Certification the Solution?" *World Rainforest Movement Bulletin* 53 (December 2001): n.p. [http://www.wrm.org.uy/deforestation/logging.html]

"Isseneru Villagers Face Risk of Mercury Contamination. *Cultural Survival News Notes*, July 19, 2001. [http://www.cs.org/main.htm]

"Is This the Way to Save Africa's Wildlife? National Wildlife Federation: Reader Follow-up." July/August, 1995. [http://www.nwf.org/internationalwildlife/ltrja95.html]

Iverson, Peter. "Taking Care of the Earth and Sky." In Alvin Josephy, ed., *America in 1492: The World of the Indian Peoples Before the Arrival of Columbus.* New York: Alfred A. Knopf, 1992, 85–118.

Jacobson, Joseph L., and Sandra W. Jacobson, "Intellectual Impairment in Children Exposed to Polychlorinated Biphenyls *in Utero*," *New England Journal of Medicine* 335:11 (September 12, 1996): 83–789.

Jaimet, Kate. "Global Warming 'Lethal' to Rare Northern Caribou." *Ottawa Citizen*, November 30, 2000, A-4.

———. "Placer Dome Blamed in 'World-calibre Disaster'; Canadian Mining Giant Says Philippine Government Blocking Cleanup." *Ottawa Citizen*, January 29, 2002, n.p.

"Japan Imports African Ivory." British Broadcasting Corporation News, July 16, 1999. [http://news.bbc.co.uk/hi/english/world/africa/newsid_396000/396123.stm

Jayaraman, Nityanand. "Norsk Hydro: Global Compact Violator." Corporate Watch, 2001. [http://www.igc.org/trac/un/updates/2001/norskhydro.html]

Johansen, Bruce. "Arctic Heat Wave." *The Progressive*, October 2001, 18–20.

———. "A Sensible, Impossible Fishing Solution." *Argus* (Seattle), April 7, 1978, 3, 12.

———. "Ward Valley: A 'Win' for Native Elders." *Native Americas* 15:2 (summer 1998): 7–8.

Johansen, Bruce, and Roberto Maestas. *Wasi'chu: The New Indian Wars*. New York: Monthly Review Press, 1979.

Johansen, Bruce E. "Exxon vs. Chippewas in Wisconsin." *Native Americas* 15:1 (spring 1998): 6.

———. *Life and Death in Mohawk Country*. Golden, Col.: North American Press/Fulcrum, 1993.

Johansen, Bruce E. "Pristine No More: The Arctic, Where Mother's Milk Is Toxic." *The Progressive*, December 2000, 27–29.

———. "The Right to One's Own Home: The Seminole Chickee Sustains Despite County Codes." *Native Americas* 13:3 (fall 1996): 44–47.

———. "U'was in Colombia Say They Will Commit Suicide if Occidental Drills." *Native Americas* 16:1 (spring 1998): 5.

Johansen, Bruce E., and Donald A. Grinde Jr. *The Encyclopedia of Native American Biography*. New York: Henry Holt, 1997.

Johnson, Guy. "Proclamation, October 4, 1774." In James Sullivan, ed., *The Papers of Sir William Johnson*. 14 vols. Albany, N.Y.: University of the State of New York, 1921–1965, 8:683–84.

"Johnston Atoll to be Used as Site for Contaminated U.S. Military Waste." Pacific Concerns Resource Center, New Zealand, May 5, 2000. [http://www.converge.org.nz/pma/jmil.htm]

Jordan, Miriam. "Brazilian Mahogany: Too Much in Demand; Illegal Logging, Exports Are Lucrative for Criminals, Disastrous for Rainforest." *Wall Street Journal*, November 14, 2001, B-1, B-4.

"Judge Rules Local Agreement on Crandon Mine Is Valid." *Milwaukee Journal-Sentinel*, March 8, 2001. [http://www.jsonline.com/WI/030801/wi-crandonmine030801174314.asp]

Kakuchi, Suvendrini. "Protests Hold Back Philippine Dam Project." Interpress Service, Corporate Watch, November 28, 2000 (in Japanese). [cwj@corpwatch.org] [http://www.jca.ax.apc.org/web-news/corpwatch-jp/132.html]

Kaltim, Jatam. "Kronologis Jebol dan Meluapnya Air Limbah di Rapak Lama, Marangkayu." *Kompas Daily* (February 16, 2000): n.p.

Kanesta, George. "No Mine, No Way—Stay Away SRP!" *The Shiwi Messenger: A Public Forum By, Of and For the Zuni Community*, August 3, 2001. [http://www.zunispirits.com/messenger/080301editorial.html]

Karen Human Rights Group. "Interviews About Shan State." July 27, 1996. [http://www.rainforestrelief.org/reports/teak_tort.html]

———. "Ye-Tavoy Railway Area: An Update." July 31, 1996. [http://www.rainforestrelief.org/reports/teak_tort.html]

Kauffman, P. *Wik, Mining and Aborigines*. Sydney, Australia: Allen & Unwin, 1999.

"Kayapo and Yanomami Battle Miners." *Native Americas* 12:1–2 (summer 1998): 14.

Keating, Tim. "Forced-labor Logging in Burma. Draft; Second in the Rainforest Relief Reports Series of Occasional Papers; In Cooperation with the Burma UN Service

Office of the National Coalition Government of the Union of Burma." June 1997. [http://www.rainforestrelief.org/reports/teak_tort.html]

Keefe, Tom, Jr. "A Tribute to David Sohappy." *Native Nations* (June–July 1991): 4–6.

Kemp, Elizabeth, ed. *The Law of the Mother: Protecting Indigenous Peoples in Protected Areas*. San Francisco: Sierra Club Books, 1993.

"Kenya: Forest Destruction for the Benefit of Government Cronies." *World Rainforest Movement Bulletin* 55 (February 2002): n.p. [wrm@wrm.org.uy]

"Kenyan Government Set to Destroy Honey-hunting Tribe." Survival International, December 5, 2001. [is@survival-international.org]

"Kenyan Honey-hunters' Land to be Carved Up?" Survival International, n.d. [http://www.survival-international.org/index2.htm]

"Kenyan President Implicated in Tribal Forest Land Grab." Survival International, February 11, 2002. [http://www.survival-international.org]

Kerebok 2:10 (May 2001): n.p. [http://www.jatam.org/xnewsletter/kk10.html]

Kettl, Paul A. " Suicide and Homicide: The Other Costs of Development." *Akwe:kon Journal* 8:4 (winter 1991): n.p.

Khandy, Vershina. [Speaking for the] Upper Khanda Band of the Evenk, April 27, 2001. Cited in "L'Auravetl'an: An Indigenous Information Center by Indigenous Peoples of Russia." United Nations Information Center, Moscow, April 27, 2001. [http://www.indigenous.ru/english/english.htm]

Kimerling, Judith. "Oil, Lawlessness, and Indigenous Struggles in Ecuador's Oriente." In Helen Collinson, ed., *Green Guerillas: Environmental Conflicts and Initiatives in Latin America and the Caribbean*. London: Latin America Bureau, 1996.

Knickerbocker, Brad. "To Some Here, Global Warming Seems Real Today." *Christian Science Monitor*, August 1, 2001, 9.

Knight, Danielle. "Amazon Indigenous Groups Oppose Infrastructure Projects." Interpress Service, January 18, 1998. [http://www.amazonwatch.org/newsroom/mediaclips/jan1898ips.html]

———. "Indigenous Groups 'at War' with U.S. Oil Giant." Interpress Service, World News, July 1999. [http://www.oneworld.org/ips2/july99/04_04_005.html]

———. "Inuit Tell Negotiators of Climate Change Impact." Interpress Service, November 16, 2000. [http://www.oneworld.org/ips2/nov00/01_44_005.html]

Kolar, Michael J. "Waste That Won't Go Away; It's Time for Washington to Store Nuclear Leftovers at Yucca Mountain." *Pittsburgh Post-Gazette*, August 15, 2001, A-13.

Korkeatkachorn, Wipaphan. "'Son of a Commoner' Faces the Assembly of the Poor." In *Focus on the Global South*. Bangkok, Thailand: 2000. [http://www.focusweb.org/publications/2000/Son%20of%20a%20commoner%20faces%20the%20Assembly%20of%20the%20Poor.htm]

Krech, Shepard, III. *The Ecological Indian: Myth and History*. New York: W. W. Norton, 1999.

Kristof, Nicholas. "For Pacific Islanders, Global Warming Is No Idle Threat." *New York Times*, March 2, 1997. [http://sierraactivist.org/library/990629/islanders.html]

Kurath, Gertrude P. *Iroquois Music and Dance: Ceremonial Arts of Two Seneca Longhouses*. Bulletin 187, Bureau of American Ethnology. Washington, D.C.: Smithsonian Institution, 1964.

LaDuke, Winona. *All Our Relations: Native Struggles for Land and Life*. Cambridge, Mass.: South End Press, 1999.

———. "Breastmilk, P.C.B.s, and Motherhood: An Interview With Katsi Cook, Mohawk." *Cultural Survival Quarterly* 17:4 (winter 1994): 43–45.

———. "Commentary: Blast from the Past: Bush's Energy Policy." *The Circle: Native American News and Arts* 22:9 (September 2001): n.p. [http://www.thecirclenews.org/092001/news2.html]

———. "Environmental Work: An Indigenous Perspective." *Akwe:kon Journal* 8:4 (winter 1991): n.p.

———. "The Future of Yellowstone Buffalo." In *Honor the Earth*. N.p.: n.d. [http://honorearth.org/buffalo/future.html]

———. "Indigenous Environmental Perspectives: A North American Primer." *Native Americas* 9:2 (summer 1992): 52–71.

———. "The Indigenous Women's Network: Our Future, Our Responsibility. Statement of Winona LaDuke, Co-Chair Indigenous Women's Network, Program Director of the Environmental Program at the Seventh Generation Fund, at the United Nations Fourth World Conference on Women. Beijing, China, August 31, 1995. [http://www.igc.org/beijing/plenary/laduke.html]

———. "Insider Essays: Our Responsibility." Electnet/Newswire, October 2, 2001. [http://www.electnet.org/dsp_essay.cfm?intID=28]

"Lake Powell Is Silting In." Environment News Service, January 15, 2002. [http://ens-news.com/ens/jan2002/2002L-01-15-09.html]

Lalonde, Michelle. "Natives to Appeal Mining Decision." *Montreal Gazette*, December 17, 2001. [http://www.nationalpost.com/]

Lama, Abraham. "Indigenous Peoples, the Invisible Victims of War." NativeNet, March 26, 1995. [http://nativenet.uthscsa.edu/archive/nl/9503/0346.html]

Lamb, David Michael. "Toxins in a Fragile Frontier." Transcript, Canadian Broadcasting Corporation News, n.d. [http://cac.ca/news/indepth/north/]

"Landmark Victory for Indians in International Human Rights Case Against Nicaragua." Press Release, Indian Law Resource Center, September 21, 2001. [http://www.indianlaw.org]

"Last Kalahari Bushmen Tortured and Facing Starvation." Survival International, May 2001. [http://www.survival.org.uk/about.htm]

Lazaroff, Cat. "Independent Review Questions Approval of Yucca Mountain." Environment News Service, January 25, 2002. [http://ens-news.com/ens/jan2002/2002L-01-28-07.html]

"Leaked Report Blasts Yucca Mountain Nuclear Waste Site." Environment News Service, November 30, 2001. [http://ens-news.com/ens/nov2001/2001L-11-30-01.html]

Lee, Susan. "Blockade of Nevada Test Site Highway." *Green Party Synthesis/Regeneration* 14 (fall 1997): n.p. [http://www.greens.org/s-r/14/14-24.html]

Lewis, Sunny. "Angry Nevadans Pack Yucca Mountain Hearing." Environment News Service, September 6, 2001. [http://ens.lycos.com/ens/sep2001/2001L-09-06-01.html]

Lickers, Henry. "Guest Essay: The Message Returns." *Akwesasne Notes*, n.s. 1:1 (spring 1995): 10–11.

"Little Progress in the Recognition and Remarcation of Indigenous Lands in Guyana." *Native Americas* 14:1 (spring 1997): 11–12.

Ljunggren, David. "Effects of Global Warming Clear in Canada Arctic." Reuters News Service, Environment News Network, April 20, 2000. [http://www.enn.com/enn-subsciber-news-archive/2000/04/04202000/reu_arctwarm_12170.asp]

Lwori, John. "Niger Delta Records 4,835 Oil Spills in 20 Years." *This Day* (Lagos, Nigeria), August 3, 2001. In *Drillbits and Tailings* 6:7 (August 31, 2001): n.p.

Lyons, Oren. "The Canandaigua Treaty: A View from the Six Nations." In G. Peter Jemison and Anna M. Schein, eds., *Treaty of Canandaigua 1794*. Santa Fe, N.M.: Clear Light, 67–75.

Macafee, Michelle. "Cree Grand Chief Defends Hydro Deal With Quebec as Negotiations Wind Down." *Ottawa Citizen*, December 18, 2001. [http://www.canada.com/news/story.asp?id={565BD3CB-8DAB-4CD2-878F-EE2F7807666B}]

Magagnini, Stephen. "Orphans of History: A Special Report by the *Sacramento Bee*," December 31, 2000. [http://www.sacbee.com/static/archive/news/projects/hmong/123100/diary.html]

"Makah Whaling Can Proceed, Judge Rules." Environment News Service, May 21, 2002. [http://ens-news.com/ens/may2002/2002L-05-21-09.html#anchor4]

"Malaysia: Self-defense Blockades from Sarawak Indigenous Peoples." *World Rainforest Movement Bulletin* (April 2002): n.p. [http://www.wrm.org.uy]

Manning, Mary. "Shoshone Leader Describes Deadly Nuclear Legacy." *Las Vegas Sun*, May 29, 1996. [http://www.lasvegassun.com/sunbin/stories/archives/1996/mar/29/504524152.html]

———. "Shoshone Move Anti-nuke Office to Nevada." *Las Vegas Sun*, January 29, 1997. [http://www.lasvegassun.com/sunbin/stories/archives/1997/jan/29/505534873.html]

"The Marcopper Toxic Mine Disaster: Philippines' Biggest Industrial Accident." 1996. [http://www.tigerherbs.com/eclectica/earthcrash/subject/mining.html]

Marsh, Michael. "Honduras is Worth More Than Gold." 2001. [http://www.menominee.com/nomining/honduras.html]

Matteo, Enzo di. "Damned Deal: Cree Leaders Call Hydro Pact Signed in Secret a Monstrous Sellout." *Now Magazine* (Toronto), February 2002. [http://www.nowtoronto.com/issues/2002-02-14/news_story.php]

Matthiessen, Peter. *Indian Country*. New York: Penguin Books, 1979.

Maveneke, T., J. Hutton, and E. Kawadza. "Campfire in Zimbabwe: Integrating African Wildlife into the Local Economy through Sustainable Use." Africa Resource Trust. N.d. [http://www.art.org.uk/articles/art_integratingwildlife.html]

May, James. "Quechans Oppose Proposed Gold Mine: Site Located on 'Spirit Trail' Where Youths Learn Tribal Conditions." *Indian Country Today*, November 7, 2001, B-1, B-2.

McGirk, Jan. "Tattooed Tribes Keep a Vigil for Kidnapped Chief." *London Independent*, June 26, 2001, 15.

McNeely, Jeffrey A. "Afterword: People and Protected Areas: Partners in Prosperity." In Elizabeth Kemf, ed., *The Law of the Mother: Protecting Indigenous Peoples in Protected Areas*. San Francisco: Sierra Club Books, 1993, 249–57.

Mercer, D. *A Question of Balance: Natural Resources Conflict Issues in Australia*. 3d ed. Sydney, Australia: The Federation Press, 2000.

"Mercury, Mining, and Mayhem: Slow Death in the Amazon." *Drillbits and Tailings* 4:10 (June 24, 1999): n.p. [http://www.moles.org/ProjectUnderground/drillbits/4_10/4.html]

"Mercury Spill Poisons Villagers Near the Yanacocha Mine in Peru." *Drillbits and Tailings* 5:11 (June 30, 2000): n.p. [http://www.moles.org/ProjectUnderground/drillbits/5_11/1.html]

"Militarization and Oil in Chiapas, Mexico." *Drillbits and Tailings* 4:16 (October 8, 1999): n.p. [http://www.moles.org/ProjectUnderground/drillbits/4_16/3.html]

"Mindex Moves in on the Peoples of Mindaro." *Drillbits and Tailings* 4:18 (November 9, 1999): n.p. [http://www.moles.org/ProjectUnderground/drillbits/4_18/4.html]

"Mine Approved for Rich Salmon River Pits Alaska against B.C." Environment News Service, June 14, 2000. [http://ens.lycos.com/ens/jun2000/2000L-06-14-05.html]

Miniclier, Kit. Nuclear-storage Plan Targeted; E.P.A. Cites Indians' Inability to Keep Water Supply Clean." Denver Post, June 17, 2001. [http://www.denverpost.com/Stories/0,1002,53%257E47549,00.html]

"Mining in Peru." Oxfam America. Global Programs: South America, n.d. [http://www.oxfamamerica.org/advocacy/mining.html]

"Mining in the South Pacific: On the Other Side of the Forests." N.d. [http://parallel.acsu.unsw.edu.au/mpi/docs/niart.html]

Minnis, Paul E., and Wayne J. Elisens, eds. Biodiversity and Native America. Norman: University of Oklahoma Press, 2000.

"Mobil Evaluates Next Step in Madre de Dios Basin." Oil and Gas Journal, April 24, 2000.

Mofina, Rick. "Study Pinpoints Dioxin Origins: Cancer-causing Agents in Arctic Aboriginals' Breast Milk Comes from U.S. and Quebec." Montreal Gazette, October 4, 2000, A-12.

"The Mohawk of Kanesatake Resist Nobium Mining." Drillbits and Tailings 7:3 (March 29, 2002): n.p. [www.moles.org]

Moody, Roger. The Gulliver File: Mines, People and Land: The Global Battleground. London: Minewatch/Pluto Press, 1992.

———. "Mining the World: The Global Reach of Rio Tinto Zinc." The Ecologist 26:2 (1996): 46–52.

Moore, Sarah. "For Shell, Nigerian Debacle Isn't the End of the Line: Danger Lurks in Ogoniland for People and Firm, but the Place Beckons." Wall Street Journal, January 10, 2002, A-10.

"Mother's Day Gathering at the Nevada Test Site." Presentation at Healing Global Wounds "Honoring the Mother" Gathering at Nevada Test Site, May 12–15, 2000 . . . [http://www.groundworkmag.org/nuke/nuke-nts-news.html]

Muggiati, André. "One Thousand Brazilian Babies Poisoned by Mercury." Environment News Service, May 20, 2003. [http://ens-news.com/ens/may2003/2003-05-20-01.asp]

Munro, Margaret. "Are the Great Lakes Safe or Sorry? The Great Lakes Appear to Have Made a Triumphant Recovery; but Looks can be Deceiving." The National Post (Canada), October 6, 2001. [http://www.nationalpost.com/features/1001/100601 lake.html]

Murillo, Mario. "Colombian Indians Indignant over Fumigations, Children Blinded." Native Americas 13:4 (winter 1996): 5.

Murphy, Dan. "The Quest for Certifiably Eco-friendly Lumber." Christian Science Monitor, August 23, 2001, 13.

———. "Why Borneo's Sun Bears Now Attack." Christian Science Monitor, August 27, 2001, 8.

"Nabarlek Uranium Mine." N.d. [http://www.urg.org.au/other_mines/nabarlek_intro.html]

"Namosi Copper Mine Proposes to Dump 100,000 Tons of Waste per Day." Drillbits and Tailings 6:5 (June 30, 2001): n.p. [http://www.moles.org/ProjectUnderground/drill bits/6_05/4.html]

"NATO and Innu Set for Showdown in Eastern Canada." Native Americas 12:3 (fall 1995): 10–11.

"Navy, Environmental Protection Agency to Clean Unexploded Ordinance off Adak Island." Environment News Service, December 14, 2001. [http://ens-news.com/ens/dec2001/2001L-12-14-09.html]

Neary, Ben. "Mine Could Affect Zuni Salt Lake." *Boulder (Colorado) Daily Camera* in *Santa Fe New Mexican*, February 5, 2001. [http://www.thedailycamera.com/news/statewest/05azuni.html]

"Nerve Gas Used in Northern Iraq on Kurds: Medical Group Proves Use of Chemical Weapons Through Forensic Analysis." Physicians for Human Rights, April 29, 1993. [http://www.phrusa.org/research/chemical_weapons/chemiraqgas2.html]

"Nevada Governor Vetoes Yucca Mountain." Environment News Service, April 8, 2002. [http://ens-news.com/ens/apr2002/2002L-04-08-09.html]

"Nevada Outraged: Yucca Mountain OK'd for Nuclear Waste." Environment News Service, January 10, 2002. [http://ens-news.com/ens/jan2002/2002L-01-10-04.html]

"New Digging: Placer Dome, Inc. Returns to Exploit Another Area of Indonesia." *Kerebok* 2:10 (May 2001): n.p. [http://www.jatam.org/xnewsletter/kk10.html]

"New Evidence Reveals Unocal's Complicity in Abuses in Burma." *Drillbits and Tailings* 5:8 (May 31, 2000): n.p. [http://groups.yahoo.com/group/graffis-l/message/11105]

"The New Gold Rush: The 1995 Philippine Mining Act Lures a New Wave of Profit-hungry Gold Diggers." *KASAMA* 10:2 (April–May–June, 1996): n.p. Solidarity Philippines Australia Network. [http://www.cpcabrisbane.org/Kasama/V10N2/Gold Rush.htm]

"Nigeria: Focus on Ogoni Oil Spill." Integrated Regional Information Networks, United Nations Office for the Co-ordination of Humanitarian Affairs, June 12, 2001. http://www.reliefweb.int/IRIN/wa/countrystories/niger/20010612.phtml

"Nigeria: Godforsaken by Oil." *World Rainforest Movement Bulletin* (March 2002): n.p. [www.wrm.org.uy]

Nikiforuk, Andrew. "Echoes of the Atomic Age: Cancer Kills Fourteen Aboriginal Uranium Workers." *Calgary Herald,* March 14, 1998, A-1, A-4. [http://www.ccnr.org/deline_deaths.html]

———. "Still a Fine Mess; The Controversy over a Philippine Mine Cleanup Rages On." *Canadian Business,* February 22, 2002, n.p.

Norrell, Brenda. "Hopi and Navajo Protest Pipeline." *Indian Country Today* in *Asheville (North Carolina) Global Report* 129 (July 5–11, 2001): n.p. [http://www.agrnews.org/issues/129/nationalnews.html]

Norton, Thomas Eliot. *The Fur Trade in Colonial New York, 1686–1776.* Madison: University of Wisconsin Press, 1974.

"Occidental Petroleum Abandons Oil Development on U'wa Land." Environment News Service, May 3, 2002. [http://ens-news.com/ens/may2002/2002L-05-03-01.html]

"Ogiek Given a Month to Quit Forest." Survival International, July 2001. [http://www.survival-international.org/index2.htm]

"Ogiek.org: Supporting the Rights of a Kenyan Indigenous Group." Ogiek Welfare Council, 2001. [http://www.ogiek.org/]

"Oil Operations in Guatemala Declared Threat to Human Rights." *Drillbits and Tailings* 5:3 (February 28, 2000): n.p. [http://www.moles.org/ProjectUnderground/drillbits/5_03/2.html]

"Oil Spill Contaminates Ecuadorian Amazon." Environment News Service, January 10, 2002. [http://ens-news.com/ens/jan2002/2002L-01-10-01.html]

"Omai Update: The Disaster Continues; Water Is More Valuable Than Gold." N.d. [http://saxakali.com/CommunityLinkups/omaiupdate799.htm]

Orr, David. "Native Americans, Environmentalists Question Proposed Lake Powell Marina: Groups Call for E.I.S. on Project." *Wild Wilderness*, November 30, 2000. [http://www.wildwilderness.org/aasg/powell.htm]

Osadolor, Kingsley. "The Rise of the Women of the Niger Delta." *London Guardian*, July 24, 2002, in *World Press Review*, October 2002, 47–48.

O'Shaughnessy, Hugh. "A Deadly Trade: How Global Battle against Drugs Risks Backfiring: The Global Battle against Narcotics Is Going Awry. Chemical Spraying of Coca Bushes Is Poisoning Colombian Villages." *London Observer*, June 17, 2001, 20.

"An Overview of Mining-related Environmental and Human Health Issues, Marinduque Island, Philippines: Observations from a Joint U.S. Geological Survey–Armed Forces Institute of Pathology Reconnaissance Field Evaluation." May 12–19, 2000. Washington, D.C.: Government Printing Office, 2000.

"Panama: Mining, Forests and Indigenous Peoples' Rights." *World Rainforest Movement Bulletin* 46 (May 2001): n.p. [http://www.wrm.org.uy/bulletin/46/Panama.html]

The Parliament of the Commonwealth of Australia. *Jabiluka: The Undermining of Process: Inquiry into the Jabiluka Uranium Mine Project.* Report of the Senate Environment, Communications, Information Technology and the Arts References Committee, Canberra, June 1999.

———. *Unlocking the Future: The Report of the Inquiry into the Reeves Review of the Aboriginal Land Rights (N.T.) Act 1976.* House of Representatives Standing Committee on Aboriginal and Torres Strait Islander Affairs, Canberra, August 1999.

Parry, Richard Lloyd. "The Hunt for Bruno Manser; In 1984, Bruno Manser Went to Live with a Tribe of Rainforest Nomads. He Dressed as a Native, Hunted with Darts, and Became the Most Famous Green Activist of his Generation; Then, Last Year, He Vanished; a Tragic Accident? Or Was the 'Wild Man of Borneo' Murdered?" *London Independent*, September 23, 2001, 18–22.

"Partners in Mahogany Crime: Amazon at the Mercy of 'Gentlemen's Agreements.'" Greenpeace, October 2001. [http://www.greenpeace.org/%7Eforests/forests_new/html/content/reports/Mahoganyweb.pdf]

P.C.B. Working Group, IPEN. "Communities Respond to P.C.B. Contamination." N.d. [http://www.ipen.org/circumpolar2.html]

"P.D. 2000 Projects." *Arctic Monitoring and Research—Project Directory.* April 11, 2001. [http://amap.no/pd2000.htm]

Pierpoint, Mary. "Reno-Sparks Fights Kitty Litter Mine; High Traffic, Dust, Potentially Toxic Minerals All Concerns." *Indian Country Today*, December 12, 2001, A-1.

People's Commission on Environment and Development in India. [http://www.pcedindia.com/peoplescomm/advocacy]

"Peoples of the Frozen North." Survival International, n.d. [http://www.survival.org.uk/siberiabg.htm]

"The People's Refusal over the Gold Mining Plan in Tahura Poboya-Paneki." *Kerebok* 2:10 (May 2001): n.p. [http://www.jatam.org/xnewsletter/kk10.html]

Perry, Tony. "U.S. Backs Tribe, Rejects Gold Mine Proposal." *Los Angeles Times,* November 17, 2000. [http://www.ienearth.org/mining_campaign_1c.html#nov17-2000]

"Peru: Illegal Loggers' Invasion of Indigenous Community's Territory." *World Rainforest Movement Bulletin* 53 (December 2001): n.p. [http://www.wrm.org.uy/deforestation/logging.html]

"Peru Villagers Poisoned after Truck Spills Mercury." Reuters News Service, June 14, 2000.

"Petaka Pembuangan Tailing ke Laut." Published by J.A.T.A.M., April 2001. In *MinergyNews*.com, April 28, 2001.

Peterson, Diane J. "Two Twin Cities Churches Protest Hydro Power Injustices." *Earthkeeping News* 9:5 (July–August 2000): n.p. [http://www.nacce.org/2000/manitoba.html]

Peterson, D. J. *Troubled Lands: The Legacy of Soviet Environmental Destruction*. Boulder, Col.: Westview Press, 1993.

"Philippine Province Bans Mining for Twenty-five Years." *Drillbits and Tailings* 7:2 (February 28, 2002): n.p. [http://www.moles.org]

"Philippines: Local People against the San Roque Dam." *World Rainforest Movement Bulletin* 42 (January 2001): n.p. [http://www.wrm.org.uy/bulletin/42/Philippines.html]

Phillip, Chief Stewart. "The Criminalization of Skwelkwek'welt Defenders." September 4, 2001.

"Picuris Pueblo Wrestles with Summo on Copper Hill and Wins." *Drillbits and Tailings* (August 7, 1998): n.p. [http://www.moles.org/ProjectUnderground/drillbits/981207/98120703.html]

"Pipeline Project Opposed by Argentinian Kollas." *World Rainforest Movement Bulletin* 18 (August, 1998): n.p. [http://www.wrm.org.uy/bulletin/5/Argentina.html]

"Polluting Mexican Refinery Told to Halve Output." Reuters News Service, May 25, 1999. [http://www.tigerherbs.com/eclectica/earthcrash/subject/mining.html]

"Population, Highways Lead to Amazon Deforestation." Environment News Service, July 8, 2002. [http://ens-news.com/ens/jul2002/2002-07-08-09.asp#anchor3]

"Portrait of Indonesian Forestry: Supply, Demand and Debt: A Call for Moratorium on Industrial Logging." Indonesian Forum on the Environment, April 24, 2001. [http://www.walhi.or.id/KAMPANYE/Moratorium.htm]

Project Underground. "More Blood Is Spoiled for Oil in the Niger Delta." *Drillbits and Tailings* 4:20 (December 11, 1999): n.p. [http://www.moles.org/ProjectUnderground/drillbits.html]

"Protecting Knowledge: Native Youth Movement Protest at Sun Peaks Resort; Interview with Amanda Soper, Spokesperson for the Native Youth Movement, and Chris Rogers, Spokesperson for Sun Peaks Resort, British Columbia. Canadian Broadcasting Corporation Radio, British Columbia Almanac, August 24, 2001.

"Protect Moorea Lagoon!" Pacific Concerns Resource Centre, February 28, 2000. [http://www2.planeta.com/mader/ecotravel/resources/rtp/rtp.html]

"Protesters Shot at Unocal Refinery in Indonesia." *Drillbits and Tailings* 5:17 (October 20, 2000): n.p. [http://www.moles.org/ProjectUnderground/drillbits/5_17/1.html]

Purdom, Tim. "Radiation Victims Rally; Marchers Complain about IOUs, Delays in Payments.' *Gallup Independent,* March 13, 2001. [http://www.cia-g.com/~gallpind/3–13–01.html#anchor1]

"Putin's Oil Politics Threaten Siberia and Sakhalin's Indigenous Peoples." *Drillbits and Tailings* 5:16 (September 30, 2000): n.p. [http://www.moles.org/ProjectUnderground/drillbits/5_16/1.html]

Pye-Smith, Charlie. "Living with Elephants." *One World* (United Kingdom), n.d. [http://www.oneworld.org/patp/pap_7_4/pye.htm]

"Quebec Admits Toxic Waste Poisons Cree." Canadian Broadcasting Corporation News On-line, October 26, 2001. [http://cbc.ca/cgi-bin/view?/news/2001/10/26/cree_tox011026]

"Quicksilver for Gold: The Poisoning of the Amazon." Greenpeace, n.d. [http://www.greenpeace.org/~thoml/mercury.html]

Rajagopal, Balakrishnan. "The Violence of Development." *Washington Post,* August 9, 2001, A-19.

Rauber, Paul. "Heat Wave." September 1997. [http://www.sierraclub.org/sierra/199709/HeatWave.html]

Rayburn, Rosalie. "Coal Mine Plan Goes to Public for Comment." *Albuquerque Journal,* June 22, 2001. [http://www.albuquerquejournal.com/news/366027news06-22-01.htm]

"A Reindeer Herder's Tale." *Survival Newsletter* 42 (2000): 8.

"Reindeer Herders Under Siege by Oil Industry in Siberia." Institute for Ecology and Action Anthropology; Report from a Fact-finding Mission to Khanty-Mansi Autonomous Region, March 18, 1997. [http://www.hartford-hwp.com/archives/56/003.html]

"Resisting Destruction: Chronology of the Lubicon Crees' Struggle to Survive." N.d. [http://www.lubiconsolidarity.ca/resisting.html]

"Resource Boom or Grand Theft?" Australia West Papua Association, Sydney, n.d. [http://www.cs.utexas.edu/users/cline/papua/deforestation.htm]

"Response to the Final Report: James Bay Cree Nation and the Pimicikamak Cree Nation." In *Dams and Development: A New Framework for Decision-making: The Report of the World Commission on Dams.* N.p.: November 16, 2000. [http://www.dams.org/report/reaction_cree.htm]

Richter, Daniel K. *The Ordeal of the Longhouse: The Peoples of the Iroquois League in the Era of European Colonization.* Chapel Hill: University of North Carolina Press, 1992.

"Rio Tinto Kelian Mine Shut Down by Community Blockade." *Drillbits and Tailings* 5:8 (May 16, 2000): n.p. [http://www.moles.org/ProjectUnderground/drillbits/5_08/1.html]

"Rio Tinto's Environmental Record in East Kalimantan." *Down to Earth.* Published by International Campaign for Ecological Justice in Indonesia, September 1999. [http://www.gn.apc.org/dte/Cklpl.htm]

"Rio Tinto's Shame File: Indonesian Landowners' Discontent Represented at Rio Tinto A.G.M." Mineral Policy Institute, May 22, 2000. [http://www.mpi.org.au/releases/rio_agm.html]

Roberts, Greg. "Mining Big Money in Irian Jaya." *Sydney Morning Herald,* April 6, 1996. In *World Press Review,* July 1996, 14–16.

Robinson, Deborah. *Ogoni: The Struggle Continues.* 2d ed. Geneva, Switzerland: World Council of Churches, 1996.

Rogers, Keith. "Nevada Anti-Nuclear Protest." *Las Vegas Review-Journal,* April 20, 1992, n.p.

———. "Political Leaders, Residents Speak out against Yucca Mountain Project." *Las Vegas Review-Journal,* September 6, 2001. [http://www.commondreams.org/headlines01/0906-01.htm]

Roslin, Alex. "Crees Revive Hydro Project." *Montreal Gazette,* January 21, 2000. [http://www.montrealgazette.com/news/pages/010121/5036705.html]

Ross, John. "Is Zapatista Rebellion Rooted in Oil?" *Earth Island Journal* 11:2 (1996): 20.

Rozell, Ned. "Alaska Science Forum: Dioxins: Another Uninvited Visitor to the North." Geophysical Institute, University of Alaska–Fairbanks, November 9, 2000. [http://www.gi.alaska.edu/ScienceForum/ASF15/1515.html]

"Russian Gas Companies Follow Receding Arctic Ice." *Drillbits and Tailings* 5:15 (September 19, 2000): n.p. [http://www.moles.org/ProjectUnderground/drillbits/5_15/1.html]

Sabaratnam, Sarah. "Plunder of Indigenous Knowledge." *New Straits Times* (Malaysia), May 17, 2001, 1.

Sagapolutele, Fili, Aldwin R. Fajardo, Eric Say, and Radio Australia. "Government Briefs." *Pacific Magazine*, March 2001. [http://cust16530.lava.net/PM/pm32001/pmdefault.cfm?articleid = 23]

Sando, Joe S. *The Pueblo Indians.* San Francisco: Indian Historian Press, 1976.

"San Roque Dam, Agno River, Philippines." International Rivers Network, 2000. [http://www.irn.org/wcd/sanroque.shtml]

Sarawak Peoples Campaign: The Penan of Sarawak. N.d. [http://www.rimba.com/spc/spcpenanmain1.html]

Sarfati, Gigi. "The Dam and the Women of Dalupirip." *KASAMA* (Solidarity Philippines Australia Network) 12:2 (April–May–June 1998): n.p. Reprinted from *CHANEG* (September–December. 1997), published by the Cordillera Women's Education and Resource Center. [http://www.cpcabrisbane.org/Kasama/V12n2/Dalupirip.htm]

Saro-Wiwa, Ken. *Genocide in Nigeria: The Ogoni Tragedy.* Port Harcourt, Nigeria: Saros International Publishers, 1992.

Schaff, Deborah. "Belize: Rainforest Destruction Threatened Mayan People." *Native Americas* 13:3 (fall 1996): 10–11.

Schetagne, R., J.-F. Doyon, and R. Verdon. *Summary Report: Evolution of Fish-mercury Levels in the La Grande Complex, Quebec (1978–1994).* Hydro-Quebec, 1997. [http://www.hydroquebec.com/environment/activites/pdf/doc_c2.pdf]

Schneider, Howard. "Facing World's Pollution in the North." *Washington Post*, September 21, 1996, A-15. [http://www.washingtonpost.com/wp-srv/inatl/longterm/canada/stories/pollution092196.htm

Schneider, Iyawata Britten. "Environmental & Human Rights Devastation in the Southwest." *Country Road News*, n.d. [http://www.countryroadchronicles.com/Articles/CountryRoadNews/devastation.html]

Schneider, Keith. "Washington Nuclear Plant Poses Risk for Indians." *New York Times*, September 3, 1990, A-9.

Seldon, Ron. "Families Sicken and Die in Mold-Plagued HUD Housing; Residents Fight Back against Substandard Construction." *Indian Country Today*, April 3, 2002, B-1, B-2.

"Senate Defeats Energy Rider Which Would Open Arctic National Wildlife Refuge." Environment News Service, December 3, 2001. [http://ens-news.com/ens/dec2001/2001L-12–03–09.html]

"Settlers Killed in Peru Jungle Clash." Associated Press, January 18, 2002.

"Sharing the Land: People and Elephants in Rural Zimbabwe." Project Campfire, Zimbabwe, 1995. [http://www.campfire-zimbabwe.org/facts_07.html]

"Shell: 100 Years Is Enough!" October, 1997. [http://www.kemptown.org/shell/rest.html]

"Shell Oil Spills Continue to Ravage Communities and the Environment in Nigeria." *Drillbits and Tailings* 6:7 (August 31, 2001): n.p. [http://groups.yahoo.com/group/protecting_knowledge/message/1770]

"Shell Says Ogoni Oil Blow-out Now under Control." Integrated Regional Information Networks, United Nations Office for the Co-ordination of Humanitarian Affairs, June 12, 2001. [http://www.reliefweb.int/IRIN/wa/countrystories/nigeria/20010507.phtml]

Shimony, Annemarie Anrod. *Conservatism among the Iroquois at the Six Nations Reserve.* 1961. Reprint. Syracuse, N.Y.: Syracuse University Press, 1994.

Sierra Club. "Hudson Bay/James Bay Watershed Eco-region." [http://www.sierraclub.org/ecoregions/hudsonbay.asp]

Silverstein, Ken. "Bangladesh: The Ambassador from Big Oil." *Earth Island Journal* (fall 1998): n.p.

———. "Gore's Oil Money." *The Nation,* May 22, 2000.

Singh, Rowena." Eco-Timber Export Brings Hope to Solomon Islands." Environment News Service, December 8, 2000. [http://ens.lycos.com/ens/dec2000/2000L-12-08-02.html]

"The Situation of Indigenous Peoples in Mindanao Culled from the Regional Reports of Lumad Organizations and Support Groups in Southern Mindanao, Far Southern Mindanao, Caraga Region and Western Mindanao." N.d. [http://www.mindanow.com/text/articles/people/intro_people.htm]

Skogland, T. "Radiocesium Concentrations in Wild Reindeer at Dovrefjell, Norway." *Rangifer* 7 (1987): 42–45.

Smith, Dean Howard. *Modern Tribal Development: Paths to Self-sufficiency and Cultural Integrity in Indian Country.* Walnut Creek, Cal.: AltaMira Press, 2000.

Sohappy, David. "This Has Been a Long Road." *Native Nations* (June–July 1991): 7–8.

Solly, Richard, Armando Pérez Araújo, and Roger Moody. "Urgent Action on Crisis Involving Exxon in Colombia." August 3, 2001. [http://www.minesandcommunities.org/Action/action9.htm]

"Solomon Islands Murder and Corruption: Logging Takes its Toll. Martin Apa Murdered; Greenpeace Calls for Investigation." Background Briefing by Greenpeace, New Zealand, December 1995. [http://www.greenpeace.org/~comms/forestry/solomo.html]

Stahl, Johannes. "Indigenous Land Rights and the Military; Army Exercises on Maasai Land." *Cultural Survival News Notes,* 2001. [http://www.cs.org/main.htm]

"Statement of Angel Valencia, Yoemem Tekia Foundation, Tucson, Arizona POPs Negotiations, March 23, 2000." *Native News* (March 27, 2000): n.p. [http://www.mail-archive.com/natnews@onelist.com/msg00572.html]

Stevens, Hazard. *The Life of Isaac I. Stevens.* 2 vols. Boston: Houghton-Mifflin, 1900.

Stevens, Jane. "Teak Forests of Burma Fall Victim to Warfare." *The Oregonian* (Portland, Oregon), March 16, 1994.

"Suriname: Logging and Tribal Rights." *World Rainforest Movement Bulletin* 53 (December 2001): n.p. [http://www.wrm.org.uy/deforestation/logging.html]

"Survival's Campaign for the Khanty. Economic Development and the Environment: The Sakhalin Offshore Oil and Gas Fields." Unpublished paper presented at Slavic Research Center, Hokkaido University, Japan, 1999. [http://src-h.slav.hokudai.ac.jp/sakhalin/eng/71/kitagawa3.html]

"Sustainable Development in the Hudson Bay: James Bay Bio-region." Unpublished paper presented at Canadian Arctic Resources Committee, Environmental Committee of Sanikiluaq, Rawson Academy of Aquatic Science, n.d. [http://www.carc. org/pubs/v19no3/2.htm]

"Sustainable Rural Development: Driven by Campfire." Project Campfire, 1995. [http://www.campfire-zimbabwe.org/facts_02.html]

Suzuki, David. "Science Matters: POP Agreement Needed to Eliminate Toxic Chemicals." December 6, 2000. [http://www.davidsuzuki.org/Dr_David_Suzuki/ Article_Archives/weekly12060002.asp]

Swamp, Jake, and John Stokes. *Thanksgiving Address*. Onchiota, N.Y. and Corrales, N.M.: Six Nations Indian Museum and the Tracking Project, 1993.

"Tabaco Attacked by the Armed Forces and by the Mining Company Intercor." *La Guajira*, August 12, 2001 [http://bargeldsparen.tripod.com/id107.htm]

Taliman, Valerie. "N.R.C. Reviews Goshute Nuke Plan." *Indian Country Today*, April 10, 2002, A-1, A-2.

Taylor, Robert. "Cree Leaders Jailed in Protest Over Power Deal." *Indian Country Today*, February 17, 2002. [http://www.indiancountry.com/?1013955409]

"Teak Is Torture and Burma's Reign of Terror; Mon, Karen and Karenni Indigenous Peoples Threatened in Burma." Rainforest Relief, March 29, 1997. [http://nativenet. uthscsa.edu/archive/nl/9703/0119.html]

"Ten Years Without Chico Mendes." *World Rainforest Movement Bulletin* 18 (August 1998): n.p. [http://www.wrm.org.uy/bulletin/18/Brazil2.html]

Thomas, R. "Future Sea-level Rise and its Early Detection by Satellite Remote Sensing." In *Effects of Changes in Atmospheric Ozone and Global Climate*, vol. 4. New York: United Nations Environment Programme/United States Environmental Protection Agency, 1986.

"The Threat of Climate Change to Arctic Human Communities." Greenpeace, 1997. [http://xs2.greenpeace.org/~comms/97/arctic/library/region/people.html]

"Thunder Storms [in the Arctic] Are Latest Evidence of Climate Change." Associated Press Canada, November 15, 2000. [http://abcnews.go.com/sections/science/Daily News/arctic_thunder001115.html]

"Tibetan Monk Tortured for Protesting Chinese Gold Mining." *Drillbits and Tailings* (January 21, 1999): n.p. [http://www.moles.org/ProjectUnderground/drillbits/990121/ 99012103.html]

"Tibet Railway to Fast-track Political and Economic Goals." Knight-Ridder News Service, July 23, 2001.

Tobin, Francis X. "Save the Narmada." University of Toledo, June 28, 2001. [http://comm-org.utoledo.edu/pipermail/announce/2001-June.txt]

Tokar, Brian. "Resisting Global Logging: Forest and Trade Activists Meet in New Zealand." *Earth Island Journal*, 16:2(Summer, 2001). [http://www.earthisland.org/ eijournal/new_articles.cfm?articleID=167&journalID=46]

Tomsho, Rupert. "Reservations Bear the Brunt of New Pollution." *Wall Street Journal*, November 29, 1990, 1.

Tooker, Elisabeth. *The Iroquois Ceremonial of Midwinter*. Syracuse, New York: Syracuse University Press, 1970.

"To the Ends of the Earth: Revenge of the Lost Tribe: the Amazon's Indigenous Peoples." N.d. [http://www.channel4.com/plus/ends/tribe4.html]

"Toxic Ash 'Burnt Offerings' Going to American Indian Lands: Environmental Injustice in Oklahoma Cherokee Nation." Indigenous Environmental Network, September 25, 2001. [http://www.ienearth.org/alerts.html#cherokee]

"Traditional Owners Concerned at More Leaks from Ranger Mine." Australian Broadcasting Corporation Indigenous News, April 24, 2002. [http://abc.net.au/news/newsitems/s538750.htm]

"Traditional Owners Statements: Statement from the Gundjehmi Aboriginal Corporation." Sustainable Energy and Anti-Uranium Service, Australia. April 1997. [http://www.sea-us.org.au/trad-owners.html]

Trelease, Allen W. *Indian Affairs in Colonial New York: The Seventeenth Century.* Ithaca, N.Y.: Cornell University Press, 1960.

"Tribe, Environmentalists Fear Opening of Uranium Mine near Grand Canyon." Associated Press, August 1, 2001. [http://www.ienearth.org/mining_campaign_1a.html#0801-tribe]

Udall, Stewart. *The Myths of August.* New York: Pantheon, 1994.

"U.N. Condemns Botswana's 'Dispossession' of Bushmen." Survival International (London), April 2, 2002. [http://www.survival-international.org]

"Unions Back Lubicons." Indigenous Environmental Network, August 16, 2000. [http://www.ienearth.org/lubicon.html#canada]

United States v. Washington. 520 F. 2d 676 (9th Cir. 1975); cert. denied 423 U.S. 1086 (1976).

Unocal Tailing Pipe Flooded; Rice Fields in Maraangkayu Contaminated." *Minergy News,* Indonesia. 2000. [http://www.minergynews.com/ngovoice/voice3.shtml]

"Untamed Wildlife Safaris: South Africa—Botswana—Namibia—Zimbabwe—Zambia—Malawi—Tanzania—Uganda—Cameroon—Mozambique." N.d. [http://www.untamedwildlife.com/program3.html]

"U.N. Warns over Indigenous Tongues." British Broadcasting Corp., February 8, 2001, 23:02 GMT. [http://news.bbc.co.uk/hi/english/sci/tech/newsid_1161000/1161406.stm]

"Uranium Mine in Australian National Park Dead." Environment News Service, September 6, 2002. [http://ens-news.com/ens/sep2002/2002-09-06-01.asp]

"Uranium Mine Poisons Indigenous Communities in India." *Drillbits and Tailings* 4:7 (June 1, 1999): n.p. [http://www.moles.org/ProjectUnderground/drillbits/990601/99060101.html]

"Uranium Miners, Families Bring Tales of Pain to Washington." *Arizona Republic,* April 15, 2000. [http://www.ienearth.org/mining_campaign_1c.html#nov17-2000]

"Uranium/Nuclear Issues and Native Communities." Indigenous Environmental Network, 2001. [http://www.ienearth.org/nuciss.html]

U'wa Traditional Authorities. "U'wa Statement [Regarding] Oxy Withdrawal from Sacred Territory." May 7, 2002. [http//www.indymedia.org]

"U'wa Update: Colombia's U'wa Tribe and Supporters Celebrate Oxy's Failure to Find Oil." Indigenous Environment Network: Project Underground, August 9, 2001. [http://www.ienearth.org/mining_campaign.html#u-wa-0809]

"The Valley of the Dammed: The Ibaloi People's Struggle against the San Roque Dam; Ibaloi Cultural Heartland under Threat, Philippines." N.d. [http://www.philsol.nl/B99/San-Roque-Dam-1.htm]

Vance, Chris, and John Shafer. "Voices of Resistance." Radio Transcript, CFUV 102 FM, [community radio station, University of Victoria, British Columbia], 1995; transcribed in 1995 and reprinted as the magazine *All That's Left is Struggle,* n.d.. [http://www.finearts.uvic.ca/~vipirg/SISIS/sov/allcree.html]

Vandevelder, Paul. "Between a Rock and a Dry Place." *Native Americas* 17:2 (summer 2000): 10–15.

———. "A Native Sense of Earth." *Native Americas* 18:1 (spring 2001): 42–49.

Vecsey, Christopher, and Robert W. Venables, eds. *American Indian Environments: Ecological Issues in Native American History.* Syracuse, N.Y.: Syracuse University Press, 1980.

Wagner, Martin. "Earth Justice Legal Defense Fund." Statement presented at Organization of American States, October 7, 1997. [http://www.moles.org/ProjectUnderground/drillbits/971007/97100703.html]

Wald, Matthew L. "Handicapping Reactors by the Numbers." *New York Times,* June 19, 2001, C-1.

Wanamaker, Tom. "Hopi Tribe Opposes Lease Extension on Water-thirsty Black Mesa Mine." *Indian Country Today,* February 13, 2002, C-1.

Wanjiru, Jennifer. "Another Ultimatum." Rights Features Service, December 27, 2001. [URL: dfn.org/focus/kenya/ultimatum.htm]

Washington v. Washington State Commercial Passenger Fishing Association. 443 U.S. 658, 686 (1979).

Wastell, David. "Fear and Loathing in Las Vegas over Underground Nuclear Dump." *London Sunday Telegraph,* May 20, 2001, 27.

Waters, Frank. *Brave Are My People: Indian Heroes Not Forgotten.* Santa Fe, N.M.: Clear Light Publishers, 1993.

"Water Supplies Alleged Contaminated by Peru Gold Mine." Reuters News Service, September 28, 2001. [http://www.planetark.org/dailynewsstory.cfm/newsid/12577/story.htm]

Watt-Cloutier, Sheila. Personal communication with author, March 28, 2001.

Weaver, Jace, ed. *Defending Mother Earth: Native American Perspectives on Environmental Justice.* New York: Maryknoll, 1996.

Weaver, Kasey. "Sacred Sites Threatened." *We Have Many Voices* 1:13 (July 21, 2001): n.p. [http://www.turtletrack.org/ManyVoices/Issue_13/SacredSites_721.htm]

Webb, Jason. "Small Islands Say Global Warming Hurting Them Now." Reuters News Service, 1998. [http://bonanza.lter.uaf.edu/~davev/nrm304/glbxnews.htm]

Webb-Vidal, Andy. "Venezuela's Deal to Help Ease Power Shortage in Brazil Sparks Controversy: Indigenous Groups Say Their Wishes Have Been Ignored in the Rush to Build the New Power Line." *London Financial Times,* August 14, 2002, n.p. (in LEXIS)

Weinberg, Bill. "Amazonia: Planning for the Final Destruction." *Native Americas* 18:3/4 (fall–winter 2001): 8–17.

———. "Flooding the Jungle." *Native Americas* 11:2 (summer 1994): 43.

———. "The Golf War of Tepoztlan." *Native Americas* 13:3 (fall 1996): 32–42

———. "La Miskita Rears Up: Industrial Recolonization Threatens Nicaraguan Rainforests: An Indigenous Response." *Native Americas* 15:2 (summer 1998): 22–32.

———. "The Shuar Federation: A Rainforest People Confront Modernity." *Native Americas* 18:3/4 (fall–winter 2001): 11.

Westlund, Mark. "Akawaio Nation in Cross-hairs of Timber Frenzy." *Native Americas* 14:1 (spring 1997): 11.

West, Maureen. "Fatal Ski-boat Fumes Sound Alarm." *Arizona Republic,* July 5, 2001, A-1.

Whalen, Jeanne. "Exxon's Russian Oil Deal Makes Other Firms Feel Lucky." *Wall Street Journal,* December 13, 2001, A-10.

Wheatley, M. A. "The Importance of Social and Cultural Effects of Mercury on Aboriginal Peoples." *Neurotoxicology* 17 (1996): 251–56.

"Who Really Owns West Java Forests?" *Jakarta Post,* October 22, 2001, n.p. (in LEXIS)

"Wichí: Fighting for Survival in Argentina." *Abya Yala News: The Journal of the South and Meso-American Rights Center.* N.d. [http://saiic.nativeweb.org/ayn/wichi.html]

Wilkinson, Charles. *Messages from Frank's Landing: A Story of Salmon, the Treaties, and the Indian Way.* Seattle: University of Washington Press, 2000.

Williams, James. "Environmental Injustice in the Pacific Islands: United States Accused of Environmental Crimes over Failure to Clean up Toxic Dump Sites." Greenpeace, August 4, 2000. [http://headlines.igc.apc.org:8080/enheadlines/965700046/index_html]

Williams, Paul. "Treaty Making: The Legal Record." In G. Peter Jemison and Anna M. Schein, eds., *Treaty of Canandaigua 1794.* Santa Fe, N.M.: Clear Light, 2000, 35–42.

"Will There be Justice for the Pataxó-Hã-Hã-Hãe?" *World Rainforest Movement Bulletin* 32 (March 2000): n.p. [http://www.wrm.org.uy/bulletin/32/Brazil.html]

Wilson, Scott. "Colombian Indians Resist an Encroaching War; Indigenous People Join to Search for Leader." *Washington Post,* June 18, 2001. [http://www.globalexchange.org/colombia/washpost061801.html]

Witzig, Ritchie. "New and Old Disease Threats in the Peruvian Amazon: The Case of the Urarina." *Abya Yala News: The Journal of the South and Meso-American Rights Center.* N.d. [http://saiic.nativeweb.org/ayn/urarina.html]

Witzig, Ritchie, and Massiel Ascencios. "Urarina Survival Update: Continued Resource Exportation and Disease Importation by Foreigners and Newly Initiated by Multinational Oil Companies." *Abya Yala News: The Journal of the South and Meso-American Rights Center.* N.d. [http://saiic.nativeweb.org/ayn/urupdate.html]

"Women and Children Begin Blockades of Pipeline Construction Crews in Threatened Ecuadorian Cloud Forest Reserve." Amazonwatch, October 11, 2001. [http://www.amazonwatch.org/newsroom/newsreleases01/oct1101_ec.html]

"Women Occupy Chevron/Texaco Facilities in the Niger Delta." *Drillbits and Tailings* 7:6 (July 31, 2002):. n.p. [http://moles.org/Order/order.mv?]

"World Bank Funds Controversial Diamond Project on Bushmen's Land." Survival International Press Release, February 17, 2003. [st2@survival-international.org]

"World Bank Quietly Insures Major Polluters." *Native Americas* 13:1 (spring 1996): 4–5.

Yablokov, Alexi, Sviatoslav Zabelin, Mikhail Lemeshev, Svetlana Revina, Galina Flerova, and Maria Cherkasova. "Russia: Gasping for Breath, Choking in Waste, Dying Young." *Washington Post,* August 18, 1991, C-3.

Yasuni National Park and Waorani Ethnicx Reserve, Ecuador. N.d. [http://www.nmnh.si.edu/botany/projects/cpd/sa/sa8.htm]

"Yemeni Tribes Blow Up Oil Pipeline." *Drillbits and Tailings* (December 7, 1998): n.p. [http://www.moles.org/ProjectUnderground/drillbits/981207/98120703.html]

Young, E. *Third World in the First: Development and Indigenous Peoples.* London: Routledge, 1995.

"Yucca Mountain: No Place for Nuclear Waste." In *Honor the Earth.* N.p.: n.d. [http://honorearth.org/nuclear/no_place.html]

"Yucca Mountain Volcanic Hazard Greater Than Thought." Environment News Service, July 31, 2002. [http://ens-news.com/ens/jul2002/2002-07-31-09.asp]

Ziman, Jenna E. "Freeport McMoran: Mining Corporate Greed." *Z Magazine,* January 1998. [http://www.zmag.org/zmag/articles/jan98ziman.htm]

Zoe, John B. "Diamonds Aren't Our Best Friend; A Negotiator for the Dogrib Nation Says a New Mine in [the Northwest Territories] will Threaten the Lifeblood of His People." *Toronto Globe and Mail,* November 2, 1999, n.p.

"Zunis Object to Coal Mine Ruling." *Gallup Independent,* August 27, 2001. [http://www.cia-g.com/~gallpind/8-27-01.html]

"Zuni: Utility Plan Would Desecrate Sacred Lake." *Salt Lake City Tribune,* June 9, 2001. [http://members.tripod.com/TopCat4/news/4jun01.txt]

INDEX

About the Author

BRUCE E. JOHANSEN is Professor of Communication and Native American studies at the University of Nebraska at Omaha. He has authored a number of books on indigenous and environmental subjects, most recently *The Global Warming Desk Reference* (Greenwood, 2001) and *The Dirty Dozen: Toxic Chemicals and the Earth's Future* (Greenwood, 2003).

About the Contributors

JOHN KAHIONHES FADDEN is associate curator of the Six Nations Indian Museum, Onchiota, New York. Fadden, who is Mohawk, is also a noted artist who has contributed covers to more than 50 books. He is a longtime Haudenosaunee culture bearer.

ROBERT W. VENABLES is a long-time senior lecturer in the American Indian Program at Cornell University. He has also edited *The Six Nations of New York: The 1892 United States Census Bulletin* (1995). With Christopher Vescey, Venables coauthored *American Indian Environments: Ecological Issues in Native American History* (1980). Venables is also well known for his free lectures to members of the Onondaga Nation in the LaFayette, New York community, near Ithaca.